Lecture Notes in Computer Science 4041

Commenced Publication in 1973
Founding and Former Series Editors:
Gerhard Goos, Juris Hartmanis, and Jan van Leeuwen

Siu-Wing Cheng   Chung Keung Poon (Eds.)

Algorithmic Aspects in Information and Management

Second International Conference, AAIM 2006
Hong Kong, China, June 20-22, 2006
Proceedings

 Springer

Volume Editors

Siu-Wing Cheng
Hong Kong University of Science and Technology, Department of Computer Science
Clear Water Bay, Hong Kong, China
E-mail: scheng@cs.ust.hk

Chung Keung Poon
City University of Hong Kong, Department of Computer Science
83 Tat Chee Avenue, Kowloon Tong, Hong Kong, China
E-mail: ckpoon@cs.cityu.edu.hk

Library of Congress Control Number: 2006927039

CR Subject Classification (1998): F.2.1-2, E.1, G.1-3, J.1

LNCS Sublibrary: SL 3 – Information Systems and Application, incl. Internet/Web
and HCI

ISSN   0302-9743
ISBN-10   3-540-35157-4 Springer Berlin Heidelberg New York
ISBN-13   978-3-540-35157-3 Springer Berlin Heidelberg New York

Springer is a part of Springer Science+Business Media

springer.com

© Springer-Verlag Berlin Heidelberg 2006
Printed in Germany

Typesetting: Camera-ready by author, data conversion by Scientific Publishing Services, Chennai, India
Printed on acid-free paper   SPIN: 11775096   06/3142   5 4 3 2 1 0

Preface

The papers contained in this volume were presented at the Second International Conference on Algorithmic Aspects in Information and Management (AAIM 2006), held on June 20–22, 2006 at the City University of Hong Kong, Hong Kong, China.

The series of AAIM conferences provides an annual international forum for the communication of research advances on algorithms pertinent to information management and management science. The first conference (AAIM 2005) was held in Xi'an, China and it is planned for the near future that conferences of the series will be held in cities in the Pacific Rim.

This volume contains 34 papers selected from a total of 263 papers submitted from places all over the world: Australia, Canada, China, France, Germany, India, Israel, Italy, Japan, Mexico, Mongolia, Netherlands, New Zealand, Poland, Singapore, South Korea, Sweden, Taiwan, Ukraine, UK and USA. In addition to the selected papers, the volume also contains two papers by the invited speakers, Allan Borodin and Ming-Yang Kao.

We thank all the people who made this meeting possible: the authors who submitted papers, the Program Committee members and external reviewers, the invited speakers, the local organizers, and the sponsors for their effort, advice and support. We also thank EasyChair (www.easychair.org) for providing the free conference software.

April 2006

Siu-Wing Cheng
Chung Keung Poon

Conference Organization

AAIM 2006 was jointly organized by the City University of Hong Kong and the Hong Kong University of Science and Technology.

Program Chairs

Siu-Wing Cheng (Hong Kong U of Science and Technology)
Chung Keung Poon (City U of Hong Kong)

Program Committee

Hee-Kap Ahn (Korean Advanced Institute of Science and Technology)
Takao Asano (Chuo U)
Amotz Bar-Noy (City U of New York)
Hans Bodlaender (U of Utrecht)
Peter Brucker (U of Osnabrueck)
Leizhen Cai (Chinese U of Hong Kong)
Jianer Chen (Texas A&M U)
Marek Chrobak (U of California at Riverside)
Rudolf Fleischer (Fudan U)
Joachim Gudmundsson (National ICT Australia)
Gregory Gutin (Royal Holloway, U of London and U of Haifa)
Wen-Lian Hsu (Academia Sinica, Taiwan)
Giuseppe F. Italiano (U of Rome "Tor Vergata")
Tao Jiang (U of California at Riverside and Tsinghua U)
Tak-Wah Lam (U of Hong Kong)
Xiang-Yang Li (Illinois Institute of Technology)
Peter Bro Miltersen (U of Aarhus)
Pat Morin (Carleton U)
Seffi Naor (Technion and Microsoft Research)
Kirk Pruhs (U of Pittsburgh)
Vijaya Ramachandran (U of Texas at Austin)
Rajeev Raman (Leicester U)
Jiri Sgall (Academy of Sciences of Czech Republic)
Paul Spirakis (U of Patras and CTI Greece)
Wing Kin Sung (National U of Singapore)
Hisao Tamaki (Meiji U)
Jan van Leeuwen (U of Utrecht)
Lusheng Wang (City U of Hong Kong)
Yinfeng Xu (Xi'an Jiaotong U)
Binhai Zhu (Montana State U)

External Reviewers

Sang Won Bae
Rezaul Alam Chowdhury
Xiaotie Deng
Andrew Goldberg
Wen-Liang Hwang
Hyunwoo Jung
Yue-Kuen Kwok
James Kwok
Eric Law
Peter Lennartz
Chi-Jen Lu
Esther Moet
Mart Molle
Andrea Pacifici
Guido Proietti
Chan-Su Shin
Maurizio A. Strangio

Organizing Committee

Matthew Chang (City U of Hong Kong)
Mordecai Golin (Hong Kong U of Science and Technology)
Jinxin Huang (Hong Kong U of Science and Technology)
Xiaohua Jia (City U of Hong Kong)
Hing Fung Ting (U of Hong Kong)
Yajun Wang (Hong Kong U of Science and Technology)

Sponsors

City University of Hong Kong
Hong Kong Pei Hua Education Foundation Limited

Table of Contents

Further Reflections on a Theory for Basic Algorithms

Allan Borodin

Department of Computer Science
University of Toronto
bor@cs.toronto.edu

1 Introduction

Can we optimally solve $Max2SAT$ in (say) time $(|F|\log|F|)$ where $|F|$ is the length of formula F. Of course, since $Max2SAT$ is NP-hard, we can confidently rely on our strongly held belief that no NP-hard problem can be solved optimally in polynomial time. But obtaining *unconditional* complexity lower bounds (even linear or near linear bounds) remains the central challenge of complexity theory. In the complementary fields of complexity theory and that of algorithm design and analysis, we ask questions such as "what is the best polynomial time approximation ratio" that can be achieved for $Max2SAT$. The best negative results are derived from the beautiful development of PCP proofs. In terms of obtaining better[1] approximation algorithms, we appeal to a variety of algorithmic techniques, including very basic techniques such as greedy algorithms, dynamic programming (with scaling), divide and conquer, local search and some more technically involved methods such as LP relaxation and randomized rounding, semi-definite programming (see [34] and [30] for an elegant presentation of these randomized methods and the concept of derandomization using conditional expectations). A more refined question might ask "what is the best approximation ratio (for a given problem such as $Max2SAT$) that can be obtained in (say) time $O(n\log n)$" where n is the length of the input in some standard representation of the problem. What algorithmic techniques should we consider if we are constrained to time $O(n\log n)$?

In order to bring some coherence to the "Design and Analysis of Algorithms", most courses and texts will organize much of the content in terms of basic "algorithmic paradigms", such as greedy algorithms, backtracking, dynamic programming, divide and conquer, local search, primal-dual, IP/LP rounding, etc. (but not etc. etc.). To this small set of paradigms, we can add randomization and sometimes very creative ways to utilize and combine these basic algorithmic approaches. Although we seem to be able to intuitively describe these basic classes of algorithms, we (in computer science) rarely attempt to *precisely* define what we mean by such terms as greedy, dynamic programming, etc. Clearly, a precise definition is required if we want to defend statements such as "there is no greedy

---

[1] For maximization (respectively, minimization) problems we will use approximation ratios ≤ 1 (resp. ≥ 1).

S.-W. Cheng and C.K. Poon (Eds.): AAIM 2006, LNCS 4041, pp. 1–9, 2006.

algorithm that provides a good approximation for problem X" or "there is no efficient dynamic programming algorithm for optimally solving problem Y".

In the context of combinatorial search and optimization problems, I will suggest some simple but precise algorithmic models for some basic algorithmic paradigms. This is not a new approach, as there were (for example) a number of important attempts to characterize greedy algorithms (e.g. in terms of matroids [16] and greedoids [25]), and characterizing dynamic programming and branch and bound (e.g. in terms of formal languages [20]). In contrast to the elegant abstraction provided by matroids, we are not attempting to (say) characterize when *the* greedy algorithm is optimal for a set system but rather (similar, for example, to some previous studies for local search [24], "branch and bound algorithms" [12] and IP/LP rounding [4]) we are trying to explore the limitations of basic (simply defined) methods in terms of approximation ratios (or time complexity vs approximation/optimality results).

This talk is based on ideas and results from a number of recent papers. In particular, I will present formulations for greedy and greedy-like algorithms [8], simple dynamic programming and backtracking [10], and basic primal-dual/local ratio algorithms [7]. As will be explained, these models are all based on giving priorities to input items and concern worst case time complexity[2] for search and optimization problems.

2   Priority Algorithms as a Model of Greedy and Greedy-Like Algorithms

With the exception of naive brute force search, greedy algorithms are arguably the simplest approach[3] for solving combinatorial optimization problems. The relation between *the* greedy algorithm for set systems and matroids was formalized by Rado [31] and Edmonds [16] with later extension to greedoids [25]. This early development did not address the use of greedy algorithms to achieve guaranteed approximation ratios but rather focused on the question of understanding when *the* greedy algorithm was optimal. Recently, k-extendible set systems are defined in [29] to help address the issue of when *the* greedy algorithm can provide a good approximation. In [8], we offered a simple model, called *priority algorithms* for greedy algorithms that can be applied in a wide variety of applications not restricted to set systems. We will briefly describe this model and some examples of well known greedy algorithms that are captured by this model. In the next section, we will use priority algorithms as the starting point for some other

---

[2] We note that the algorithmic models can be applied to any measure of complexity (e.g. space complexity, time vs space, average case or smoothed analysis).

[3] On a conceptual level, local search algorithms are perhaps equally simple. Obviously, simplicity is in the eyes of the beholder and ignoring the complexity of optimally solving an LP relaxation, one can easily argue that IP/LP relaxations are also conceptually very simple. But whatever one's experience and intuition, greedy algorithms are certainly considered to be conceptually simple.

"priority based" paradigms, namely "simple dynamic programming", "simple backtracking" and "simple primal dual" algorithms.

The priority model (and the extensions that will follow) relies on the assumption that we represent an input instance as a set of "locally defined input items", each item represented in some "natural" way. For a scheduling problem (such as interval or job scheduling to maximize profit, makespan minimization, etc.) the choice and representation of an "input item" is usually quite natural and not an issue. Namely, for scheduling problems, an input item is a "job" and each job is represented by the parameters of that job (e.g., the duration, deadline, value, etc. of the job). For other applications, such as graph theory there is a choice of whether to represent the items as edges or as vertices, and having done so there is a choice of how much information to provide within the representation of (say) a vertex. For the well known greedy Kruskal and Prim MST algorithms, the input items are edges, represented by their weights and their end points. For greedy vertex cover approximation algorithms the input items are the vertices, each vertex represented by its weight and its list of adjacent vertices[4]. Similarly, for the CNF-SAT problem, we can think of the clauses as the items or the propositional variables as the items. In the latter case, a variable could be represented by a full description of each clause in which it appears.

Of course there is no one "correct way" to define what is an input item and exactly how much "local information" should be included in the representation of an input item. For example, with regard to the interval selection problem, it seems that the most natural representation would be that each input interval is an item and is represented by a triple (s, f, v) where s (respectively, t and v) is the start (respectively, finish and value) of the interval. But one could also represent the input as an interval graph, or combining these representations by representing each interval I by the tuple (s, f, v, L) where L is a list of the intervals that intersect I. But, of course, this representation or the representation as an interval graph could result in a representation of size $\Omega(n^2)$ for an input instance having n intervals. This would seem to defeat the purpose of greedy algorithms which are utilized because of their efficiency.

After agreeing on the nature of the input representation, we are able to define a priority algorithm as formulated in [8]. The basic idea is that a priority algorithm is a *one-pass* algorithm in which the input is processed one input item at a time. The order or priority in which input items are "considered" is determined by a "local ordering". When an input item is considered, the algorithm makes an irrevocable decision about this item. The nature of a problem usually determines the set of allowable decisions (e.g. accept/reject, schedule on particular machine, etc.). But what is an *allowable ordering* of the input items? We impose the priority condition that if inputs I_j and I_k are in the input sets $\mathcal{I}' \subseteq \mathcal{I}$ and I_j has higher priority than I_k in $\mathcal{I}$ then that priority is maintained in the subset $\mathcal{I}'$. In particular then, any function $f : \mathcal{I} \to \mathbf{R}$ induces an allowable ordering by ordering input items in (say) non-decreasing (or non-increasing) order of their f

---

[4] In fact, as discussed in [9], it is usually sufficient to represent a vertex by its list of adjacent edges.

value. Any function (including functions of arbitrarily high complexity or even non computable functions) that takes an input item given by its representation and produces a real number can use this real value as the priority of an item. The only other distinction to be made about the ordering is whether the ordering is *fixed* initially (before any item is considered) or if the ordering is *adaptive* in that the priority of an item can depend on the items previously considered.

In either the fixed or adaptive case, we emphasize that priority algorithms do not impose any explicit complexity limitations as the ordering and decisions being made can be of arbitrary complexity. It is only the "syntactic structure" of a priority algorithm that limits its power. Finally, we view priority algorithms as being "greedy-like" and reserve the term *greedy* for those priority algorithms which make "greedy (irrevocable) decisions" for each input item in the sense of making a decision as if this is the last input item and the decision must minimize/maximize the given objective function[5].

We claim that the priority model seems to capture almost all algorithms that we commonly consider as greedy algorithms. The following are examples of fixed order priority algorithms: Kruskals MST, the maximal matching algorithm for unweighted vertex cover, optimal scheduling of unit profit intervals (ordering intervals according to non-decreasing finishing times), Graham's "online" and LPT greedy approximation algorithms for minimizing makespan on identical machines. We also claim that for the exact $MaxkSAT$ problem (where each clause has exactly k literals), the naive randomized algorithm (independently set each variable to true/false with probability 1/2) can be derandomized to be a (online) priority algorithm (where the input items are the propositional variables represented by the description of the clauses in which they appear). This is similar to exercise 16.6 in [34] which indicates how to turn the naive randomized algorithm for MaxCut (independently place each vertex in S or $\bar{S}$ with probability 1/2) into an (online) greedy algorithm. The following are examples of adaptive order priority algorithms: the H_n approximation greedy set cover algorithm, Prim's MST, Dijkstra's shortest path algorithm (for digraphs where all edge costs are non-negative)[6], Huffman optimal prefix trees[7], various greedy algorithms for weighted vertex cover (see, for example, [13]). It also can be shown that the best (to date) polynomial time computable approximation ratio for the uncapacitated facility location problem is a greedy priority algorithm [27].

---

[5] This "live for today" definition of greediness may not always make sense in settings where the input items are not "isolated"; for example, in a graph problem, if the items are the vertices, we may already know that a given vertex v is present since it is adjacent to a vertex already considered but v itself has not yet been considered. Our view is that in the context of approximation algorithms the distinction between greedy priority and (non-greedy) priority algorithms is not an essential distinction beyond the historical importance of the term. For optimal algorithms, each decision must be greedy by definition.

[6] Here we view Dijkstra's algorithm as computing the optimal tree from a single source to all other vertices.

[7] Here we consider the input items to be nodes of the prefix tree, with leaf nodes representing the keys to be coded and internal nodes representing subtrees.

The priority algorithm model includes online algorithms (where the ordering of input items can be assumed to be dictated by an adversary) and as in the case of online algorithms one can derive negative results on approximation ratios (called the competitive ratio in the online setting). For example, it can be shown that weighted interval scheduling cannot have a c-approximation priority algorithm for any constant c (see [8]). Other negative results for priority algorithms can be found in [8, 32, 15, 3, 10, 9].

But have we captured everything that one would tend to call greedy? In [17], a 4-approximation algorithm called "greedy" is given for the weighted interval scheduling problem. The algorithm is a one pass algorithm in which rejections are irrevocable but acceptances are revocable. The condition that must be satisfied is that after each interval is considered, the partial solution being constructed is feasible. To incorporate such revocable acceptances, an appropriate extension of the priority model has been introduced and studied in [21].

3   Using Priorities Beyond Priority Algorithms

Obviously (and by design) the priority algorithm framework is a very limited algorithmic model. However, a number of other conceptually simple algorithms can also be viewed in terms of priorities given to input items. We briefly consider the stack model of [7], and the BT model of [10].

3.1   The Stack Model as a Model of Simple Primal Dual Algorithms

One of the most interesting developments in approximation algorithms has been the use of primal dual algorithms as pioneered in [1] and [18]. In many cases, primal dual algorithms can be realized as greedy algorithms (for example, as in the greedy approximation algorithms [23, 19] for the uncapacitated facility location problem) and also can be used to analyze known greedy algorithms (i.e. using the method of dual fitting). Simply stated, in a one pass primal dual algorithm, dual variable are increased until at least one constraint becomes tight. When each constraint corresponds to an input variable, then the fact that a constraint becomes tight can be used to make a decision about the corresponding variable. If our input representation has enough information to determine that a constraint has become tight then we can view the primal dual algorithm as an adaptive priority algorithm. For some applications of the primal dual method, a second clean up phase is also required (e.g. for covering problems to remove items not needed for a feasible solution and for packing problems to insure feasibility). The primal dual framework has been shown to be equivalent to the local ratio method [5, 6]. The local ratio method is used in [7] to motivate a stack model which attempts to model primal dual algorithms where the second clean up phase is a simple "popping" of a stack of items that were pushed onto the stack in the first phase (i.e. when dual constraints became satisfied). To be more precise, in a stack algorithm, the priority framework is used to decide on the order in which to consider input items and an "irrevocable accept/reject decision" is replaced by a decision to push the item onto the stack or else reject it. Once again, as in the one pass priority model, no explicit complexity considerations are introduced. Some negative results are obtained in [7] for adaptive ordering stack algorithms (applied to the set cover and Steiner tree problems) and for fixed order stack algorithms (applied to packing problems).

3.2 The BT Model as a Model of Simple Dynamic Programming and Simple Backtracking

For the knapsack problem where every item has a value v and a weight w, it is shown in [10] that no priority algorithm can achieve a constant approximation. In particular, the natural "cost effective" greedy algorithm which orders items by non-increasing v/w does not have a constant approximation ratio. However, either the item having largest value or the cost effective greedy solution will achieve a 1/2 approximation. We can view this "1/2" approximation as a "more permissive" priority algorithm where we are allowed to maintain two partial solutions as we consider each item. More generally, we can think of taking each possible subset of the k largest valued items and then extending each such partial solution using the greedy most cost effective ordering for the remaining items. This results in a $k - 1/k$ approximation algorithm (i.e. a PTAS) at a cost of maintaining 2^k partial solutions [33]. In fact, the well known FPTAS algorithms for knapsack based on dynamic programming and scaling [22, 26, 28] can be implemented so as to achieve a $1 - \epsilon$ approximation at the cost of maintaining polynomial (in $(1/\epsilon)$) many partial solutions. These algorithms for the knapsack problem and the well known optimal dynamic programming algorithm for weighted interval scheduling are called *simple dynamic programming* algorithms in [35]. Such simple dynamic programming (DP) algorithms can be nicely formulated as BT algorithms [10]. Simply stated, BT algorithms are "priority based" algorithms which generate a computation tree of partial solutions by considering one item at a time and branching on a set of decisions being made for each item. As in the more basic priority framework, the order in which items are considered can be determined by a fixed ordering or an adaptive ordering where the item considered depends on previously seen items but not on the decisions taken; i.e. the same item is being considered at all nodes on any given level of the tree. Moreover, in BT algorithms the ordering can be *fully adaptive* in the sense that the item being considered at any node of the computation tree can depend on the path of item (i.e. on the items already seen and the decisions taken along the path). In this way, BT algorithms also become a model for "simple backtracking".

In the BT framework, it is no longer a question of studying the best approximation factor. For now, by allowing an exponential size tree, we can optimally solve any NP-optimization problem. Rather, we study the tradeoff between the width of the BT (i.e. the largest number of nodes at any level = the largest number of partial solutions being maintained) and the approximation ratio, or we can study the width required for computing optimal solutions (or finding a solution in a search problem). Or one can study the related measure of "depth first size" instead of width to better capture the sense of backtracking. Within the BT model and for the different types of ordering , a number of results are derived for interval scheduling, knapsack, Max2SAT, vertex cover, and the 2SAT and 3SAT search problems [10].

In what sense, do BT algorithms only capture "simple" DP and "simple" backtracking? The BT model seems to capture those DP algorithms where the "DP recursion" is based on an induction on the number of items in the (partial) solution. In contrast, consider a standard DP algorithm for least cost paths in a graph with negative edges but no negative cycles (or similarly for the longest path in a DAG). Here the recursion is based on the number of edges in a path. The standard DP algorithm for edit distance between two strings is based on the sum of the two prefix lengths. And there are also so-called "non-serial" DP applications such as the standard optimal algorithms for constructing binary search trees, and matrix chain products (see, for example, [14]). With regard to backtracking, the fully adaptive BT model does not allow information to be shared between different paths on the tree[8]. Recently, [11] extend the BT model from computation trees to DAGS to provide a model that can capture (serial) non-simple DPs such as in the least cost path algorithm.

4 Where Is This All Going?

Keeping in mind Samuel Johnson's (1709-1784) warning that "All theory is against freedom of the will; all experience for it", the reader may well wonder by now what is the goal of this type of research. An algorithm designer is not restricted to any small set of algorithmic paradigms no matter how well they may capture many useful algorithms. Why try to formalize some kinds of restrictive thinking? But these basic simple kinds of algorithms we have been discussing are often the first approach that is used by "practitioners" and it is what we teach in our courses. If for no other than pedagogical reasons, it seems worthwhile to be able to give some clarity to terms we use in our algorithm design courses. But, moreover, as we sometimes religiously profess, by carefully understanding the limitations of basic methods and our inability to derive good negative results for a particular problem, we may be led to new algorithms. This has been the case in (for example) online algorithms and we believe similarly that new simple offline approximation algorithms can and will be derived by a more careful understanding of these simple algorithmic design methods.

If indeed, the difficulty of providing significant negative results is any indication, there are great opportunities to derive new algorithms within the simple algorithmic methods we have been studying. We have found it difficult to produce significant negative results in many settings even for priority algorithms, especially with regard to unweighted graph theoretic problems. For the more expressive BT model, although we do have some strong results concerning the (width or depth first size) complexity required for optimality, we only have one approximation lower bound (somewhat matching the FPTAS for the knapsack

---

[8] By definition, fixed order and adaptive order BT algorithms implicitly know what has happened on each path.

problem) for polynomial width BT algorithms using an adaptive ordering. We are also unable to obtain results for the stack model with adaptive ordering when applied to packing problems. Developing a much more expressive framework that can model more general dynamic programming, backtracking, and primal dual algorithms and yet be amenable to analysis is a major challenge. Would such a framework for dynamic programming include divide and conquer algorithms as a special case? Can we show that dynamic programming cannot optimally solve graph matching problems? How much power can be added to these models by allowing randomization? Only one paper [2] thus far has considered priority algorithms allowing randomization. And so far, all of our efforts toward defining precise algorithmic definitions have been restricted priority based algorithms for search and optimization problems. Can we develop a more inclusive theory?

References

1. A. Agrawal, P. Klein, and R. Ravi. When trees collide: An approximation algorithm for the generalized steiner problem on networks. *SICOMP*, 24:440–465, 1995.
2. S. Angelopoulos. Randomized priority algorithms. In *Proceedings of 1st Workshop on Approximation and Online Algorithms*, 2003.
3. S. Angelopoulos and A. Borodin. On the power of priority algorithms for facility location and set cover. *Algorithmica*, 40(4):271–291, 2004.
4. S. Arora, B. Bollobás, and L. Lovász. Proving integrality gaps without knowing the linear program. In *Proceedings of the 43rd Annual IEEE Conference on Foundations of Computer Science*, pages 313–322, 2002.
5. R. Bar-Yehuda, A. Bendel, A. Freund, and D. Rawitz. Local ratio: A unified framework for approxmation algorithms in memoriam: Shimon even 1935-2004. *Computing Surveys*, 36:422–463, 2004.
6. R. Bar-Yehuda and D. Rawitz. On the equivalence between the primal-dual schema and the local ratio technique. In *4th International Workshop on Approximation Algorithms for Combinatorial Optimization Problems, APPROX*, pages 24–35, 2001.
7. A. Borodin, D. Cashman, and A. Magen. How well can primal-dual and local-ratio algorithms perform? In *Proceedings of the 32nd International Colloquium on Automata, Languages and Programming (ICALP)*, 2005.
8. A. Borodin, M. N. Nielsen, and C. Rackoff. (Incremental) priority algorithms. *Algorithmica*, 37(4):295–326, 2003.
9. Allan Borodin, Joan Boyar, and Kim S. Larsen. Priority Algorithms for Graph Optimization Problems. In *Second Workshop on Approximation and Online Algorithms*, volume 3351 of *Lecture Notes in Computer Science*, pages 126–139. Springer-Verlag, 2005.
10. M. Alekhnovich A. Borodin, J. Buresh-Oppenheim, R. Impagliazzo, A. Magen, and T. Pitassi. Toward a model for backtracking and dynamic programming. In *Proceedings of Computational Complexity Conference (CCC)*, pages 308–322, 2005.
11. J. Buresh-Oppenheim, S. Davis, and R. Impagliazzo. A formal model of dynamic programming algorithms. Manuscript in preparation, 2006.
12. V. Chvátal. A greedy heuristic for the set covering problem. *Mathematics of Operations Research*, 4(3):233–235, 1979.
13. Kenneth L. Clarkson. A modification of the greedy algorithm for vertex cover. *Information Processing Letters*, 16:23–25, 1983.

14. T. Cormen, C. Leiserson, R. Rivest, and C. Stein. *Introduction to Algorithms, Second Edition (Page 1015)*. MIT Press, Cambridge, Mass., 2001.
15. S. Davis and R. Impagliazzo. Models of greedy algorithms for graph problems. In *Proceedings of the 15th ACM-SIAM Symposium on Discrete Algorithms*, 2004.
16. Jack Edmonds. Matroids and the greedy algorithm. *Mathematical Programming*, 1:127–136, 1971.
17. T. Erlebach and F.C.R. Spieksma. Interval selection: Applications, algorithms, and lower bounds. *Technical Report 152, Computer Engineering and Networks Laboratory, ETH*, October 2002.
18. M. X. Goemans and D.P. Williamson. A general approximation technique for constrained forest problems. *SICOMP*, 24:296–317, 1995.
19. S. Guha and S. Khuller. Greedy strikes back: Improved facility location algorithms. In *Proceedings of the 9th ACM-SIAM Symposium on Discrete Algorithms*, pages 649–657, 1998.
20. P. Helman. A common schema for dynamic programming and branch and bound algorithms. *Journal of the Association of Computing Machinery*, 36(1):97–128, 1989.
21. S.L. Horn. One-pass algorithms with revocable acceptances for job interval selection. *MSc Thesis, University of Toronto*, 2004.
22. O. Ibarra and C. Kim. Fast approximation algorithms for the knapsack and sum of subset problems. *JACM*, 4:463–468, 1975.
23. K. Jain and V.V. Vazirani. Approximation algorithms for metric facility location and k-median problems using the primal-dual schema and lagrangian relaxation. *Journal of the ACM*, 48(2):274–296, 2001.
24. S. Khanna, R. Motwani, M. Sudan, and U. Vazirani. On syntactic versus computational views of approximability. *SIAM Journal on Computing*, 28:164–a91, 1998.
25. B. Korte and L. Lovász. Mathematical structures underlying greedy algorithms. *Lecture Notes in Computer Science*, 117:205–209, 1981.
26. E. L. Lawler. Fast approximation algorithms for knapsack problems. In *Proc. 18th Ann. Symp. on Foundations of Computer Science*, Long Beach, CA, 1977. IEEE Computer Society.
27. M. Mahdian, J. Ye, and J. Zhang. Improved approximation algorithms for metric facility location problems. In *Proceedings of the 5th International Workshop on Approximation Algorithms for Combinatorial Optimization Problems (APPROX)*, pages 229–242, 2002.
28. A. Marchetti-Spaccamela. Personal communication as stated in [10], 2004.
29. J. Mestre. Greedy in approximation algorithms. *unpublished manuscript*, 2006.
30. Rajeev Motwani and Prabhakar Raghavan. *Randomized Algorithms*. Cambridge University Press,, 1995.
31. R. Rado. A theorem on independence relations. *Quart. Jorunal of Mathematics*, 13:83–89, 1942.
32. Oded Regev. Priority algorithms for makespan minimization in the subset model. *Information Processing Letters*, 84(3):153–157, Septmeber 2002.
33. S. Sahni. Approximate algorithms for the 0-1 knapsack problem. *JACM*, 1:115–124, 1975.
34. V. V. Vazirani. *Approximation algorithms*. Springer-Verlag New York, Inc., 2001.
35. G. Woeginger. When does a dynamic programming formulation guarantee the existence of a fully polynomial time approximation scheme (FPTAS)? *INFORMS Journal on Computing*, 12:57–75, 2000.

Algorithmic DNA Self-assembly

Ming-Yang Kao

Department of Electrical Engineering and Computer Science
Northwestern University
Evanston, IL 60208, USA
kao@cs.northwestern.edu

Abstract. Self-assembly is the ubiquitous process by which objects autonomously assemble into complexes. This phenomenon is common in nature and yet is poorly understood from mathematical and programming perspectives. It is believed that self-assembly technology will ultimately permit the precise fabrication of complex nanostructures. Of particular interest is DNA self-assembly. Double and triple crossover DNA molecules have been designed that can act as four-sided building blocks for DNA self-assembly. Experimental work has been done to show the effectiveness of using these building blocks to assemble DNA crystals and perform DNA computation. With these building blocks (called *tiles*) in mind, researchers have considered the power of the *tile* self-assembly model.

The tile assembly model extends the theory of Wang tilings of the plane by adding a natural mechanism for growth. Informally, the model consists of a set of four sided Wang tiles whose sides are each associated with a type of glue. The bonding strength between any two glues is determined by a *glue function*. A special tile in the tile set is denoted as the *seed* tile. Assembly takes place by starting with the seed tile and attaching copies of tiles from the tile set one by one to the growing seed whenever the total strength of attraction from the glue function meets or exceeds a fixed parameter called the *temperature*.

Algorithmic DNA self-assembly is both a form of nanotechnology and a model of DNA computing. As a computational model, algorithmic DNA self-assembly encodes the input of a computational problem into DNA patterns and then manipulates these patterns to produce new DNA patterns that encode the desired output of the computational problem. As a nanotechnology, algorithmic DNA self-assembly aims to design tiles with carefully chosen glue types on their four sides. Two tiles are said to be of different types if their sides have different glue types. Useful tile types are nontrivial to design but relatively easy to duplicate in large quantity. A key design challenge for algorithmic DNA self-assembly is to use only a small number of different tile types to assemble a target nanostructure.

This talk will survey recent results in algorithmic DNA self-assembly and discuss future research directions.

S.-W. Cheng and C.K. Poon (Eds.): AAIM 2006, LNCS 4041, p. 10, 2006.
© Springer-Verlag Berlin Heidelberg 2006

Online Scheduling on Parallel Machines with Two GoS Levels*

Yiwei Jiang

Department of Mathematics, Zhejiang University, Hangzhou 310027, P.R. China
mathjyw@yahoo.com.cn

Abstract. This paper investigates the online scheduling problem on parallel and identical machines with a new feature that service requests from various customers are entitled to many different grade of service (GoS) levels. Hence each job and machine are labelled with the GoS levels, and each job can be processed by a particular machine only when the GoS level of the job is not less than that of the machine. The goal is to minimize the makespan. In this paper, we consider the problem with two GoS levels. It assumes that the GoS levels of the first k machines and the last $m - k$ machines are 1 and 2, respectively, and every job has a GoS level of 1 alternatively or 2. We first prove the lower bound of the problem under consideration is at least 2. Then we discuss the performance of algorithm AW presented in [2] for the problem and show it has a tight bound of $4 - 1/m$. Finally, We present an approximation algorithm with a competitive ratio of $\frac{12+4\sqrt{2}}{7} \approx 2.522$.

Keywords: Online algorithm, Competitive analysis, Parallel machine scheduling, Grade of Service.

1 Introduction

We study the problem of online scheduling on identical parallel machines with a new feature that service requests from various customers are entitled to many different *grade of service* (GoS) levels. The goal is to minimize the *makespan* under the constraint that all requests are satisfied. This problem is first proposed by Hwang et al. [1] and is motivated by the following scenario. In service industry, the service providers often have special customers, such as, gold, silver, or platinum members who are more valued than the regular members. Those special members are usually entitled to premium services, and hence, some kind of differentiated service policy must be implemented by the service provider. One simple scheme for providing differentiated service is to label machines and jobs with pre-specified GoS levels and allow each job to be processed by a particular machine only when the GoS level of the job is not less than that of the machine. In effect, the processing capability of the machines labelled with high GoS levels

---

* Research supported by National Natural Science Foundation of China (10271110)and Natural Science Foundation of Zhejiang Province (Y605316).

tends to be reserved for the jobs with high GoS levels. In such situation, assigning jobs to the machines becomes a parallel machine scheduling problem with a special eligibility constraint.

Formally, this problem can be described as follows. We are given a sequence $\mathcal{J} = \{p_1, p_2, \ldots, p_n\}$ of independent jobs with positive processing time, which must be processed onto m identical machines $M_1, M_2, \cdots, M_m$. We identify jobs with their processing time. Each job p_j is labelled with the GoS level of $g(p_j)$, and each machine M_i is also labelled with the GoS level of $g(M_i)$. M_i is allowed to process job p_j only if $g(M_i) \leq g(p_j)$. The objective is to minimize the makespan, i.e., the maximum completion time of all jobs. This problem is called *parallel machine scheduling with GoS eligibility* [1].

We consider online algorithms in this paper, hence, we assume that jobs arrive in *online over list*, that is to say, jobs arrive one by one and the jobs are required to be scheduled irrevocably onto machines as soon as they are given, without any knowledge of the jobs that will arrive later. If we have full information on the job data before constructing a schedule, it is called *offline*. Algorithms for an online/offline problem are called *online/offline algorithms*.

The performance of an online algorithm is measured by its *competitive ratio*. For a job sequence $\mathcal{J}$ and an algorithm A, let $c^A(\mathcal{J})$ (or shortly c^A) denote the makespan produced by A and let $c^*(\mathcal{J})$ (or shortly c^*) denote the optimal makespan in an off-line version. Then the competitive ratio of A is defined as the smallest number C such that for any $\mathcal{J}$, $c^A(\mathcal{J}) \leq Cc^*(\mathcal{J})$. An online problem has a *lower bound* ρ if no online deterministic algorithm has a competitive ratio smaller than ρ. An online algorithm is called *optimal* if its competitive ratio matches the lower bound.

Clearly, the offline version of the problem is NP-hard. Note that Lenstra et al. [3] propose a binary search algorithm based on linear programming with makespan no more than 2 times the optimum for the most general problem of unrelated parallel machine scheduling which certainly covers the problem under consideration. Recently, for the offline version of the problem, Hwang et al. [1] present an algorithm $LG - LPT$ with makespan no more than 5/4 for $m = 2$ and $2 - 1/(m-1)$ for $m \geq 3$ times the optimum. For the online version, Azar et al. [2] present an online algorithm with competitive ratio $\log_2 2m$ for any m. In particular, the competitive ratio turns into 2 for $m = 2$. All the results are for non-preemptive algorithms.

In most service industry, the service provider always divides their customers into two parts such as VIP and regular members. So, in this paper, we consider the online version of the problem on m identical machines with two GoS levels, that is, $g(M_i)$ is equal to 1 or 2 for all $1 \leq i \leq m$, and $g(p_j)$ is also equal to 1 or 2 for all $1 \leq j \leq n$. Without loss of generality, we assume that $g(M_i) = 1$ for $i = 1, \cdots, k$ and $g(M_i) = 2$ for $i = k + 1, \cdots, m$, where $0 \leq k \leq m$. If $k = 0$ or $k = m$, then the problem can be reduced to the classical parallel machine scheduling with the objective of minimizing the makespan, which has been well studied in the literature. Therefore, we suppose $1 \leq k \leq m - 1$ in this paper. Especially, for $k = 1$ and $m = 2$, Jiang et al. in [4] presented an optimal online algorithm with a competitive ratio of 5/3. For the off-line version of the problem under consideration, Zhou et al.[5] present an algorithm with worst-case ratio of $\frac{4}{3} + (\frac{1}{2})^r$, where r is the desired number of iteration.

For our result, we first prove that the lower bound of the considered problem is at least 2. Then we consider the upper bound of the problem. We show that the competitive ratio of algorithm AW proposed in [2] for this problem has a tight bound of $4 - 1/m$. Finally, we proposed an online algorithm with a competitive ratio of $\frac{12+4\sqrt{2}}{7} \approx 2.522$.

The rest of the paper is organized as follows. Section 2 gives some basic notation and the lower bound of the problem. Section 3 discuss the performance of the algorithm AW. Section 4 present a new online algorithm. Finally, section 5 contains some remarks.

2   Preliminary

To simplify the presentation, the following notation and definitions are required in the remainder of the paper.

- T_j The total size of the first j jobs.
- T_{ji} The total size of the jobs with GoS level of i in the first j jobs, $i = 1, 2$.
      Then $T_j = T_{j1} + T_{j2}$.
- $p_j^{\max}$ The largest job size in the first j jobs.
- L_j^i The completion time of machine M_i at time $j \geq 0$ (i.e., the moment right after the j-th job is scheduled) in an online algorithm A, $i = 1, 2, \cdots, m$.
- L_j^A The current makespan yielded by algorithm A at time j.
- L_j^* The optimal makespan at time j.
      It clearly follows that $c^A = L_n^A$ and $c^* = L_n^*$.
- U_i The set of the jobs with the GoS level of i, $i = 1, 2$.
- V_i The set of the machines with the GoS level of i, $i = 1, 2$.
      Then $V_1 = \{M_1, \cdots, M_k\}$ and $V_2 = \{M_{k+1}, \cdots, M_m\}$.
- S_p The starting time of the job p.
- C_p The completion time of the job p.

Let $C_j^0 \doteq \max\{p_j^{\max}, \frac{T_j}{m}, \frac{T_{j1}}{k}\}$, then we have a lower bound of optimal makeapan as described as following Theorem.

Theorem 1. *The optimal makespan of the problem $L_j^* \geq C_j^0$ at any time $j \geq 1$.*

Proof. It is clear that the optimal makespan satisfies $L_j^* \geq \max\{p_j^{\max}, \frac{T_j}{m}\}$ at any time j. By the definition of the problem, all the jobs in set U_1 only can be processed on the machines in V_1, which implies that the optimal makespan $L_j^* \geq \frac{T_{j1}}{k}$ according to the definition of T_{j1}.   □

Theorem 2. *The competitive ratio of any non-preemptive online algorithm is at least 2.*

Proof. We use adversary method to establish the result. Let $k = 4$ and $m = 16$, that is, the numbers of machines in V_1 and V_2 are 4 and 12, respectively. Assume that there exists an online algorithm A with competitive ratio C. The first 16 jobs with $p_i = 1$ and $p_i \in U_2$ arrive, $i = 1, \cdots, 16$. If there are at least two jobs to be processed on the same machine by algorithm A, then no more job arrives. It follows that $c^A = 2$ and $c^* = 1$ which implies that $C \geq 2$. Therefore we assume that the algorithm schedules these jobs onto different machines. In other words, every machine exactly has a job. Then 7 jobs with $p_i = 2$ and $p_i \in U_2$ arrive, $i = 17, \cdots, 23$.

We claim that these seven jobs must be processed on the different machines. Otherwise, no more job arrives and there are at least two jobs to be processed on the same machine, which, together with the assignment of the first 16 jobs, follows that $c^A \geq 5$. It is easy to obtain that $c^* = 2$, which yields that $C \geq 5/2 > 2$. More detailed assignment of these 7 jobs is distinguished by the following two cases.

Case 1. Suppose that there is at least one job to be processed on the first 4 machines. It yields that there exists at least one machine with the load of 3. Then the last 4 jobs with $p_i = 3$ and $p_i \in U_1$ arrive, $i = 24, \cdots, 27$, which must be processed on the first 4 machines. Thus we can conclude that there is at least one machine with load of 6. While the optimal makespan is 3 by scheduling the last 4 jobs on the machines in V_1 and the first 23 jobs on the machines in V_2. It follows that $C \geq 6/3 = 2$.

Case 2. Suppose that all these seven jobs are processed on the last 12 machines. That is, there are exactly 7 machines in V_2 with load of 3. Then 6 jobs with $p_i = 3$ and $p_i \in U_2$ arrive, $i = 24, \cdots, 29$. If all these jobs are processed on the last 12 machines, then no more job arrives. Therefore, we can obtain that the makespan yielded by A is at least 6. However, it is not hard to obtain that the optimal makespan is 3. It follows that $C \geq 6/3 = 2$. Hence, we assume there is at least one job p_i for some $24 \leq i \leq 29$ to be processed on the first 4 machines. It means that there is at least one of the first 4 machines has a load of $1 + 3 = 4$. Then the last 4 jobs with $p_i = 4$ and $p_i \in U_1$ arrive, $i = 30, \cdots, 33$. It is clear that the last 4 jobs must be processed on the first 4 machines, which implies that the makespan yielded by A is at least $4 + 4 = 8$. While we can obtain that the optimal makespan is 4 by scheduling the last 4 jobs on the machines in V_1 and the rest jobs on the machines in V_2.

In summary, we have shown that the competitive ratio of any online algorithm is at least 2. $\square$

3   Algorithm AW

In this section, we show that the competitive ratio of algorithm AW proposed in [2] for this problem has a tight bound of $4 - 1/m$. The main idea of the algorithm AW is to assign the arriving job to the machine with currently minimum load. Therefore, for the problem under consideration, the algorithm AW can be described as follows:

Algorithm AW**:**

For any $1 \le j \le n$, if $p_j \in U_2$, schedule p_j onto the machine with the minimum load at time $j-1$ in all m machines (i.e., $V_1 \bigcup V_2$). Otherwise, if $p_j \in U_1$, schedule p_j onto the machine with the minimum load at time $j-1$ in the first k machines (i.e., V_1).

Let job p_l be the job which determined the makespan produced by AW. It yields that $c^{AW} = S_{p_l} + p_l$.

Lemma 1. *If p_l is assigned to the machine with minimum load in m machines at time $l-1$, then we have $C_{p_l} = S_{p_l} + p_l \le (2 - \frac{1}{m})c^*$.*

Proof. It is clear that the minimum load is S_{p_l} in m machines at time $l-1$. Then we conclude that $S_{p_l} \le \frac{T_n - p_l}{m} \le \frac{T_n - p_l}{m}$, which together with Theorem 1, leads that $C_{p_l} = S_{p_l} + p_l \le \frac{T_n - p_l}{m} + p_l = \frac{T_n}{m} + (1 - \frac{1}{m})p_l \le (2 - \frac{1}{m})c^*$. □

Theorem 3. *The algorithm AW has a competitive ratio of $4 - 1/m$.*

Proof. Two cases are considered with respect to the GoS level of p_l.

Case 1. $g(p_l) = 2$, i.e., $p_l \in U_2$. By the algorithm rule, p_l is assigned to the machine with minimum load at time $l-1$ in $V_1 \bigcup V_2$. It is easy to obtain $c^{AW} = S_{p_l} + p_l \le (2 - \frac{1}{m})c^*$ from Lemma 1.

Case 2. $g(p_j) = 1$. i.e., $p_l \in U_1$. It is clear that, from step 3 of the algorithm, p_l is assigned to the machine with minimum load in V_1 at time $l-1$. If no jobs from U_2 to be processed on machines in V_1, that is, all jobs processed on machines in V_1 are from U_1, then by Theorem 1 we have

$$C_{p_l} = S_{p_l} + p_l \le \frac{T_{l1} - p_l}{k} + p_l = \frac{T_{n1}}{k} + (1 - \frac{1}{k})p_l \le (2 - \frac{1}{m})c^*.$$

Therefore, we assume there exist some jobs from U_2 to be processed on machines in V_1 and let q be the job with maximum completion time in those jobs. It means that all jobs processed on the machines in V_1 after the completion time of q (i.e., C_q) are from U_1. From Lemma 1, we can obtain that $C_q = S_q + q \le (2 - 1/m)c^*$. If $S_{p_l} \le C_q$, then we obtain that $c^{AW} = S_{p_l} + p_l \le C_q + p_n^{\max} \le (3 - 1/m)c^*$. Suppose $S_{p_l} > C_q$. By the definitions of q and T_{j1}, we conclude that $T_{n1} \ge T_{l1} \ge p_l + k(S_{p_l} - C_q)$. It follows that $S_{p_l} < \frac{T_{n1}}{k} + C_q$. According to Theorem 1, we obtain that

$$c^{AW} = S_{p_l} + p_l \le \frac{T_{n1}}{k} + C_q + p_n^{\max} \le c^* + (2 - \frac{1}{m})c^* + c^* = (4 - \frac{1}{m})c^*. \quad □$$

We next show the bound is tight. Let ε be a sufficiently small number and $m = 2k^2 + k$. Consider an instance in arriving order of $A_1, A_2, A_3, A_4, A_5, A_6, p_n$ as follows:

- A_1 A set of jobs in U_2 with total size of m. The size of every job is ε.
- A_2 A set of k jobs in U_1. The size of every job is ε.

- A_3 A set of $m - k$ jobs in U_2. The size of every job is 2ε.
- A_4 A set of k jobs in U_2. The size of every job is 1.
- A_5 A set of $k - 1$ jobs in U_1. The size of every job is 1.
- A_6 A set of k jobs in U_1. The size of every job is $1/k$
- The last job $p_n = 1$ in U_1.

Clearly, we have $U_1 = A_2 \bigcup A_5 \bigcup A_6 \bigcup \{p_n\}$ and $U_2 = A_1 \bigcup A_3 \bigcup A_4$.

According to the rule of the algorithm AW, the assignment of the instance can be shown as Figure 1, from which we can obtain that the makespan of AW is $4 + \varepsilon$.

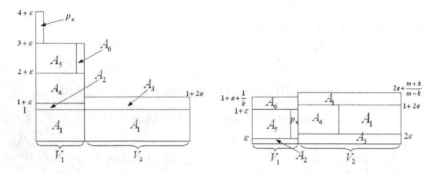

Fig. 1. The schedule of AW **Fig. 2.** A feasible schedule

It is not difficult to obtain a feasible schedule as shown as Figure 2. It schedules all jobs in U_i onto the machines in V_i, $i = 1, 2$. For jobs in U_1, every machine in V_1 exactly contains one of the jobs in A_2, one of the jobs in $A_5 \bigcup \{p_n\}$ and one of the jobs in A_6. It follows that the load of each machine is $\varepsilon + 1 + \frac{1}{k}$. For jobs in U_2, we assign all $m - k$ jobs in A_3 onto $m - k$ machines in V_2 exactly, and all k jobs in A_4 to different machines in V_2. Finally, we can assign those small jobs in A_1 such that the load of each machine in V_2 is identical so far as ε is sufficiently small. Since the total size of all jobs in U_2 is $m + k + 2(m - k)\varepsilon$, we have the load of each machine in V_2 is $2\varepsilon + \frac{m+k}{m-k} = 2\varepsilon + \frac{2k^2+2k}{2k^2} = 2\varepsilon + 1 + \frac{1}{k}$. It is clear that $c^* \leq \max\{\varepsilon + 1 + \frac{1}{k}, 2\varepsilon + 1 + \frac{1}{k}\} = 2\varepsilon + 1 + \frac{1}{k}$.

Hence,

$$\frac{c^{AW}}{c^*} \geq \frac{4 + \varepsilon}{2\varepsilon + 1 + \frac{1}{k}} \rightarrow 4. \quad (\varepsilon \rightarrow 0, \quad k \rightarrow +\infty)$$

4   A New Algorithm HA

In this section, we present a new online algorithm with a competitive ratio of $2 + \alpha$, where $\alpha = \frac{4\sqrt{2}-2}{7}$. We partition our algorithm into two parts with respect to the ratio of m and k. The first part considers the cases $\frac{m}{k} \leq 1 + \alpha$ or $\frac{m}{k} \geq 1 + \frac{1}{\alpha}$. The second part of the algorithm is presented for the case $1 + \alpha < \frac{m}{k} < 1 + \frac{1}{\alpha}$. The detailed can described as follows:

Algorithm HA
Part 1

1. If $\frac{m}{k} \leq 1 + \alpha$, assign the arriving job to the machine with the current minimum load in V_1.
2. If $\frac{m}{k} \geq 1 + \frac{1}{\alpha}$ and the arriving job belongs to U_1, assign it to the machine with the current minimum load in V_1.
3. If $\frac{m}{k} \geq 1 + \frac{1}{\alpha}$ and the arriving job belongs to U_2, assign it to the machine with the current minimum load in V_2.

Part 2 (For the case $1 + \alpha < \frac{m}{k} < 1 + \frac{1}{\alpha}$)

1. Let $j = 1$.
2. If $p_j \in U_1$, go to 3. Otherwise, go to 4.
3. Schedule p_j onto the machine with the minimum load at time $j - 1$ in V_1, go to 5.
4. Let $L_{j-1}^t = \min_{k+1 \leq i \leq m} L_{j-1}^i$. If $p_j + L_{j-1}^t \leq (2 + \alpha)C_j^0$, then schedule p_j onto the machine M_t, go to 5. Otherwise, go back to 3.
5. Let $j = j + 1$. If no new job arrives, stop. Otherwise, go back to 2.

Remark 1. It is clear that Part 1 of the above algorithm is based on a greedy idea in some sense. If $\frac{m}{k} \leq 1 + \alpha$, all jobs are schedule on the first k machines (i.e., V_1) and the machines in V_2 are not used . On the other hand, if $\frac{m}{k} \geq 1 + \frac{1}{\alpha}$, the algorithm assigns the jobs in U_1 and U_2 to the machines in V_1 and V_2, respectively.

Remark 2. For Part 2, the main idea is that we assign the jobs in U_2 to the machines in V_2 as many as possible, in other wards, we would not assign them to the machines in V_1 unless the sum of the current job size and the minimum load in V_2 is too large. And the jobs in V_1 just are assigned to the machine in V_1.

Let p_l be the job which determined the makespan. Then $c^{HA} = p_l + S_{p_l}$.

Theorem 4. *Part 1 of algorithm HA achieves a competitive ratio of $2 + \alpha - \frac{1}{m}$.*

Proof. By the algorithm rule, if $\frac{m}{k} \leq 1 + \alpha$, all jobs are scheduled onto the first k machines. It is obvious that $S_{p_l} \leq \frac{T_n - p_l}{k}$ due to the assignment of p_l. It follows that

$$c^{HA} \leq p_l + \frac{T_n - p_l}{k} \leq p_l + \frac{(1 + \alpha)(T_n - p_l)}{m}$$

$$\leq (1 - \frac{1}{m})p_n^{\max} + (1 + \alpha)\frac{T_n}{m} \leq (2 + \alpha - \frac{1}{m})c^*.$$

The last inequality above holds because of Theorem 1.

If $\frac{m}{k} \geq 1 + \frac{1}{\alpha}$, we conclude that all jobs in U_1 are scheduled on the machines in V_1 and jobs in U_2 on the machines in V_2. Suppose $p_l \in U_1$. Then by Theorem 1, $c^{HA} = p_l + S_p \leq p_l + \frac{T_{n1} - p_l}{k} = (1 - \frac{1}{k})p_n^{\max} + \frac{T_{n1}}{k} \leq (2 - \frac{1}{k})c^*$. On the other hand, if $p_l \in U_2$, it is easy to obtain that

$$c^{HA} = p_l + S_p \le p_l + \frac{T_{n2} - p_l}{m - k}$$

$$\le p_l + \frac{T_n - p_l}{m - \frac{m}{1+1/\alpha}} \le (1 - \frac{1+\alpha}{m})p_l + (1+\alpha)\frac{T_n}{m}$$

$$< (1 - \frac{1}{m})p_n^{\max} + (1+\alpha)c^* \le (2+\alpha - \frac{1}{m})c^*. \qquad \square$$

Note that, if $\alpha = 1$, Part 1 can be regarded as an independent and complete algorithm with a competitive ratio of $3 - 1/m$. Clearly, it is better than the algorithm AW though both them seem like greedy algorithms. The difference between them is that the algorithm AW only considers the current minimum load (which assigns the current job to a machine with the currently minimum load) while Part 1 considers both the currently minimum load and the ratio of m and k.

In the remainder of this section, we focus on discussing the performance of Part 2 of HA. We begin with the following lemma.

Lemma 2. *If* $p_l \in U_2$, *we have* $c^{HA} \le (2+\alpha)c^*$.

Proof. Note that the algorithm schedules all jobs in U_2 by step 3 or alternatively by step 4. If p_l is scheduled by step 4, that is, it is scheduled onto machine M_t with minimum load, then we obtain that $c^{HA} = S_{p_l} + p_l = L_{l-1}^t + p_l \le (2+\alpha)C_j^0 \le (2+\alpha)c^*$. If p_l is scheduled by step 3, then we conclude that

$$p_l + L_{l-1}^t > (2+\alpha)C_j^0. \qquad (1)$$

Then we claim that p_l is scheduled onto the machine with minimum load (denoted by $M_{t'}$) in $V_1 \bigcup V_2$ at time $l - 1$. In fact, by the rule of step 3, it is obvious that the load of the machine $M_{t'}$ is minimum in V_1 at time $l - 1$. Clearly, the load of machine M_t is minimum in V_2 by the definition of M_t. Then we only need to show that $L_{l-1}^{t'} \le L_{l-1}^t$. Otherwise, suppose $L_{l-1}^{t'} > L_{l-1}^t$, that is, M_t is the machine with minimum load in $V_1 \bigcup V_2$. Therefore, by Lemma 1, we can obtain that $S_{p_l} + p_l = L_{l-1}^t + p_l \le (2 - 1/m)C_j^0$, which contradicts the inequality (1). So $M_{t'}$ is the machine with minimum load in $V_1 \bigcup V_2$, which, together with Lemma 1, yields that

$$c^{HA} = S_{p_l} + p_l = L_{l-1}^{t'} + p_l \le (2 - \frac{1}{m})c^* < (2+\alpha)c^*. \qquad \square$$

We next show that the desired result still holds when the job $p_l \in U_1$.

Lemma 3. *If* $p_l \in U_1$ *and there are no jobs in* U_2 *to be processed on the machines in* V_1, *then we have* $c^{HA} \le (2+\alpha)c^*$.

Proof. By the assumption, we can conclude that all jobs in U_i are scheduled on the machines in V_i, $i = 1, 2$. As $p_l \in U_1$ and Theorem 1, we obtain that

$$c^{HA} = S_{p_l} + p_l \le \frac{T_{l1} - p_l}{k} + p_l \le \frac{T_{n1}}{k} + (1 - \frac{1}{k})p_l \le (2 - \frac{1}{k})c^* < (2+\alpha)c^*. \qquad \square$$

Now we are ready to consider the case that there are many jobs from U_2 to be processed on the machines in V_1. And let job p_x be the last completion one in those jobs. That is to say, after the time C_{p_x}, all jobs processed on machines in V_1 are from U_1. We first give a lemma on the completion time of p_x.

Lemma 4. *The following inequalities hold:*
(i) $L_{x-1}^i > (1+\alpha)C_x^0$, $i = k+1, \cdots, m$;
(ii) $C_{p_x} \leq (2 + \alpha - \frac{m}{k}\alpha)C_x^0$.

Proof. (i) As p_x belongs to U_2 and is scheduled on the machine in V_1 by step 3, we can obtain that $p_x + L_{x-1}^t > (2+\alpha)C_x^0$. It yields that $L_{x-1}^i > (2+\alpha)C_x^0 - p_x$ for every $i = k+1, \cdots, m$. The result can obtained directly from $p_x \leq p_x^{\max} \leq C_x^0$.

(ii) It is clear that $S_{p_x} \geq 0$. Since p_x is scheduled onto the machine with minimum load in V_1, which, together with (i) and $T_x \leq mC_x^0$ from Theorem 1, leads

$$S_{p_x} \leq \frac{T_x - p_x - \sum_{i=k+1}^m L_{x-1}^i}{k}$$
$$\leq \frac{T_x - p_x - (m-k)(1+\alpha)C_x^0}{k}$$
$$\leq \frac{mC_x^0 - (m-k)(1+\alpha)C_x^0}{k} = (1 + \alpha - \frac{m}{k}\alpha)C_x^0.$$

Note that $m/k < 1 + 1/\alpha$, which implies that $(1 + \alpha - \frac{m}{k}\alpha)C_x^0 > 0$. It follows that $0 \leq S_{p_x} < (1+\alpha - \frac{m}{k}\alpha)C_x^0$. Hence, $C_{p_x} = S_{p_x} + p_x \leq (2+\alpha - \frac{m}{k}\alpha)C_x^0$ with $p_x \leq C_x^0$. □

Parallel to Lemma 3, we have the following lemma:

Lemma 5. *If $p_l \in U_1$ and there are some jobs in U_2 to be processed on the machines in V_1, then we have $c^{HA} \leq (2+\alpha)c^*$.*

Proof. Let $\triangle = S_{p_l} - C_{p_x}$, then we have $c^{HA} = p_l + S_{p_l} = p_l + \triangle + C_{p_x}$. Two cases are considered according to the value of $\triangle$.

Case 1. $\triangle > 0$. By the definition of C_{p_x} and S_{p_l}, we conclude that all machines in V_1 is busy between time C_{p_x} and S_{p_l} and all jobs processed on these machine in this time interval are from U_1. It implies that $T_{n1} \geq k\triangle$, which follows that $c^* \geq \frac{T_{n1}}{k} \geq \triangle$ due to Theorem 1. If $C_{p_x} \leq \alpha\triangle$, then we have

$$c^{HA} = p_l + \triangle + C_{p_x} \leq p_n^{max} + \triangle + \alpha\triangle \leq (2+\alpha)c^*.$$

Now we turn to consider the case that $C_{p_x} > \alpha\triangle$. Since $c^{HA} = p_l + \triangle + C_{p_x} \leq c^* + \triangle + C_{p_x}$, we only need to show that $\triangle + C_{p_x} \leq (1+\alpha)c^*$, i.e.,

$$\triangle + C_{p_x} - (1+\alpha)c^* \leq 0. \tag{2}$$

By the assignment of p_l, it is clear that $L_{l-1}^i \geq S_{p_l} = \triangle + C_{p_x}$ for any $1 \leq i \leq k$. Together with Lemma 4(i), we have

$$T_n \geq \sum_{i=1}^k L_{x-1}^i + \sum_{i=k+1}^m L_{x-1}^i \geq k(\triangle + C_{p_x}) + (m-k)(1+\alpha)C_x^0,$$

which follows that $c^* \geq \frac{T_n}{m} \geq \frac{k(\triangle + C_{p_x}) + (m-k)(1+\alpha)C_x^0}{m}$ by Theorem 1. Hence the left member of the inequality (2)

$$\leq \triangle + C_{p_x} - (1+\alpha)\frac{k(\triangle + C_{p_x}) + (m-k)(1+\alpha)C_x^0}{m}$$

$$= \frac{(m - (1+\alpha)k)(\triangle + C_{p_x}) - (m-k)(1+\alpha)^2 C_x^0}{m}. \tag{3}$$

Recall that $m/k > 1 + \alpha$ and $C_{p_x} > \alpha\triangle$, which, together with Lemma 4(ii), leads that

$$(3) \leq \frac{(m - (1+\alpha)k)(\frac{1}{\alpha}C_{p_x} + C_{p_x}) - (m-k)(1+\alpha)^2 C_x^0}{m}$$

$$= \frac{(m - (1+\alpha)k)C_{p_x} - (m-k)(1+\alpha)\alpha C_x^0}{m\alpha}(1+\alpha)$$

$$\leq \frac{(m - (1+\alpha)k)(2 + \alpha - \frac{m}{k}\alpha)C_x^0 - (m-k)(1+\alpha)\alpha C_x^0}{m\alpha}(1+\alpha)$$

$$= [-\alpha(\frac{m}{k})^2 + (2+\alpha)\frac{m}{k} - 2(1+\alpha)]\frac{(1+\alpha)C_x^0}{\frac{m}{k}\alpha}.$$

Hence, to obtain (2), we only need to show that $-\alpha(\frac{m}{k})^2 + (2+\alpha)\frac{m}{k} - 2(1+\alpha) \leq 0$, which holds because that $(2+\alpha)^2 - 4\alpha \cdot 2(1+\alpha) = 0$ with $\alpha = \frac{4\sqrt{2}-2}{7}$.

Case 2. $\triangle \leq 0$, i.e., $S_{p_l} \leq C_{p_x}$. If $S_{p_l} \leq L_{x-1}^t$, since p_l is assigned to the machine with minimum load in V_1, we can obtain that $S_{p_l} \leq (T_n - p_l)/m$. It follows that $c^{HA} = p_l + S_{p_l} \leq T_n/m + (1 - /m)p_l \leq (2 - 1/m)c^* < (2+\alpha)c^*$. Now we assume that $S_{p_l} > L_{x-1}^t$. It yields that the load of each machine is at least L_{x-1}^t, which, together with Lemma 4(i), deduces that $c^* \geq L_{x-1}^t \geq (1+\alpha)C_x^0$. By the assumption and Lemma 4(ii), we obtain that $c^{HA} = p_l + S_{p_l} \leq p_l + C_{p_x} \leq c^* + (2 + \alpha - \frac{m}{k}\alpha)C_x^0$. Note that $\frac{m}{k} > 1 + \alpha$, hence,

$$\frac{c^{HA}}{c^*} \leq \frac{c^* + (2 + \alpha - \frac{m}{k}\alpha)C_x^0}{c^*}$$

$$\leq 1 + \frac{2 + \alpha - \frac{m}{k}\alpha}{(1+\alpha)} < 1 + \frac{2 + \alpha - (1+\alpha)\alpha}{(1+\alpha)} < 2 + \alpha. \qquad \square$$

Now we give the main Theorem of this section:

Theorem 5. *Part 2 of the algorithm HA has a competitive ratio of $2+\alpha$. Thus, we have $c^{HA} \leq (2+\alpha)c^*$ for all k and m.*

Proof. It is a direct result due to Lemmas 2, 3 and 5. Together with Theorem 4, we have $c^{HA} \leq (2+\alpha)c^*$ for all k and m. $\qquad \square$

5   Conclusion and Open Problems

In this paper, we studied the online scheduling problem on parallel machines with two GoS levels. We proved the lower bound of the problem is at least 2.

The upper bound was also considered. We first analyzed the performance of algorithm *AW* presented in [2]. Then we presented an online algorithm *HA* which greatly improved the bound of *AW*.

The results of this paper suggest a number of problems deserving further study. An important and natural open question is to design an optimal online algorithm. As this paper only consider the case of two GoS levels, it can be very interesting to extend the result to general m GoS levels case. In addition, it is also worthy studying the preemptive version of the problem as presented in [4].

References

1. H. Hwang, S. Chang, K. Lee, Parallel machine scheduling under a grade of service provision, *Computer & Operations Research*, 2004, 31, 2055-2061.
2. Y. Azar, J. Naor, R. Rom, The competitiveness of on-line assignments, *Journal of Algorithms*, 1995, 18, 221-37.
3. J. K. Lenstra, D. B. Shmoys, N. E. Tardos, Approximation algorithms for scheduling unrelated parallel machines, *Mathematical Programming*, 1990, 46, 259-271.
4. Y. W. Jiang, Y. He, C. M. Tang, Optimal online algorithms for scheduling on two identical machines under a grade of service, *Journal of Zhejiang University SCIENCE*, 2005, 7(3), 309-314.
5. P. Zhou, Y. He, Y. W. Jiang, Parallel machines scheduling with two GoS levels, Technical Report, Zhejiang University, 2005.(in Chinese)

Online Dial-A-Ride Problem with Time-Windows Under a Restricted Information Model

Fanglei Yi[1], Yinfeng Xu[1,2], and Chunlin Xin[1]

[1] School of Management, Xi'an Jiaotong University,
Xi'an, ShaanXi 710049, P.R. China
fangleiyi@163.com
[2] The State Key Lab for Manufacturing Systems Engineering,
Xi'an, ShaanXi 710049, P.R. China
yfxu@mail.xjtu.edu.cn

Abstract. In online dial-a-ride problem with time-windows, requests for rides consist of two points in a metric space, a source and a destination. One server with some finite capacity is required to transports a specified amount of goods for requests from the sources to the destinations. Calls for rides come in while the server is travelling. Each request also specifies a deadline. If a request is not be served by its deadline, it will be called off. The server travels at unit speed in the metric space and the goal is to plan the motion of the server in an online way so that the maximum number of requests (or the maximum quantity of goods) is met by the deadlines of the requests. Usually it is assumed that the server knows the complete information on the ride when the requests are presented. We study this problem under a restricted information model. At the release time of one request, only the information on the source is presented. The server does not have the information on the destination until it reaches the source of the request. This models, e.g. the taxi problem, or elevator problem. We study the problem in the *uniform* metric space and *K-constrained* metric space. We perform competitive analysis of two deterministic strategies in the two types of metric spaces. The competitive ratios of the strategies are obtained. We also prove a lower bound on the competitive ratio of any deterministic algorithm of Z for the *uniform* metric space and of KZ for the *K-constrained* metric space, where Z denotes the capacity of the server and K denotes the diameter of the metric space.

1 Introduction

The class of Dial-A-Ride Problem (DARP) has been studied extensively in the area of operations research, management science, and combinatorial optimization because of its usefulness to the logistics and transportation industry. In DARP, there are servers that travel in some metric space to serve requests for rides. Each ride is characterized by two points in the metric space, a *source*, the starting point of the ride, and a *destination*, the ending point of the ride. The problem is to

S.-W. Cheng and C.K. Poon (Eds.): AAIM 2006, LNCS 4041, pp. 22–31, 2006.

design routes for the servers through the metric space, such that all requested rides are made and some optimality criterion is met. To meet real life needs, many new side constraints have been added to the problem. One useful extension is the Dial-A-Ride Problem with Time-Windows (DARPTW). Each request specifies a deadline. If a request is not be served by its deadline, it will be called off. The goal is to plan the motion of servers so that the maximum number of requests is met by their deadlines. In the natural online setting, requests for rides are presented over time while the servers are enroute serving other rides. And the servers do not know any information on the future requests at any time until they are presented. In other words, the time flows while decisions are made and executed in the online setting of the problem.

Traditionally, when a certain request is presented, all the information on the request becomes known. That is, both the source and the destination of the ride are specified completely upon presentation at the release time. However, in many practical situations complete specification of the rides is not realistic. Often only the source of the ride is presented at the release time of the request. The server is not able to get the information on the destination of the ride until it arrives at the source. This models, e.g. the taxi problem, or elevator problem. So this model is called the *restricted information model*.

In this paper we study the Online Dial-A-Ride Problem with Time-Windows under the Restricted Information Model (ODARPTWRIM). There is a server which has some finite capacity, travelling in an unit speed in a metric space to serve a set of requests for ride. Calls for rides come in with some amount of goods while the server is travelling. Each request specifies a deadline. If the server does not arrive at the source of the request by its deadline, the request will be called off. At the release time of one request, only the information on the source is presented. The server does not have the information on the destination until it reaches the source of the request. The goal is to plan the motion of the server in an online way so that the maximum quantity of goods is transported by the deadlines of the requests.

The online DARP and in general vehicle routing and scheduling problems have been widely studied for more than three decades (see [1] for a survey on the subject). Most previous researches on online routing problems focused on the objectives of minimizing the makespan [2, 3, 4], the weighted sum of completion times [2, 5], and the maximum/average flow time [6, 1]. In the paper [7, 8, 9], results on the online k-taxi scheduling problem have been presented, in which a request consists of two points (a *source* and a *destination*) on a graph or in a metric space. Subsequently, a similar problem, online k-truck scheduling problem has been studied in [10]. Both of them (online k-taxi/truck scheduling) assumed that k servers (taxies or trucks) are all free when a new service request occurs, and the goal is to minimize the total distance travelled by servers. [2] studied the online DARP in which calls for rides come in while the server is travelling. The authors also considered two different cases, where the server has infinite capacity and where the server has finite capacity. The first results of the online DARP under a restricted information model have been presented in [11] in which the objective function is to minimize the time by which the server has executed all the rides and is back in the origin. All of these previous work assumed that the requests could wait for any length of time until the server completed them. That is, they did not consider the constraints of the time window.

Most previous works on the DARP with time window constraints are the offline point of view. The input is known completely beforehand. For some related works, please refer to [12, 13]. In reference [14], we presented the first results on the online DARP with time windows. We study the problem again in this paper under the restricted information model .

We perform competitive analysis of deterministic algorithm for ODARPTW-RIM in two types of metric space, respectively. In the first metric space, uniform metric space, we present an online FCFS (First-Come-First-Serve) algorithm. We prove that FCFS is $2Z$-competitive, where Z denotes the server's capacity. And in the second metric space, K-constrained metric space, another online algorithm, Greedy Algorithm (abbr.GR), is shown which has a $2KZ$-competitive ratio, where K is the diameter of the metric space. We also prove a lower bound on the competitive ratio of any deterministic algorithm of Z for the uniform metric space and of KZ for the K-constrained metric space.

The rest of this paper is organized as follows. In section 2 we give some definitions and notations. Section 3 contains two online algorithms and the discussions of their performances. In section 4 we present lower bounds for the problem in two types of metric space. Section 5 concludes this paper.

2   Definitions and Notations

Let $\mathcal{M} = (X, d)$ be a metric space with n points which is induced by an undirected unweighted graph $G = (V, E)$ with $V = X$, i.e., for each pair of points from the metric space $\mathcal{M}$ we have $d(x, y)$ that equals the shortest path length in G between vertices x and y. We consider two types of metric space in this paper, the *uniform metric space* and the *K-constrained metric space*. In the uniform metric space, the distance between any two points is unit length. It can be considered as a special metric space that is induced by a complete graph with unit edge weights. And in the K-constrained metric space, $\frac{d_{max}}{d_{min}} = K$, where $d_{max} = \max d(x, y), d_{min} = \min d(x, y), x \neq y, x, y \in V$. Without loss of generality, we assume that d_{min} equals 1 and d_{max} is K long in the K-constrained metric space. We call K the *diameter* of the metric space which can be considered the maximum time required to travel between the two farthest points in the metric space. Note that in the uniform metric space, $K = 1$. An instance of the basic ODARPTW in the metric space $\mathcal{M}$ consists of a sequence $R = (r_1, r_2, \cdots, r_m)$ of *requests*. Each request is a quadruple $r_i = (t_i, z_i, a_i, b_i) \in \mathcal{R} \times \mathbf{N} \times X \times X$ with the following meaning: $t_i \in \mathcal{R}$ is the time that request r_i is released and $z_i \in \mathbf{N}$ is the quantity of goods that needs to be transported by the server; $a_i \in X$ and $b_i \in X$ are the source and destination, respectively, between which the goods corresponding to request r_i is to be transported. The capacity of the server is finite, denoted by constant Z. That is, the upper bound of the goods loaded by the server is Z units. We assume the goods of the requests is partible. If a given request has overmany goods in the sense of the current capacity of the server, it can be divided into several partitions. The server is allowed to load some partitions of them. And the rest of them can be considered as the goods of new requests. In this paper we assume that $z_i \leq Z, \forall i \in \mathbf{N}$. It is also assumed that the sequence $R = (r_1, r_2, \cdots, r_m)$ of requests is given in order of non-decreasing release times, that is, $0 \leq t_1 \leq t_2 \leq \cdots \leq t_m$. A request is said to be *accepted request* if the corresponding object is picked up by the server at source, and a request is said to be *completed request* if the corresponding object is transported to the destination. We do not allow *preemption*: it is not allowed to drop an accepted request at any other place than its destination. This means, once a request is accepted, it will not be called off.

Definition 1. *[11] Under the restricted information model only the source a_i of r_i is revealed at time t_i. The destination of the ride becomes known only at picking up the ride in the source.*

In this paper, we consider the following assumptions for ODARPTWRIM as we did in [14]: a) The speed of the server is constant 1. This means that, the time it takes to travel from one point to another is exactly the distance between the two points; b) The window sizes for all requests are uniform, denoted by T. To make sure that the problem is feasible, we assume that $T \geq d_{min}$ in the K-constrained metric space and $T \geq 1$ in the uniform metric space.

We evaluate the quality of online algorithms by competitive analysis [1-3], which has become a standard yardstick to measure the performance. In competitive analysis, the performance of an online algorithm is compared to the performance of the optimal off-line algorithm, which knows about all future jobs. An algorithm $\mathcal{A}$ for ODARPTWRIM is called α-*competitive* if for any instance R the number of goods completed (transported) by $\mathcal{A}$ is at least $1/\alpha$ times the number of goods completed by an optimal off-line algorithm OPT.

3   Algorithms for Two Types of Metric Space

In this sections we firstly study the problem in a general metric space. We prove that there is no competitive deterministic online algorithm for a non-constrained metric space if the time windows $T > 0$ is arbitrary. Then we study the problem in two types of metric space, the uniform metric space and the K-constrained metric space, respectively. We propose the FCFS algorithm for the uniform metric space and the Greedy algorithm for the K-constrained metric space. The performance guarantees of the two algorithms for the problem are shown in this section.

Proposition 1. *If the time windows $T > 0$ is arbitrary, there is a metric space in which no deterministic online algorithm can obtain a constant competitive ratio for ODARPTWRIM.*

Proof. Consider a metric space which contains a line. At time 0, two request with one unit of goods is presented on the line. One request requires the server to load the goods at position T and deliver it to position $2T$, where T is the time window of requests. And the other requires the server to transport the goods from position $-T$ to position $-2T$. Note that the requests will be called off at time T if the server does not arrive at the sources by time T. If the online server does not immediately leave for the requests, the sequence stops and the server can not serve any one of the requests. Otherwise, we assume without loss of generality that at time T the online server reaches position $-T$.The off-line server arrives at position T at time T and delivers the goods to the destination $2T$ at time $2T$. Then, from time $t = 2T$ onwards a request is presented with the source position t and the destination position $t + T$ at each time $(2 + i)T$ ($i = 0, 1, 2, \ldots$). The off-line server can serve all the requests whereas the online server is not able to complete any of them.   □

3.1 FCFS Algorithm in the Uniform Metric Space

The FCFS algorithm works as follows. The online server always goes to the source of the request firstly that has been presented for the longest time among all yet unserved requests, getting the information on the destination and picking up the goods, then delivering them to the destination. If there are not unserved requests that can be *accepted* by their deadlines after completing a certain request, the server remains at its current point and waits for new requests to occur.

Theorem 1. *In the uniform metric space, FCFS algorithm is $2Z$-competitive for ODARPTWRIM, where Z is the capacity of the server.*

Proof. Given any input sequence R, it can always be divided into such m maximal sub-sequences, $R = (R_1, R_2, \ldots, R_m)$ that in each sub-sequence $R_i = (r_{i,1}, r_{i,2}, \ldots)(1 \leq i \leq m)$, the online server works continuously i.e. it serves constantly some requests in R_i one after another.

We will first analyze an arbitrary sub-sequence R_i and then extend the result to the whole sequence R. We will show that in each sub-sequence OPT can not transport (or pick up) more than $2Z$ times as many goods as FCFS does which is followed by the theorem.

Denote by t_1 the released time of the first request $r_{i,1}$ in R_i. Let $r_{i,l} \in R_i$ and $r_{i,l'} \in R_i$ be the last requests that FCFS and OPT served, respectively. Let t_{FCFS} be the time when FCFS reaches the ride destination of $r_{i,l}$, and t_{OPT} denote the time when OPT reaches the ride destination of $r_{i,l'}$. We define t^* which makes sure that $(t_{OPT} - t^*)/2$ is an integral and $t_1 \leq t^* \leq t_1 + 2$. There are two possible cases for t_{FCFS} and t_{OPT}.

Case 1. FCFS completes $r_{i,l}$ no later than time t_{OPT}, i.e., $t_{FCFS} \leq t_{OPT}$. It can be proved that $t_{OPT} - t_{FCFS} < 2$. Since the sub-sequence R_i is maximal and the server works continuously in R_i, no new requests which belongs to R_i can be presented at time $t \geq t_{FCFS}$. And the total time needed to pick up the goods of one request at the source and deliver them to the destination is not more than 2 units of time in the unit metric space. Thus, in any interval $(t, t+2]$ for $t = t^*, t^* + 2, \ldots, t_{OPT} - 2$, the server of FCFS can pick up at least one unit of goods and/or deliver it to the destination while the OPT's server can not pick up more than $2Z$ units of goods at the sources and/or deliver them to the destinations due to the finite capacity of Z. Also, in the interval $(t_1, t^*]$ the FCFS's server can accept at least one request which has one unit of goods, and the OPT's server can not accept more than 2 requests (with $2Z$ units of goods) because $t^* - t_1 \leq 2$. We notice that the number of the goods which is accepted (picked up) by the server is as many as the one that is transported (delivered) since we do not allow preemption. So we can say that in any interval of $(t_1, t^*]$ and $(t, t+2]$ for $t = t^*, t^* + 2, \ldots, t_{OPT} - 2$, OPT can not transport more than $2Z$ times as many goods as FCFS does.

Case 2. FCFS still works after time t_{OPT}, i.e., $t_{FCFS} > t_{OPT}$. Obviously, in each interval $(t, t+2]$ for $t = t^*, t^* + 2, \ldots, t_{OPT} - 2$, OPT can transport at most $2Z$ times as many goods as FCFS does. It also holds in the interval $(t_1, t^*]$ with the same reasoning in case 1.

This completes the proof.   □

3.2   Greedy Algorithm in the K-Constrained Metric Space

We present the GR as follows. At any time t when the server arrives at one point in the metric space, it finds the source of the request which has the least time path from its current position among all the outstanding requests that it can reach by their deadlines, picking up the goods and getting the information on the destinations, until it is overweighted or there are no outstanding requests. Then it delivers the goods of the accepted requests to their destinations.

Theorem 2. *In the K-constrained metric space, the Greedy algorithm is $2ZK$-competitive for ODARPTWRIM, where Z is the capacity of the server and K denotes the diameter of the metric space.*

Proof. The proof of theorem 1 goes through with a few changes. Given any input sequence R, we can always divide it into such m sub-sequences $R = (R_1, R_2, \ldots, R_m)$ that in each sub-sequence $R_i = (r_{i,1}, r_{i,2}, \ldots)(1 \leq i \leq m)$, the GR's server is empty when it begins to serve the first request $r_{i,1}$ of R_i. According to the algorithm of GR, under the circumstance of that the online server is not overweighted, the server always picks up the goods of requests if there are some outstanding requests existing or delivers the accepted goods to the destinations otherwise. So in each sub-sequence the online server will works continuously. As we did in theorem 1, we need only to show that OPT can transport at most $2ZK$ times as many goods as GR does for each sub-sequence.

Denote by t_1 the released time of the first request $r_{i,1}$ in R_i. Let $r_{i,l} \in R_i$ and $r_{i,l'} \in R_i$ be the last requests that GR and OPT served, respectively. Let t_{GR} be the time when GR reaches the ride destination of $r_{i,l}$, and t_{OPT} denote the time when OPT reaches the ride destination of $r_{i,l'}$. We define t^* such that

$(t_{OPT} - t^*)/2K$ is an integral and $t_1 \leq t^* \leq t_1 + 2K$, where K is the diameter of the metric space. There are two possible cases for t_{GR} and t_{OPT} in the same way. Note that in case 1, $t_{OPT} - t_{GR} < 2K$ since the total time needed to pick up the goods of one request at the source and deliver them to the destination is not more than $2K$ units of time in the K-constrained metric space. The rest of the proof holds unchanged.   □

Corollary 1. *GR is $2Z$-competitive for ODARPTWRIM in the uniform metric space.*

Though GR has the same competitive ratio as FCFS does in the uniform metric space, we notice that FCFS is simpler than GR and the FCFS's server may carry less goods than GR's server during the working time.

4   Lower Bounds

In this section we derive lower bounds on the competitive ratio of any deterministic online algorithm for serving the requests in the version of the problem in two types of metric space, respectively. The results are obtained by considering the optimal algorithm as an adversary that specifies the request sequence in a way that the online algorithm performs badly.

Theorem 3. *In the uniform metric space, no deterministic online algorithm can obtain a competitive ratio less than Z for ODARPTWRIM, where Z is the capacity of the server.*

Proof. We consider an arbitrary deterministic algorithm $\mathcal{A}$ and an adversary $(\mathcal{AD})$ which constructs an input request sequence so that $\mathcal{A}$ will not achieve a competitive ratio small than Z. At time 0 both of their servers locate at the origin. Set the size of time window T is 1. $\mathcal{AD}$ will present the requests in steps. In the first step, $\mathcal{AD}$ ensures that the server of itself and the server of $\mathcal{A}$ are not at the same point in the metric space.

Step 1. Two requests with different positions of sources and destinations from each other are presented at time 0, each with one unit goods. That is, $r_1 = (0, 1, a_1, b_1)$ and $r'_1 = (0, 1, a'_1, b'_1)$ where $a_1 \neq a'_1 \neq b_1 \neq b'_1$. If the server of $\mathcal{A}$ does not leave immediately for the requests, then $\mathcal{AD}$ stops the sequence and let its server go to any one of points a_1 and a'_1, completing one request with 1 unit goods. Otherwise, we assume without loss of generality that $\mathcal{A}$'s server goes to a'_1, then the server of $\mathcal{AD}$ goes to a_1 for serving the request r_1. So $\mathcal{A}$ can serve at most 1 unit goods while $\mathcal{AD}$ is able to complete at least 1 unit goods. Go to the step 2.

Step 2. After step 1, there are two possible position for the server of $\mathcal{A}$.
    P1. The server of $\mathcal{A}$ is at a vertex point different from the position of $\mathcal{AD}$.
    P2. $\mathcal{A}$'s server is on an edge.
    If P1 holds, $\mathcal{AD}$ releases another two requests which have the same characteristic to the ones that are presented in step 1. If $\mathcal{A}$'s server does not immediately

go to one of points of the sources, it does not serve any one of these two requests, whereas $\mathcal{AD}$'s server can complete at least one request. Otherwise, after a short period of time $0 < \Delta t < 1$, $\mathcal{AD}$ releases a request r^* with Z units goods whose source point is not incident to the edge on which $\mathcal{A}$'s server is. Hence, the $\mathcal{A}$'s server can not pick up the goods by the deadline of request r^*, whereas $\mathcal{AD}$'s server can do that by remaining in its position for Δt time and then going to the source of r^*.

If $P2$ holds, $\mathcal{A}$'s server is in the interior of and edge. Hence, it can not reach any vertex point which is not incident to this edge in 1 unit time. Now $\mathcal{AD}$ presents a request with Z units goods whose point of source is not incident to the edge on which $\mathcal{A}$'s server is. Thus, $\mathcal{A}$'s server is not able to serve the request, whereas $\mathcal{AD}$ can complete these Z units goods.

So in this step the server of $\mathcal{A}$ can complete at most 1 unit goods, and $\mathcal{AD}$'s server is able to serve at leat Z units goods.

In step 2, the adversary can arrange to make its server stay at a point different from the position of $\mathcal{A}$'s server when presenting the requests.

Step 3. Repeating step 2 for M times.

Denote by $|\mathcal{AD}|$ and $|\mathcal{A}|$ the total number of goods completed by $\mathcal{AD}$ and $\mathcal{A}$ respectively. We get

$$\frac{|\mathcal{AD}|}{|\mathcal{A}|} \geq \frac{ZM + 1}{M + 1}$$

As M grows, the right hand side gets arbitrarily close to Z. So the theorem holds.   □

Theorem 4. *In the K-constrained metric space, no deterministic online algorithm can obtain a competitive ratio less than KZ for ODARPTWRIM, where Z is the capacity of the server and K denotes the diameter of the metric space.*

Proof. The proof follows the proof of theorem 3 closely. Let $T = K$. In step 1 the way in which $\mathcal{AD}$ presents the requests is the same to theorem 3.

In step 2, when $P1$ holds, $\mathcal{AD}$ then presents a request r^* with 1 unit goods of which the source point a^* is K units length away from the current position of $\mathcal{A}$'s server. If the server of $\mathcal{A}$ does not go to a^* immediately, $\mathcal{AD}$ stop the sequence and let its server to complete the request r^* while $\mathcal{A}$'s server can not serve the request. Otherwise, $\mathcal{AD}$ releases K requests, $r_i (i = 1, 2, \ldots, K)$, each with Z units goods just after the $\mathcal{A}$'s server leaves. The source points $a_i (i = 1, 2, \ldots, K)$ and the destination points $b_i (i = 1, 2, \ldots, K)$ of the requests have the following characteristic. Each request for ride needs only 1 unit time to complete,.i.e., $d(a_i, b_i) = 1 (i = 1, 2, \ldots, K)$ and $a_i = b_{i-1}$ for $i = 2, 3, \ldots, K$. The source point of the first request is the current position of $\mathcal{AD}$'s server. And none of the source points of K requests is incident to the edge on which $\mathcal{A}$'s server is. Thus, the $\mathcal{A}$'s server can not serve any one of the K requests, whereas $\mathcal{AD}$'s server can complete all of them with KZ units goods. The rest of the proof remains the same.   □

5 Conclusions

In this paper we discuss the online dial-a-ride problems with time windows under the restricted information model which is occurring in a wide variety of practical settings. It is an important issue since in practice complete information is often lacking[11]. We present two online algorithm for two types of metric space. For upper bounds, we analyze the performance of FCFS in the uniform metric space and of GR in the K-constrained metric space respectively. We also give the analysis for lower bounds on the competitive ratio of any deterministic algorithm for the problem in different metric spaces.

It is worth the whistle that the uniform metric space is in fact a special case of the K-constrained metric space with $K = 1$. It is shown in corollary 1 that GR has the same competitive ratio as FCFS does in the uniform metric space. However, we feel that FCFS is a simpler algorithm than GR. And more important, the FCFS's server may carry less goods than GR's server during the working time which implies that FCFS may cost less energy for per unit of goods than GR does in practice.

An interesting extension of the problem considered in this paper may take into account the non-uniform time windows. That is, each of the requests has a different size of time window. It would also be interesting to study other particular metric spaces (such as trees, cycles, etc.) to see if better bounds can be obtained. Other possible extension of the problem is to crew scheduling in which more than one server is used to serve the requests. All of these can be further investigated.

Acknowledgements

This work was partly supported by a grant from the National Science Fund of China (No.70471035) and the National Science Fund of China for Distinguished Young Scholars (No.70525004).

References

1. Krumke S.O., Laura L., Lipmann M., and Marchetti-Spaccamela et al. Non-abusiveness helps: An $o(1)$-competitive algorithm for minimizing the maximum flow time in the online traveling salesman problem. *Lecture Notes in Computer Science*, (2002) 200–214.
2. Feuerstein E. and Stougie L. On-line single server dial-a-ride problems. *Theoretical Computer Science*, **268**(1), (2001) 91–105.
3. Ascheuer N., Krumke S.O., and Rambau J. Online dial-a-ride problems: Minimizing the completion time. *Lecture Notes in Computer Science*, (2000) 639–650.
4. Ausiello G., Feuerstein E., Leonardi S., Stougie L., and Talamo M. Algorithms for the on-line traveling salesman. *Algorithmica*, **29**(4), (2001) 560–581.
5. Krumke S.O., de Paepe W.E., Poensgen D., and Stougie L. News from the online traveling repairman. *Theoretical Computer Science*, **295**, (2003) 279–294.

6. Hauptmeier D., Krumke S.O., and Rambau J. The online dial-a-ride problem under reasonable load. *Lecture Notes in Computer Science*, (2000) 125–136.
7. Xu Y.F. and Wang K.L. Scheduling for on-line taxi problem and competitive algorithms. *Journal of Xi'an Jiao Tong University*, **31**(1), (1997) 56–61.
8. Xu Y.F., Wang K.L., and Zhu B. On the k-taxi problem. *Journal of Information*, **2**, (1999) 429–434.
9. Xu Y.F., Wang K.L., and Ding J.H. On-line k-taxi scheduling on a constrained graph and its competitive algorithm. *Journal of System Engineering(P.R. China)*, **4**, (1999).
10. Ma W.M., XU Y.F., and Wang K.L. On-line k-truck problem and its competitive algorithm. *Journal of Global Optimization*, **21**(1), (2001) 15–25.
11. Lipmann M., Lu X., de Paepe W.E., and Sitters R.A. On-line dial-a-ride problems under a restricted information model. *Lecture Notes in Computer Science*, **2461**, (2002) 674–685.
12. Psaraftis H.N. An exact algorithm for the single vehicle many-to-many dial-a-ride problem with time windows. *Transportation Science*, **17**, (1983) 351–357.
13. Diana M. and Dessouky M.M. A new regret insertion heuristic for solving large-scale dial-a-ride problems with time windows. *Trasportation Reserch Part B*, **38**, (2004) 539–557.
14. Yi F.L. and Tian L. On the online dial-a-ride problem with time windows. *Lecture Notes in Computer Science*, **3521**, (2005) 85–94.

Online Scheduling with Hard Deadlines on Parallel Machines[*]

Jihuan Ding[1,2] and Guochuan Zhang[1]

[1] Department of Mathematics, Zhejiang University
Hangzhou 310027, China
{dingjihuan, zgc}@zju.edu.cn
[2] College of Operations Research and Management Science,
Qufu Normal University, Rizhao 276826, China

Abstract. In this paper, motivated by on-line admission control in the hard deadline model, we deal with the following scheduling problem. We are given m identical machines (multi-streams). All jobs (requests) have identical processing time. Each job is associated with a release time and a deadline, neither of which is known until the job arrives. As soon as a job is available, we must immediately decide if the job is accepted or rejected. If a job is accepted, then it must be completed no later than its deadline. The goal is to maximize the total number of jobs accepted. The one-machine case has been extensively studied while little is known for multiple machines. Our main result is deriving a nontrivial optimal online algorithm with competitive ratio $\frac{3}{2}$ for the two-machine case by carefully investigating various strategies. Deterministic lower bounds for the general case are also given.

1 Introduction

Problem statement. We deal with the following real-time scheduling problem with online admission control. We are given m identical machines. Each job j arrives at its release time r_j that is not known in advance. Upon arrival of job j its deadline d_j is revealed. All jobs have equal processing times of 1. Preemption is not allowed. At any time when some machine is idle, we have to decide whether to start an "accepted" job or not, and if so, to choose which one, based only on the information on the jobs released so far. Those jobs that can not be scheduled to meet their deadlines will be lost (not processed at all). The objective is to maximize the number of completed jobs, i.e., the number of jobs meeting their deadlines. We also call this objective throughput maximization.

To evaluate online algorithms we adopt the standard measure of *competitiveness*. An online algorithm is *c-competitive* if on every input instance the number of early jobs by the algorithm is at least $\frac{1}{c}$ times that of an optimum schedule.

Previous work. The throughput maximization problem on a single machine has been extensively studied in the literature. Goldman *et al.* [6] gave a lower

---

[*] This work has been supported by NSFC (60573020).

S.-W. Cheng and C.K. Poon (Eds.): AAIM 2006, LNCS 4041, pp. 32–42, 2006.

bound of $\frac{4}{3}$ on the competitive ratio of randomized algorithms and the tight bound of 2 for deterministic algorithms for the online problem. They further proved that a greedy algorithm is a $\frac{3}{2}$-competitive if $d_j - r_j \geq 2$ for all jobs j, which implies that the lower bound of 2 can be beaten if the jobs have sufficiently large slack. Along this line, Goldwasser [3] made a parameterized extension of this result: if $d_j - r_j \geq \lambda$ for all jobs j, where $\lambda > 0$ is a real number, then the competitive ratio is $1 + \frac{1}{\lambda}$. In 2003 Goldwasser and Kerbikov [4] extended the previous results [3] under a reasonable assumption called *immediate notification*. This assumption requires the scheduler to determine whether to accept job j immediately at its arrival. If the job is accepted it must be completed by this deadline.

Chrobak *et al.* [2] considered randomization and restarts for online scheduling of equal-length jobs. They gave a $\frac{5}{3}$-competitive randomized algorithm. For the restart model (allowed to abort a job during execution and an aborted job can be restarted and completed later), they presented an optimal $\frac{3}{2}$-competitive algorithm.

In contrast, to our best known, there are few results for the throughput maximization problem on parallel machines. The offline problem on m machines can be solved in polynomial time [1]. For the online case Lee [7] considered the problem of maximizing the sum of the length of accepted jobs on m machines, where each job of length L can be delayed for at least $k \cdot L$ for $k < 1$ before it is started, while still meeting its deadline. He presented an $O(\log(1/k))$-competitive randomized algorithm for m machines where $m = 1, 2, \ldots, O(\log(1/k))$. For $m \geq 2$ a $[m + 1 + m \cdot (1/k)^{(1/m)}]$-competitive deterministic algorithm was derived.

Our results. We first derive simple deterministic lower bounds for m parallel machines. Then we show that the competitive ratio of a simple algorithm is exactly two. Our main result is an optimal non-trivial on-line algorithm with competitive ratio of 3/2 for the two-machine case (very recently Goldwasser and Pedigo [5] obtained the same ratio independently). Moreover all our results are still valid under the assumption of *immediate notification*.

2   Lower Bounds for $m \geq 2$ Machines

Theorem 1. *For on-line scheduling of unit-length jobs on m machines to maximize the number of jobs completed, the competitive ratio of any deterministic online algorithm is not smaller than*

$$R = \begin{cases} 3/2 & \text{if } m = 2 \\ 6/5 + 1/(5k) & \text{if } m = 3k \\ (6k+2)/(5k+1) & \text{if } m = 3k+1 \\ (6k+3)/(5k+2) & \text{if } m = 3k+2 \end{cases}, k \geq 1.$$

Proof. We only prove the lower bound for $m = 2$. At time 0, a job with deadline 5 comes. For any algorithm A, let $t(0 \leq t \leq 4)$ denote the start time of the job (If the algorithm does not start the job by time 4, the job will not be scheduled. It results in an unbounded competitive ratio.) Then two jobs with deadline $t+1+x$ come at time $t + x(0 < x < 1)$. So one job is lost and the algorithm accepts two jobs. But the optimal schedule can accept all the three jobs. So the competitive ratio is at least $\frac{3}{2}$ for $m = 2$.

3   Online Algorithms

For an instance I, let σ and σ^* denote the schedule produced by an algorithm A and the optimum schedule respectively. For given schedules σ^* and σ, we say job i *blocks* job j if a) $j \in \sigma^*, j \notin \sigma$ and b) job i is processed in σ on the same machine with j in σ^* and $s_i \le s_j^* < s_i + 1$, where s_i is the start time of job i in σ and s_j^* is the start time of job j in σ^*.

Simple Algorithm LS. Before we start scheduling jobs all machine loads $L_i = 0$ for $i = 1, 2, \ldots, m$. As a job j is coming check if there exists a machine that is able to complete the job by its deadline. If such a machine, say M_i, exists, schedule the job j at time $s_j = \max\{r_j, L_i\}$ and set $L_i = s_j + 1$. If such a machine does not exist, discard job j (job j is thus lost).

Clearly the Simple Algorithm satisfies *immediate notification*. The following theorem can be easily shown with the similar analysis as in [6]. For the sake of completeness we also give a proof.

Theorem 2. *The competitive ratio of algorithm LS is 2.*

Proof. Consider the following instance with $2m$ jobs. The first m jobs are available at time zero, each of which has a deadline $\frac{5}{2}$. And the other m ones with a common deadline $\frac{3}{2}$ come at time $\frac{1}{2}$. Using algorithm LS the second m jobs will be lost, while the optimal schedule, that the first m jobs are processed immediately after the second m jobs are completed at time $\frac{3}{2}$, enable all jobs to meet their deadlines. Thus we have $R_{LS} \ge 2$.

Let σ and σ^* denote schedules produced by algorithm LS and a given optimal schedule respectively. For each job $J_i \in \sigma^*$, if $J_i \in \sigma$, then assign $\frac{1}{2}$ of J_i to J_i; if $J_i \notin \sigma$, then some job $J_j \in \sigma$ must block J_i, we assign $\frac{1}{2}$ of J_j to J_i, because each job can block at most one job. So the number of jobs in σ is at least half that of σ^*. So we have $R_{LS} \le 2$.

Note that algorithm LS is simple but not greedy. However, even if the algorithm becomes greedy in the following sense the ratio of two is still tight: always pick the available job with the smallest deadline, and always assign the job to the machine with the least load. To improve the algorithm we are suggested to leave some space in the schedule for the future jobs. Basically we have to answer two questions below.

Q1. How to decide if the incoming jobs can be accepted and make an immediate notification?

Q2. How to manage a schedule so that all accepted jobs will not be lost and there is some idle space for the future jobs? More precisely, for how long an accepted job can wait and when it should be processed?

To prepare our answers to the above questions we need some preliminaries.

EDF (Earliest Deadline First) Rule. Re-number the available jobs in nondecreasing order of their deadlines. Always pick the first job from the remaining job list and schedule the job on the machine with the smallest load (the completion time of the last job on this machine); remove the job from the job list. Continue the process till the job list is null.

Online Admission Control. Let $P(t)$ denote the set of the accepted jobs that have not been started at time t. Set $P(t) = \emptyset$ if $t < 0$. If some new jobs arrive, then we test the new jobs one by one in the nondecreasing order of their deadlines, namely, for a newly arrived job set $J(t) = \{j_1, j_2, \ldots, j_k\}$, $d_1 \geq d_2 \geq \ldots d_k$, where d_i is the deadline of job j_i, we first "put" j_1 into $P(t)$ and use EDF to pre-schedule all the jobs in $P(t)$ on the two machines (they may have already completed/started some old jobs before the jobs in $P(t)$ are put). If all the jobs meet their deadlines then we accept j_1 (and update $P(t)$ by adding j_1), otherwise reject job j_1 (remove it from $P(t)$). Continue this process until all jobs in $J(t)$ has been tested.

In *Online Admission Control* a new job is tested immediately. It is either accepted (but may be processed later) or rejected. Thus an online algorithm with *Online Admission Control* satisfies the immediate notification. Thus we have answered question **Q1**. In the next subsection we will present an improved online algorithm for the case $m = 2$, where question **Q2** will be answered.

3.1   Algorithm MEDF for Two Machines

If at time t some new jobs arrive, we update $P(t)$ according to the *Online Admission Control*. For the job set $P(t)$, we pre-schedule $P(t)$ at time t with EDF rule and let C_j denote the completion time of job j for each job $j \in P(t)$. If for every job $j \in P(t)$ we have $d_j - C_j > 1$, then we call $P(t)$ an *open* job group; Otherwise, we call it an *impatient* job group. For the jobs that satisfy $d_j - C_j \leq 1$, we call them *critical* jobs at time t.

In the following algorithm $MEDF$, at any time t, we always use $L_1(t)$ and $L_2(t)$) to denote the earliest time points not smaller than t at which either machine can start processing next jobs. Assume that $L_1(t) \leq L_2(t)$. The machine of $L_1(t)$ is called the Min machine while the other is called Max machine.

Algorithm MEDF (Modified EDF):

0. Set $t = 0$.
1. If $P(t) = \emptyset$, then go to **5**.
2. At time t, if both the machines are idle, then start the ED job in $P(t)$ to one of the two machines and remove this job from $P(t)$. $L_1(t) := t$, $L_2(t) := t+1$.
3. Pre-schedule $P(t)$ according to EDF rule at time t and let $M_1(t)(\subset P(t))$ denote the set of jobs arranged to the machine Min. Let C_j be the completion time of job j for each $j \in P(t)$. Then compute $\beta' = \min_{j \in P(t)}\{d_j - C_j - 1\}$ and $\beta = \min_{j \in M_1(t)}\{d_j - C_j - 1\}$.

3.1: If $\beta' > 0$, then we do not arrange any job at time t. Once some new jobs come in $(t, \min\{L_2(t), t + \beta\}]$, then update $P(t)$; Otherwise replace t with $\min\{L_2(t), t + \beta\}$. Go to **1**.

3.2: If $\beta' < 0$, then find the last critical job and denote by $U(t)$ the job set that contains all the jobs $(\in P(t))$ with deadline not larger than the deadline of the last critical job. If $U(t)$ cannot be completed by either machine, then arrange the ED job in $P(t)$. $L_1(t) := L_2(t)$, $L_2(t) := L_1(t) + 1$. If some jobs arrive in $(t, L_2(t)]$ then update $P(t)$. Replace t with $L_2(t)$ and go to **1**. If $U(t)$ can be completed on one machine then compute $\gamma = \min_{j \in M_1(t) \backslash U(t)}\{d_j - C_j - 1\}$. Go to **4**.

4. $U(t)$ can be completed on one machine.

4.1: $U(t)$ cannot be completed on the machine Max. Pre-schedule $U(t)$ on the machine Min according to EDF Rule and let C_j' denote the completion time of job j for each job $j \in U(t)$. Then compute $\alpha = \min_{j \in U(t)}\{d_j - C_j'\}$,

4.1.1 If $\alpha = 0$ or $\alpha > 0$ but $\min_{j \in U(t)}\{d_j\} < L_2(t) + 1$, we start the ED job in $U(t)$ on the machine Min. If some jobs arrive in $(t, L_2(t)]$ then update $P(t)$. Replace t with $L_2(t)$. Go to **1**.

4.1.2 If $\alpha > 0$ and $\min_{j \in U(t)}\{d_j\} \geq L_2(t) + 1$, we do not arrange any job at time t. Once some new jobs arrive in $(t, t + \min\{\alpha, \gamma\}]$ then update $P(t)$. Go to **1** . If no job arrives in $(t, t + \min\{\alpha, \gamma\}]$, then replace t with $t + \min\{\alpha, \gamma\}$. Go to **1**.

4.2: $U(t)$ can be completed on the machine Max, then we do not arrange any job at time t. Once some new jobs arrive in $(t, \min\{L_2(t), t + \gamma\}]$, update $P(t)$ and go to **1**. Otherwise, replace t with $\min\{L_2(t), t + \gamma\}$. Go to **1**.

5. If no more jobs arrive, stop. Otherwise let t be the arrival time of the next incoming job. Update $P(t)$ and go to **1**.

An Example. There are thirteen jobs in total. One job j_0 with very large deadline arrives at time zero. At time $x(0 < x < 1)$, eight jobs with deadline 7 arrive. At time $3 + x$, two jobs with deadline $4 + x$ come. At time $4 + x$, two jobs with deadline $5 + x$ come. For this instance, it is obvious that the algorithm accepts nine jobs, but the optimal schedule accepts all the thirteen jobs (see Figure 1 and Figure 2).

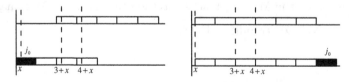

Fig. 1. The MEDF schedule   **Fig. 2.** An optimal schedule

3.2   Analysis of Algorithm *MEDF*

We will prove that algorithm $MEDF$ has a competitive ratio of $3/2$ with the help of a minimum counterexample. Let I be a counterexample with the minimum number of jobs. Let σ and σ^* be the schedule produced by the algorithm $MEDF$ and an optimum schedule on I, respectively. According to $MEDF$, it is obvious that at any time t smaller than the makespan of the schedule σ, at least one machine is busy. Otherwise if at some time t both machines are idle, then either the jobs arriving before time t or the jobs arriving after time t forms a job list conflicting with competitive ratio $3/2$. It is a contradiction with the assumption of a minimum counterexample. Let j_0 be the first job arranged in the schedule σ.

Let $[s_i, C_{i,\min}]$, $i = 1, \ldots, k$, be the i-th period during which both machines are busy in σ (see Figure 3). Let f_i be the number of jobs starting before time $C_{i,\min}$ in σ. Denote by $C_{i,\max}$ the largest completion time among the jobs starting before time $C_{i,\min}$. Let $\mathcal{A}_i, \mathcal{B}_i$ denote the job sets in which all the jobs arrive before time $C_{i,\min}$ and are started at or after $C_{i,\min}$ in σ and in σ^*, respectively. Moreover assume that $|\mathcal{A}_i| = a_i$ and $|\mathcal{B}_i| = b_i$. Without loss of generality let $\mathcal{A}_i = \{j_1, j_2, \ldots, j_{a_i}\}$ and the deadlines of the jobs in $\mathcal{A}_i$ satisfy $d_1 \geq d_2 \geq \ldots \geq d_{a_i}$. We also sort the jobs in $\mathcal{B}_i \backslash \{j_0\} = \{j'_1, \ldots, j'_{|\mathcal{B}_i \backslash \{j_0\}|}\}$ in non-increasing order of their deadlines, i.e., $d'_1 \geq d'_2 \geq \ldots \geq d'_{|\mathcal{B}_i \backslash \{j_0\}|}$. Consider those jobs with release times smaller than $C_{i,\min}$ in I and assume that σ^* accepts L_i more such jobs than σ does. Then the number of jobs released before time $C_{i,\min}$ that are accepted in σ^* is exactly $f_i + a_i + L_i$. Thus $(f_i + a_i + L_i - b_i)$ jobs are started before time $C_{i,\min}$ in σ^*. Check these jobs and let c_1 and c_2 be the largest completion times among them on the two machines, respectively, where $c_1 \leq c_2$. We further define $C'_{i,\min} = \max\{C_{i,\min}, c_1\}$ and $C'_{i,\max} = \max\{C_{i,\min}, c_2\}$ in σ^*. For the sake of convenience, let $l_i = \frac{1}{2}[f_i - (a_i - b_i)]$, $i = 1, \ldots, k$. Let x_i, $i = 1, \ldots, k$, denote the number of jobs which are started before time $C_{i,\min}$ but completed after time s_i in σ.

We will prove that $L_i \leq l_i$ for $1 \leq i \leq k$ by induction. Then for the last busy period we obtain $L_k \leq l_k = \frac{1}{2}[f_k - (a_k - b_k)]$. Finally we can achieve $L_k \leq \frac{1}{2}(f_k + a_k)$ with the analysis of the values of a_k and b_k. In the following we first give some properties of the algorithm.

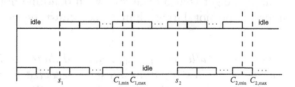

Fig. 3. Illustration of the MEDF schedule

Lemma 1. *Let C_0 be the makespan of the schedule σ, then at most one job, namely the first job j_0 arranged by the algorithm in σ, which has deadline larger than $C_0 + 1$.*

Proof. We prove it by contradiction. We assume $j_*(\neq j_0)$ is the last job with deadline larger than $C_0 + 1$. Let s_* be the starting time of the job j_*, it is easy to verify that at time s_* only j_* is available. Now we construct another instance I' from instance I. We change the release time and the deadline of the job j_* to

s_* and $+\infty$ in instance I'. Any other jobs in I' are the same as the job in I. It is obvious that for instance I and I', the algorithm generates the same schedule. And any optimal schedule for instance I' accepts at least the same number of jobs as the optimal schedule for instance I. So I' is a counterexample too. We can easy to verify that either the jobs with arrival time smaller than s_* in I' or the jobs with arrival time not smaller than s_* in I' construct a counterexample. It contradicts with the minimum counterexample I.

Lemma 2. *Let $C_{0,\min} = 0$, then for any $t_0 \in [C_{i-1,\min}, s_i)$, $i = 1, \dots, k$, while one of the two machines is idle in σ all the jobs with arrival time t_0 must be accepted by the algorithm.*

Proof. Suppose that $t_0 \in [C_{i-1,\min}, s_i)$ and one machine is idle during $[C_{i-1,\min}, s_i)$ in σ. We consider the job set $P(t_0)$ after the deletion of the job arranged at time t_0 by the algorithm if at time t_0 a job is really arranged. We prove that j_* must be accepted by the algorithm if job j_* comes at time t_0 for such $P(t_0)$. Therefore j_* must be accepted by the algorithm before the job is arranged by the algorithm at time t_0. We have $|U(t_0)| \leq 2$. Otherwise both machines should be busy from t_0. If $P(t_0)$ is an open job group at time t_0, then it is obvious that the algorithm can accept job j_*. If $P(t_0)$ is an impatient job group at time t_0.

Case 1. $|U(t_0)| = 1$. Then the deadline of the job in $U(t_0)$ is not smaller than $L_2(t_0) + 1$. The i-th job in $P(t_0) \backslash U(t_0)$ must have a deadline larger than $L_2(t_0) + (i+1)/2 + 1$ when i is odd and $t_0 + i/2 + 2$ when i is even. So job j_* is accepted by the algorithm.

Case 2. $|U(t_0)| = 2$. Let j_1, j_2 be the jobs in $U(t_0)$. Then we have $d_1 \geq L_2(t_0)+1$ and $d_2 > t_0 + 2$. The i-th job in $P(t_0) \backslash U(t_0)$ must have deadline larger than $t_0 + \frac{i+1}{2} + 2$ when i is odd and $L_2(t_0) + \frac{i}{2} + 2$ when i is even. So job j_* can be accepted by the algorithm.

Remark. By Lemma 2, we get that all the jobs with deadline not smaller than $C_{i,\min} + 1$ and arrival time not larger than $C_{i,\min}$ must be accepted by the algorithm.

Lemma 3. *If one machine is idle in $[C_{i-1,\min}, s_i)$ and there are jobs coming at time s_i, then the job with smallest deadline must be accepted by the algorithm. b). If $L_2(s_i) > s_i$ and only jobs with deadline not smaller than $L_2(s_i) + 1$ come in $[s_i, L_2(s_i)]$, then at least one job with deadline smaller than $C_{i,\min} + 1$ can be accepted by the algorithm if such jobs come in $[s_i, L_2(s_i)]$.*

Proof. Let $r < s_i$ be the largest arrival time over all the jobs arrived before s_i. There must exist a $t_0 \in (r, s_i)$ such that for any $t \in (t_0, s_i)$ we have $P(t) = P(t_0)$ and $U(t) = U(t_0)$ (see Figure 4). It is obvious that $|U(t_0)| \leq 2$. Otherwise both machines should be busy from time t_0. If we schedule all the jobs in $P(t_0)$ using EDF at time s_i, then we must have $\min_{j \in P(t_0) \backslash U(t_0)} \{d_j - C_j\} \geq 1$.

a) If $|U(t_0)| = 0$, then it is obvious that the job with smallest deadline can be accepted by the algorithm if some jobs arrive at time s_i. If $|U(t_0)| = 1$, then the job in $U(t_0)$ must have deadline not smaller than $L_2(s_i)+1$. We can schedule one newly arrived job with smallest deadline at time s_i on one machine and arrange the job in $U(t_0)$ on the other machine at time $L_2(s_i)$. If $|U(t_0)| = 2$, let j_1, j_2 be the jobs in $U(t_0)$, Then we have $d_1 \geq L_2(s_i)+1$ and $d_2 \geq s_i+2$. We can schedule one newly arrived job with smallest deadline at time s_i on one machine and the job with smaller deadline in $U(t_0)$ at time $L_2(s_i)$ on another machine. For the jobs in $P(t_0)\backslash U(t_0)$, they must be completed before their deadlines by EDF algorithm. Therefore if some jobs arrive at time s_i, then the job with smallest deadline can be accepted by the algorithm.

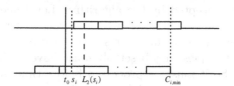

Fig. 4. The case $L_2(s_i) > s_i$ in the MEDF schedule

b) If only those jobs with deadline not smaller than $C_{i,\min}+1$ come in $[s_i\,,L_2(s_i)]$ and we schedule the jobs in $P(L_2(s_i))$ at time $L_2(s_i)$ by the $MEDF$ algorithm then the time $C^*_{i,\min}$ when one machine is idle must not larger than $C_{i,\min}$. If $|U(t_0)| \leq 1$, then for the jobs in $P(L_2(s_i))$, with deadline smaller than $C_{i,\min}+1$, started before $C^*_{i,\min}$ we have $d_j - C_j \geq 1$. For the jobs, with deadline not smaller than $C_{i,\min}+1$, started before $C^*_{i,\min}$, they can be delayed one by one. So if some jobs with deadline smaller than $C_{i,\min}+1$ come in $[s_i\,,L_2(s_i)]$, then at least one of them can be accepted. If $|U(t_0)| = 2$, note that the job j_1 in $U(t_0)$ must have been arranged at time s_i, we can schedule one job with deadline smaller than $C_{i,\min}+1$ at time $L_2(s_i)$ on one machine and the job j_2 at time s_i+1 on another machine. The remaining jobs must be completed before their deadlines by EDF algorithm. So at least one such job can be accepted by the algorithm.

Lemma 4. *For the jobs in $\mathcal{A}_i$ and $\mathcal{B}_i$, $i = 1,\ldots,k$, we have $d_j \geq d'_j, j = 1, 2, \ldots, \min\{a_i, |\mathcal{B}_i\backslash\{j_0\}|\}$.*

Proof. (By induction) We first prove that $d_1 \geq d'_1$; Otherwise $d_1 < d'_1$. According to Lemma 2 all the job with deadline at least d'_1 must be accepted by the algorithm. Let j_* be the last job, with deadline at least d'_1, which is started before $C_{i,\min}$ in σ. Let s_* be the starting time of the job j_*. If $s_* \geq s_i$, then all the jobs in job set $\mathcal{A}_i$ must have deadline at least d'_{t+1}, it is a contradiction. So it must be processed before time s_i(see Figure 5). According to the algorithm, $P(s_*)$ contains only job j_*. Let $I'_1 = \{j|r_j \leq s_*\}$. Now we consider $I'_1\backslash j_*$. According to the algorithm j_* cannot affect the configuration of schedule of the algorithm on job set $I'_1\backslash j_*$, so if $I'_1\backslash j_*$ is a counterexample then we derive a smaller counterexample than I, it is a contradiction; if it is not, then we construct I' from I. We change the release time of job j_* to s_* and omit all the jobs arrived before s_*; For the jobs with release time larger than s_* in I', they are the same as the jobs in I. Then we get a smaller counterexample I' than I. It is a contradiction.

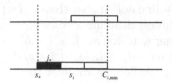

Fig. 5. The case that $s_* < s_i$

Now suppose that $d_1 \geq d'_1, \ldots, d_t \geq d'_t$, then we show that $d_{t+1} \geq d'_{t+1}$. Otherwise $d_{t+1} < d'_{t+1}$. According to the algorithm all the job with deadline at least d'_{t+1} must be accepted by the algorithm. Let j_* be the last job, with deadline at least d'_{t+1}, which is started to be processed before $C_{i,\min}$ in the schedule produced by the algorithm. Let s_* be the starting time of the job j_*. If $s_* \geq s_i$, then all the jobs in job set $\mathcal{A}_i$ must have deadline at least d'_{t+1}, it is a contradiction. So it must be processed before time s_i. According to the algorithm, $P(s_*)$ must be an open job group. Otherwise, any job in $\mathcal{A}_i$ must have deadline at least d'_{t+1}, it is a contradiction. For the optimal schedule, at least $|P(s_*)|$ jobs arrived at or before s_* are processed at or after $C_{i,\min}$. Let $I'_1 = \{j | r_j \leq s_*\}$. Now we consider $I'_1 \backslash P(s_*)$. According to the algorithm, $P(s_*)$ cannot affect the configuration of schedule of the algorithm on job set $I'_1 \backslash P(s_*)$, so if $I'_1 \backslash P(s_*)$ is a counterexample we derive a smaller counterexample than I, it is a contradiction; if it is not, then we construct I' from I. We change the release time of jobs in $P(s_*)$ into s_* and omit all the jobs arrived before s_*; For other jobs in I', it is the same as the jobs in I. Then we get a smaller counterexample I' than I. It is a contradiction.

Lemma 5. *If $a_i = 0$, $i = 1, \ldots k$, then $|\mathcal{B}_i \backslash \{j_0\}| = 0$.*

Proof. We prove this lemma by contradiction. If $|\mathcal{B}_i \backslash \{j_0\}| \geq 1$, then there exists a job $j' \neq j_0$ in $\mathcal{B}_i \backslash \{j_0\}$, with deadline at least $d'_j \geq C_{i,\min} + 1$, which is processed at or after time $C_{i,\min}$ in σ^*. According to the algorithm all the jobs with deadline at least $C_{i,\min} + 1$ must be accepted by the algorithm. Let j_* be the last job with deadline at leat $d'_j \geq C_{i,\min} + 1$ which is started before time $C_{i,\min}$ in σ. Let s_* be the starting time of job j_*. It is obvious that $P(s_*)$ contains just one job j_*.

If $s_* < C_{i,\min} - 1$(see Figure 6), then according to the algorithm, $P(s_*)$ contains only one job and it is an open job group at s_*. Let $I'_1 = \{j | r_j \leq s_*\}$. Now we consider $I'_1 \backslash P(s_*)$. According to the algorithm $P(s_*)$ cannot affect the configuration of the schedule of the algorithm on job set $I'_1 \backslash P(s_*)$, so if $I'_1 \backslash P(s_*)$ is a counterexample we derive a smaller counterexample than I, it is a contradiction; if it is not, then we construct I' from I. We change the release time of jobs in $P(s_*)$ into s_* and delete all the jobs arrived before s_*; For other jobs in I', they are the same as the jobs in I. Then we get a smaller counterexample I' than I. It arrives at a contradiction.

If $s_* = C_{i,\min} - 1$(see Figure 7), then at time $s_* = C_{i,\min} - 1$ there must exist two jobs that have not been arranged. According to the algorithm the other job must have deadline at least $C_{i,\min} + 1$, so another job should be arranged at time $C_{i,\min}$ when j_* is completed. It is a contradiction.

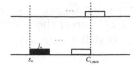 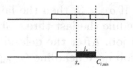

Fig. 6. The case that $s_* < C_{i,\min} - 1$ **Fig. 7.** The case that $s_* = C_{i,\min} - 1$

We further give two lemmas below. The omitted proofs will be included in the full paper.

Lemma 6. $L_1 \leq l_1$. Furthermore, if $L_1 = l_1$, then $C'_{1,\min} > C_{1,\max}$; if $L_1 = l_1 - \frac{1}{2}$, then $C'_{1,\max} \geq C_{1,\max}$.

Lemma 7. $L_i \leq l_i$ for $i = 1, \ldots, k$. Furthermore, if $L_i = l_i$, then $C'_{i,\min} > C_{i,\max}$; if $L_i = l_i - \frac{1}{2}$, then $C'_{i,\max} \geq C_{i,\max}$.

Theorem 3. The competitive ratio of algorithm $MEDF$ is $\frac{3}{2}$.

Proof. Now we consider the last period. By Lemma 7, we have $L_k \leq l_k$. It is easy to verify $a_k \leq 3$. Otherwise if we schedule the last a_k jobs at $C_{k,\min}$ using EDF, then the last job is not a critical job. So after the deletion of the last job from I, it is still a counterexample. It is a contradiction. So we have $a_k \leq 3$.

Case 1. If $a_k = 0$, $L_k \leq \frac{1}{2}(f_k + 1)$ by Lemmas 5 and 7. If f_k is even, then the theorem holds; If it is odd, we will prove that $L_k \leq \frac{1}{2}f_k$. Suppose that it is not true, $L_k = \frac{1}{2}(f_k + 1)$. If $L_{k-1} = l_{k-1}$, then we have $C'_{k-1,\min} > C_{k-1,\max}$. In this period $\frac{1}{2}(f_k + 1) - l_{k-1}$ jobs are blocked. We have $x_k \geq \frac{1}{2}(f_k + 1) - l_{k-1}$. If $x_k = \frac{1}{2}(f_k + 1) - l_{k-1}$, all these x_k jobs must be accepted by the optimal schedule, otherwise $L_k < \frac{1}{2}(f_k + 1) - 1 \leq \frac{1}{2}f_k$. Because $b_k = 1$, all these x_k jobs must arrive before s_k. According to the algorithm, these jobs should be started before s_k in the schedule σ. It is a contradiction. If $x_k = \frac{1}{2}(f_k + 1) - l_{k-1} + 1$, we can derive a similar contradiction with the same analysis. Thus $L_k \leq \frac{1}{2}f_k$. It is easy to verify $L_k \leq \frac{1}{2}f_k$ for the case that $L_{k-1} \leq l_{k-1} - \frac{1}{2}$. Therefore $L_k \leq \frac{1}{2}f_k$.

Case 2. If $a_k = 1$, then the last job j_1 must have deadline $d_1 = C_{k,\max} + 1$. Otherwise, if $d_1 > C_{k,\max} + 1$, then we consider the job j_2 that is started at time $C_{k,\max} - 1$. By the algorithm $d_2 < C_{k,\min} + 1$. If not, then $\alpha > 0$ and $d_2 \geq C_{k,\min} + 1$. Job j_2 should not have been started at $C_{k,\max} - 1$. So we have $d_2 < C_{k,\min} + 1$. By the algorithm after the deletion of job j_1, it is still a counterexample. It is a contradiction. Thus $d_1 = C_{k,\max} + 1$. If $L_k = l_k$, then $C'_{k,\min} > C_{k,\max}$. We get $b_k \leq 2$ and $L_k \leq \frac{1}{2}(f_k + 1)$; If $L_k \leq l_k - \frac{1}{2}$, then $b_k \leq 3$. Hence $L_k \leq \frac{1}{2}(f_k + 1)$.

Case 3. If $a_k = 2$, then the last job must have deadline $C_{k,\max} + 2$. Otherwise if we schedule the last two jobs at $C_{k,\min}$ using EDF, then the last job is not a critical job. After the deletion of the last job, it is still a counterexample. It is a contradiction. So the last job have deadline $C_{k,\max} + 2$. If $L_k = l_k$, then we have $C'_{k,\min} > C_{k,\max}$. Therefore $b_k \leq 4$ and $L_k \leq \frac{1}{2}(f_k + 2)$; If $L_k \leq l_k - \frac{1}{2}$, then we have $b_k \leq 5$ and thus prove $L_k \leq \frac{1}{2}(f_k + 2)$.

Case 4. If $a_k = 3$, then the last job must have deadline $C_{k,\min} + 3$. Otherwise if we schedule the last three jobs at $C_{k,\min}$ using EDF, then the last job is not a critical job. After the deletion of the last job, it is still a counterexample. So the last job have deadline $C_{k,\min} + 3$. If $L_k = l_k$, then we have $C'_{k,\min} > C_{k,\max}$ by Lemma 7. Thus $b_k \leq 5$, so $L_k \leq \frac{1}{2}(f_k + 3)$; If $L_k \leq l_k - \frac{1}{2}$, we get $b_k \leq 7$. So it is obvious that $L_k \leq \frac{1}{2}(f_k + 3)$.

Summarizing the above case analysis, we show that the competitive ratio of the algorithm is $3/2$.

4 Final Remarks

We consider the problem scheduling jobs of unit-length online to maximize the number of early jobs. Other than the single machine case in which a greedy algorithm can achieve the best ratio of two, the case that $m \geq 2$ becomes much more complex. It is easy to have an online algorithm with competitive ratio of two. To improve the bound we design a non-trivial optimal $3/2$-competitive online algorithm for $m = 2$. Our algorithms satisfy the assumption of immediate notification. An obvious question is to improve the Simple Algorithm LS for the general case of m machines. Before closing the paper we propose a stronger property than *immediate notification*, called *immediate decision*. As a job is coming the scheduler is asked to decide immediately whether to accept the job or not, if yes when to start it (its deadline must be met). Note that the algorithm LS has this property but the algorithm $MEDF$ does not. It is interesting to see an optimal online algorithm with *immediate decision* even for $m = 2$.

References

1. P. Baptiste, P. Brucker, S. Knust, and V. Timkovsky. Ten notes on equal-excution-time scheduling: at the frontiers of solvability in polynomial time. *4OR* 2:111-127, 2004.
2. M. Chrobak, W. Jawor, J. Sagll, T. Tichy. Online schedulig of equal-length jobs: randomization and restarts help. In *Proc. 31st International Colloquium on Automata, Languages, and Programming (ICALP)*, LNCS 3142, pages 358-370, 2004.
3. M.H. Goldwasser. Patience is a virtue: the effect of slack on competitiveness for admission control. In *Proc. 10th ACM-SIAM Symposium on Discrete Algorithms (SODA)*, pages 396-405, 1999.
4. M.H. Goldwasser and B. Kerbikov. Admission control with immediate notification. *Journal of Scheduling* 6: 269-285, 2003.
5. M.H. Goldwasser and M. Pedigo, Online, non-preemptive scheduling of equal-length jobs on two identical machines. To appear in SWAT 2006.
6. S. Goldman, J. Parwatikar, and S. Suri. On-line scheduling with hard deadlines. *Journal of Algorithms* 34: 370-389, 2000.
7. J.-H. Lee. Online deadline scheduling: multiple machines and randomization. In *Proc. the Fifteenth Annual ACM Symposium on Parallel Algorithms (SPAA)*, pages 21-25, 2003.

Maximizing the Throughput of Multiple Machines On-Line*

Jae-Hoon Kim

Department of Computer Engineering,
Pusan University of Foreign Studies, Busan 608-738, Korea
jhoon@pufs.ac.kr

Abstract. We study the nonpreemptive online scheduling of jobs with deadlines and weights. The goal of the scheduling algorithm is to maximize the total weight of jobs completed by their deadlines. As a special case, the weights may be given as the processing times of jobs, where the job instance is said to have *uniform value density*.

Most previous work of nonpreemptively scheduling jobs online concentrates on a single machine and uniform value density. For the single machine, Goldwasser [6] shows a matching upper bound and lower bound of $(2 + \frac{1}{\kappa})$ on the best competitive ratio, where every job can be delayed for at least κ times its processing time before meeting its deadline. This paper is concerned with multiple machines. We provide a $(7 + 3\sqrt{\frac{1}{\kappa}})$-competitive algorithm defined on multiple machines. Also we consider arbitrary value density, where jobs have arbitrary weights. We derive online scheduling algorithms on a single machine as well as on multiple machines.

1 Introduction

The problem of scheduling jobs with deadlines has extensively been studied in the literature. In this problem, a set of jobs with release times, processing times, and deadlines is given as a input. The output is a schedule of jobs in which the jobs are completed by their deadlines. This paper is concerned with an online environment, where jobs arrive over time, and their processing times and deadlines are revealed at the release times. The online scheduling algorithms have to make decisions with no information about the future arrivals of jobs.

The job has a weight which represents its importance or a gain of a system when serving it. The total gain of a system is the sum of weights of jobs which are completed by their deadlines. So the goal of a scheduling algorithm is to maximize the gain of the system. In many cases, the weight of the job may be given as its processing time.

The scheduling algorithms are divided into preemptive and nonpreemptive. The preemptive algorithm can abort a job in its execution and resume it later,

---

* This work was supported by the Korea Research Foundation Grant funded by the Korean Government(KRF-2005-003-D00277).

S.-W. Cheng and C.K. Poon (Eds.): AAIM 2006, LNCS 4041, pp. 43–52, 2006.

but the nonpreemptive algorithm cannot stop a job after accepting it. In a field of online deadline scheduling, there are a large number of researches for the preemptive algorithms both on a single machine [1, 2, 4] and on multiple machines [3]. In this paper, we will focus our attention on the nonpreemptive algorithms. In this case, a little number of results on a single machine are known [5, 6, 12], but there are few ones on multiple machines. We are interested in multiple machines.

In case the weight of a job is given as its processing time, Goldwasser [6] provided an optimal online algorithm on a single machine for an instance parameterized by a term, called a patience. This paper can be considered as an generalization of [6] to multiple machines, which was proposed as one of further studies in citegoldwasser. In case the weight of a job is given arbitrarily, to the best of our knowledge, there is no previously known work for the nonpreemptive scheduling.

Notation. In an input instance $\mathcal{I}$, each job J_j has release time r_j, processing time p_j, deadline d_j, and weight w_j. The *expiration time* x_j is defined by $d_j - p_j$, which means the latest time when the job J_j can be started while still meeting its deadline. Then we define the *slack* s_j of J_j by $x_j - r_j$. We assume all jobs in $\mathcal{I}$ satisfy the following inequality: $s_j \geq \kappa p_j$, for a nonnegative constant κ, called *patience*. The weight of a job divided by its processing time, namely, $\frac{w_j}{p_j}$, is called the *value density* of the job and denoted by λ_j. The *importance ratio* of an input instance $\mathcal{I}$, denoted by λ, is the ratio of the largest value density to the smallest value density. When the weights of all jobs are given as their processing times, the importance ratio is 1 and $\mathcal{I}$ is said to have *uniform value density*. For convenience, we assume the smallest processing time is 1 and we denote the largest processing time as Δ.

Related work. Most previous work considers the case when the instance has uniform value density. In this problem, the online algorithms will maximize the total processing time of the completed jobs. For nonpreemptive scheduling, the problem was initiated by Lipton and Tomkins [12]. They studied only the instance where all jobs have zero slack, that is, the job should be accepted as soon as it arrives, if not, it must be rejected. This problem is called *online interval scheduling*. The authors provide an $O((\log \Delta)^{1+\epsilon})$-competitive randomized algorithm, and a lower bound of $\Omega(\log \Delta)$ on the competitive ratio of any randomized online algorithm. Goldman *et. al.* [5] eliminate the assumption of zero slack, allowing jobs to have arbitrary slacks, specifically, $\kappa = 0$. They show there is also an $O(\log \Delta)$-competitive randomized algorithm and for jobs with an equal processing time, there is a 2-competitive algorithm. In [6], Goldwasser introduced the patience κ. For a job instance with an equal processing time, he shows a simple $(1 + \frac{1}{\lfloor \kappa \rfloor + 1})$-competitive algorithm and the ratio is best possible for any deterministic online algorithm. For arbitrary processing times, he shows there is a $(2 + \frac{1}{\kappa})$-competitive algorithm and the ratio is also best possible.

The above results all deal with the problem of a single machine. There is few work about multiple machines, following the outcomes on the single machine. In [11], Lee considers the problem of multiple machines to obtain a randomized algorithm on a single machine. The randomized algorithm simulates a randomly-chosen machine, envisaging the schedule of the deterministic algorithm for multiple machines. In [7], it is shown that a simple greedy algorithm has the same upper bounds as those in [6] on its competitive ratios even if it is defined on multiple machines.

For arbitrary value density, *i.e.*, the case where each job has an arbitrary value of a weight, there is a little work only for the preemptive scheduling. Koren and Shasha [10] provide an optimal online algorithm with its competitive ratio having a matching upper bound and lower bound of $(1 + \sqrt{\lambda})^2$. Also they extend the problem to the multiple machines in [9]. In [8], the resource augmentation model is studied, where the throughput of the online algorithm on multiple machines is compared with that of the offline optimal algorithm on a single machine.

Our results. For uniform value density, we present a $(7 + 3\sqrt{\frac{1}{\kappa}})$-competitive algorithm defined on multiple machines. It is an improvement on the upper bound of $O(\frac{1}{\kappa})$ given in [7]. For arbitrary value density, we first derive a lower bound on the competitive ratio of any online algorithm on a single machine. Moreover, we provide a simple algorithm with the competitive ratio matching the lower bound. Also we propose an online algorithm for multiple machines.

Throughout the paper, for a set S of jobs, $\|S\|$ denotes the total weight of jobs in S. Also for a job J, $p(J)$, $w(J)$, and $sl(J)$ denotes the processing time of J, the weight of J, and the slack of J, respectively.

2   Uniform Value Density

In this section, we consider the case when the weights of jobs are given as their processing times. We will present an $O(\sqrt{\frac{1}{\kappa}})$-competitive online algorithm, defined on multiple machines, where the number of machines, denoted by m, is at least two. The machines are divided into classes so that each class consists of two machines. Of course, when m is odd, the last class has only one machine. Specifically, the class c_i contains the machine m_{2i-1} and m_{2i}, for $i = 1, \ldots, \lceil \frac{m}{2} \rceil$. For the convenience of analysis, the algorithm does not use the machine in the last class when m is odd.

There is a pool of jobs, waiting to be scheduled. If a job in a pool cannot meet its deadline anymore, that is, it is not scheduled until its expiration time, then it is canceled from the pool. We will adopt an identical algorithm to each class. At any time in the execution of the algorithm, only one machine in the class is *active*, that is, a job can be scheduled only on the active machine in the class. For a class c_i, we define a variable *active-machine$_i$* to denote its active machine. Assume at time t, a job J is given in c_i, to be scheduled. If the *active-machine$_i$* is idle, schedule J on it. Otherwise, we consider the job J' which the *active-machine$_i$* currently processes in c_i. Then we compare the processing times of J and J'. In case $p(J) \geq C \cdot p(J')$, if the other machine M than *active-machine$_i$* in c_i, called the *inactive* machine, is idle at t, then update *active-machine$_i$* to M and schedule

J on M. If there is a job J'' processed on M at t, then put J into the pool. But we consider the time s when the job J'' is completed after t. If J can be scheduled at s without violating its deadline, that is, its expiration time is greater than or equal to s, then we say that J is reserved from J' at s and the time s is called a *reservation time* of J. In case $p(J) \leq C \cdot p(J')$, put J into the pool. Thus with the independent algorithms of the classes, we can design the master algorithm A. Whenever a job arrives, a job is completed on *active-machine$_i$* in a class c_i or a reservation time is met, A makes decisions. When a job J arrives, J is given to the classes subsequently from c_1. After J is given to c_i, the above algorithm is performed in c_i, and if J is not scheduled in c_i and put into the pool, then J is given to the next class c_{i+1}, if any. When a job is completed on *active-machine$_i$* in a class c_i, choose the job with the longest processing time in the pool and schedule it on *active-machine$_i$*. Lastly, when a reservation time s is met, if the time s is defined in a class c_i and M is the inactive machine in c_i, then choose the job with the longest processing time among all the jobs which are reserved at s in the pool, if any. If such a job exists, then update *active-machine$_i$* to M and schedule it on M.

We consider specific periods in the execution time of the algorithm A. The period in which a job is processed on *active-machine$_i$* for every class c_i is called a *busy* period. Then the whole period in the execution of A is divided into busy periods and the other periods, called *loose* periods. At any time in a loose period, there is an *active-machine$_i$* for some class c_i on which no job is processed. Then we first pay attention to the jobs with expiration times lying in a loose period, called *patient* jobs. We can prove that A schedules all patient jobs. Note that no job arrives during any loose period.

Lemma 1. *All patient jobs are scheduled in the execution of the algorithm A.*

Proof. Assume that J is a patient job with expiration time lying in a loose period (s, t) and not scheduled by A. At time s, J would remain in the pool since its expiration time is after s. Then J could be scheduled on an idle *active-machine$_i$* for some class c_i. It is a contradiction. □

The jobs other than the patient jobs are called *urgent* jobs, which have expiration times lying in busy periods. All jobs are classified into urgent jobs and patient jobs. For the sake of convenience, we consider only the urgent jobs that are scheduled by the offline optimal algorithm OPT but rejected by our algorithm A. We call them *missing* jobs. If a missing job J has its expiration time lying in a busy period $\mathcal{I}$, then J arrives in $\mathcal{I}$, because if J arrived before $\mathcal{I}$, J would be scheduled in a loose period by A. Also the missing job J will be scheduled and rejected within $\mathcal{I}$ by OPT and A, respectively.

Here we can derive a simple upper bound on the competitive ratio of A. Let J be a missing job with expiration time lying in a busy period $\mathcal{I}$. Then J arrives in $\mathcal{I}$. Thus J may be completed within $\mathcal{I}$ or if not, from the definition of the patience κ, we can see that $p(J)$ is at most $\frac{1}{\kappa}$ times the length of the interval $\mathcal{I}$. Also, if a job is scheduled by OPT and not a missing job, then from lemma 1, it is scheduled by A. Consequently, let $\mathcal{O}$ and $\mathcal{A}$ be the set of jobs scheduled by OPT and A, respectively, and let $\mathcal{M}$ be the set of missing jobs. Then $||\mathcal{O}|| \leq ||\mathcal{M}|| + ||\mathcal{O} \setminus \mathcal{M}|| \leq 3(1 + \frac{1}{\kappa})||\mathcal{A}|| + ||\mathcal{A}|| = (4 + \frac{3}{\kappa})||\mathcal{A}||$, because in A, there are at least $\lceil \frac{m}{2} \rceil - 1$ machines processing a job in a busy period. Thus we obtain the upper bound of $(4 + \frac{3}{\kappa})$ on the competitive ratio of A. From now on, we will try to derive an upper bound of $O(\sqrt{\frac{1}{\kappa}})$. Since $\kappa \geq \sqrt{\kappa}$ if $\kappa \geq 1$, hereafter, we assume that $0 < \kappa < 1$.

Let J be a job given at time t in a class c_i in A. If there is a job J' currently processed on $active\text{-}machine_i$ such that $p(J) \geq C \cdot p(J')$ and there is also a job J'' processed at t on the inactive machine M, then we can prove that J is reserved at s if C is given as $\sqrt{\frac{1}{\kappa}}$, where the time s is the completion time of J''. Since $\sqrt{\frac{1}{\kappa}} > 1$, we can easily see that the inactive machine M is also idle whenever $active\text{-}machine_i$ is idle. When J' is scheduled at time u, there may be two cases: If J' was chosen from the pool after the completion of a job, then M was idle at time u. It is a contradiction. If J' was scheduled, satisfying that $p(J') \geq \sqrt{\frac{1}{\kappa}}p(J'')$, then we can see that $p(J) \geq \sqrt{\frac{1}{\kappa}}p(J') \geq \frac{1}{\kappa}p(J'')$. Therefore the slack of J is sufficiently large to satisfy that $sl(J) \geq \kappa p(J) \geq p(J'')$. It says that J can be scheduled after J'' on M, that is, J is reserved at the time s when J'' is completed.

Lemma 2. *Let J be a job given in a class c_i at time t in A. If a job J' is currently processed on $active\text{-}machine_i$ such that $p(J) \geq C \cdot p(J')$ and a job J'' is processed at t on the inactive machine M, then J is reserved at s if C is given as $\sqrt{\frac{1}{\kappa}}$, where the time s is the completion time of J''.*

Let J be a missing job, arriving in a busy period $\mathcal{I}$. Then J is scheduled on a machine m_j within $\mathcal{I}$ by OPT. Let m_j belong to a class c_i. If J starts to be processed at time s, then in A, we consider the job which the $active\text{-}machine_i$ in the class c_i processes at time s. It is called the blocking job of J and it is denoted by b^J. Note that several missing jobs may have a same blocking job. Here we will show that if missing jobs have a same blocking job, then the total processing time of the missing jobs is bounded by the processing times of the blocking job and a job reserved from the blocking job.

Lemma 3. *If missing jobs $J_1, \ldots, J_n$, scheduled on a same machine and arranged in order of their start times in the optimal schedule, have a same blocking job b, then,*

$$\sum_{i=1}^{n} p(J_i) \leq (1 + \sqrt{\frac{1}{\kappa}})p(b) + p(\tilde{b}),$$

where $\tilde{b}$ is reserved from b and scheduled at the reservation time if such a job exists.

Proof. Let $C = \sqrt{\frac{1}{\kappa}}$. If the block job b is processed during $[s, t]$. Then only the last missing job J_n can be completed after time t. So, $\sum_{i=1}^{n-1} p(J_i) \leq p(b)$. If

J_n arrives before s, then we consider the last time u when J_n is checked by A. Note that in A, a job is checked at its arrival time, its reservation time, or a completion time of a job. Let a be the job processed on the *active-machine$_i$* at u, where b is processed in a class c_i. Since J_n is a missing job, we can see that $w(J_n) \leq Cw(a)$. Then, $w(J_n) \leq Cw(b)$ since $w(a) \leq w(b)$. If J_n arrives after s, then from the lemma 2, $w(J_n) \leq w(b)$ if $p(J_n) \geq Cp(b)$.   □

Here we can see that the total processing time of the missing jobs is bounded by the throughput of A. Let $C = \sqrt{\frac{1}{\kappa}}$ and $\mathcal{M}$ be the set of all missing jobs. Also let $\mathcal{B}$ be the set of all blocking jobs and $\tilde{\mathcal{B}}$ be the set of all jobs reserved from blocking jobs and scheduled at the reservation times. Then directly from the lemma 3, we can see that $||\mathcal{M}|| \leq 2(1 + C)||\mathcal{B}|| + 2||\tilde{\mathcal{B}}||$ if m is even. In the case when m is odd, the jobs scheduled on the last machine in the schedule of OPT have the jobs scheduled on the class c_{n-1} as the blocking jobs, defined similarly to the above, where $n = \lceil \frac{m}{2} \rceil$. Then we can also see that $||\mathcal{M}|| \leq 3(1 + C)||\mathcal{B}|| + 3||\tilde{\mathcal{B}}||$.

Corollary 1. *Let $\mathcal{M}$ be the set of all missing jobs and let $\mathcal{A}$ be the set of all jobs scheduled by A. Then,*

$$||\mathcal{M}|| \leq 3(2 + \sqrt{\frac{1}{\kappa}})||\mathcal{A}||.$$

As a consequence of the above lemmas, we show that A is $O(\sqrt{\frac{1}{\kappa}})$-competitive.

Theorem 1. *Let $\kappa > 0$ be the patience of a job instance. Then the nonpreemptive online algorithm A defined on multiple machines is $(7 + 3\sqrt{\frac{1}{\kappa}})$-competitive.*

Proof. Let $\mathcal{A}$ and $\mathcal{O}$ be the set of all jobs scheduled by A and OPT, respectively, and let $\mathcal{M}$ be the set of all missing jobs. Also let $\mathcal{U}$ and $\mathcal{P}$ be the set of all urgent jobs and the set of all patient jobs, respectively. Then all jobs in $(\mathcal{O} \cap \mathcal{U}) \setminus \mathcal{M}$ are scheduled by A and from the lemma 1, all jobs in $\mathcal{P}$ are also scheduled by A. So from the corollary 1,

$$||\mathcal{O}|| \leq ||\mathcal{M}|| + ||\mathcal{O} \setminus \mathcal{M}|| \leq (7 + 3\sqrt{\frac{1}{\kappa}})||\mathcal{A}||.$$ □

3   Jobs with Arbitrary Weights

In this section, we will study about an online scheduling for jobs with weights. Before we consider the multiple machines case, we will deal with the single machine. In both case, each job J_i has a constant number w_i, representing its importance and a value density $\lambda_i = \frac{w_i}{p_i}$. Also we consider the importance ratio λ, the ratio of the maximum value density to the minimum value density. Similarly to the previous section, we define busy periods, loose periods, urgent jobs, and patient jobs. Then the lemma 1 still holds in both case.

3.1   Single Machine

First we are concerned with the difficulty of the problem. We derive a lower bound on the competitive ratio of any online algorithm on a single machine.

Theorem 2. *Let $\kappa > 0$ and λ be the patience of the job instance and the importance ratio, respectively. Then the competitive ratio of any online algorithm on a single machine is at least $1 + \lambda(1 + \frac{1}{\kappa})$.*

Proof. Let $\alpha = \kappa + 1$ be an integer. At time 0, there arrives a job J_1 with the processing time of $\alpha + 2\epsilon$, an infinite deadline, and the weight of $\alpha + 2\epsilon$. Then the job J_1 must be scheduled at some time t by any online algorithm B so that it may have a bounded competitive ratio. At time $t + \epsilon$, α jobs with the processing time of 1, the deadline of $t + \alpha + \epsilon$, and the weight of λ arrive. Also at the same time, there arrives another job J_2 with the processing time of $1 + \frac{1}{\kappa}$ and the weight of $(1 + \frac{1}{\kappa})\lambda$. The job J_2 has the deadline of $t + \alpha + (1 + \frac{1}{\kappa}) + \epsilon$. Then all the given jobs satisfy the definition of the patience κ. Also we see that the α jobs and the job J_2 have their expiration times of at most $t + \alpha + \epsilon$. So they cannot be scheduled due to J_1. But OPT can schedule all the jobs, delaying the first job J_1 after the schedule of the other jobs. Thus the competitive ratio of B is at least $1 + \frac{\lambda\alpha}{\alpha + 2\epsilon}(1 + \frac{1}{\kappa})$. By choosing sufficiently small ϵ, the lower bound becomes $1 + \lambda(1 + \frac{1}{\kappa})$. $\qquad\square$

Here we present an optimal online algorithm with the competitive ratio to meet the lower bound. It is a simple greedy algorithm G which at any idle time of the machine, schedules any job having already arrived. For convenience, we choose the job with the largest weight. The schedule of G is divided into busy periods and the other ones. The analysis can be performed independently in each busy period and we will concentrate on only one busy period. So all jobs are assumed to arrive in the busy period $[0, \ell]$. The urgent jobs and the patient jobs are defined to have expiration times in $[0, \ell]$ and after time ℓ, respectively. The set of all urgent jobs and the set of all patient jobs are denoted by $\mathcal{U}$ and $\mathcal{P}$, respectively.

Theorem 3. *Let $\kappa > 0$ and λ be the patience of the job instance and the importance ratio, respectively. The greedy algorithm G is $1 + \lambda(1 + \frac{1}{\kappa})$-competitive.*

Proof. Let $\mathcal{O}$ and $\mathcal{G}$ be the set of all jobs scheduled by OPT and G, respectively. The jobs scheduled by G in $[0, \ell]$ are denoted by $J_1, \ldots, J_n$, in order. Let J_i have the processing time p_i and the value density v_i, $i = 1, \ldots, n$. Then, $||G|| = \sum_i v_i p_i \geq v_{min}\ell$, where v_{min} is the smallest value density in the job instance.

We also consider the jobs $O_1, \ldots, O_m$ in $\mathcal{U} \cap \mathcal{O}$ in order. Each job O_i has the processing time p_i^o and the value density v_i^o. Then only the last job O_m may be completed after time ℓ. Assume that O_m is completed after time ℓ. Then, $\sum_{i=1}^{m-1} p_i^o \leq \ell$ and $\kappa p_m^o \leq \ell$. Thus, $||\mathcal{U} \cap \mathcal{O}|| = \sum_i v_i^o p_i^o \leq v_{max}(\ell + \frac{\ell}{\kappa})$, where v_{max} is the largest value density. So the upper bound on the competitive ratio of G is derived as follows: $||\mathcal{O}|| = ||\mathcal{P} \cap \mathcal{O}|| + ||\mathcal{U} \cap \mathcal{O}|| \leq ||G|| + \lambda(1 + \frac{1}{\kappa})||G|| = (1 + \lambda(1 + \frac{1}{\kappa}))||G||$. $\qquad\square$

3.2 Multiple Machines

To obtain an algorithm A^w, we will slightly modify the algorithm A given in the previous section. In A, we compared the processing times of J and J', for the arriving job J and the currently processed job J'. But in A^w, we compare their weights, that is, we use the inequality $w(J) \geq C \cdot w(J')$. When the job J arrives at time t, if $w(J) \geq C \cdot w(J')$ and the inactive machine M is idle, then update *active-machine$_i$* to M and schedule J on M. If there is also a processed job J'' on M, then put J into the pool and if J'' is completed at time s while J' is still processed at s, we say that J is reserved from J' at s and s is the reservation time of J. When a job is completed at t on *active-machine$_i$*, we choose the job J with the largest weight in the pool. If M is idle at t, then we schedule J on *active-machine$_i$*. If there is a job J'' processed at time t on M, then we schedule J on *active-machine$_i$* only if $w(J) \geq C \cdot w(J'')$. Otherwise, we just update *active-machine$_i$* to M. When a reservation time s is met, choose the job with the largest weight among all jobs reserved at s, if any and if such a job exists, update *active-machine$_i$* to M and schedule the chosen job on M. Here we choose C as $\max\{1, \sqrt{\frac{\lambda}{\kappa}}\}$. As in the previous section, we can define a missing job J and its blocking job b^J in the same meaning. Then the following lemmas are derived.

Lemma 4. *Let J be a job given in a class c_i at time t in A^w. If a job J' is currently processed on active-machine$_i$ such that $w(J) \geq C \cdot w(J')$ and a job J'' is processed at t on the inactive machine M, then J can be scheduled at time s, where s is the completion time of J''.*

Proof. There are two cases in which J' is scheduled. One is that J' arrives after the start time of J'' and satisfies $w(J') \geq \sqrt{\frac{\lambda}{\kappa}} w(J'')$, and the other is that J' is scheduled after a completion of a job. In the latter case, J' also satisfies $w(J') \geq \sqrt{\frac{\lambda}{\kappa}} w(J'')$. Thus we can see that $w(J) \geq \frac{\lambda}{\kappa} w(J'')$. From the definition of the patience κ, $sl(J) \geq \kappa p(J) \geq p(J'') \lambda \frac{w(J'')/p(J'')}{w(J)/p(J)} \geq p(J'')$. It says that J can be scheduled after the completion of J''.   □

Lemma 5. *If missing jobs $J_1, \ldots, J_n$, scheduled on the same machine and arranged in order of their start times in the optimal schedule, have the same blocking job b, then,*

$$\sum_{i=1}^{n} w(J_i) \leq (1 + \Delta)(Cw(b) + w(\tilde{b})),$$

where $\tilde{b}$ is the job scheduled next to b in the class if such a job exists.

Proof. First we see that $\sum_{i=1}^{n-1} p_i \leq p(b)$. Let J_j be a missing job for $j = 1, \ldots, n$ and let s be the time when J_j starts to be processed in the optimal schedule. We consider the last time t when J_j is checked by A^w before s. Let a be the job processed at t in the class. If a is not b, then we can see that $w(J_j) \leq Cw(a)$. Since $w(a) \leq w(b)$, $w(J_j) \leq Cw(b)$. If a is exactly b, then either $w(J_j) \leq Cw(b)$ or $w(J_j) \leq w(\tilde{b})$. Therefore, $w(J_n) \leq Cw(b) + w(\tilde{b})$ and

$$\sum_{i=1}^{n-1} w(J_i) \leq \sum_{i=1}^{n-1} w(J_i)p_i \leq \sum_{i=1}^{n-1} p_i(Cw(b) + w(\tilde{b})) \leq p(b)(Cw(b) + w(\tilde{b}))$$

$$\leq \Delta(Cw(b) + w(\tilde{b})). \qquad \square$$

Corollary 2. *Let* $\mathcal{M}$ *be the set of all missing jobs and let* $\mathcal{A}$ *be the set of all jobs scheduled by* A. *Then,*

$$\|\mathcal{M}\| \leq 3(1 + \Delta)(C + 1)\|\mathcal{A}\|.$$

From corollary 2, an upper bound on the competitive ratio of A^w is given.

Theorem 4. *Let* $\kappa > 0$ *and* λ *be the patience of the job instance and the importance ratio, respectively, and let* Δ *be the largest processing time. Then* A^w *is* $(1 + 3(1 + \Delta)(\max\{1, \sqrt{\frac{\lambda}{\kappa}}\} + 1))$-*competitive.*

From theorem 4, in case $\kappa \geq \lambda$, *i.e.*, κ is sufficiently large, A^w is $(7 + 6\Delta)$-competitive. If $\Delta = 1$, *i.e.*, each job has an equal processing time, then A^w is 13-competitive. In case $\kappa \leq \lambda$, for the equal processing time, A^w is $(7 + 6\sqrt{\frac{\lambda}{\kappa}})$-competitive.

References

[1] S. Baruah, G. Koren, D. Mao, B. Mishra, A. Raghunathan, L. Rosier, D. Shasha, and F. Wand. On the competitiveness of on-line task real-time task scheduling. *Journal of Real-Time Systems*, 4(2):124–144, 1992.

[2] S. Baruah, G. Koren, B. Mishra, A. Raghunathan, L. Rosier, and D. Shasha. On-line scheduling in the presence of overload. In *Proc. of the 32nd Annual Symposium on Foundation of Computer Science*, pages 100–110, 1991.

[3] Bhaskar DasGupta and Michael A. Palis. Online real-time preemptive scheduling of jobs with deadlines on multiple machines. *Journal of Scheduling*, 4:297–312, 2001.

[4] Juan A. Garay, Joseph Naor, Bulent Yener, and Peng Zhao. On-line admission control and packet scheduling with interleaving. In *Proc. of the 21st IEEE IN-FOCOM*, 2002.

[5] S. Goldman, J. Parwatikar, and S. Suri. On-line scheduling with hard deadlines. In *Proc. of the Workshop on Algorithms and Data Structures (WADS)*, volume 1272 of *Lecture Notes in Computer Science (LNCS)*, pages 258–271. Springer-Verlag, 1997.

[6] Micheal H. Goldwasser. Patience is a virtue: The effect of slack on competitiveness for admission control. *Journal of Scheduling*, 6:183–211, 2003.

[7] Jae-Hoon Kim and Kyung-Yong Chwa. On-line deadline scheduling on multiple resources. In *Proc. of the 7th Annual Int. Computing and Combinatorics Conference (COCOON)*, volume 2108 of *Lecture Notes in Computer Science (LNCS)*, pages 443–452. Springer-Verlag, 2001.

[8] C. Y. Koo, T. W. Lam, T. W. Ngan, and K. K. To. Extra processors versus future information in optimal deadline scheduling. In *Proc. of the 14th ACM Annual Symposium on Parallel Algorithms and Architectures*, pages 55–764, 2002.

[9] Gilad Koren and Dennis Shasha. Moca : A multiprocessor on-line competitive algorithm for real-time system scheduling. *Theoretical Computer Science*, 128(1):75–97, 1994.

[10] Gilad Koren and Dennis Shasha. D^{over} : An optimal on-line scheduling algorithm for overloaded real-time systems. *SIAM Journal of Computing*, 24(2):318–339, 1995.

[11] Jae-Ha Lee. On-line deadline scheduling: multiple machines and randomization. In *Proc. of the 15th Annual ACM Symposium on Parallel Algorithms*, pages 19–23, 2003.

[12] R. Lipton and A. Tomkins. Online interval scheduling. In *Proc. of the Fifth Annual ACM-SIAM Symposium on Discrete Algorithms*, pages 302–311, 1994.

Lattice Embedding of Direction-Preserving Correspondence over Integrally Convex Set

(Extended Abstract)

Xi Chen[1] and Xiaotie Deng[2]

[1] Department of Computer Science, Tsinghua University,
Beijing, P.R. China
`xichen00@mails.tsinghua.edu.cn`
[2] Department of Computer Science, City University of Hong Kong,
Hong Kong SAR
`deng@cs.cityu.edu.hk`

Abstract. We consider the relationship of two fixed point theorems for direction-preserving discrete correspondences. We show that, for any space of no more than three dimensions, the fixed point theorem [4] of Iimura, Murota and Tamura, on integrally convex sets can be derived from Chen and Deng's fixed point theorem [2] on lattices by extending every direction-preserving discrete correspondence over an integrally convex set to one over a lattice. We present a counter example for the four dimensional space. Related algorithmic results are also presented for finding a fixed point of direction-preserving correspondences on integrally convex sets, for spaces of all dimensions.

1 Introduction

A recent work on discrete fixed point introduced by Iimura [4] has attracted a series of work on related problems. Iimura, Murota and Tamura [6] improved the original proof of Iimura. Chen and Deng presented an alternative discrete fixed point theorem for general domain with a matching algorithmic bound for all finite dimensions [2]. In [8], Laan, Talman and Yang presented an iterative algorithm for the zero point problem. Friedl, Ivanyosy, Santha and Verhoeven obtained a $\sqrt{n}$ upper bound for the dimension two Sperner problem [7], thus a matching bound when combined with the lower bound of Crescenzi and Silvestri [3].

These problems are closely related. The matching bound of Friedl, Ivanyosy, Santha and Verhoeven for the Sperner problem is in some sense a mirror result of an earlier work of Hirsch, Papadimitriou and Vavasis on 2D approximate fixed point [5]. In addition, the higher dimensional query complexity for the Sperner problem of Friedl, Ivanyosy, Santha and Verhoeven, i.e., with query time linear in the separation number of the skeleton graph of the manifold and the size of its boundary, compares closely with the upper bound of Chen and Deng [2], for the query complexity of finding a discrete fixed point.

In this work, we set to understand the relationship between the discrete fixed point theorem of Iimura, Murota and Tamura, and the discrete fixed point

S.-W. Cheng and C.K. Poon (Eds.): AAIM 2006, LNCS 4041, pp. 53–63, 2006.

theorem of Chen and Deng. In both cases, the discussion focuses on direction-preserving correspondences. The main differences are the restriction of the domains for which the theorems could apply. Murota, Iimura and Tamura consider a domain which is integrally convex. Informally, a point in the convex hull of the domain can be represented by a convex combination of integral points in the domain within unit distance from it. The work of Chen and Deng allows the domain not to be convex at all. Moreover, the result of Murota, Iimura and Tamura restricts the correspondence to be bounded in the domain. A more general boundary condition for correspondence is presented in Chen and Deng [2]. It is therefore natural to believe that the work of Murota, Iimura and Tamura can be derived from the seemingly more general version of Chen and Deng.

Indeed, for dimension two and three, we confirm it by embedding an integrally convex set in a lattice so that the bounded correspondence on the integrally convex set can be extended to a bounded and direction-preserving function on the lattice. Any fixed point of this function leads to a fixed point of the original correspondence. Therefore, a claim of existence of a fixed point on the lattice leads to a claim of existence of a fixed point in the integrally convex set. Such a direct extension, however, does not carry over to higher dimensions. We derive an interesting counter example for four-dimensional space.

There is another unsettled issue for the discrete fixed point theorem of Murota, Iimura and Tamura, that of algorithmic issues. In [8], Laan, van der Talman, and Yang presented an iterative algorithm which is shown to terminate with a fixed point. Our extension theorem for two and three dimensional spaces directly answers this problem and derives a matching algorithmic bound. For higher constant dimensional spaces, we need to refine the domain to derive an algorithmic solution.

In section 2, we define a fixed point problem called $\mathbf{FPC}^d$. Previous results are then reviewed in section 3. We formalize the concept of function extension mechanism in section 4. After presenting positive results for both two and three dimension spaces, we derive a counter example for the four-dimensional space in section 5. Section 6 gives a sketch of an algorithm to solve problem $\mathbf{FPC}^d$, for spaces of all dimensions, which implies a matching bound for the time complexity of $\mathbf{FPC}^d$. Finally, we conclude in section 7 with discussions on the difference between the two approaches.

2 Definition of Problem $\mathbf{FPC}^d$

In this section, we will define a fixed point problem called $\mathbf{FPC}^d$. It originates from the fixed point theorem of Iimura, Murota and Tamura [4, 6] concerning direction-preserving correspondences on integrally convex sets.

Definition 1. *Let X be a nonempty finite subset of $\mathbb{Z}^d$ and $\Gamma : X \rightrightarrows X$ be a nonempty-valued correspondence (that is, for every $x \in X$, $\Gamma(x) \subset X$).*

A point $x \in X$ is said to be a fixed point of Γ if $x \in \Gamma(x)$.

For each $x \in X$, let $\tau(x) \in \overline{\Gamma(x)}$ denote the projection of x onto $\overline{\Gamma(x)}$, i.e.,

$$||\tau(x) - x||_2 = \min_{y \in \overline{\Gamma(x)}} ||y - x||_2$$

where $\|y - x\|_2 = (\sum_{i=1}^{d}(y_i - x_i)^2)^{1/2}$.

Definition 2. *A correspondence* $\Gamma : X \rightarrowtail X$ *where* $X \subset \mathbb{Z}^d$ *is said to be direction-preserving on* X *if for all* $x, x' \in X$ *with* $\|x - x'\|_\infty \leq 1$, *we have* $(\tau_i(x) - x_i)(\tau_i(x') - x_i') \geq 0$ *for every* $1 \leq i \leq d$.

We now define two classes of convex sets in $\mathbb{Z}^d$, integrally convex sets and discretely convex sets, which play important roles in the fixed point theorem.

Definition 3. *A finite set* $X \subset \mathbb{Z}^d$ *is integrally convex if for all points* $y \in \overline{X}$, $y \in \overline{X \cap N(y)}$, *where* $N(y) = \{ z \in \mathbb{Z}^d \mid \|z - y\|_\infty < 1 \}$.

Definition 4. *A finite set* $X \subset \mathbb{Z}^d$ *is discretely convex if* $X = \overline{X} \cap \mathbb{Z}^d$.

Theorem 1 (Theorem of Iimura, Murota and Tamura [6]). *Let* $X \subset \mathbb{Z}^d$ *be a nonempty integrally convex set. For every nonempty, discretely convex-valued and direction-preserving correspondence* Γ *from* X *to itself, there must exist a fixed point* $x^* \in X$ *such that* $x^* \in \Gamma(x^*)$.

In brief, the task of the fixed point problem $\mathbf{FPC}^d$ is to find a fixed point of correspondence Γ which satisfies all the conditions in Theorem 1. Formally speaking, the input includes both the set X and correspondence Γ. Here X is described by all the extreme points of convex set $\overline{X}$. This representation of X is succinct, according to the following lemma.

Lemma 1. *For every* $d \geq 1$, *there exists an integer* N_d *such that, for all integrally convex sets* $X \subset \mathbb{Z}^d$, *the number of extreme points of* $\overline{X}$ *is less than* N_d.

On the other hand, correspondence Γ looks like a black box to algorithms. We only consider algorithms which are based on correspondence evaluations. Such an algorithm should behaves as follows: It makes up a test point $r_1 \in X$, sends it to the black box and receives $\tau(r_1)$. Based on r_1 and $\tau(r_1)$, it computes a new test point r_2 and evaluate $\tau(r_2)$. It continues until a fixed point of Γ is reached. We assume that each evaluation of τ takes one step.

Diameter of the integrally convex set X, that is, $n = \max_{x,y \in X} \|x - y\|_\infty$, is taken as the input size of $\mathbf{FPC}^d$. We are interested in the time complexity $T_d(n)$ of problem $\mathbf{FPC}^d$. Our main result is stated in the following theorem.

Theorem 2. *For every constant* $d \geq 2$, $T_d(n) = \Theta(n^{d-1})$.

Problem $\mathbf{FPC}^d$ is closely related to problems $\mathbf{DFP}^d$ and $\mathbf{AFP}^d$ [2].

3   Previous Results on Fixed Point Problems

In this section, we review both the problem definitions and algorithmic results in [2]. For every $1 \leq k \leq d$, we use e^k to denote the kth unit vector of $\mathbb{Z}^d$. Here $e_k^k = 1$ and $e_i^k = 0$ for all $1 \leq i \neq k \leq d,$.

Definition 5. *For all $p < q \in \mathbb{Z}^d$, $A_{p,q} = \{\, r \in \mathbb{Z}^d \mid \forall\, 1 \leq i \leq d,\ p_i \leq r_i \leq q_i \,\}$. Its boundary is defined as $B_{p,q} = \{\, r \in A_{p,q} \mid \exists\, 1 \leq i \leq d,\ r_i = p_i \text{ or } q_i \,\}$.*

Definition 6. *Function $f : S \to \{\, 0, \pm e^1, \pm e^2 \dots \pm e^d \,\}$ where $S \subset \mathbb{Z}^d$ is said to be direction-preserving if for all $r^1, r^2 \in S$ which satisfy $\|r^1 - r^2\|_\infty \leq 1$, we have $\|f(r^1) - f(r^2)\|_\infty \leq 1$.*
   When $S = A_{p,q}$, f is said to be bounded if $r + f(r) \in A_{p,q}$ for all $r \in B_{p,q}$.

It is proved in [2] that any function f which is both bounded and direction-preserving has a zero point $r^* \in A_{p,q}$ such that $f(r^*) = 0$. The task of problem **DZP**^{d} is to find such a point in $A_{p,q}$. To get information of f, algorithms make up test points and evaluate f at these points. Similarly, we use $T_d^1(n)$ to denote the time complexity of **DZP**^{d}, where $n = \max_{1 \leq i \leq d}(q_i - p_i)$.

Definition 7. *Map $\mathcal{G} : E^d = [0,1]^d \to \mathbb{R}^d$ satisfies a Lipschitz condition with constant M if $\|\mathcal{G}(x) - \mathcal{G}(y)\|_\infty \leq M\|x - y\|_\infty$ for all $x, y \in E^d$.*
   We use $L_{M,d}$ to denote the set of all those maps $\mathcal{F} : E^d \to E^d$ such that $\mathcal{G}(x) = \mathcal{F}(x) - x$ satisfies a Lipschitz condition with constant M.

By Brouwer's fixed point theorem, every map $\mathcal{F} \in L_{M,d}$ has a fixed point $x^* \in E^d$ such that $\mathcal{F}(x^*) = x^*$. Given a map $\mathcal{F} \in L_{M,d}$ and $\epsilon > 0$, the output of problem **AFP**^{d} is an approximate fixed point $x^* \in E^d$ with error bounded by ϵ. More exactly, x^* should satisfy $\|\mathcal{F}(x^*) - x^*\|_\infty \leq \epsilon$. Similarly, $\mathcal{F}$ looks like a black box to algorithms, which can only be accessed by evaluations. We use $T_d^2(M, \epsilon)$ to denote the time complexity of problem **AFP**^{d}.

Theorem 3 ([2]). *For every constant $d \geq 2$,*

$$T_d^1(n) = \Theta(n^{d-1}) \quad and \quad T_d^2(M, \epsilon) = \Theta\left(\left(\frac{M}{\epsilon}\right)^{d-1}\right).$$

In fact, the lower bound of $T_d(n)$ in Theorem 2 can be easily derived from the lower bound of $T_d^1(n)$ above.

4   Extension Mechanism for Low Dimensional Spaces

In this section, we focus on a natural idea to solve problem **FPC**^{d}. First, we formalize the concept of function extension mechanism $\mathcal{M}^d$. Its existence gives an algorithm for **FPC**^{d} with time complexity $O(n^{d-1})$. $\mathcal{M}^2$ and $\mathcal{M}^3$ are then constructed and we get the upper bound in Theorem 2 for cases $d = 2$ and 3.

4.1   Definition of Function Extension Mechanism $\mathcal{M}^d$

The discrete approach presented in this section is based on the existence of algorithms for problem **DZP**^{d} with time complexity $O(n^{d-1})$. Let $A_{p,q}$ be the smallest set that contains X which is the domain of Γ and τ. A function extension mechanism $\mathcal{M}^d$ extends map τ to be a direction function f from $A_{p,q}$ to $\{\, 0, \pm e^1, \pm e^2 \dots \pm e^d \,\}$ which is both bounded and direction-preserving. We can use any algorithm for problem $\mathbf{DZP}^d$ to find a zero point of f. Properties of $\mathcal{M}^d$ guarantee that, given a zero point of f, one can find a fixed point of map τ (and thus, correspondence Γ) very efficiently.

Definition 8. *Given an input pair (X, Γ) of $\mathbf{FPC}^d$, if the integrally convex set X is non-degenerate, that is, $\overline{X}$ is a d-polytope in $\mathbb{R}^d$, then function extension mechanism $\mathcal{M}^d = (\mathcal{A}^d, \mathcal{B}^d)$ for d-dimensional space constructs a direction function f from $A_{p,q}$ to $\{\, 0, \pm e^1, \dots \pm e^d \,\}$.*

The following five properties should be satisfied:

- P_1. *Function f is both bounded and direction-preserving on $A_{p,q}$;*
- P_2. *For every $r \in A_{p,q}$, algorithm $\mathcal{A}^d$ takes (r, X, τ) as input and computes $f(r)$ with $O(1)$ (d is viewed as a constant here) running time;*
- P_3. *For every $r \in X$, $f(r) = 0$ if and only if $\tau(r) = r$;*
- P_4. *For evert $r \in X$ such that $\tau(r) \neq 0$, $f(r) \cdot (\tau(r) - r) > 0$;*
- P_5. *For every zero point r of f such that $r \notin X$, algorithm $\mathcal{B}^d$ takes (r, X, τ) as input and computes a fixed point $r' \in X$ of τ with $O(1)$ running time.*

Clearly, once we find a mechanism $\mathcal{M}^d$ for d-dimensional space, we get an algorithm for $\mathbf{FPC}^d$ with time complexity $O(n^{d-1})$ (if X is degenerate, then we exhaustively check every point in X, since $|X| \leq n^{d-1}$). From now on, we always assume that X is non-degenerate.

4.2 Function Extension Mechanism $\mathcal{M}^2$ for Case $d = 2$

$\mathcal{M}^2$ is closely related to a map ψ from $A_{p,q}$ to X. For every $r \in X$, $\psi(r) = r$. Otherwise, $\psi(r) = \widetilde{r}$ where

$$|r_1 - \widetilde{r}_1| = \min_{r' \in X,\ r_2 = r'_2} |r_1 - r'_1|.$$

The construction of function f is described in figure 1.

Properties P_2, P_3 and P_4 are easy to verify. For property P_5, if $r \notin X$ and $f(r) = 0$, then $f(r') = 0$ where $r' = \psi(r)$. With the succinct representation of X, $r' = \psi(r)$ can be computed in $O(1)$ time. Proof of the following lemma is available in the full version [1].

Lemma 2. *f constructed by $\mathcal{M}^2$ is both bounded and direction-preserving.*

Function Extension Mechanism $\mathcal{M}^2$

1: **for any** $r \in X$
2:    **if** $\tau(r) = 0$ **then** $f'(r) = 0$
3:    **else if** $\tau_2(r) \neq 0$ **then** $f'(r) = \text{sign}\,(\tau_2(r))\,e^2$
4:    **else** $f'(r) = \text{sign}\,(\tau_1(r))\,e^1$
5: **for any** $r \in A_{p,q}$, $f(r) = f'(\psi(r))$

Fig. 1. Details of the Function Extension Mechanism $\mathcal{M}^2$

4.3   Function Extension Mechanism $\mathcal{M}^3$ for Case $d = 3$

Behavior of the mechanism $\mathcal{M}^3$ is similar to $\mathcal{M}^2$, while the details are a little more complicated. First, we divide $A_{p,q}$ into three pairwise disjoint sets, X, S_1 and S_2 where

$$S_1 = \{\, r \notin X,\ r \in A_{p,q} \mid \exists\, r' \in X, r_2 = r_2'\ \text{and}\ r_3 = r_3'\,\},$$
$$S_2 = \{\, r \notin X \cup S_1,\ r \in A_{p,q} \mid \exists\, r' \in X \cup S_1, r_1 = r_1'\ \text{and}\ r_3 = r_3'\,\}.$$

We then define two maps. ψ^1 is from $X \cup S_1$ to X. For all $r \in X$, $\psi^1(r) = r$. For all $r \in S_1$, $\psi^1(r) = \widetilde{r}$ where

$$|r_1 - \widetilde{r}_1| = \min_{r' \in X,\ r_2 = r_2',\ r_3 = r_3'} |r_1 - r_1'|.$$

Map ψ^2 is from $A_{p,q}$ to $X \cup S_1$. For all $r \in X \cup S_1$, $\psi^2(r) = \psi^1(r)$. For all point $r \in S_2$, $\psi^2(r) = \widetilde{r}$ where

$$|r_2 - \widetilde{r}_2| = \min_{r' \in X \cup S_1,\ r_1 = r_1',\ r_3 = r_3'} |r_2 - r_2'|.$$

Given a map τ, $\mathcal{M}^3$ first convert it into a direction function f' from X to $\{0, \pm e^1, \pm e^2, \pm e^3\}$. After extending f' to be f'' on $X \cup S_1$ using map ψ^1, we employ map ψ^2 to extend f'' onto $A_{p,q}$. The difficulty here is that, to keep the direction-preserving property, we must be careful when dealing with some boundary points of X.

Definition 9. *Point $r \in X$ is said to be a left (or right) boundary point of X if $(r_1 - 1, r_2, r_3) \notin X$ (or $(r_1 + 1, r_2, r_3) \notin X$). We use L_X (or R_X) to denote the set of left (or right) boundary points of X.*

From the definition of integrally convex sets, we get the following lemma.

Lemma 3. *For all points $r^1, r^2 \in L_X$ (or R_X) which satisfy $|r_2^1 - r_2^2| \le 1$ and $|r_3^1 - r_3^2| \le 1$, we have $|r_1^1 - r_1^2| \le 2$.*
   Furthermore, if $|r_1^1 - r_1^2| = 2$, then $|r_2^1 - r_2^2| = |r_3^1 - r_3^2| = 1$.

Definition 10. *Pair (r^1, r^2) where $r^1, r^2 \in L_X$ (or $r^1, r^2 \in R_X$) is said to be a bad pair of X if $|r_2^1 - r_2^2| = |r_3^1 - r_3^2| = 1$ and $|r_1^1 - r_1^2| = 2$. We use B_X to denote the set of bad pairs of X.*
   $r \in X$ is said to be bad if there exists $r' \in X$ such that $(r, r') \in B_X$.

Each bad pair (r^1, r^2) of X gives a supporting hyperplane $H_{r^1, r^2} = (u, a)$ of $\overline{X}$ where $|u_i| = 1$, for all $1 \le i \le 3$. For example, if $r^1 = (0, 0, 0)$ and $r^2 = (2, 1, 1)$ are two left boundary points, then one can prove both $(1, 1, 0)$ and $(1, 0, 1)$ belong to L_X. These points together define a hyperplane $H_{r^1, r^2} = (-1, 1, 1, 0)$. With $H_{r^1, r^2} = (u, a)$, we define $S_{r^1, r^2} = \{-u_1 e^1, -u_2 e^2, -u_3 e^3\}$.
   On the other hand, for a bad point $r \in X$, there might be more than one point r' such that $(r, r') \in B_X$. We define $S_r = \bigcap_{(r, r') \in B_X} S_{r, r'}$ which has the following property.

Lemma 4. *For every bad point $r \in X$, $1 \le |S_r| \le 3$.*

If $S_r = \{+e^k\}$ (or $S_r = \{-e^k\}$) where $1 \le k \le 3$, then $r_k = \min_{r' \in X} r'_k$ (or $r_k = \max_{r' \in X} r'_k$). Furthermore, if $k \ne 1$, then there are exactly two points $r' \in X$ such that $(r, r') \in B_X$.

If $|S_r| > 1$ and $\tau(r) \ne r$, then there exists a unit vector $ce^k \in S_r$ such that $ce^k \cdot (\tau(r) - r) > 0$.

For every bad point r such that $S_r = \{ce^k\}$ where $k \ne 1$ and $|c| = 1$, we define vectors $v_L, v_R \in \{\pm e^1, \pm e^2, \pm e^3\}$ based to the value of $\tau(r)$ and the shape of X around r. Only the case for $c = 1$ and $k = 3$ is described below, as other cases are similar.

> **Case 1:** $(r, r^L), (r, r^R) \in B_X$ where $r^L = (r_1 - 2, r_2 - 1, r_3 - 1)$, $r^R = (r_1 + 2, r_2 + 1, r_3 - 1)$. If $\tau_3(r) < 0$, then $v_L = v_R = -e^3$. Otherwise, we have $\tau_1(r) = \tau_2(r)$. If $\tau_1(r) > 0$, then $v_L = +e^1$ and $v_R = +e^2$, or else $v_L = -e^2$ and $v_R = -e^1$.

> **Case 2:** $(r, r^L), (r, r^R) \in B_X$ where $r^L = (r_1 - 2, r_2 + 1, r_3 - 1)$, $r^R = (r_1 + 2, r_2 - 1, r_3 - 1)$. If $\tau_3(r) < 0$, then $v_L = v_R = -e^3$. Otherwise, we have $\tau_1(r) = -\tau_2(r)$. If $\tau_1(r) > 0$, then $v_L = +e^1$ and $v_R = -e^2$, or else $v_L = +e^2$ and $v_R = -e^1$.

In both cases, we have $v_L \cdot (\tau(r) - r) > 0$, $v_L \in S_{r, r^L}$, $v_R \cdot (\tau(r) - r) > 0$ and $v_R \in S_{r, r^R}$. Details of the mechanism $\mathcal{M}^3$ are described in figure 2. Similarly, properties P_2, P_3, P_4 and P_5 are easy to verify. Proof of the following lemma is available in the full version.

Function Extension Mechanism $\mathcal{M}^3$

1: **for any** $r \in X$
2:      **if** $\tau(r) = 0$ **then** $f'(r) = 0$
3:      **else if** r is a bad point of X and $|S_r| > 1$ **then**
4:          there must exist k such that $ce^k \in S_r$ and $c\tau_k(r) > 0$, set $f'(r) = ce^k$
5:      **else** let k be the largest integer satisfies $\tau_k(r) \ne 0$, set $f'(r) = \operatorname{sign}(\tau_k(r)) e^k$
6: **for any** $r \in X \bigcup S_1$
7:      **if** $r \in X$ **then** $f''(r) = f'(r)$
8:      **else if** $f'(\psi^1(r)) = 0$ **then** $f''(r) = 0$
9:      **else if** $\psi_1^1(r) = \min_{r' \in X} r'_1$ **then** $f''(r) = +e^1$
10:      **else if** $\psi_1^1(r) = \max_{r' \in X} r'_1$ **then** $f''(r) = -e^1$
11:      **else if** $r' = \psi^1(r)$ is a bad point of X and $S_{r'} = \{ce^k\}$ where $k \ne 1$ **then**
12:          **if** $r_1 < r'_1$ **then** $f''(r) = v_L$
13:          **else** $f''(r) = v_R$
14:      **else** $f''(r) = f'(\psi^1(r))$
15: **for any** $r \in A_{p,q}$, $f(r) = f''(\psi^2(r))$

Fig. 2. Details of the Function Extension Mechanism $\mathcal{M}^3$

Lemma 5. *f constructed by $\mathcal{M}^3$ is both bounded and direction-preserving.*

5   A Counter Example for 4-Dimensional Space

Although function extension mechanism $\mathcal{M}^d$ does exist for cases $d = 2$ and 3, we find great difficulty in designing $\mathcal{M}^d$ for higher dimensional spaces. In this section, we construct a set of maps S in the 4-dimensional space and prove the non-existence of mechanism $\mathcal{M}^4$.

The domain of maps in S is

$$X = \left\{ r \in \mathbb{Z}^4 \;\middle|\; \forall\, 1 \le i \le d,\ r_i \ge 0 \text{ and } r_1 + r_2 + r_3 + r_4 \le n \right\}$$

which can be divided into layers $X = X_1 \cup X_2 \ldots \cup X_n = Y \cup Z$. Here set $X_i = \{r \in X \mid r_4 = i\}$, $Y = X_n \cup X_{n-1} \ldots \cup X_{n-5}$ and $Z = X - Y$. For every $r \in Z$, we construct a map τ_r as follows, and $S = \{\tau_r \mid r \in Z\}$.

For every two maps $\tau_r, \tau_{r'} \in S$, $\tau_r(p) = \tau_{r'}(p)$ for all $p \in Y$. Values of τ, where $\tau \in S$, on the first four layers X_n, X_{n-1}, X_{n-2} and X_{n-3} are described in figure 3. In this figure, an arrow ce^k on point r means $\Gamma(p) = \{p + ce^k\}$ and $\tau(p) = p + ce^k$. For every $p \in X_{n-4}$, if $||p - (2,0,0,n-4)||_\infty > 1$, then $\tau(p) = p - e^4$. If $p = (2,0,0,n-4)$, then $\tau(p) = p - e^1$. Otherwise, $\tau(p) - p = \tau((p_1, p_2, p_3, p_4+1)) - (p_1, p_2, p_3, p_4+1)$. Finally, $\tau(p) = p - e^4$ for every $p \in X_{n-5}$,.

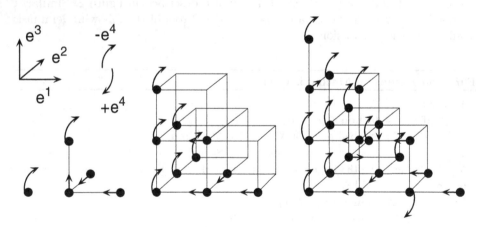

Fig. 3. A Counter Example

Values of τ_r on Z are described as follows. For every $p \in Z$, if $p = r$, then $\tau_r(p) = p$. Otherwise, we have two cases. If $||r||_1 > ||p||_1$ where $||r||_1 = \sum_{i=1}^{4} r_i$, letting k be an integer such that $r_k > p_k$, then $\tau_r(p) = p + e^k$. If $||r||_1 \le ||p||_1$, letting k be an integer such that $r_k < p_k$, then $\tau_r(p) = p - e^k$. One can prove the following property of maps in S.

Lemma 6. $\tau_r : X \to X$ *is direction-preserving and r is its only fixed point.*

Now we prove the non-existence of mechanism $\mathcal{M}^4$. Let's make a reduction to absurdity, considering that there exists a mechanism $\mathcal{M}^4 = (\mathcal{A}^d, \mathcal{B}^d)$, however, satisfies all the five properties $P_1, P_2 \ldots P_5$, then for every map $\tau_r \in S$, it constructs a direction-preserving function f_r. By property P_4, we have $f_r(r') = \tau_r(r') - r'$ for every $r' \in X$. Since f_r is direction-preserving, we must have $f_r(r^*) = 0$ where $r^* = (1, 1, 1, n-2)$.

Let's pick a map $\tau_r \in S$ arbitrarily and run $\mathcal{B}^d$ with input (r^*, X, τ_r). After constant steps, it should output a fixed point r' of τ_r according to P_5. By Lemma 6, we have $r' = r$. This means that maps in S can be recognized within constant steps, which contradicts with the fact that $|S| = \Theta(n^4)$. As a result, our assumption is wrong and no such mechanism exists.

6   An Algorithm for Problem $\mathbf{FPC}^d$

In this section, we briefly describe an algorithm for $\mathbf{FPC}^d$ and prove the upper bound in Theorem 2, for spaces of all dimensions.

Definition 11. *For every point $r \in \mathbb{Z}^d$, we define a hypercube $C_{r,n} \subset \mathbb{R}^d$ as*

$$C_{r,n} = \{ x \in \mathbb{R}^d \mid r_i \leq x_i \leq r_i + n, \text{ for all } 1 \leq i \leq d \}.$$

Let (Γ, X) be an input instance of $\mathbf{FPC}^d$, then we use $C_{r,n}$ to denote the smallest hypercube containing X. Starting from Γ, we build a map $\mathcal{F}$ from $C_{r,n}$ to itself. Details of the construction can be found in the full version. We give the following lemmas without proof.

Lemma 7. *Given an input instance (Γ, X) of problem $\mathbf{FPC}^d$, for every point $x \in C_{r,n}$, $\mathcal{F}(x)$ can be computed in $O(1)$ time.*

Lemma 8. *For every constant $d \geq 2$, there exists a constant D_d such that, for every input instance (Γ, X) of problem $\mathbf{FPC}^d$, map $\mathcal{F}$ belongs to $L_{D_d, d}$.*

Lemma 9. *For every point $x^* \in \overline{X}$ such that $\|\mathcal{F}(x^*) - x^*\|_\infty < 1/(d+1)^2$, there must exist a fixed point of correspondence Γ in $N(x^*) \cap X$. Recall that $N(x^*) = \{ r \in \mathbb{Z}^d \mid \|r - x^*\|_\infty < 1 \}$.*

Lemma 10. *For every $x \in C_{r,n}$ such that $\|\mathcal{F}(x) - x\|_\infty < 1/(d^{1/2}(d+1)^2)$, point $x^* = \Psi_{\overline{X}}(x)$ must satisfy $\|\mathcal{F}(x^*) - x^*\|_\infty < 1/(d+1)^2$. Here $\Psi_{\overline{X}}$ is the projection onto $\overline{X}$ where $\|x - \Psi_{\overline{X}}(x)\|_2 = \min_{y \in \overline{X}} \|x - y\|_2$.*

$\mathcal{F}$ can be scaled to be a map $\mathcal{F}'$ from $E^d = [0, 1]^d$ to itself as follows. For every point $x \in E^d$, $\mathcal{F}'(x) - x = (\mathcal{F}(nx + r) - (nx + r))/n$.

The reason we build $\mathcal{F}$ and $\mathcal{F}'$ is to find a fixed point of Γ. By Lemma 8, one can prove that the new map $\mathcal{F}'$ also belongs to $L_{D_d, d}$, thus we can use an algorithm for $\mathbf{AFP}^d$ to compute an $\epsilon = 1/(d^{1/2}(d+1)^2 n)$ approximate fixed point x of $\mathcal{F}'$, and $x^* = nx + r$ must be an $1/(d^{1/2}(d+1)^2)$ approximate fixed

---

Algorithm for Problem FPC^d

1: Let (Γ, X) be the input instance of problem **FPC**^{d}
2: Let $\mathcal{F}$ and $\mathcal{F}'$ be the two maps constructed
3: Use an algorithm for **AFP**^{d} to find an ϵ approximate fixed point x of $\mathcal{F}'$
4: compute $x^* = nx + r$
5: **if** $x^* \in \overline{X}$, **then**
6:     query Γ for every point in $N(x^*) \cap X$ and output a fixed point of Γ
7: **else**
8:     compute $x' = \Psi_{\overline{X}}(x^*)$
9:     query Γ for every point in $N(x') \cap X$ and output a fixed point of Γ
10: **endif**

Fig. 4. The Algorithm for Fixed Point Problem **FPC**^{d}

point of $\mathcal{F}$. Lemma 9 and 10 together show that, once we get x^*, a fixed point of Γ can be located easily.

The algorithm is described in figure 4. Let's analyze its time complexity. For every test point $x \in E^d$ which is queried by the **AFP**^{d} algorithm, constant steps are sufficient to compute $\mathcal{F}'(x)$ according to Lemma 7. By Theorem 3, the time used by the **AFP**^{d} algorithm in line 3 is $O\left((D_d/\epsilon)^{d-1}\right) = O\left(n^{d-1}\right)$. This gives us the upper bound of time complexity $T_d(n)$ in Theorem 2.

7   Concluding Remarks

In this paper, we described two different approaches to solve the discrete fixed point problem **FPC**^{d}. In the discrete approach, we try to extend map τ to be a direction-preserving function f on lattice $A_{p,q}$. In the continuous approach, we construct a Lipschitz map $\mathcal{F}_3$ from $C_{r,n}$ to itself. While the former only works for low dimensional spaces, the latter solves problem **FPC**^{d} for spaces of all dimensions. But how does the algorithm for problem **AFP**^{d} work? Actually, it samples map $\mathcal{F}_3$ with a suitable interval, builds a direction function which is both bounded and direction-preserving, and employs an algorithm for problem **DZP**^{d} to find an zero point which is also an approximate fixed point of $\mathcal{F}_3$ [2].

Thus in both approaches, we construct (explicitly or implicitly) a bounded and direction-preserving function on some lattice. The difference is that, the lattice of the continuous approach has much higher density than the one of the discrete approach. While no function extension mechanism exists for high dimensional spaces, we can always construct a direction-preserving function on a denser lattice implicitly using the continuous method.

References

1. Xi Chen and Xiaotie Deng. Lattice Embedding of Direction-Preserving Correspondence Over Integrally Convex Set (Full version). *manuscript, available at* http://www.cs.cityu.edu.hk/~deng/.

2. Xi Chen and Xiaotie Deng. On algorithms for discrete and approximate brouwer fixed points. In *STOC 2005*, pages 323–330.
3. P. Crescenzi and R. Silvestri. Sperner's lemma and robust machines. *Comput. Complexity*, 2(7):163–173, 1998.
4. T. Iimura. A discrete fixed point theorem and its applications. *J. of Mathematical Economics*, 7(39):725–742, 2003.
5. C. Papadimitriou M.D. Hirsch and S. Vavasis. Exponential lower bounds for finding brouwer fixed points. *J. Complexity*, (5):379–416, 1989.
6. K. Murota, T. Iimura, and A. Tamura. Discrete fixed point theorem reconsidered. *J. of Mathematical Economics*, to appear.
7. M. Santha, K. Friedl, G. Ivanyos, and F. Verhoeven. On the black-box complexity of sperner's lemma. In *FCT*, 2005.
8. D. Talman, G. Laan, and Z. Yang. Solving discrete zero point problems. In *Tinbergen Institute Discussion Papers*, 2004.

Linear Programming Polytope and Algorithm for Mean Payoff Games*

Ola Svensson[1,**] and Sergei Vorobyov[2]

[1] IDSIA, Istituto Dalle Molle di Studi sull'Intelligenza Artificiale, Lugano,
Switzerland
[2] Information Technology Department, Uppsala University, Sweden
Sergei.Vorobyov@it.uu.se

Abstract. We investigate LP-polytopes generated by mean payoff
games and their properties, including the existence of tight feasible solutions of bounded size. We suggest a new associated algorithm solving a
linear program and transforming its solution into a solution of the game.

1 Introduction

The goal of this paper is to investigate linear programming formulations for
mean payoff games (MPGs) [8, 9], a well-known problem in NP ∩ coNP, with an
open P-membership status. Recently combinatorial randomized subexponential
algorithms for linear programming were successfully applied for solving several
kinds of games [5, 6, 2, 1, 3]. However, to our knowledge, there are no previous
attempts at investigating LP-formulations for MPGs and associated polyhedra,
except recent work [2, 1, 3] representing some infinite games as instances of the
new so-called *controlled linear programming problem (CLPP)*. In contrast, LP-
formulations and relaxations are well studied and understood for the overwhelm-
ing majority of combinatorial optimization problems [13].

Several naturally arising questions we address and solve are as follows.

1. Is it possible to describe/approximate solutions to an MPG by linear constraints, i.e., as a polyhedron or a polytope?
2. Do "real" solutions to MPGs lie inside this polytope, how can they be characterized, are they vertices of this polytope?

We present several surprisingly interesting and simple properties, classify-
ing feasible solutions of the MPG-polytopes and giving new insights into the
combinatorial structure of the problem. Based on these, we describe a new
MPG-solving algorithm, which solves a linear program and then transforms (if
necessary) an optimal solution into a solution of the game by "tightening".

---

* Research supported by the grants from the Swedish Scientific Council and the Foundation for International Cooperation in Research and Higher Education.
** Partially supported by the Swiss National Science Foundation Project 200020-
109854.

More specifically, we represent an MPG by a linear system with a *totally unimodular* matrix, defining a nonempty integral MPG-polytope. Some vertices of this polytope represent the so-called "*tight*" feasible solutions, which solve the corresponding MPG. We suggest a new algorithm for finding tight solutions based on minimizing a simple linear function and "tightening" an optimal solution. In contrast to the TSP-polytope, for which no LP-description is to be expected, MPG-polytopes are easily characterized, and any game can be solved by optimizing a single (but unknown) linear function over such polytope. This provides for a certain reduction in the size of the search space and is suggestive for a new class of algorithms.

Combinatorial optimization and linear programming seem to be very productive tools for solving games. In [6] we generalized the *shortest paths* problem to the *controlled* or *longest* shortest paths problem, and used it together with combinatorial linear programming for solving MPGs. Combinatorial structures underlying iterative improvement for games are explored in [5]. A related line of research concerns applications of the *Linear Complementarity Problem* (LCP) [10, 7], a nonlinear optimization theory we recently successfully applied to solving several classes of infinite games and P-matrix Generalized LCPs [4, 14].

2 Preliminaries

2.1 Mean Payoff Games

We start by recalling basic definitions about mean payoff games (MPGs) and then introduce the 0-mean partition problem, to which all other problems for MPGs are polynomially reducible. The 0-mean partition problem is convenient for the linear programming formulations and simplifies descriptions of different algorithms. We further show that simplifying restrictions to ergodic MPGs (all vertices have the same value), ergodic bipartite MPGs (players strictly alternate moves), and ergodic complete bipartite MPGs (the game graph is complete bipartite) can be done without loss of generality.

A *mean payoff game (MPG)* is a two-player game, played on a finite directed edge-weighted graph $G = (V, E, w)$, where the set of vertices V is partitioned into two nonempty sets $V_{\max}$, $V_{\min}$, every vertex has at least one outgoing edge (no sinks or leaves), and the weight function w is integer-valued.

We assume throughout the paper that $n = |V|$ is the number of vertices of the game graph G, $n_{\max} = |V_{\max}|$, $n_{\min} = |V_{\min}|$, and W is the maximal absolute edge weight; thus $w : E \to \{-W, \ldots, W\}$.

Given an MPG, a play develops in the following way. Initially, a pebble is placed in some vertex v_0 and players MAX and MIN start constructing an infinite sequence of edges $\{(v_i, v_{i+1})\}_{i=0}^{+\infty}$. If the pebble is in a vertex $v_i \in V_{\max}$ then MAX selects an outgoing edge from v_i and moves the pebble to its destination vertex v_{i+1}, otherwise MIN makes the analogous choice and move.

Players MAX and MIN are adversaries, the first one wants to maximize, whereas the second one wants to minimize, respectively, the values

$$\liminf_{k\to\infty} \frac{1}{k}\sum_{i=0}^{k-1} w(v_i, v_{i+1}), \quad \text{and} \quad \limsup_{k\to\infty} \frac{1}{k}\sum_{i=0}^{k-1} w(v_i, v_{i+1}). \tag{1}$$

It turns out that MPGs are solvable in *pure positional* strategies for both players, and every vertex has a *value* $\nu(v)$ [8, 9]. This value is equal to both limits in (1), and both players can secure it by applying these strategies. Moreover, when one player fixes his pure positional strategy, an optimal counterstrategy of his adversary is polynomial time computable. Consequently the problem whether the value of a vertex is above/below a certain threshold is in NP∩coNP.

An MPG is called *bipartite* if $E \subseteq (V_{max} \times V_{min}) \cup (V_{min} \times V_{max})$, i.e., players strictly alternate moves. A bipartite MPG is *complete* if $E = (V_{max} \times V_{min}) \cup (V_{min} \times V_{max})$, and *incomplete* otherwise. As usual, the weight of a cycle is the sum of edge weights along the cycle.

2.2   0-Mean Partition Problem for MPGs

In this paper we concentrate on the following restricted problem, which polynomially subsumes the problem of computing values of MPGs (as well as other problems, as ergodic partitioning, finding optimal strategies). It also simplifies the the algorithms, structure and properties of LP-representations.

0-MEAN PARTITION PROBLEM FOR MPGS.

Given: a bipartite MPG G without 0-weight cycles.
Find: a partition of vertices of G into sets $G_{>0}$ and $G_{\leq 0}$ of vertices with positive and nonpositive values.   □

Restricting to this problem, with the additional constraints as stated, is no loss of generality. We summarize it in the following two propositions.

Proposition 1. *Finding values of MPGs is polynomial time reducible to the 0-mean partitioning problem.*

Proof. For an arbitrary MPG, adding a constant k to every edge weight adds k to every vertex value; multiplying every edge weight by a constant k multiplies every vertex value by k. This is because values are defined by mean values of optimal cycles wrt positional strategies, and because every cycle mean changes by additive or multiplicative constant, respectively. Therefore, partitioning with a rational mean threshold reduces to 0-mean partitioning.

Values of an MPG vertices are rationals with numerators and denominators up to nW and n, respectively. If a value is known to belong to an interval of length $\leq 1/n^2$, then it is uniquely determined (the smallest difference between two values is $\frac{1}{n-1} - \frac{1}{n}$). Bisecting the range $[-W, W]$ with rational thresholds, polynomially many in n and $\log W$ times, each time invoking the partition algorithm, we may uniquely determine the value of a vertex [9, 15, 6].   □

Proposition 2. *In the 0-mean partition problem the following assumptions can be done without loss of generality:*

1. *the game graph has no 0-weight cycles,*
2. *the game graph is bipartite,*
3. *the values of all vertices are of the same sign;*
4. *the game graph is complete bipartite (with $|V_{max}| = |V_{min}|$).*

Every reduction from general MPGs to a restricted case is polynomial.

Proof. Given an arbitrary MPG, consider the following chain of reductions.

1. Multiplying every edge weight by $n+1$ and subtracting one does not change signs of positive- and negative-weight cycles, but 0-weight cycles (if any) become negative-weight. The 0-mean partition remains the same.

2. The straightforward solution is to introduce a vertex of the opposite player between two vertices of the same player. This, however, may increase the number of vertices quadratically. A more economic solution, leading to just a linear increase in the number of vertices is depicted in the figure below (the outgoing edge from the new vertex gets weight 0). Note that this transformation may actually change means of cycles, but not 0-means partitions, which is enough for our purpose of computing values.

3. Let v be an arbitrary vertex of a bipartite MPG G without 0-weight cycles. Construct G' by adding a new backward edge from every vertex $u \neq v$ of G to v of weight $-M$ for an edge from a MAX vertex and of weight $+M$ for a MIN vertex, where $M = (n-1)W + 1$. Suppose a play in G' starts from v. If MAX can secure a positive value of v in G, he can use the same strategy as in G never using new edges. If MIN never uses his new edges, then the value is the same as in G. But if MIN is the first to use his heavy edge back to v, the cycle thus formed has a mean $\geq [(n-1)W + 1 - (n-1)W]/n > 0$ (and we refer to the equivalence of finite and infinite MPGs [8]). The case when v has negative value in G is symmetric. Now suppose a play starts in any other vertex $v' \neq v$. Then each player can reach v and then follow the same strategy as he uses from v. The signs of means in *infinite* plays thus formed, one starting from v', the other from v, are the same (the initial finite path does not matter (this argument also depends on the equivalence between finite and infinite MPGs [8]). Hence, in G' values of all vertices are of the same sign.

4. Let $M = (n-1)W + 1$. Add all missing edges between V_{max} and V_{min} of weight $-M$ or $+M$ depending on whether an edge leaves a MAX or a MIN vertex. This makes the graph complete bipartite and preserves the signs of all values. We can also assume both partitions have the same number of vertices.   □

Remark 1. In the above chain of reductions the numbers of vertices and edges grow just linearly in the number of vertices n. In contrast, maximum absolute weights in each of 1, 3, and 4 are multiplied by n, resulting in the overall weight multiplication by n^3. Our algorithm operates on bipartite MPGs without 0-weight cycles. Thus, assumptions 1, 2 cost us a factor of n in the weight increase.

2.3 Longest Shortest Paths (LSP)

The Longest Shortest Paths problem has previously been successfully applied to solve MPGs in *randomized subexponential* time [6, 5]. Here we will use it to prove the existence of small tight feasible solutions of the MPG-generated systems of linear constraints (Section 4).

THE LONGEST SHORTEST PATH PROBLEM.

Given: a weighted digraph (without 0-weight cycles) with a sink and a set of *controlled* vertices.
Find: a selection of *exactly one* edge from each controlled vertex maximizing the lengths of the shortest paths from each vertex to the sink.   □

3   LP Formulations for MPGs

The definition below does not assume that an MPG is bipartite nor complete.

Definition 1 (Linear Slack Constraints). *For an MPG G let S_G, called* slack constraints, *be the following system of linear constraints:*

1. for every edge $x \xrightarrow{w} v$ with $x \in V_{\max}$ write constraints

$$x = v + w_{xv} + s_{xv}, \tag{2}$$
$$s_{xv} \geq 0, \tag{3}$$

where s_{xv} is a MAX *slack variable for the edge;*
2. similarly, for every edge $y \xrightarrow{w} v$ with $y \in V_{\min}$ write constraints

$$y + s'_{yv} = v + w'_{yv}, \tag{4}$$
$$s'_{yv} \geq 0, \tag{5}$$

where s'_{yv} is a MIN *slack variable for the edge.*   □

We adopt the convention that primed s' and w' denote MIN slacks and weights of edges outgoing from MIN vertices. In the sequel we will freely identify edges with their corresponding equality constraints.

The simple LP-formulation above allows one to derive many interesting MPG properties to be discussed below. We start with the simplest, but useful

Proposition 3. *For a cycle in an MPG G let w_i, s_i, s'_i (for $i \in I$) be weights of all edges, all* MAX *slacks, and all* MIN *slacks on the cycle. Then S_G implies*

$$\sum_{i \in I} w_i + \sum_{i \in I} s_i - \sum_{i \in I} s'_i = 0. \tag{6}$$

Proof. Just sum up left- and right-hand sides of the equalities corresponding to edges on the cycle.   □

This proposition partially explains why the bipartite requirement is useful. Indeed, whenever a positive weight cycle traverses only MAX vertices in G, or a negative weight cycle traverses only MIN vertices, the system S_G is infeasible, because (6) cannot be satisfied.

With the introductory purpose of explaining the usefulness of linear slack constraints, let us temporarily assume complete bipartiteness. Say that a solution to a linear slack system is *tight for* MAX (for MIN, resp.), if for every MAX vertex (MIN vertex, resp.) at least one outgoing edge has slack zero (we call such edges *tight*). The following proposition shows that tight solutions determine the winner.

Proposition 4. *If the system of slack constraints has a tight solution for*

1. MAX, *then* MAX *can enforce a nonnegative cycle in the corresponding MPG from every vertex;*
2. MIN, *then* MIN *can enforce a nonpositive cycle in the corresponding MPG from every vertex.*

Proof. Let MAX use any tight edges with zero slacks as his strategy. Then, by (6), for every cycle that MIN can create the sum of edge weights on the cycle is nonnegative. The proof of the second claim is analogous.   □

The next section addresses the existence of (tight) solutions for the MPG linear slack constraints and their relation to determining the winner. Now we introduce MPG-polyhedra.

Definition 2 (MPG Polyhedron). *An MPG-polyhedron is the feasible set of the linear slack constraints corresponding to an MPG; see Definition 1.*   □

We have seen above that some MPGs may induce empty polyhedra. The next section shows that bipartite MPGs always have nonempty polyhedra. Here we state simple properties of MPG-polyhedra.

Proposition 5. *An MPG-polyhedron has no vertices.*

Proof. Suppose (x, y, s) is a vertex. Then $(x + \alpha 1, y + \alpha 1, s)$ (1 is a vector of ones, $\alpha \in \mathbb{R}$) is also a feasible solution to slack constraints. Thus, any MPG-polyhedron, with each point contains a line, hence has no vertices.   □

In Section 4 we introduce additional bounding constraints and an MPG-polyhedron becomes an MPG-polytope (bounded polyhedron), with vertices.

Another useful property of MPG-polyhedra is their integrality.

Proposition 6. *For any MPG-polyhedron P one has $conv(P) = conv(P_I)$.*

Proof. Any MPG-generated linear slack system can be written as $[A\ I](x, y, s)^T = b$, where the entries in A correspond to x and y variables, and the identity matrix corresponds to the slacks. Every row of A has *exactly* one +1 and one −1 entry and is thus totally unimodular. Totally unimodularity for $[A\ I]$ follows directly, since it is preserved when adding a column with at most one nonzero, being ± 1 [12, p. 280]. By [12, Thm. 19.1, p. 266], the polyhedron $\{v|\ [A\ I]\ v \leq b\}$ is integral whenever b is integral. Duplicating a row and multiplying a row by -1 preserve total unimodularity and the polyhedron $\{v|\ [A\ I]\ v = b\}$ is integral.   □

As a consequence, any linear function over an MPG-polyhedron with finite optimum, has an integral optimum. Moreover, optimizing any linear function over an MPG-polyhedron can be done in *strongly* polynomial time, because the constraint matrix consists of 0 and ± 1 entries.

4   Existence of Tight Solutions

In this section we consider linear slack systems corresponding to bipartite (not necessarily complete) MPGs and show that they possess tight feasible solutions of bounded size. We first generalize the notion of tightness, introduced (for the case of complete bipartite MPGs) in the previous section.

Definition 3 (Tight Solution). *Given a linear slack system S_G obtained from a bipartite MPG G without 0-weight cycles, say that a solution to S_G is tight if there is a partition of vertices of G into sets X and N such that:*

1. *every MIN vertex in X has a tight edge to X;*
2. *every MAX edge from X leads to X;*
3. *every MAX vertex in N has a tight edge to N;*
4. *every MIN edge from N leads to N.*   □

(Note that in the case of an ergodic MPG, e.g., a complete bipartite MPG, either X or N should be necessarily empty.)

   A tight solution to a slack system gives the 0-mean partitioning for the associated MPG as shows the following

Proposition 7. $G_{>0} = N$ *and* $G_{\leq 0} = X$.

Proof. If a play starts in N, then MAX may just use his tight edges to stay in N. When a cycle is eventually formed, by (6), the sum of weights on the cycle is positive (there are no 0-weight cycles); hence, the mean is also positive.

   Symmetrically, if a play starts in X, then MIN just uses his tight edges to stay in X. When a cycle is eventually formed, by (6), the sum of weights on the cycle is nonpositive; hence, the mean is also nonpositive.   □

Here comes the main result of this section. Although there are well-known general bounds on some feasible solution to a system of linear constraints (if it exists) [12, Ch. 10], our bounds for MPG-generated constraints are stronger. We also show that tight solutions of bounded size always exist. In Section 5 we prove related results for complete bipartite MPGs.

Theorem 1 (Tight Solution Existence). *A linear slack system of every bipartite MPG without 0-weight cycles always has a tight solution with integral components of absolute value $O(nW)$, where n is the number of vertices and W is the maximal absolute edge weight.*

Proof. Add *retreat* edges, of weight 0, from all MAX vertices to the sink (new vertex), and of weight $M = (2n - 1)W + 1$ from all MIN vertices to the sink. The resulting graph determines an instance of the *Longest Shortest Paths* (LSP) problem [6]. In this instance optimal positional strategies of both players create no cycles, because each cycle is either positive or negative, which one of the players always wants to avoid (and can due to bipartiteness). Thus all optimal plays end up in the sink, through a 0- or M-weight retreat edge. The *unique* [6] solution (with all components finite, because every cycle is broken by one of the players selecting to retreat) determines a feasible solution to the linear slack system. Optimal edges for both players have associated slacks equal zero. Moreover, by the properties of the shortest paths [6] and optimality for both players, the following conditions are satisfied for every edge (v, u) of the game graph, because $d(v)$, $d(u)$ are shortest path distances:

$$d(v) \leq w(v, u) + d(u), \text{ if } v \in V_{\min}, \tag{7}$$

$$d(v) \geq w(v, u) + d(u), \text{ if } v \in V_{\max}. \tag{8}$$

These conditions ensure that all slacks are *nonnegative*. Moreover, at least one slack per vertex is zero, since $d(v)$ are defined by shortest paths.

Let the required sets X and N be as follows:

1. N is the set of vertices starting from which MAX can force a play into a MIN vertex from which MIN retreats through the retreat edge with weight M, when both players can use tight edges only;
2. X is the set of vertices starting from which MIN can force a play into a MAX vertex from which MAX retreats through the retreat edge with weight 0, when both players can use tight edges only.

The graph on tight edges is acyclic, bipartite, spanning all vertices of the game graph, with leaves being vertices selecting retreat edges. Therefore, N and X form a partition, which can be easily computed, after topological sorting, by dynamic programming. We have to show that MAX has no edges (including non-tight) from X to N and MIN has no edges (including non-tight) from N to X (see Definition 3).

Since X and N do not intersect and shortest distances inside them are defined by tight edges, the choice of the weights for the retreat edges implies the bounds on the values of MAX and MIN vertices in X and N summarized in the table.

	X	N
MAX	$[0, (n-1)W]$	$[nW + 1, 2nW + 1]$
MIN	$[-W, (n-1)W]$	$[nW + 1, (2n-1)W + 1]$

In the left column, the common upper bound is explained by the fact that the longest path in X may traverse at most $n - 1$ edges of weight at most W. The lower bounds 0 and $-W$ in the left column are due to the MAX retreat and to bipartiteness: the best MIN can do is to go to the 0-value vertex via a $-W$ edge. In the right column, the common lower bound is because the shortest path in N is through the M-weighted retreat and at most $n - 1$ edges of weight $-W$. The upper bound for a MIN variable is due to the retreat weight, and for a MAX variable it is just W larger.

To show that MAX has no edges from X to N, assume, toward a contradiction, that MAX has an edge from $v \in X$ to $u \in N$. The bound from the table above together with (8) imply $w(v, u) < -W$, a contradiction, since W is the maximal absolute edge weight. A similar argument shows that MIN cannot have edges from N to X.

Now delete the sink and retreat edges to return to the original game. All equalities in the associated linear slack system are satisfied. This solution is tight as shown above. Note that some 0 slacks for some variables can disappear (in the vertices where a retreat was taken).

Since a slack s is always equal $s = x - y \pm w$, from the table above we conclude that all slacks are at most $O(nW)$. $\square$

Remark 2. We can thus impose additional bounding constraints for all variables in the linear slack systems from Definition 1. The feasible set becomes a polytope with vertices, which we call an *MPG-polytope*.

Proposition 8. *An MPG-polytope of a bipartite game always has at least one vertex, which is a tight solution.*

Proof. Consider a tight solution, which exists by Theorem 7. Minimize the sum of slacks, which are zero in the tight solution, over the MPG-polytope. Obviously, the value of the optimum will be zero. Furthermore, the optimal solution can be attained in a vertex of the polytope. $\square$

Proposition 11 shows a simple form of a linear target function for a complete bipartite MPG with an optimum attained in a tight solution.

Corollary 1. *Vertices of an MPG-polytope of a bipartite game are integral.* $\square$

5   MPGs on Complete Bipartite Graphs

In this section we assume that MPGs are played on complete bipartite graphs $K_{p,p}$. Thus the number of vertices $n = 2p$. We use a convention that x_i, y_i denote variables associated to the i-th vertex of MAX and MIN respectively, s_{ij} and s'_{ij} denote slacks for MAX and MIN edges, and w_{ij}, w'_{ij} denote edge weights of MAX and MIN. Slack equality constraints (2) and (4) in this case are (for $1 \leq i, j \leq p$):

$$x_i = y_j + w_{ij} + s_{ij}, \tag{9}$$
$$y_i + s'_{ij} = x_j + w'_{ij}. \tag{10}$$

5.1 Invariant Properties

Proposition 9. *Every solution to a linear slack constraint system S_G obtained from a complete bipartite MPG G satisfies the invariant*

$$\sum_{ij} s_{ij} - \sum_{ij} s'_{ij} = -\sum_{ij} (w_{ij} + w'_{ij}).$$

Proof. Sum up all equalities (9) and (10). This gives $\sum s'_{ij} = \sum s_{ij} + \sum (w_{ij} + w'_{ij})$, since each variable x_i, y_i appears in the left- and right-hand sides of (9), (10) the same number of times. $\qquad\square$

The following proposition shows that one can optimize any of the several linear functions over the MPG-polytope. They happen to possess the same optimal solutions, i.e., are equivalent.

Proposition 10. *For any complete MPG-generated S_G the following functions are similar up to scaling and a constant additive term:*

$$1) \sum_{i,j} s_{ij}, \quad 2) \sum_{i,j} s'_{ij}, \quad 3) \sum_{i,j} s_{ij} + \sum_{i,j} s'_{ij}, \quad 4) \sum_{i} x_i - \sum_{i} y_i.$$

Proof. Equivalence of 1-3 follows from Proposition 9. To prove equivalence of 1 and 4, we use the fact that $s_{ij} = x_i - y_j + w_{ij}$. Thus

$$\sum_{ij} s_{ij} = (x_1 - y_1 + w_{11}) + (x_1 - y_2 + w_{12}) + \ldots + (x_1 - y_n + w_{1n}) +$$
$$(x_2 - y_1 + w_{21}) + (x_2 - y_2 + w_{22}) + \ldots + (x_2 - y_n + w_{2n}) +$$
$$\vdots$$
$$(x_n - y_1 + w_{n1}) + (x_n - y_2 + w_{n2}) + \ldots + (x_n - y_n + w_{nn})$$
$$= n(\sum_i x_i - \sum_i y_i) + c, \text{ where } c \text{ is a constant.} \qquad\square$$

5.2 Complete Bipartite MPGs as Linear Programs

The next proposition asserts that there is always a simple linear target function over the feasible polytope of a complete bipartite MPG with the optimum, which solves the game.

Proposition 11. *Let S_G be a linear slack system obtained from a complete bipartite MPG. Then there exist vectors $a, b \in \mathbb{N}^p$ such that $\sum_i a_i = \sum_i b_i = p$ and the optimal solution to S_G with the objective function $\min \sum_i a_i x_i - \sum_i b_i y_i$ has either a tight solution for MAX or for MIN and thus solves the corresponding MPG. Moreover, one of the vectors a, b consists of ones only.*

Proof. Suppose MAX has a winning strategy, hence a tight solution. Then the sum of the slacks corresponding to his optimal edges (tight), taken one per vertex, $\sum_{(i,j) \in I} s_{ij}$ has minimal solution 0. But this sum is equal $\sum_{(i,j) \in I} (x_i - y_j - w_{ij}) = \sum_{i-1}^{n} x_i - \sum_{j=1}^{n} b_j y_j + C$, where b_j counts how many times y_j is selected as a destination of some MAX optimal edge. The proof, when MIN has a winning strategy is symmetric. $\qquad\square$

As a consequence, for a complete bipartite MPG, the corresponding slack polytope has a vertex solving the game (which also follows by Proposition 8). We state two other simple corollaries.

Corollary 2. *The problem of deciding the winner for a complete MPG reduces to the problem of determining:*

1. *the number of* MAX *vertices that play, in a winning positional strategy, to the* MIN *vertex y_i, for each i, if* MAX *has a winning strategy, or*
2. *the number of* MIN *vertices that play, in a winning positional strategy, to the* MAX *vertex x_i, for each i, if* MIN *has a winning strategy.* □

Corollary 3. *If* MAX *has a winning strategy where every* MAX *vertex selects an unique* MIN *vertex. The game is solvable with the objective function* $\min \sum_i x_i - \sum_i y_i$. *The case for* MIN *is symmetric.* □

5.3   Search Space

Proposition 11 allows one to somewhat reduce the search space of all positional strategies in a complete bipartite MPG.

Proposition 12. *The problem of finding vectors a, b such that it is possible to recover the winning player from the optimal solution to S_G with objective function $\min \sum_i a_i x_i - \sum_i b_i y_i$ has strictly smaller search space than deciding the optimal strategy of one player.*

Proof. In a complete MPG G played on the graph $K_{p,p}$ both players have p^p number of strategies.

Consider the problem of finding vectors a, b recovering the winning player from an optimal solution to S_G with objective function $\min \sum_i a_i x_i - \sum_i b_i y_i$ (as explained in the proof of Proposition 11).

If MAX has a winning strategy, we can assume $a = 1$. It remains to find the correct b_i's. Any vector b with p nonnegative integer components summing up to p can be represented by a word of $p-1$ zeros (bucket separators) and p ones, i.e., p buckets and p items. The number of possible ways to distribute the items are $(2p-1)!/(p!(p-1)!) = \binom{2p-1}{p} = O(2^{2p})$.

Similarly, if MIN is winning the number of ways to select the vector a is $O(2^{2p})$. Thus, the number of different meaningful objective functions are bounded by $O(2^{2p})$, which is $o(p^p) = o(2^{p \log p})$. □

6   0-In-Out Property

In this section we only assume that MPGs are bipartite, but not necessarily complete. Consider the following interesting

Definition 4 (0-in-out property). *Say that a solution to an MPG-generated system of slack constraints satisfies the* 0-in-out *property if*

$$\forall i \in V_{\max} \, \exists j \in V_{\min}(s_{ij} = 0 \vee s'_{ji} = 0) \wedge \forall i \in V_{\min} \, \exists k \in V_{\max}(s'_{ik} = 0 \vee s_{ki} = 0).$$

Informally, it stipulates that every vertex has at least one incoming or outgoing 0-slack (tight) edge. Two propositions below summarize interesting relations between tight solutions to systems of slack constraints, solutions minimizing $\sum x_i - \sum y_i$,[1] and solutions with the 0-in-out-property.

Proposition 13. *Every solution to an MPG-generated system of slack constraints, which minimizes $\sum x_i - \sum y_i$, possesses the 0-in-out property.*

Proof. An x_i with nonzero slacks on all outgoing and incoming edges can be decreased thus diminishing the target value. Similarly, a y_i with nonzero slacks on all outgoing and incoming edges can be increased thus diminishing the target value.   □

Proposition 14. *For every* MAX- *or* MIN-*tight solution to an MPG-generated system of slack constraints there corresponds a tight solution satisfying the 0-in-out property with a smaller or equal target value $\sum x_i - \sum y_i$.*

Proof. A MAX-tight solution has 0-in-out property satisfied for all MAX vertices. If the property is not satisfied for a vertex y_i, then its value can be increased, keeping the tightness, and decreasing the target value. The proof for the MIN-tight solutions is completely similar.   □

6.1   Minimizing Slacks Does Not Give Tight Solutions

Thus, both: 1) tight solutions (modified, if necessary as explained in the proof of Proposition 14) and 2) solutions minimizing $\sum x_i - \sum y_i$, satisfy the 0-in-out property. A natural challenging question is: *whether tight solutions can always be found among minimizing $\sum x_i - \sum y_i$?* This plausible conjecture, if true, would allow us to limit the search for tight solutions among those minimizing $\sum x_i - \sum y_i$. Unfortunately, this promising conjecture fails, as demonstrated by the counterexample in Figure 1.

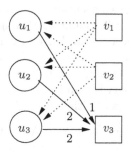

Fig. 1. A complete bipartite MPG where the dotted edges have weight -2 and the edges that are not in the figure have weight 0

---

[1] Recall that x_i, y_i are variables associated with the i-th vertex of MAX and MIN.

By Proposition 10, for any systems of linear slack constraints corresponding to complete bipartite MPGs, minimizing the objective function $\sum_i x_i - \sum_i y_i$ is equivalent to minimizing the objective functions $\sum\{\text{MAX slacks}\}$, $\sum\{\text{MIN slacks}\}$, and $\sum\{\text{All slacks}\}$.

It is easy to see that the value of $\min \sum_i x_i - \sum_i y_i$, when MAX uses his winning strategy (always plays to v_3) is 7, because $u_1 = v_3 + 1$, $u_2 = v_3 + 2$, $u_3 = v_3 + 2$, $v_2 = u_1 - 2$, $v_1 = u_1 - 2$. Letting all MAX variables equal 2 and all MIN variables equal 0 is also feasible, but then u_1 has no tight outgoing edges. Thus, a MAX- or MIN-tight solution can have a larger value than the minimal value of the objective function $\sum_i x_i - \sum_i y_i$.

7   Slacks Update "Tightening" Algorithm

Despite the fact (described by the previous counterexample) that there may be no tight solutions (solving MPGs) among those minimizing $\sum x_i - \sum y_i$, the idea to start from such a solution and transform it into a tight one seems quite tempting. We now develop this idea and describe an algorithm for finding tight solutions for MPG-generated systems of slack constraints, and thus solves MPGs by Proposition 7. The algorithm applies to systems obtained from bipartite (not necessarily complete) MPGs without 0-weight cycles. The proof of correctness and the intuitions underlying the algorithm go in parallel with its description.

The Algorithm starts by finding a solution to slack constraints minimizing $\sum x_i - \sum y_i$ (in strongly polynomial time). By Proposition 13, every vertex has at least one (incoming or outgoing) tight edge.

Main Loop. Let X_0 and N_0 be the sets of MAX and MIN vertices without tight outgoing edges. If one of these sets is empty, the 0-mean partition is found (Proposition 7). Temporarily delete all non-tight edges. Let X be the set of vertices starting from which MIN can force a play into X_0, and N be the set of vertices from which MAX can force a play into N_0. (Both sets may be easily computed in polynomial time, as shown below.)

We claim that X and N form a partition of the game vertices. Indeed, every vertex is an endpoint (source or destination) of at least one tight edge. Note also that the graph induced by tight edges is acyclic (this follows from Proposition 3, because a cycle with all slacks 0 should be 0-weight, absent by assumption). Topologically sort it, and proceed from leaves (which are either in $X_0 \subseteq X$ or in $N_0 \subseteq N$) backwards in the topological order as follows. For a MAX vertex v with all successors already decided to be in X or N, put v to N if it has a tight edge to N, and to X otherwise, and symmetrically for a MIN vertex. This classifies all vertices as members of either X or N. At this stage:

- there are no tight MAX edges from X to N, by definition of X; equivalently, all MAX edges from X to N, denote them $E_{\max}(X, N)$, are non-tight;
- there are no tight MIN edges from N to X, by definition of N; equivalently, all MIN edges from N to X, denote them $E_{\min}(N, X)$, are non-tight;
- note that there may exists tight MAX edges from N to X, as well as tight MIN edges from X to N.

Terminate? If the set of edges $E_{\max}(X, N) \cup E_{\min}(N, X)$ is empty, the 0-mean partition is found: $G_{\leq 0} = X$ and $G_{>0} = N$ (see Proposition 7), and the algorithm terminates. (Both X, N may be nonempty if the graph is not complete bipartite.)

Update. Let $\delta > 0$ be the *minimal* slack assigned to edges in $E_{\max}(X, N) \cup E_{\min}(N, X)$ (all such edges are non-tight; see above). Now, either 1) *increase* the values of all vertices in N by δ, or 2) *decrease* the values of all vertices in X by δ. This does not violate any constraints, and preserves the property that every vertex has at least one in- or outgoing tight constraint/edge. Indeed, all constraints corresponding to edges from X to X and from N to N remain satisfied (since we increase or decrease the values of variables in both sides of constraints by the same δ). Proceed to the Main Loop. □

Note that in the Update step: a) at least one non-tight edge in $E_{\max}(X, N) \cup E_{\min}(N, X)$ becomes tight, but b) all tight edges in $E_{\max}(N, X) \cup E_{\min}(X, N)$, if any, become non-tight. Therefore, we unfortunately do not have monotonic increase of the set of tight edges. However, once a vertex obtains a tight outgoing edge, it keeps at least one such edge forever. Thus, the set of vertices possessing tight edges monotonically increases. Consequently, the sets X_0 and N_0 may only *decrease (monotonicity)*. Every increase, in the Update step, of values of vertices in N decreases the positive slacks of all edges leaving vertices in N_0 and going to X, and the positive slacks of all edges leaving vertices in X_0 and going to N (there is always at least one such edge; otherwise the algorithm terminates. (The *decrease* case 2) is analogous.) Therefore, after pseudopolynomially many steps at least one vertex in $X_0 \cup N_0$ will obtain a tight edge and will leave the set $X_0 \cup N_0$ forever. We summarize the above argument in the following

Theorem 2. *The described algorithm is pseudopolynomial, $O(|G| \cdot n \cdot W)$, where G is the size of the game graph, n the number of its vertices, and W is the largest absolute edge weight.* □

Note, retrospectively, that this algorithm is similar in spirit to the iterated potential transformation algorithm of [9] (proved exponential in [9] and pseudopolynomial in [11]). Our algorithm is based on completely different principles. Moreover, our proof and the algorithm description are considerably simpler.

8   Conclusions

The idea to describe MPGs by linear constraints and investigate the associated polytopes using linear programming methods appears natural and useful. It reveals simple algebraic properties of MPG-polytopes and allows for a new transparent LP-based algorithm for solving MPGs. In a forthcoming paper we will present further properties of MPG-polytopes and a dual algorithm, which allow for a faster convergence to a tight solution.

References

1. H. Björklund, O. Nilsson, O. Svensson, and S. Vorobyov. Controlled linear programming: Boundedness and duality. Technical Report DIMACS-2004-56, DIMACS: Center for Discrete Mathematics and Theoretical Computer Science, Rutgers University, NJ, December 2004. http://dimacs.rutgers.edu/TechnicalReports/.
2. H. Björklund, O. Nilsson, O. Svensson, and S. Vorobyov. The controlled linear programming problem. Technical Report DIMACS-2004-41, DIMACS: Center for Discrete Mathematics and Theoretical Computer Science, Rutgers University, NJ, September 2004.
3. H. Björklund, O. Svensson, and S. Vorobyov. Controlled linear programming for infinite games. Technical Report DIMACS-2005-13, DIMACS: Center for Discrete Mathematics and Theoretical Computer Science, Rutgers University, NJ, April 2005.
4. H. Björklund, O. Svensson, and S. Vorobyov. Linear complementarity algorithms for mean payoff games. Technical Report DIMACS-2005-05, DIMACS: Center for Discrete Mathematics and Theoretical Computer Science, Rutgers University, NJ, February 2005.
5. H. Björklund and S. Vorobyov. Combinatorial structure and randomized subexponential algorithms for infinite games. *Theoretical Computer Science*, 349(3):347–360, 2005.
6. H. Björklund and S. Vorobyov. A combinatorial strongly subexponential strategy improvement algorithm for mean payoff games. *Discrete Applied Mathematics*, 2006. Accepted, to appear. Preliminary version in MFCS'04, Springer Lecture Notes in Computer Science, vol. 3153, pp. 673-685, and DIMACS TR 2004-05.
7. R. W. Cottle, J.-S. Pang, and R. E. Stone. *The Linear Complementarity Problem*. Academic Press, 1992.
8. A. Ehrenfeucht and J. Mycielski. Positional strategies for mean payoff games. *International Journ. of Game Theory*, 8:109–113, 1979.
9. V. A. Gurvich, A. V. Karzanov, and L. G. Khachiyan. Cyclic games and an algorithm to find minimax cycle means in directed graphs. *U.S.S.R. Computational Mathematics and Mathematical Physics*, 28(5):85–91, 1988.
10. K. G. Murty and F.-T. Yu. *Linear Complementarity, Linear and Nonlinear Programming*. Heldermann Verlag, Berlin, 1988.
11. N. Pisaruk. Mean cost cyclical games. *Mathematics of Operations Research*, 24(4):817–828, 1999.
12. A. Schrijver. *Theory of Linear and Integer Programming*. John Wiley and Sons, 1986.
13. A. Schrijver. *Combinatorial Optimization*, volume 1-3. Springer, 2003.
14. O. Svensson and S. Vorobyov. A subexponential algorithm for a subclass of P-matrix generalized linear complementarity problems. Technical Report DIMACS-2005-20, DIMACS: Center for Discrete Mathematics and Theoretical Computer Science, Rutgers University, NJ, June 2005.
15. U. Zwick and M. Paterson. The complexity of mean payoff games on graphs. *Theor. Comput. Sci.*, 158:343–359, 1996.

Atomic Routing Games on Maximum Congestion

Costas Busch and Malik Magdon-Ismail

Rensselaer Polytechnic Institute, Dept. of Computer Science, Troy, NY 12180, USA
{buschc, magdon}@cs.rpi.edu

Abstract. We study atomic routing games on networks in which players choose a path with the objective of minimizing the *maximum congestion* along the edges of their path. The social cost is the global maximum congestion over all edges in the network. We show that the *price of stability* is 1. The *price of anarchy*, PoA, is determined by topological properties of the network. In particular, $PoA = O(\ell + \log n)$, where ℓ is the length of the longest path in the player strategy sets, and n is the size of the network. Further, $\kappa - 1 \leq PoA \leq c(\kappa^2 + \log^2 n)$, where κ is the length of the longest cycle in the network, and c is a constant.

1 Introduction

A fundamental issue in the management of large scale communication networks is to route the packet traffic so as to optimize the network performance. Our measure of network performance is the worst bottleneck (most used link) in the system. We model network traffic as finite, unsplittable packets (atomic flow) [22, 26], where each packet's path is controlled independently by a selfish player. The Nash equilibrium (NE) is a natural outcome for a game with selfish players – a stable state in which no player can unilaterally improve her situation. In the recent literature, the *price of anarchy* (*PoA*) [15, 24] and the *price of stability* (*PoS*) [1, 2] have become prevalent measures of the quality of the equilibria of uncoordinated selfish behavior relative to coordinated optimal behavior. The former quantifies the worst possible outcome with selfish agents, and the latter measures the minimum penalty in performance required to ensure a stable equilibrium outcome.

We study routing games with N players corresponding to N source-destination pairs of nodes on a network G. The strategy set available to each player is a set of edge-simple paths from the player's source to the destination (typically the strategy set consists of all edge-simple paths in G). A *pure* strategy profile is a selection of a single path (strategy) by each player from her respective strategy set. We study pure Nash equilibria. In our context, a pure strategy profile corresponds to a *routing* **p**, a collection of paths, one for each player. We refer to Nash equilibra in this context as *Nash-Routings*. A routing **p** causes *congestion* in the network: the congestion C_e on an edge e is the number of paths in **p** that use this edge; the congestion C_{p_i} of a path $p_i \in$ **p** is the maximum congestion

S.-W. Cheng and C.K. Poon (Eds.): AAIM 2006, LNCS 4041, pp. 79–91, 2006.

over all edges on the path; the congestion C of the network is the maximum congestion over all edges in the network. The *dilation* D is the maximum path length in **p**.

Since a packet is to be delivered along each player's path, a natural choice for social cost is the maximum delay incurred by a packet. The packets can be scheduled along the paths in **p** with maximum delay $O(C + D)$ [6, 17, 18, 23, 25]. In heavily congested networks, $C \gg D$, and the maximum delay of a packet is governed by the congestion C. Thus, the network congestion is an appropriate social cost – this choice for the social cost is often referred to as the *maximum social cost* [4, 5, 15, 27].

Consider player i with path $p_i \in$ **p**. It is shown in [3] that player i's packet can be delivered in time $\tilde{O}(C_{p_i} + |p_i|)$, where $|p_i|$ is the path length (this holds for all players simultaneously). In congested networks, $C_{p_i} \gg |p_i|$, and so it is appropriate to use C_{p_i} as the player cost, along her chosen path. This choice of player cost is typically referred to as the *maximum player cost*. The maximum player cost is is appropriate since this is what governs the delay experienced by that player in a highly congested network [3]. In the literature it is common to use the sum player cost (instead of the maximum) [7, 9, 14, 15, 21, 27]. However, the sum of congestions does not govern the packet delays, since when a packet waits for a particular congested edge to clear of other packets, the other congested edges in its path can be cleared simultaneously. It is the maximum player and social costs that are appropriate metrics for atomic routing games.

1.1   Contributions

We give the first comprehensive analysis of routing games with maximum player and social cost. We study the quality of pure Nash-Routings with respect to the price of stability and anarchy.

In our first result, we establish that there exist *optimal* Nash-Routings where the social cost (congestion) is equal to the optimal coordinated cost; in other words, $PoS = 1$ (the price of stability expresses the ratio of the optimal social cost in the Nash-Routing with the optimal coordinated cost). We also show that any *best response dynamic*, a sequence of best response moves of players, converges to a Nash-Routing in a finite amount of time. Thus, we can easily obtain Nash-Routings, starting from arbitrary initial routings.

Theorem 1. *For every routing game:*
(i) There is a pure Nash-Routing which is optimal ($PoS = 1$).
(ii) Every best response dynamic converges to a Nash-Routing in finite time.

We continue by examining the quality of the worst case Nash-Routings. The price of anarchy, PoA, expresses the ratio of the social cost in the worst-case Nash-Routing to the optimal coordinated cost. We bound the price of anarchy in terms of topological properties of the network. The next result bounds the price of anarchy for arbitrary instances of routing games in terms of the maximum path-lengths in the strategy sets:

Theorem 2. *For any routing game where the strategy sets of the payers have paths with length at most ℓ, $PoA < 2(\ell + \log n)$.*

Theorem 2 gives good bounds for the price of anarchy for networks where it is natural to use paths with short length. For example in the Hypercube and Butterfly [16], if we choose bit-fixing paths, then $\ell = O(\log n)$, which implies that $PoA \leq c \log n$, for some constant c.

Our next result characterizes the worst case Nash-Routing in terms of the longest cycle of the network. For a graph G, the *edge-cycle number* $\kappa_e(G)$ is the length of the longest edge simple cycle in G; we will drop the dependence on G when the context is clear.

Theorem 3. *For any undirected graph G with edge-cycle number κ_e,*
(i) there exists a routing game for which $PoA \geq \kappa_e - 1$;
(ii) for any routing game, $PoA \leq c(\kappa_e^2 + \log^2 n)$, for some constant c.

Let m denote the number of edges in the network. Since $\kappa_e \leq m$, we have that $PoA \leq c \cdot m^2$. In graphs with Euler cycles, $\kappa_e = m$. Therefore, Theorem 3 implies that $m - 1 \leq PoA \leq c \cdot m^2$ (we use c to represent a generic constant).

The lower bound of Theorem 3 (part i) is obtained by constructing a game instance where the players have their sources and destination on the largest cycle. To prove the upper bound of Theorem 3 (part ii), we use Theorem 2. For 2-connected graphs, every pair of nodes has two edge-disjoint paths connecting them (Menger's theorem [32]), from which we establish that $\ell \leq c \cdot \kappa_e^2$. The cycle upper bound follows immediately by using Theorem 2.

If the graph G is not 2-connected, then the relation $\ell \leq c \cdot \kappa_e^2$ may not hold. To obtain the result for a general graph G, we decompose G into a tree of 2-connected components. We show that if in G the Nash-Routing has network congestion C, then there is some 2-connected component G' which has congestion $C' \approx C$. At the same time the players in G' are in a *partial* Nash-Routing, where many of them are locally optimal. A generalization of Theorem 2 to partial Nash-Routings, helps to establish the upper-bound of Theorem 3.

1.2 Related Work

General congestion games were introduced and studied in [22, 26]. The application of game theory in computer science, specifically the introduction of the price of anarchy was introduced in [15]. Since then, many models have been studied, categorized by: the topology of the network; the nature of the player and social costs; the nature of the traffic (atomic or splittable); the nature of the strategy sets; the nature of the equilibria studied (pure or mixed). A brief taxonomy of some relevant existing results, according to the kind of flow (atomic or splittable) and equilibria (mixed or pure), and according to the social cost SC and player cost pc (sum or maximum), are shown in the following two tables.

	Atomic Flow	Splittable Flow
Pure	[4, 19, 26], [31]*, **Our Work**	[27, 28, 29, 30]
Mixed	[7, 8, 9, 11, 12, 13, 14, 15, 20, 21, 24]*	[5], [10]*

	Max SC	Sum SC	Other SC	**
Max pc	**Our Work**	–	–	[19]
Sum pc	[4, 5, 27] [7, 8, 9, 10, 11, 14, 15, 21, 24]*	[4, 28, 29, 30], [13, 31]*	[12, 20]*	[19, 26]

($*$: A Specific network model is used, eg. parallel links, or specific player strategy sets. $**$: Results on existence or convergence to equilibrium, as opposed to quality of equilibria).

Typically, the research in the literature has focused on computing upper and lower bounds on the price of anarchy. The vast majority of the work on maximum social cost has been for parallel link networks, with only a few recent results on general topologies [4, 5, 27]. Essentially, all of the work has focused on the sum player cost, which corresponds to the sum of the edge congestions on a path (as opposed to the maximum edge congestion on the path, which we consider here).

The only result which has a brief discussion of the maximum player cost is [19] where the authors focus on parallel link networks, but also give some results for general topologies. In [19], the main content is to establish the existence of pure Nash-Routings. We present a systematic study of pure Nash-Routings in atomic routing games. Pure equilibria with atomic players and maximum player cost introduces essentially combinatoric conditions for the equilibria, in contrast to infinitely splittable flow, or mixed equilibria, which can be characterised by Wardrop-type equilibrium conditions.

Outline of Paper. In Section 2 we give some basic definitions. We prove Theorem 1 in Section 3. We continue with the proof of Theorem 2 in Section 4. The lower bound of Theorem 3 is proven in Section 5. In the same section we prove the upper bound of Theorem 3 for 2-connected graphs. We give the general version of the upper bound in Section 6. We conclude in Section 7. Some of the technical proofs have been omitted for space considerations, and will be presented in a full version of this paper.

2   Definitions

An instance $\mathcal{R}$ of a *routing (congestion) game* is a tuple $(\mathbf{N}, G, \{\mathcal{P}_i\}_{i \in \mathbf{N}})$, where $\mathbf{N} = \{1, 2, \ldots, N\}$ are the players, $G = (V, E)$ is an undirected connected graph with $|V| = n$, and $\mathcal{P}_i$ is a collection of *edge-simple* paths. Each path in $\mathcal{P}_i$ is a path in G that has the same source $s_i \in V$ and destination $t_i \in V$; each path in $\mathcal{P}_i$ is a pure strategy available to player i. A pure strategy profile $\mathbf{p} = [p_1, p_2, \cdots, p_N]$ is a collection of pure strategies (paths), one for each player, where $p_i \in \mathcal{P}_i$. We refer to a pure strategy profile as a *routing*. On a finite network, a routing game is necessarily a finite game.

For any routing $\mathbf{p}$ and any edge $e \in E$, the *edge-congestion* $C_e(\mathbf{p})$ is the number of paths in $\mathbf{p}$ that use edge e. For any path p, the *path-congestion* $C_p(\mathbf{p})$ is the maximum edge congestion over all edges in p, $C_p(\mathbf{p}) = \max_{e \in p} C_e(\mathbf{p})$. The *network congestion* is the maximum edge-congestion over all edges in E, $C(\mathbf{p}) = \max_{e \in E} C_e(\mathbf{p})$. The *social or global cost* $SC(\mathbf{p})$ is the network congestion, $SC(\mathbf{p}) = C(\mathbf{p})$. The *player or local cost* $pc_i(\mathbf{p})$ for player i is her path-congestion, $pc_i(\mathbf{p}) = C_{p_i}(\mathbf{p})$. When the context is clear, we will drop the dependence on $\mathbf{p}$ and use C_e, C_p, C, SC, pc_i.

We use the standard notation $\mathbf{p}_{-i}$ to refer to the collection of paths $\{p_1, \cdots, p_{i-1}, p_{i+1}, \cdots, p_N\}$, and $(p_i; \mathbf{p}_{-i})$ as an alternative notation for $\mathbf{p}$ which emphasizes the dependence on p_i. Player i is *locally optimal* in routing $\mathbf{p}$ if $pc_i(\mathbf{p}) \leq pc_i(p_i'; \mathbf{p}_{-i})$ for all paths $p_i' \in \mathcal{P}_i$. A routing $\mathbf{p}$ is in a Nash Equilibrium ($\mathbf{p}$ is a *Nash-routing*) if every player is locally optimal. Nash-routings quantify the notion of a stable selfish outcome. A routing $\mathbf{p}^*$ is an optimal pure strategy profile if it has minimum attainable social cost: for any other pure strategy profile $\mathbf{p}$, $SC(\mathbf{p}^*) \leq SC(\mathbf{p})$.

We quantify the quality and diversity of the Nash-routings by the *price of stability* (*PoS*) and the *price of anarchy* (*PoA*) (sometimes referred to as the coordination ratio). Let $\mathbf{P}$ denote the set of distinct Nash-Routings, and let SC^* denote the social cost of an optimal routing $\mathbf{p}^*$. Then,

$$PoS = \inf_{\mathbf{p} \in \mathbf{P}} \frac{SC(\mathbf{p})}{SC^*}, \qquad PoA = \sup_{\mathbf{p} \in \mathbf{P}} \frac{SC(\mathbf{p})}{SC^*}.$$

3 Existence of Optimal Nash-Routings

The goal in this section is to establish Main Theorem 1. For routing $\mathbf{p}$, the *congestion vector* $\mathbf{C}(\mathbf{p}) = [m_0(\mathbf{p}), m_1(\mathbf{p}), m_2(\mathbf{p}), \ldots]$, where each component $m_k(\mathbf{p})$ is the number of edges with congestion k. Note that $\sum_k m_k(\mathbf{p}) = m$, where m is the number of edges in the network. The social cost (network congestion) $SC(\mathbf{p})$ is the maximum k for which $m_k > 0$. We define a lexicographic total order on routings as follows. Let $\mathbf{p}$ and $\mathbf{p}'$ be two routings, with $\mathbf{C}(\mathbf{p}) = [m_0, m_1, m_2, \ldots]$, and $\mathbf{C}(\mathbf{p}') = [m_0', m_1', m_2' \ldots]$. Two routings are equal, written $\mathbf{p} =_c \mathbf{p}'$, if and only if $m_k = m_k'$ for all $k \geq 0$; $\mathbf{p} <_c \mathbf{p}'$ if and only if there is some k^* such that $m_{k^*} < m_{k^*}'$ and $\forall k > k^*$, $m_k \leq m_k'$.

Let $(\mathbf{N}, G, \{\mathcal{P}_i\}_{i \in \mathbf{N}})$ be an instance of a routing game. Since there are only finitely many routings (as a player's path may use any edge at most once), there exists at least one minimum routing w.r.t. the total order $<_c$. There may be many distinct routings all of which are minimum (and equal to each other). Let $\mathbf{p}^*$ be a minimum routing (which exists); then, for all routings $\mathbf{p}$, $\mathbf{p}^* \leq_c \mathbf{p}$. Every minimum routing is optimal; indeed, if $SC(\mathbf{p}) < SC(\mathbf{p}^*)$ for some other routing $\mathbf{p}$, then the maximum k for which $m_k(\mathbf{p}) > 0$ is smaller than the corresponding k for $\mathbf{p}^*$, contradicting the fact that $\mathbf{p}^* \leq_c \mathbf{p}$.

Lemma 1. *Every minimum routing (at least one exists) is optimal.*

A greedy move is available to player i if she can obtain a lower path congestion by changing her current path from p_i to p_i' – the greedy move takes the original routing $(p_i; \mathbf{p}_{-i})$ to $(p_i'; \mathbf{p}_{-i}')$ in which p_i is replaced by p_i'.

Lemma 2. *If a greedy move by any player takes $\mathbf{p}$ to $\mathbf{p}'$, then $\mathbf{p}' <_c \mathbf{p}$.*

Thus, a greedy move decreases the number of high congestion edges, by transferring the congestion to lower congestion edges. Since there are only a finite number of routings, every best response dynamic is finite. By Lemma 2, no player can have an available greedy move at a minimum routing, as this would contradict the minimality of the routing. Hence,

Lemma 3. *Every minimum routing is an optimal Nash-routing.*

Hence, $PoS = 1$. Theorem 1 now follows from Lemmas 2 and 3.

4   Path Length Bound on Price of Anarchy

Here, we prove Theorem 2. In order to do so we will use the *edge-expansion process*, that we introduce here. Before we describe this technique we need to give some necessary definitions.

Let $\mathcal{R} = (\mathbf{N}, G, \{\mathcal{P}_i\}_{i \in \mathbf{N}})$ be an instance of a routing game. Let $\mathcal{P} = \bigcup_{i \in \mathbf{N}} \mathcal{P}_i$. The *path-length* of $\mathcal{R}$ is $\ell = \max_{p \in \mathcal{P}} |p|$. A *path-cut* for player i is a set of edges E_i such that every path in $\mathcal{P}_i$ must use at least one of the edges in E_i. The congestion of a path-cut $C(E_i)$ is the minimum congestion of any edge in E_i, $C(E_i) = \min_{e \in E_i} C_e$. If player i is locally optimal with congestion pc_i, then every alternative path for that player must have congestion at least $pc_i - 1$.

Lemma 4. *Let $\mathbf{p} = [p_1, p_2, \cdots, p_N]$ be a routing for which player i is locally optimal. Then, there is a path-cut E_i for player i with congestion $C(E_i) \geq pc_i - 1$.*

4.1   Edge-Expansion Process

If only some players are locally optimal in a routing $\mathbf{p}$, then $\mathbf{p}$ is a *partial* Nash-Routing (a Nash-Routing is a special case of a partial Nash-Routing). The edge expansion process applies to any partial Nash-Routing.

Suppose routing $\mathbf{p}$ has network congestion C, and suppose that at least one player is locally optimal with player cost C. Let $\mathcal{E}_0$ be the set of edges with congestion $C_0 = C$ that are used by at least one locally optimal player, and let Π_0 be the set of these locally optimal players that use at least one edge in $\mathcal{E}_0$. By Lemma 4, each player in Π_0 has a path-cut with congestion at least $C_0 - 1$. Let $\mathcal{E}_1$ denote the union of $\mathcal{E}_0$ with all these path-cuts of every player in Π_0. Thus, $\mathcal{E}_0 \subseteq \mathcal{E}_1$ and each edge in $\mathcal{E}_1$ has congestion at least $C_1 = C_0 - 1$. Let Π_1 denote the set of locally optimal players whose paths in $\mathbf{p}$ use at least one edge in $\mathcal{E}_1$. Note that $\Pi_0 \subseteq \Pi_1$. Each player in Π_1 has player cost at least C_1, since every edge in $\mathcal{E}_1$ has congestion at least C_1.

We repeat this process as follows. Suppose that for $i \geq 1$, edge set $\mathcal{E}_i$ has been constructed as the union of $\mathcal{E}_{i-1}$ with path cuts for the players in Π_{i-1}, thus every edge in $\mathcal{E}_i$ has congestion at least $C_i = C_{i-1} - 1 = C - i$. We now construct Π_i, the set of locally optimal players whose paths use at least one edge in $\mathcal{E}_i$; every player in Π_i has player cost at least C_i. By Lemma 4, each player in Π_i has a path-cut with congestion $C_i - 1$, and we construct $\mathcal{E}_{i+1}$ to be the union of $\mathcal{E}_i$ with all these path-cuts of the players in Π_i.

Using this inductive construction, we obtain a sequence of edge sets, $\mathcal{E}_0 \subseteq \mathcal{E}_1 \subseteq \mathcal{E}_2, \cdots$, with $C(\mathcal{E}_j) \geq C_j = C - j$, and corresponding to each edge set, a set of locally optimal players $\Pi_0 \subseteq \Pi_1 \subseteq \Pi_2 \cdots$. We continue this inductive construction up to edge set $\mathcal{E}_s$ which is the first set for which $|\mathcal{E}_s| \leq 2|\mathcal{E}_{s-1}|$. We will refer to this process as the *edge-expansion process*.

4.2   Edge-Expansion Properties

Since $|\mathcal{E}_i| \leq \frac{1}{2}n^2$ and each expansion at least doubles the size of the edge set,

Lemma 5. $|\mathcal{E}_s| \geq 2^{s-1}$ *and* $1 \leq s < 2\log n$.

In routing $\mathbf{p}$, let $F(C') \subseteq \mathbf{N}$ denote the set of non-locally optimal players with player cost at least C'. We now establish a relationship between the congestion of a partial Nash-Routing and the optimal routing.

Lemma 6. $C < 2\ell \cdot (C^* + F(C - 2\log n)) + 2\log n$.

Proof. From the edge-expansion process, each edge in $\mathcal{E}_{s-1}$ has congestion at least C_{s-1}. Let M be the number of times edges in $\mathcal{E}_{s-1}$ are used by the paths in $\mathbf{p}$. Then, $M > C_{s-1} \cdot |\mathcal{E}_{s-1}|$. By construction, in $\mathbf{p}$, the congestion in each of the edges of $\mathcal{E}_{s-1}$ is caused only by the players in $A = \Pi_{s-1} \cup B$, where $B \subseteq F(C_{s-1})$ contains the non locally optimal players that use edges in $\mathcal{E}_{s-1}$. Since path lengths are at most ℓ, each player in A can use at most ℓ edges in $\mathcal{E}_{s-1}$. Hence, $C_{s-1} \cdot |\mathcal{E}_{s-1}| < M \leq \ell \cdot |A|$. Since, $|A| \leq |\Pi_{s-1}| + |F(C_{s-1})|$, we obtain, $C_{s-1} < \frac{\ell}{|\mathcal{E}_{s-1}|} \cdot (|\Pi_{s-1}| + |F(C_{s-1})|)$. We now bound $|\Pi_{s-1}|$.

$\mathcal{E}_s$ contains a path-cut for every player in Π_{s-1}, and every such player must use at least one edge in $\mathcal{E}_s$ in any routing, including the optimal routing $\mathbf{p}^*$. Thus, edges in $\mathcal{E}_s$ are used at least $|\Pi_{s-1}|$ times, hence some edge is used at least $|\Pi_{s-1}|/|\mathcal{E}_s|$ times, by the pigeonhole principle. Hence, $C^* \geq |\Pi_{s-1}|/|\mathcal{E}_s|$ (note that $|\mathcal{E}_s| > 0$). By the definition of s, $|\mathcal{E}_s| \leq 2|\mathcal{E}_{s-1}|$. Hence, $|\Pi_{s-1}| \leq 2|\mathcal{E}_{s-1}|C^*$, and $C_{s-1} < 2\ell \cdot \left(C^* + \frac{|F(C_{s-1})|}{2|\mathcal{E}_{s-1}|} \right)$. Since $C_{s-1} = C - (s-1)$ and $2|\mathcal{E}_{s-1}| \geq 2^s$ (Lemma 5), we obtain $C < 2\ell \cdot \left(C^* + \frac{|F(C-s+1)|}{2^s} \right) + s - 1$. To conclude, $2^s \geq 2$, and note that $C'' < C'$ implies $F(C') \subseteq F(C'')$, hence $|F(C')|$ is non-increasing in C'. Thus $|F(C-s+1)| \leq |F(C - 2\log n)|$.

Since in a Nash-Routing, $F(C') = 0$, $\forall C' > 0$, by dividing the result of Lemma 6 with C^*, we obtain Theorem 2.

5   Basic Cycle Bounds on Price of Anarchy

Here, we first give the lower bound (part i) of Theorem 3 for the price of anarchy; we then prove the upper bound (part ii) of Theorem 3, for the special case of 2-connected graphs. The next result establishes the lower bound of Theorem 3.

Lemma 7. *For any graph G, there is a routing game with $PoA \geq \kappa_e(G) - 1$.*

Proof. Let $Q = e_1, \ldots, e_{\kappa_e}$ be an edge simple cycle with length κ_e. We construct a routing game with κ_e players, where player i corresponds to edge $e_i = (u_i, v_i)$

in Q, that is, the source of i is $s_i = u_i$ and the destination $t_i = v_i$. The strategy set of i is the collection of all edge simple paths from s_i to t_i.

There are two special paths in the strategy set of player i, the *forward path* which is composed solely of the edge (u_i, v_i), and the *backward path* which consists of the remaining edges of cycle Q. Since Q is edge simple, if every player uses his forward path $C = 1$. Thus, the optimal social cost is 1. If on the other hand, all the players use their backward paths (backward routing $\bar{\mathbf{p}}$), then player i uses every edge in Q except e_i exactly once. Thus, the congestion on every edge in Q is $N - 1 = \kappa_e - 1$. Hence, if $\bar{\mathbf{p}}$ is a Nash-Routing, then $PoA \geq \kappa_e - 1$.

We will show that $\bar{\mathbf{p}}$ is a Nash-Routing by contradiction. Suppose that some player k is not locally optimal – so player k has lower congestion for some other path p. Since every edge on Q has congestion $\kappa_e - 1$ in routing $\bar{\mathbf{p}}$, at least $\kappa_e - 2$ players other than player k use every edge on Q. Thus, if p uses any edge on Q, then $pc_k(p; \bar{\mathbf{p}}_{-k}) = \kappa_e - 1$, which does not improve its cost, so we conclude that p does not use any edge on Q. Therefore, p has length at least 2 (since $p \neq e_k$ and G is not a multi-graph). Thus, replacing $e_k \in Q$ by p results in a new edge simple cycle Q' that is strictly longer than Q, a contradiction. Thus, $\bar{\mathbf{p}}$ is a Nash-Routing.

We now continue with the upper bound on the price of anarchy. A graph G is k-*connected* if its minimum edge-cut has size at least k. By Menger's theorem [32], G is k-connected if and only if there are at least k edge-disjoint paths between every two nodes. Let L be the longest path length in G.

Lemma 8. *If G is 2-connected, then $\kappa_e(G) \geq \sqrt{2L} - \frac{3}{2}$.*

The proof relies on the observation that the longest path p must have at least $\sqrt{L}$ edges in common with the largest cycle q, since otherwise, we would be able to construct a larger cycle by combing pieces of p and q.

Lemma 8 bounds the longest path length in G with respect to $\kappa_e(G)$. Theorem 2 bounds the price of anarchy in terms of the longest path ℓ in the players' strategy sets. Since $\ell \leq L$, we obtain the following result, with proves the upper bound of Theorem 3 for 2-connected graphs:

Lemma 9. *For any routing game on a 2-connected graph G, $PoA \leq c(\kappa_e{}^2(G) + \log n)$, for some constant c.*

6   Cycle Upper Bound for General Graphs

We now prove the upper bound (part ii) of Theorem 3 for general graphs. We will bound the price of anarchy with respect to the square of the longest cycle. The main idea behind the result is that any Nash-Routing in G can be mapped to a partial Nash-Routing on some 2-connected subgraph of G. In this partial Nash-Routing, many players are locally optimal, and we can apply Lemma 6 in combination with Lemma 7 to obtain the result.

6.1   Canonical Subgraphs

Consider an arbitrary connected graph $G = (V, E)$. A *subgraph* $G' = (V', E')$ of G contains a subset of the nodes, $V' \subseteq V$, and a subset of the edges $E' \subseteq E$, where each edge in E' is incident with two nodes in V'. We say that G' is an *induced* subgraph by the node set V' if E' contains all the edges in E that are incident with a pair of vertices in V'. We say that two subgraphs are *adjacent* if the intersection of their node sets is non-empty. The *union* of two subgraphs $G' = (V', E')$ and $G'' = (V'', E'')$ is $\hat{G} = (V' \cup V'', E' \cup E'')$.

We will focus on 2-connected subgraphs. It is easy to verify that G contains a 2-connected subgraph if and only if it is not a tree. A 2-connected subgraph G' is *maximal* if there is no larger 2-connected subgraph $G'' = (V'', E'')$ that contains G', so if G'' is 2-connected, then $E' \not\subseteq E''$. Let $A_1, \ldots, A_\alpha$ be all the maximal 2-connected subgraphs of G, where $\alpha \geq 1$, and $A_i = (V_{A_i}, E_{A_i})$. Any two subgraphs A_i and A_j, $i \neq j$, are node-disjoint since otherwise their union would be 2-connected, which contradicts their maximality.

Therefore, we can construct from G two subgraphs A and B, where A consists of $A_1, \ldots, A_\alpha$, while B consists of the remaining edges in G: $A = (V_A, E_A)$ and $B = (V_B, E_B)$, where $E_A = \bigcup_{i=1}^{\alpha} E_{A_i}$, $E_B = E - E_A$, and V_A and V_B are the nodes adjacent to the edges in E_A and E_B, respectively. Note that graphs A and B are edge-disjoint, however, they may have common nodes. Subgraph B consists also of one or more disjoint maximal connected components (each containing at least two nodes), which we will denote $B_1, \ldots, B_\beta$. (Graph A consists of connected components $A_1, \ldots, A_\alpha$.) We refer to the A_i as the *type-a canonical subgraphs* of G and the B_i as the *type-b canonical subgraphs* of G. One can show:

Lemma 10. *Every type-b subgraph is a tree. Any pair of type-a and type-b subgraphs can have at most one common node.*

We now define a simple bipartite graph $H = (V_H, E_H)$ that represents the structure of G. In $V_H = \{a_1, \ldots, a_\alpha, b_1, \ldots, b_\beta\}$, the nodes a_i, b_j correspond to the the type-a canonical subgraph A_i and the type-b canonical subgraph B_j respectively. The edge $(a_i, b_j) \in E_H$ if and only if the canonical subgraphs A_i and B_j are adjacent (have a common node). The bipartition for H is $(\mathcal{A}, \mathcal{B})$, where $\mathcal{A} = \{a_1, \ldots, a_\alpha\}$ and $\mathcal{B} = \{b_1, \ldots, b_\beta\}$. The nodes in H inherit the same type as their corresponding canonical subgraph in G. Since G is connected, it follows immediately that H is connected too. Further, we have:

Lemma 11. *Graph H is a tree.*

6.2   Canonical Subpaths

A node in G can belong to at most one type-a subgraph and one type-b subgraph, since no two canonical subgraphs of the same type are adjacent. If a node is a member of one canonical subgraph, then its type is the type of the subgraph. If the node belongs to two canonical subgraphs then it is of type-a (we assign it to the type-a canonical subgraph). An edge belongs to exactly one canonical subgraph and inherits the type of that subgraph.

Let $p = v_1, v_2, \ldots, v_k$, $k > 1$, be an edge-simple path in G. We can write p as a concatenation of subpaths $p = q_1 q_2 \cdots q_k$, where $|q_i| > 0$, $\forall i$, with the following properties: (i) the subpaths are edge disjoint; (ii) all the nodes of a subpath q_i are in the same subgraph and have the same type, which will also be the type and subgraph of q_i; (iii) the types of the subpaths alternate, i.e. the types of q_i and q_{i+1} are different; (iv) There is no type-a subpath with one node (any type-a subpath with one node can be merged with two adjacent type-b subpaths in the same type-b subgraph). We refer to the q_i as the *canonical subpaths* of p. Note that there is a unique canonical subpath decomposition for path p.

Since type-b subgraphs are trees and graph H is a tree, an arbitrary path in G can form cycles only inside type-a canonical subgraphs (in the respective type-a canonical subpaths). As a consequence, a path from a source node to a destination node follows a unique sequence of type-b edges (the union of all the edges in the type-b subpaths). Thus, we can obtain the following crucial result on paths that connect the same endpoints in G.

Lemma 12. *Any two edge-simple paths from nodes s to t in G use the same sequence of type-b edges.*

6.3   Subgames in Canonical Subgraphs

Consider a routing game $\mathcal{R} = (\mathbf{N}, G, \{\mathcal{P}_i\}_{i \in \mathbf{N}})$ in G. Let $\mathbf{p}$ be a routing with network congestion C. Let $\mathbf{p}^*$ denote an optimal routing for $\mathcal{R}$ with congestion C^*. An immediate consequence of Lemma 12 is that every path in $\mathbf{p}$ uses the same type-b edges as its corresponding path in $\mathbf{p}^*$, hence

Lemma 13. *Any type-b edge e has the same congestion in $\mathbf{p}$ and $\mathbf{p}^*$, i.e.* $C_e(\mathbf{p}) = C_e(\mathbf{p}^*) \leq C^*.$

By Lemma 13, all the edges in $\mathbf{p}$ with congestion higher that C^* must occur in type-a subpaths.

Lemma 14. *For path p, if $C_p(\mathbf{p}) > C^*$, then p must have a type-a subpath q with $C_q(\mathbf{p}) = C_p(\mathbf{p})$.*

Suppose now that $\mathbf{p}$ is an arbitrary Nash-Routing which has network congestion C. For a type-a subgraph Λ, let $\mathbf{p}_\Lambda = \{p_1, \ldots, p_\gamma\}$ denote the paths in $\mathbf{p}$ that use edges in Λ, and denote the respective users as $\mathbf{N}_\Lambda$, where $|\mathbf{N}_\Lambda| = \gamma$. Let $Q_\Lambda = \{q_1, \ldots, q_\gamma\}$ denote the type-a canonical subpaths of the paths in $\mathbf{p}_\Lambda$ that are in Λ (q_i is a subpath of p_i).

In subgraph Λ, we define a new routing game $\mathcal{R}_\Lambda = (\mathbf{N}_\Lambda, \Lambda, \{\mathcal{P}_i^\Lambda\}_{i \in N_\Lambda})$, where $\mathcal{P}_i^\Lambda$ contains all the type-a subpaths of $\mathcal{P}_i$ that are in Λ and have the same source and destination as q_i. We refer to $\mathcal{R}_\Lambda$ as the *subgame* of $\mathcal{R}$ for subgraph Λ. Q_Λ is a possible routing for $\mathcal{R}_\Lambda$. If q_i is locally optimal for player i in Λ, we say that its corresponding path p_i in G is *satisfied* in subgame $\mathcal{R}_\Lambda$. In other words, if path p_i is satisfied in $\mathcal{R}_\Lambda$, player i does not wish to change the choice q_i in Λ. Every player with high player cost (higher than C^*) must be satisfied in a type-a subgraph, since otherwise it would violate Lemma 14. Thus:

Lemma 15. *If player i has path p_i and $pc_i > C^*$, then player i is satisfied in some subgame $\mathcal{R}_\Lambda$ in a type-a subgraph Λ, and player i has congestion pc_i in Λ.*

6.4   Main Result

Consider routing game $\mathcal{R} = (\mathbf{N}, G, \{\mathcal{P}_i\}_{i \in \mathbf{N}})$ in G and a Nash-Routing $\mathbf{p}$ with congestion $C(\mathbf{p}) = C$. Lemma 15, implies that each user is satisfied in some type-a subgraph (not necessarily the same). In any type-a subgraph, the resulting routing in the subgame may be a partial Nash-Routing, since some users may not be satisfied in it. We first show that there is a subgraph with high congestion where the number of unsatisfied players is bounded. For a canonical type-a subgraph Λ, let $F_\Lambda(C')$ denote the set of non-locally optimal players in the subgame $\mathcal{R}_\Lambda$ whose congestion in $\mathcal{R}$ is at least C'. We will use C_Λ to denote the congestion in the canonical subgraph Λ. We have:

Lemma 16. *Suppose that $C(\mathbf{p}) > C^* + x(1 + \log n)$ for some $x > 0$. Then, there is a type-a canonical subgraph Λ with congestion $C_\Lambda \geq C - x \log n$ and $|F_\Lambda(C_\Lambda - x)| \leq 2C^*$.*

By combining Lemma 6 and Lemma 16 we obtain the following result which establishes the upper bound of Theorem 3.

Lemma 17. $PoA \leq c \cdot (\kappa_e^2(G) + \log^2 n)$, *for some constant c.*

Proof. Let $x = 2 \log n$. If $C \leq C^* + x(1 + \log n)$, then there is nothing to prove because $C/C^* \leq 1 + 2 \log n(1 + \log n)/C^* \leq c \log^2 n$, for some generic constant c. So, suppose that $C > C^* + x(1 + \log n)$. By Lemma 16, there exists a type-a subgraph Λ such that $C_\Lambda \geq C - 2 \log^2 n$ and $|F_\Lambda(C_\Lambda - 2 \log n)| \leq 2C^*$. By applying Theorem 6 to the subgame $\mathcal{R}_\Lambda$ we obtain,

$$C_\Lambda < 2\ell \cdot (C_\Lambda^* + F_\Lambda(C_\Lambda - 2 \log n')) + 2 \log n',$$

where ℓ is the length of the longest edge-simple path in the player strategy sets in $\mathcal{R}_\Lambda$, n' is the number of nodes in Λ and C_Λ^* is the optimal congestion for the subgame $\mathcal{R}_\Lambda$. Note that $n' \leq n$, and the subgame $\mathcal{R}_\Lambda$ cannot have a higher optimal congestion than the full game $\mathcal{R}$, hence $C^* \geq C_\Lambda^*$. Since $|F_\Lambda|$ is monotonically non-increasing ($F_\Lambda(C') \subseteq F_\Lambda(C'')$ for $C'' < C'$), we have that:

$$C - 2 \log^2 n < 2\ell \cdot (C^* + F(C_\Lambda - 2 \log n)) + 2 \log n \leq 2\ell \cdot (C^* + 2C^*) + 2 \log n.$$

From Lemma 8, $\ell \leq c\kappa_e^2(\Lambda) \leq c\kappa_e^2(G)$, and so $C \leq c \cdot (\kappa_e^2(G)C^* + \log^2 n)$. After dividing by C^*, we obtain the desired result.

7   Discussion

We believe that the price of anarchy upper bound can be improved. Specifically, we leave open the following conjecture: *for any routing game, $PoA \leq \kappa_e - 1$.*

An interesting future direction is to obtain similar results when the latency functions at each link are more general and not necessarily identical. We conclude by noting that all our results have been stated for paths that are edge-simple. Specifically the strategy set for the players is a set of *edge-simple* paths and the social and player costs are the maximum *edge-congestion* in the network and player path respectively. Exactly analogous results can be obtained for strategy sets containing *node-simple* paths with the social and player costs being the maximum *node-congestion* in the network and player path respectively. In this case, the bounds on the price of anarchy are in terms of the node-cycle number (the length of the longest node-simple cycle).

References

1. E. Anshelevich, A. Dasgupta, J. Kleinberg, , E. Tardos, T. Wexler, and T. Roughgarden. The price of stability for network design with fair cost allocation. In *Proc. FOCS*, 2004.
2. E. Anshelevich, A. Dasgupta, E. Tardos, and T. Wexler. Near optimal network design with selfish agents. In *Proc. STOC*, 2003.
3. P. Berenbrink and C. Scheideler. Locally efficient on-line strategies for routing packets along fixed paths. In *10th ACM-SIAM Symposium on Discrete ALgorithms (SODA)*, pages 112–121, 1999.
4. G. Christodoulou and E. Koutsoupias. The price of anarchy of finite congestion games. In *Proc. STOC*, 2005.
5. J. R. Correa, A. S. Schulz, and N. E. Stier Moses. Computational complexity, fairness, and the price of anarchy of the maximum latency problem. In *Proc. 10th Conf. on Integer Programming and Combinatorial Optimization (IPCO)*, 2004.
6. R. Cypher, F. Meyer auf der Heide, C. Scheideler, and B. Vöcking. Universal algorithms for store-and-forward and wormhole routing. In *In Proc. of the 28th ACM Symp. on Theory of Computing*, pages 356–365, 1996.
7. A. Czumaj, P. Krysta, and B. Vöcking. Selfish traffic allocation for server farms. In *Proc. STOC*, 2002.
8. A. Czumaj and B. Vöcking. Tight bounds for worst-case equilibria. In *Proc. SODA*, 2002.
9. D. Fotakis, S. Kontogiannis, E. Koutsoupias, M. Mavronicolas, and P. Spirakis. The structure and complexity of Nash equilibria for a selfish routing game. In *Proc. ICALP*, 2002.
10. D. Fotakis, S. Kontogiannis, and P. Spirakis. Selfish unsplittable flows. In *Proc. ICALP*, 2004.
11. M. Garing, T. Lücking, M. Mavronicolas, and B. Monien. Computing nash equilibria for scheduling on restricted parallel links. In *Proc. STOC*, 2004.
12. M. Garing, T. Lücking, M. Mavronicolas, and B. Monien. The price of anarchy for polynomial social cost. In *Proc. MFCS*, 2004.
13. M. Garing, T. Lücking, M. Mavronicolas, B. Monien, and M. Rode. Nash equilibria in discrete routing games with convex latency functions. In *Proc. ICALP*, 2004.
14. E. Koutsoupias, M. Mavronicolas, and P. Spirakis. Approximate equilibria and ball fusion. In *Proc. SIROCCO*, 2002.
15. E. Koutsoupias and C. Papadimitriou. Worst-case equilibria. In *Proc. STACS*, 1999.

16. F. T. Leighton. *Introduction to Parallel Algorithms and Architectures: Arrays - Trees - Hypercubes.* Morgan Kaufmann, San Mateo, 1992.

17. F. T. Leighton, B. M. Maggs, and S. B. Rao. Packet routing and job-scheduling in $O(congestion + dilation)$ steps. *Combinatorica*, 14:167–186, 1994.

18. T. Leighton, B. Maggs, and A. W. Richa. Fast algorithms for finding O(congestion + dilation) packet routing schedules. *Combinatorica*, 19:375–401, 1999.

19. L. Libman and A. Orda. Atomic resource sharing in noncooperative networks. *Telecomunication Systems*, 17(4):385–409, 2001.

20. T. Lücking, M. Mavronicolas, B. Monien, and M. Rode. A new model for selfish routing. In *Proc. STACS*, 2004.

21. M. Mavronicolas and P. Spirakis. The price of selfish routing. In *Proc. STOC*, 2001.

22. D. Monderer and L. S. Shapely. Potential games. *Games and Economic Behavior*, 1996.

23. R. Ostrovsky and Y. Rabani. Universal $O(congestion+dilation+\log^{1+\varepsilon} N)$ local control packet switching algorithms. In *Proceedings of the 29th Annual ACM Symposium on the Theory of Computing*, pages 644–653, New York, May 1997.

24. C. H. Papadimitriou. Algorithms, games and the internet. In *Proc. STOC*, pages 749–753, 2001.

25. Y. Rabani and É. Tardos. Distributed packet switching in arbitrary networks. In *Proceedings of the Twenty-Eighth Annual ACM Symposium on the Theory of Computing*, pages 366–375, Philadelphia, Pennsylvania, 22–24 May 1996.

26. R. W. Rosenthal. A class of games possesing pure-strategy Nash equilibria. *International Journal of Game Theory*, 1973.

27. T. Roughgarden. The maximum latency of selfish routing. In *Proc. SODA*, 2004.

28. T. Roughgarden. Selfish routing with atomic players. In *Proc. SODA*, 2005.

29. T. Roughgarden and Éva Tardos. How bad is selfish routing. *Journal of the ACM*, 49(2):236–259, March 2002.

30. T. Roughgarden and Éva Tardos. Bounding the inefficiency of equilibria in nonatomic congestion games. *Games and Economic Behavior*, 47(2):389–403, 2004.

31. S. Suri, C. D. Tóth, and Y. Zhou. Selfish load balancing and atomic congestion games. In *Proc. SPAA*, 2004.

32. D. B. West. *Introduction to Graph Theory.* Prentice Hall, Upper Saddle River, NJ, U.S.A., 2001.

Equilibrium Distribution of Advertising Prices

Qianqin Chen[1,*], Wei-Guo Zhang[2], and Guoliang Kuang[1]

[1] School of Commerce and Economy, South China University of Technology,
Guangzhou, 510640, P.R. China
marty@scut.edu.cn
[2] School of Business Administration, South China University of Technology,
Guangzhou, 510641, P.R. China

Abstract. This paper formulates a model of advertising prices in which a homogeneous product is not intended for sales at conventional stores. The product is sold by means of advertising instead. Applications of this model can be found on numerous sales activities include, for example, insurance companies, television shopping channels and Internet e-tailers who advertise their products and prices by sending e-mails to potential buyers or by means of popup windows. This paper makes endogenous both firm advertising and price strategies in the model.

1 Introduction

Stigler's (1961) seminar article motivates a growing body of theoretical literature investigating firm price strategy and consumer behavior with incomplete information. See, for example, Varian (1980), Salop and Stiglitz (1982), Rob (1985), Janssen and Moraga (2000), Hopkins and Seymour (2000), and Morgan and Sefton (2001).

Most of these models are based on the crucial assumption that buyers are freely endowed with perfect information regarding the number and locations of sellers. As a result, buyers may gather price information by visiting one seller after another. However, this assumption is unnecessarily true in reality.

There may be occasions that consumer search is impossible or costly. Butters (1977) considers the market for a homogeneous good, in which sellers randomly allocate advertisements indicating their price and location among buyers, and buyers have no other means of gathering information about sellers. Butters' (1977) assumption rises to the occasion that firms rely their sales heavily on advertising, and that "buyers' behavior is passive; they simply order at the lowest price of which they are aware".

Butters' (1977) model is helpful in analyzing firm strategies when sellers are also responsible for the provision of price and firm information. In particular, insurance companies sell their products by visiting customers, and the possibility that buyers visit insurance companies is negligible. Butters (1977) is also applicable to other sales activities, for example, Internet e-tailers who advertise

---

* Correspondence author.

S.-W. Cheng and C.K. Poon (Eds.): AAIM 2006, LNCS 4041, pp. 92–101, 2006.

their products and prices with popup windows or by sending e-mails to potential customers, and sales representatives who sell their products through personal relation with their clients.

Following Butters (1977), Kessner and Polborn (2000) propose a model of sales for heterogeneous products. Buyers differ in their preference for quality. Kessner and Polborn (2000) derive an equilibrium price distribution density that is related to the density of consumers. Kessner and Polborn (2000) also apply their model on a test of price dispersion in the German life insurance market. They suggest that it is important to distinguish between price dispersion and market failure, and that "if price dispersion is due to heterogeneous qualities, different prices are no indication of market failure".

This paper is concerned with the price distribution of a homogeneous good. The following assumptions distinguish this paper from Butters (1977). Butters (1977) assumes that a firm may advertise different prices to different buyers. This paper assumes that each firm charges all buyers the same price. This assumption seems more realistic since charging all customers the same price for an identical good is prevalent in practice. Butters (1977) assumes that each customer buys exactly one unit of the good. This paper relaxes Butters' (1977) restriction by assuming that individual demand is elastic.

This paper yields different conclusions in comparison with Butters (1977). In Butters (1977), neither the number of firms nor the number of consumers appears in the equilibrium price distribution function. This paper derives an equilibrium price distribution that is related to the number of firms.

The remainder of this paper is organized as follows. Section 2 formulates the model. Section 3 derives the equilibrium distribution of advertising prices. Section 4 illustrates the model with a numerical example. Section 5 contains concluding remarks.

2   The Basic Model

In the market for a homogeneous good, there exist $n \geq 2$ identical sellers and $l \geq 2$ identical buyers. Let $N = \{1, 2, \cdots, n\}$ denote the set of sellers, and $L = \{1, 2, \cdots, l\}$ denote the set of buyers. Both sellers and buyers are assumed to be risk neutral. Transactions are conducted on the take-it-or-leave-it basis.

All firms have identical marginal production costs $c \geq 0$. Let $p \geq 0$ be the price paid by the consumer. The individual demand function is given by

$$x = x(p) \geq 0 \tag{1}$$

There exists a unique $p_m \in (0, \infty)$ such that

$$p_m = \arg \max_{p \geq 0} \{(p - c) x(p)\} \tag{2}$$

Firms randomly allocate advertisements indicating their prices and locations among buyers. There is no way for firms to direct their advertisements to any specific buyers. The cost of sending an advertisement is $b \in [0, \infty)$. Each firm charges the same price to all customers. The probability that a buyer receives the advertisement from a firm is $q \in (0,1)$, and q is identical for all firms. One might think of a model in which each firm randomly drops h advertisements into buyers' mailboxes. If $h < l$, then $q = h/l < 1$.

Receiving advertisements entails no cost on buyers. A buyer receives an advertisement from a firm is independent to the buyer receives advertisements from other firms. Additionally, a buyer may purchase the good only if he receives at least one advertisement. If two or more firms advertise the same lowest price, a consumer buys the good by choosing a firm at random.

The above assumptions are basically following Butters (1977). One can imagine a multi-period model, in which at the beginning of each period firms randomly allocate advertisements among buyers. At the end of each period, those buyers who receive at least one advertisement purchase the good from the seller who advertised the lowest price.

Definition: Let $F_q(p)$ denote the distribution of advertising prices, where $q = h/l$ is the probability that a representative buyer receives the advertisement from a representative firm, $h \in [0, l]$ is the number of advertisements sent by a representative firm, and $E[\pi(p)]$ is the expected profit of a representative firm who charges the price $p \in [\underline{p}, \overline{p}]$, where $[\underline{p}, \overline{p}] \subset [0, \infty)$ is the support of $F_q(p)$. If $F_q(p)$, $h^* \in [0, l]$ and $E[\pi(p)]$ satisfy the following conditions:

1. $E[\pi(p)] = \overline{\pi}$ is a constant for every $(p, h) \in [\underline{p}, \overline{p}] \times \{h^*\}$; and
2. $E[\pi(p)] \leq \overline{\pi}$ for every $(p, h) \in [0, \infty) \times [0, l]$.

Then $\{F_q(\cdot), h^*, \overline{\pi}\}$ is a symmetric mixed-strategy equilibrium of the above pricing game, $F_q(p)$ and h^* are equilibrium price and advertising strategies of firms, respectively.

3   Analysis

For ease of exposition, assume for the present that the advertising cost $b = 0$ and that $q = h/l$ is given. These assumptions will be relaxed later. For any $j \in N$, if firm j sells the good for price p to a customer, then the transaction brings firm j the following profit

$$\pi(p) = (p - c)x(p) \tag{3}$$

Let ξ be the number of firms whose advertisements successfully reach a representative buyer. Since there are a total of n firms, ξ has a Binomial distribution with parameters n and q. For every $m = 0, 1, \cdots, n$, let α_i denote the probability that a representative buyer receives advertisements from m firms. Then

$$\alpha_m = P(\xi = m) = C_n^m b^m (1 - b)^{n-m} \tag{4}$$

Suppose the distribution of market price is $F_q(\cdot)$ on the support $[\underline{p}, \overline{p}] \subset [0, \infty)$. For any $j \in N$ and $k \in L$, the probability that buyer k purchases the product from firm j is

$$\delta_p = \sum_{m=1}^{n} \alpha_m \frac{C_{n-1}^{m-1}}{C_n^m} \left[1 - F_q\left(p\right)\right]^{m-1} = \frac{1}{n} \sum_{m=1}^{n} m\alpha_m \left[1 - F_q\left(p\right)\right]^{m-1} \quad (5)$$

where p is the price advertised by firm j, $\alpha_m C_{n-1}^{m-1}/C_n^m$ is the probability that buyer k receives the advertisement from firm j conditional on buyer k receives advertisements from m firms, $\left[1 - F_q\left(p\right)\right]^{m-1}$ is the probability that p is the lowest of the m prices. Substituting (4) in (5) yields

$$\delta_p = \frac{1}{n} \sum_{m=1}^{n} m C_n^m q^m \left(1 - q\right)^{n-m} \left[1 - F\left(p\right)\right]^{m-1} = q\left[1 - qF\left(p\right)\right]^{n-1} \quad (6)$$

Let η be the number of customers who buy the product from firm j. Since there are a total of l buyers in the market, η has a Binomial distribution with parameters l and δ_p. If firm j advertises price p and all other $n-1$ firms independently advertise prices according to $F\left(\cdot\right)$, then the expected profit for firm j is

$$E\left[\pi\left(p\right)\right] = \sum_{i=0}^{l} iP\left(\eta = i\right)\pi\left(p\right) = \sum_{i=0}^{l} iC_l^i \delta_p^i \left(1 - \delta_p\right)^{l-i} \pi\left(p\right)$$
$$= l\delta_p \pi\left(p\right) = lq\left[1 - qF_q\left(p\right)\right]^{n-1} \pi\left(p\right) \quad (7)$$

where $\pi\left(p\right)$ is given by (3). By definition, if $\{F_q\left(\cdot\right), h, \overline{\pi}\}$ is a non-degenerate mixed-strategy equilibrium, then evaluating (7) at $p = \overline{p}$ yields

$$E\left[\pi\left(p\right)\right] = \overline{\pi} = E\left[\pi\left(\overline{p}\right)\right] = lq\left(1 - q\right)^{n-1} \pi_q\left(\overline{p}\right) \quad (8)$$

for any $p \in \left[\underline{p}, \overline{p}\right]$, where $\left[\underline{p}, \overline{p}\right] \subset \left[0, \infty\right)$ is the support of $F_q\left(\cdot\right)$. It follows from (2) that $p_m > 0$ is the highest price an optimizing firm would ever charge. Hence $\overline{p} = p_m$. Substituting $\overline{p} = p_m$ in (8) yields the expected profit

$$\overline{\pi} = E\left[\pi\left(p\right)\right] = lq\left[1 - qF_q\left(p_m\right)\right]^{n-1} \pi\left(p_m\right) = lq\left(1 - q\right)^{n-1} \pi_q\left(p_m\right) \quad (9)$$

Inserting (9) in (8) and solving for $F_q\left(p\right)$ yields the equilibrium price distribution

$$F_q\left(p\right) = \frac{1}{q} \left\{1 - \left(1 - q\right) \left[\frac{\pi\left(p_m\right)}{\pi\left(p\right)}\right]^{\frac{1}{n-1}}\right\} \quad (10)$$

for every $p \in \left[\underline{p}, p_m\right]$. Substituting $F_q\left(\underline{p}\right) = 0$ in (10) and solving for $\underline{p}$ yields

$$\underline{p} = \pi^{-1} \left[\left(1 - q\right)^{n-1} \pi\left(p_m\right)\right] \quad (11)$$

This establishes the following proposition.

Proposition 1. Let $\{F_q\left(\cdot\right), h, \overline{\pi}\}$ be a symmetric mixed-strategy equilibrium of the pricing game. If the equilibrium advertising strategy h is given and the advertising cost is $b = 0$, then $F_q\left(\cdot\right)$ is given by (10) and (11), and $\overline{\pi}$ is given by (9).

If the advertising cost is $b > 0$ and advertising strategy h is given, then equation (8) can by rewritten as follows

$$\tilde{\pi} \equiv \overline{\pi} - hb \equiv E\left[\pi\left(p\right)\right] - hb = lq\left[1 - qF_q\left(p\right)\right]^{n-1} \pi\left(p\right) - hb \quad (12)$$

for every $p \in [\underline{p}, p_m]$. Solving for $F_q(\cdot)$ in (12) yields (10). This proves the following proposition.

Proposition 2. Let $\{F_q(\cdot), h, \tilde{\pi}\}$ be a symmetric mixed-strategy equilibrium of the pricing game. If the equilibrium advertising strategy h is given and the advertising cost is $b > 0$, then $F_q(\cdot)$ is given by (10) and (11), and $\tilde{\pi}(q)$ is given by

$$\tilde{\pi}(q) = lq(1-q)^{n-1}\pi(p_m) - hb = lq\left[(1-q)^{n-1}\pi(p_m) - b\right] \qquad (13)$$

Note that, in Proposition 2, the expected profit $\tilde{\pi}(q)$ is a function in q. To determine the equilibrium advertising strategy h^*, one only has to solve

$$q^* \equiv \arg\max_{q \in [0,a]} \tilde{\pi}(q) = \arg\max_{q \in [0,a]} lq\left[(1-q)^{n-1}\pi(p_m) - b\right] \qquad (14)$$

Differentiating (13) with respect to q and setting the derivative equal zero yields the necessary condition

$$(1-nq^*)(1-q^*)^{n-2}\pi(p_m) - b = 0 \qquad (15)$$

Since advertising cost $b \geq 0$, (15) implies $q^* \leq 1/n$. Taking derivative of (15) yields the sufficient condition

$$-(n-1)(1-q^*)^{n-3}(2-nq^*)\pi(p_m) \leq 0 \qquad (16)$$

which gives $q^* \leq 2/n$. It follows that, if (15) holds, then (16) holds. Therefore, (15) is also sufficient for (14). If (15) and (16) hold, then the optimal advertising strategy is $h^* = q^*l$. However, to prove h^* is optimal is more difficult than to derive it.

Proposition 3. Let $h^* \in (0, l)$ and $q^* = h^*/l$. If (15) and (16) hold, then $\{F_{q^*}(\cdot), h^*, \pi^*\}$ is a symmetric mixed-strategy of the pricing game, the equilibrium price distribution is given by

$$F_{q^*}(p) = \frac{1}{q^*}\left\{1 - (1-q^*)\left[\frac{\pi(p_m)}{\pi(p)}\right]^{\frac{1}{n-1}}\right\} \qquad (17)$$

for every $p \in [\underline{p}^*, p_m]$, where

$$\underline{p}^* = \pi^{-1}\left[(1-q^*)^{n-1}\pi(p_m)\right] \qquad (18)$$

and the expected profit is given by

$$\pi^* = lq^*(1-q^*)^{n-1}\pi(p_m) - h^*b = lq^*\left[(1-q^*)^{n-1}\pi(p_m) - b\right] \qquad (19)$$

where $\pi(p)$ is given by (3).

Proof of Proposition 3. In the pricing game described in Section 2, a Nash equilibrium is an n-tuple of strategies $(s_1, s_2, \cdots, s_n)$ such that, for every $j \in L$, if all other firms $L \setminus \{j\}$ play $(s_1, s_2, \cdots, s_{j-1}, s_{j+1}, \cdots, s_n)$, then firm j has no incentives to deviate his strategy s_j, where $s_j = (F_{q^*}(\cdot), h^*)$ for all $j \in L$.

To prove h^* is an optimal advertising strategy in a Nash equilibrium, one has to start with asymmetric strategies.

Suppose that firm $j \in L$ decides to send $g \leq l$ advertisements among buyers, and that all other $n - 1$ firms $L \setminus \{j\}$ each sends $h \leq l$ advertisements among buyers. The probability that a buyer receives the advertisement from firm j is $a = g/l$. And the probability that a buyer receives the advertisement from firm $s \in L \setminus \{j\}$ is $q = h/l$. Let ξ_{n-1} be the number of advertisements a representative buyer receives from firms belong to $L \setminus \{j\}$. ξ_{n-1} has a Binomial distribution with parameters $n - 1$ and q. For every $m = 0, 1, \cdots, n - 1$, let τ_m denote the probability that a representative buyer receives advertisements from m firms belong to $L \setminus \{j\}$. Then

$$\tau_m = P(\xi_{n-1} = m) = C_{n-1}^m q^m (1 - q)^{(n-1)-m} \tag{20}$$

for every $m = 0, 1, \cdots, n-1$. Let ξ_j be the number of advertisements a representative buyer receives from firm j. Then $\xi_j = 0$ with probability $1 - a$ and $\xi_j = 1$ with probability a. Let ξ be the total number of advertisements a representative buyer receives. Then

$$\tilde{\alpha}_m = P(\xi = m) = \frac{nb(1-a) - m(a-q)}{n} C_{n-1}^m q^{m-1} (1 - q)^{(n-1)-m} \tag{21}$$

for every $m = 0, 1, \cdots, n - 1$. And

$$\tilde{\alpha}_n = P(\xi = n) = aq^{n-1} \tag{22}$$

Suppose that firm j advertises price p and that all other $n - 1$ firms independently advertise prices according to $F(\cdot)$. For every $k \in L$, the probability that buyer k purchases the product from firm j is given by (5). Substituting $\tilde{\alpha}_m$ for α_m in (5) yields

$$\tilde{\delta}_p = \left(\frac{1}{1-q}\right) [1 - qF_q(p)]^{n-2} \left(\frac{1}{n}\right) \\ \{nq(1 - a)[1 - qF_q(p)] + (a - q)[1 + nqF_q(p)]\} \tag{23}$$

Note that, if $a = q$, then (23) simply reduces to (6). Let $\tilde{\eta}$ be the number of customers who buy the product from firm j. Since there are a total of l identical buyers in the market, $\tilde{\eta}$ has a Binomial distribution with parameters l and $\tilde{\delta}_p$. Therefore, the expected profit of firm j is given by

$$\tilde{\pi}(a, p) \equiv E[\pi(p)] - hb \equiv \sum_{i=0}^{l} i P(\eta = i) \pi(p) - hb$$

$$= \sum_{i=0}^{l} i C_l^i \tilde{\delta}_p^i \left(1 - \tilde{\delta}_p\right)^{l-i} \pi(p) - hb = l\tilde{\delta}_p \pi(p) - hb \tag{24}$$

$$= l\left[[n(1-q)]^{-1} [1 - qF_q(p)]^{n-2} \{nq(1 - a)[1 - bF_q(p)] \\ + (a - q)[1 + nqF_q(p)]\} \pi(p) - ab \right]$$

where the last equality follows from $a = h/l$. Let $[\underline{p}, p_m] \subset [0, \infty)$ be the support of $F_q(\cdot)$. If $\{F_q(\cdot), h, \tilde{\pi}\}$ is an equilibrium, then $\tilde{\pi}(a, p) = \tilde{\pi}(a)$ for all $p \in [\underline{p}, p_m]$. That is

$$
\tilde{\pi}(a) = lq \left[(1-q)^{n-1} \pi(p_m) - b \right]
$$
$$
+ (a-q) l \left(\tfrac{1}{n} \right) \left[(1+nq^2)(1-q)^{n-3} \pi(p_m) - nb \right] \tag{25}
$$

Note that the first term in (25) is the expected profit of a representative firm, which is given by (13). The difference $a - q$ of advertising strategies only affects $\tilde{\pi}(a)$ through the second term in (25). Denote

$$
A(q) = l \left(\frac{1}{n} \right) \left[(1+nq^2)(1-q)^{n-3} \pi(p_m) - nb \right] \tag{26}
$$

If $A(q) > 0$, then firm j has incentives to increase a; and the reverse is true if $A(q) < 0$. Firm j knows that all other firms can see this just as well as he does. If firm j thinks a^* is optimal, he might expect $q^* = a^*$ as well. This justifies a symmetric solution to the game. The second term can be dropped from (25). It follows from (15) and (16) that q^* solves (14) is optimal for all firms. Hence, $h^* = q^* l$ is an optimal advertising strategy for all firms. Equations (17) through (19) follow by Proposition 2. This completes the proof of Proposition 3.

Varying the number $n \geq 2$ of firms in (19), denote

$$
V(n) \equiv \max_{q \in [0,a]} lq \left[(1-q)^{n-1} \pi(p_m) - b \right] = lq^* \left[(1-q^*)^{n-1} \pi(p_m) - b \right] \tag{27}
$$

where q^* is given by (14). Note that, by varying the number $n \geq 2$ of firms in (14), q^* is a function of n. It follows by the Envelope Theorem that

$$
V'(n) = lq^* (1-q^*)^{n-1} \pi(p_m) \ln(1-q^*) \leq 0 \tag{28}
$$

where the inequality follows by the fact that $q^* \in [0,1]$. Since the optimal expected profit is a decreasing function in n, the expected profit approaches zero as more and more firms enter the market. The solution to the following equation set with respect to (n, q^*) gives the number of firms in the long run:

$$
lq^* \left[(1-q^*)^{n-1} \pi(p_m) - b \right] = 0
$$
$$
(1 - nq^*)(1 - q^*)^{n-2} \pi(p_m) - b = 0 \tag{29}
$$

where the first equation refers to zero expected profit, and the second refers to optimal condition. Unfortunately, there does not exist an explicit solution to (29). A numerical solution to (29) is derived in the next section.

4   Numerical Example

Consider the model in which the individual demand function is $x(p) = 5 - p$, and the marginal cost is $c = 1$. It follows that

$$
\pi(p) = (p - c) x(p) = (p - 1)(5 - p) = -p^2 + 6p - 5
$$
$$
p_m = \arg\max_{p \geq 0} \pi(p) = \arg\max_{p \geq 0} \{-p^2 + 6p - 5\} = 3 \tag{30}
$$
$$
\pi(p_m) = -p_m^2 + 6p_m - 5 = 4
$$

The inverse individual profit function is given by

$$p = \pi^{-1}\left(\hat{\pi}\right) = \frac{1}{2}\left[6 - \sqrt{36 - 4\left(5 + \hat{\pi}\right)}\right] \tag{31}$$

for any $\hat{\pi} \in [0, 4]$. Suppose that there are $n = 4$ firms and $l = 100$ consumers. It costs a firm $b = 0.1$ to send an advertisement. If each firm randomly delivers $h = 50$ advertisements among consumers, then the probability that a consumer receives the advertisement from a firm is $q = h/l = 50/100 = 1/2$, and the equilibrium distribution of advertising prices is given by

$$F_q\left(p\right) = \frac{1}{q}\left\{1 - \left(1 - q\right)\left[\frac{\pi\left(p_m\right)}{\pi\left(p\right)}\right]^{\frac{1}{n-1}}\right\} = 2\left\{1 - \frac{1}{2}\left(\frac{4}{-p^2 + 6p - 5}\right)^{\frac{1}{3}}\right\} \tag{32}$$

for every $p \in \left[\underline{p}, p_m\right] = \left[\underline{p}, 3\right]$, where the lowest price a representative firm will charge is

$$\underline{p} = \pi^{-1}\left[\left(1 - q\right)^{n-1}\pi\left(p_m\right)\right] = \tfrac{1}{2}\left(6 - \sqrt{14}\right) \tag{33}$$

The expected profit of a representative firm is

$$\tilde{\pi} = lq\left[\left(1 - q\right)^{n-1}\pi\left(p_m\right) - b\right] = 100\left(\frac{1}{2}\right)\left[\left(1 - \frac{1}{2}\right)^{4-1}4 - 0.1\right] = 20 \tag{34}$$

However, $\{F_q\left(\cdot\right), h, \tilde{\pi}\}$ is unnecessarily a Nash equilibrium of the pricing game. To find a Nash equilibrium, one can solve problem (14) of the model

$$q^* \equiv \arg\max_{q \in [0,a]} lq\left[\left(1 - q\right)^{n-1}\pi\left(p_m\right) - b\right] = 0.2392 \tag{35}$$

If $q^* = 0.2392$, the equilibrium distribution of advertising prices is

$$F_{q^*}\left(p\right) = \frac{1}{0.2392}\left\{1 - \left(1 - 0.2392\right)\left(\frac{4}{-p^2 + 6p - 5}\right)^{\frac{1}{3}}\right\} \tag{36}$$

for every $p \in \left[\underline{p}, p_m\right] = \left[\underline{p}, 3\right]$, where

$$\underline{p}^* = \pi^{-1}\left[\left(1 - q^*\right)^{n-1}\pi\left(p_m\right)\right] = 1.5038 \tag{37}$$

The expected profit of a representative firm is

$$\pi^* = lq^*\left[\left(1 - q^*\right)^{n-1}\pi\left(p_m\right) - b\right] = 39.7420 \tag{38}$$

One can solve the following equation set to determine the number of firms in the long run

$$\begin{aligned} lq^*\left[\left(1 - q^*\right)^{n-1}\pi\left(p_m\right) - b\right] &= 100q^*\left[\left(1 - q^*\right)^{n-1}4 - 0.1\right] = 0 \\ \left(1 - nq^*\right)\left(1 - q^*\right)^{n-2}\pi\left(p_m\right) - b &= \left(1 - nq^*\right)\left(1 - q^*\right)^{n-2}4 - 0.1 = 0 \end{aligned} \tag{39}$$

A feasible way to solve (39) numerically is employ the following iterative method

$$
\begin{aligned}
q_{k+1}^* &= \frac{1}{n_k} - \frac{1}{n_k\left(1-q_k^*\right)^{n_k-2}\pi(p_m)} \\
n_{k+1} &= \frac{\ln b - \ln \pi(p_m)}{\ln\left(1-q_{k+1}^*\right)} + 1
\end{aligned}
\tag{40}
$$

for $k = 1, 2, \cdots$, where $(q_0^*, n_0) = (0.1, 4)$ is given. Three steps of iteration give $n = 47190$, $q^* = 1.82 \times 10^{-5}$ and $\pi^* = 8.54 \times 10^{-9}$.

5   Conclusion

This paper derives the equilibrium price distribution in the market for a homogeneous good, where buyers are passive in information gathering. The game described in this paper can be referred to as a temporal price dispersion model. In Varian's (1980) model of sales, temporal price dispersion arises when firms randomize their prices so as to price discriminate against uninformed customers.

As shown in Section 2, the model in this paper can be viewed as the same pricing game being repeated in discrete time periods. At the beginning of each period, each firm selects a price according to the equilibrium distribution and delivers advertisements among buyers. At the end of each period, those consumers who receive at least one advertisement buy the good at the lowest price they are aware of. Price dispersion characterized in this paper is persistent.

Standard economics textbooks tell us that, in a market with complete information, the market will reduce to complete competition model, as the numbers of buyers and sellers approach infinity. In a market with incomplete information, this rule unnecessarily holds.

The equilibrium price distributions derived by Butters (1977) and Kessner and Polborn (2000) are irrelevant to either the number of buyers or the number of sellers. In this paper, the equilibrium price distribution is related to the number of sellers, but the number of buyers does not affect the equilibrium price distribution. The equilibrium price distribution derived in this paper permits comparative statics with variable firm numbers.

Acknowledgements

The correspondence author gratefully acknowledges financial support from Ministry of Education of P. R. China under grant no. 05JC70097 and South China University of Technology under grant no. G05N7040580. Authors are all grateful for funding from National Social Science Foundation grant 05BJY056, Provincial Natural Science Foundation grant 04020108 and Provincial Social Science Foundation grant 04GC102.

References

1. Ed Hopkins and Robert M. Seymour, "The Stability of Price Dispersion under Seller and Consumer Learning", Working Paper, 2000
2. Ekkehard Kessner and Mattias K. Polborn, "A New Test of Price Dispersion", German Economic Review, 2000, 1(2), p221-237

3. George J. Stigler, "The Economics of Information", Journal of Political Economy, 69, June 1961, p213, 25p
4. Gerard R. Butters, "Equilibrium Distributions of Sales and Advertising Prices", Review of Economic Studies, Oct 77, Vol. 44 Issue 138, p465, 27p
5. Hal R. Varian, "A Model of Sales", The American Economic Review, Sep 1980, Vol. 70 Issue 4, p651-659
6. John Morgan and Martin Sefton, "Information Externalities in a Model of Sales", Economics Bulletin, 2001(4), p1, 5p
7. Maarten C. W. Janssen and Jose Luis Moraga, "Consumer Search and the Size of Internet Markets", Tinbergen Institute Discussion Paper 2000
8. Rafael Rob, "Equilibrium Price Distributions", Review of Economic Studies, July 1985, Vol. 52 Issue 170, p487, 18p
9. Steven C. Salop and Joseph E. Stiglitz, "The Theory of Sales: A Simple Model of Equilibrium Price Dispersion with Identical Agents", The American Economic Review, Dec 1982, Vol. 72 Issue 5, p1121, 10p

Finding Faithful Boyce-Codd Normal Form Decompositions

Henning Koehler

Massey University, Department of Information Systems
Private Bag 11222, Palmerston North, New Zealand
h.koehler@massey.ac.nz

Abstract. It is well known that faithful (i.e. dependency preserving) decompositions of relational database schemas into Boyce-Codd Normal Form (BCNF) do not always exist, depending on the set of functional dependencies given, and that the corresponding decision problem is NP-hard. The only algorithm to guarantee both faithfulness and BCNF (if possible) proposed so far in [Os79] is a brute-force approach which always requires exponential time. To be useful in practice, e.g. in automated design tools, we require more efficient means.

In this paper we present an algorithm which always finds a faithful BCNF decomposition if one exists, and which is usually efficient, and exponential only in notorious cases.

1 Introduction

We begin by introducing some basic terms from relational database theory. A *relation schema* $R = \{A_1, A_2, \ldots, A_n\}$ is a set of *attributes*. A *relation* r over a schema R is a set of tuples (where each tuple represents one data item), and each element of the tuple corresponds to one attribute in R.

With each relation schema we associate a set F of *functional dependencies* (FD). A FD on R is an expression of the form $X \rightarrow Y$ (read "X *determines* Y") where X and Y are subsets of R. We say that a FD $X \rightarrow Y$ *holds* on a relation r over R if every pair of tuples in r that coincides on all attributes in X also coincides on all attributes in Y. We call a FD $X \rightarrow Y$ *trivial* if $Y \subseteq X$, or, equivalently, if it holds on every relation.

A set F of FDs over R *implies* a FD $X \rightarrow Y$, written $F \vDash X \rightarrow Y$, if $X \rightarrow Y$ holds on every relation r over R for which all FDs in F hold. We say that F implies a set G of FDs if F implies every FD in G. If F and G imply each other, we call G a *cover* of F (and vice versa).

A relation schema R is in *Boyce-Codd Normal Form* (BCNF) w.r.t. a set F of FDs on R if and only if for every non-trivial FD $X \rightarrow Y \in F$ the left hand side (LHS) X is a *key* for R, i.e., $X \rightarrow R \in F^+$, where

$$F^+ := \{ X \rightarrow Y \mid X, Y \subseteq R, F \vDash X \rightarrow Y \}.$$

This normal form is desirable as it prevents redundancy and update anomalies caused by such redundancy [MR87].

S.-W. Cheng and C.K. Poon (Eds.): AAIM 2006, LNCS 4041, pp. 102–113, 2006.

A standard approach to achieve BCNF (or at least the weaker third normal form (3NF)) is to decompose R into several smaller relation schemas $R_i \subseteq R$, such that each R_i is in BCNF. For this, we need to define what FDs hold on each R_i. The *projection* of a set F of FDs onto a subschema $R_i \subseteq R$ is

$$F[R_i] := \{ X \to Y \in F \mid XY \subseteq R_i \}$$

For each R_i, the set of FDs on R_i that are implied by F, is then $F^+[R_i]$. When we ask whether R_i is in BCNF, it is w.r.t. this set $F^+[R_i]$.

To ensure that the decomposed schema can hold the same data as the original schema, we must ask for a decomposition that has the *lossless join* property, that is to say if we project some relation over R onto the R_i and then join them back together, we get back the original relation. Furthermore, the decomposition should be dependency preserving or *faithful*, i.e., $(\bigcup F_i)^+ = F^+$, where F_i is a cover for $F^+[R_i]$ (when describing the decomposition, we usually want to represent $F^+[R_i]$ by a smaller cover for it).

Unfortunately, there does not always exist a faithful BCNF decomposition[1] - consider e.g. the schema $R = ABC$ with FDs

$$F = \{ AB \to C, C \to B \}.$$

The well-known decomposition algorithms ([MR87], [TF80]) produce a lossless BCNF decomposition, which however is not always faithful. While this is sometimes unavoidable, there are cases where they do not produce a faithful decomposition even though one exists (example 1).

Example 1. Consider the schema $CLRT$ containing the attributes C=Course, L=Lecturer, R=Room, T=Time with the functional dependencies

$$F = \{ C \to L, CT \to R, LT \to C, RT \to C \}$$

The only FD in F for which the left hand side is not a key is $C \to L$, so the algorithm from [MR87] produces the decomposition

$$R' = \left\{ \begin{array}{l} (CL, \{ C \to L \}), \\ (CRT, \{ CT \to R, RT \to C \}) \end{array} \right\}$$

The missing FD $LT \to C$ is not implied by $\{ C \to L \} \cup \{ CT \to R, RT \to C \}$, thus the decomposition is not faithful.

On the other hand, the popular synthesis algorithm in its various forms ([BD79], [MR87], [LL99]) produces a faithful decomposition, but the resulting relations R_i need not be in BCNF. And again, there are cases where a faithful BCNF decomposition exists, but the synthesis algorithm does not find one:

Example 2. Consider again the schema $CLRT$ with the FDs

$$F = \{ C \to L, CT \to R, LT \to C, RT \to C \}$$

---

[1] As the condition of being lossless does not conflict with BCNF or faithfulness, we usually will not mention it.

If we synthesize a decomposition by projecting on the attributes involved in each FD in F and eliminate contained sets, we get the decomposition

$$R' = \left\{ \begin{array}{l} (CLT, \{C \to L, LT \to C\}), \\ (CRT, \{CT \to R, RT \to C\}) \end{array} \right\}$$

While this decomposition is clearly faithful, the subschema $CLT, \{C \to L, LT \to C\}$ is not in BCNF as the left hand side C of $C \to L$ is not a key for CLT.

Since the schema CLT is not in BCNF, the information who is lecturer of a course is stored multiple times (once for each lecture time). Thus, in order to change the lecturer of a course, multiple tuples need to be updated. These problems are avoided by the BCNF decomposition CL, CRT from example 2, but there we lost $LT \to C$ which prevented us from creating tables where a lecturer is supposed to give different courses at the same time.

The question of whether a faithful BCNF decomposition exists has been shown to be NP-hard [BB79], so we cannot hope to always find one (if it exists) in polynomial time. In the following we will present an algorithm which does always find a faithful BCNF decomposition if it exists, and computes a faithful decomposition into 3NF otherwise. While the runtime of this algorithm is exponential in the worst case, it appears to be efficient in practise.

2   Linear Resolution and Atomic Closure

The central idea of our approach is to compute all "minimal" FDs in F^+. While the number of such FDs can grow exponentially in the number of attributes and FDs, it often turns out to be reasonably small. The main contribution of this paper is an efficient algorithm for computing the set of all "minimal" FDs.

In the following R denotes a relation schema with FD set F. The variables X, Y, Z shall denote subsets of R, while A, B, C denote single attributes.

Definition 1. (i) *A FD* $X \to A$ *is called* singular.
(ii) *A non-trivial singular FD* $X \to A \in F^+$ *is called* atomic, *if and only if for all* $Y \subsetneq X$ *we have* $Y \to A \notin F^+$.
(iii) *The* atomic closure $\overline{F}$ *of* F *is the set of all atomic FDs in* F^+.
(iv) *A set* $G \subseteq \overline{F}$ *is called* canonical cover *if* $G^+ = F^+$ *and for all* $H \subsetneq G$ *we have* $H^+ \neq F^+$.

Example 3. Consider the set of FDs

$$F = \{A \to B, AB \to C, BC \to AD\}$$

(i) The FDs $A \to B$ and $AB \to C$ are singular but $BC \to AD$ is not.
(ii) While $A \to B$ is atomic, $AB \to C$ is not, since $A \to C \in F^+$.
(iii) For the atomic closure of F we get

$$\overline{F} = \{A \to B, A \to C, BC \to A, BC \to D, A \to D\}$$

(*iv*) The canonical covers of F are

$$\{A \rightarrow B, A \rightarrow C, BC \rightarrow A, BC \rightarrow D\},$$
$$\{A \rightarrow B, A \rightarrow C, BC \rightarrow A, A \rightarrow D\}$$

Checking whether a FD $X \rightarrow Y$ is implied by a set F of functional dependencies is easy: We can do so by computing the *closure* $\overline{X}$ of the left hand side X, which is the set of all attributes determined by X:

$$\overline{X} := \left\{ A \in R \mid X \rightarrow A \in F^+ \right\}$$

Once computed, we only need to check whether the right hand side Y is a subset of $\overline{X}$. Computing $\overline{X}$ can be done quickly using the well-known closure algorithm:

Algorithm. "closure"

```
X̄ := X
while ∃X' → Y ∈ F with X' ⊆ X, Y ⊄ X̄ do
    X̄ := X̄Y
end
```

However, creating all possible FDs on R and testing whether they are implied by F is inefficient. In order to compute $\overline{F}$ efficiently, we need some method for deriving new FDs. In our approach we use a single derivation rule, namely the resolution rule

$$\frac{X \rightarrow A \quad AY \rightarrow B}{XY \rightarrow B}$$

which is easily checked to be sound, i.e. the FDs at the top imply the FD at the bottom. We can then derive new FDs implied by a given set of FDs by applying the resolution rule multiple times:

$$\frac{A \rightarrow B \quad \dfrac{AB \rightarrow C \quad BC \rightarrow D}{AB \rightarrow D}}{A \rightarrow D}$$

The derivation tree above derives $A \rightarrow D$ from the FDs $A \rightarrow B, AB \rightarrow C, BC \rightarrow D$. Note that the derivation tree is *right-linear*, i.e. the left branch always ends in a leaf (a FD in F, rather than an arbitrary sub-tree, in this case $A \rightarrow B$ and $AB \rightarrow C$). We shall refer to such derivation trees as *linear resolution trees*.

The following theorem, which shows that such derivations are possible in general, is central, as it allows us to create a fast algorithm for computing $\overline{F}$.

Theorem 1. *Let F be a set of singular FDs. Then every atomic FD in $\overline{F}$ can be derived from F using the resolution rule*

$$\frac{X \rightarrow A \quad AY \rightarrow B}{XY \rightarrow B} \tag{1}$$

This result still holds if we restrict ourselves to derivations where the substituting FDs $X \rightarrow A$ lie in F, i.e. for every atomic FD $X_i \rightarrow A_i \in \overline{F}$ there exists a linear resolution tree deriving $X_i \rightarrow A_i$ from F.

Proof: Let $X \rightarrow A \in \overline{F}$, and thus $A \in \overline{X} \setminus X$. We use the known fact that the "closure" algorithm works. For any run of the "closure" algorithm let $X_i \rightarrow A_i \in F, i = 1 \ldots k$ be the FDs $X' \rightarrow Y$ used to compute $\overline{X}$, in that order. We start our derivation with $X_k \rightarrow A (= A_k)$, and then successively use $X_i \rightarrow A_i$ for $i = k - 1, \ldots, 1$ in the resolution rule (1):

$$\frac{X_i \rightarrow A_i \quad U_{i+1} \rightarrow A}{U_i \rightarrow A}$$

provided $A_i \in U_{i+1}$. In this, the derived left hand sides U_i have the form

$$U_k = X_k, U_i = \begin{cases} X_i \, (U_{i+1} \setminus A_i) \text{ if } A_i \in U_{i+1} \\ U_{i+1} \qquad\qquad \text{else} \end{cases}$$

It is easy to see that $U_j \subseteq X \cup \{A_1, \ldots, A_{j-1}\}$, and in particular $U_1 \subseteq X$. Since $X \rightarrow A$ is atomic, we get $U_1 = X$, thus we have indeed constructed the derivation we wanted. ∎

Note that the intermediate FDs $U_i \rightarrow A$ during the derivation need not be atomic, as U_i need not be minimal. As example 4 shows, this is unavoidable.

Example 4. Consider the set

$$F = \{A \rightarrow B, A \rightarrow C, BC \rightarrow D\}$$

The atomic FD $A \rightarrow D$ cannot be derived using (1) without intermediate non-atomic FDs: The only possible applications of the resolution rule are

$$\frac{A \rightarrow B \quad BC \rightarrow D}{AC \rightarrow D} \quad \text{and} \quad \frac{A \rightarrow C \quad BC \rightarrow D}{AB \rightarrow D},$$

and neither $AC \rightarrow D$ nor $AB \rightarrow D$ are atomic.

Since we are only interested in atomic FDs, we reduce the left hand side of any FD we derive. As the reduced FD implies the derived one, this still allows us to derive all atomic FDs, and it can reduce the number of possible derivation sequences considerably. This is all we need to turn theorem 1 into an efficient algorithm (that we shall name *linear resolution*) which computes the atomic closure $\overline{F}$ of any set of functional dependencies F. Note that we can easily compute a canonical cover of F by splitting up FDs by their right hand side attributes, reducing the left hand sides and removing redundant FDs [MR87].

Algorithm. "linear resolution" to compute $\overline{F}$

```
compute a canonical cover F' of F
F̄ := F'
for all Y → B ∈ F̄ do
   for all X → A ∈ F' with A ∈ Y, B ∉ X do
      // derive (XY \ A) → B by rule (1)
      find U ⊆ (XY \ A) with U → B atomic
      if U → B ∉ F̄ then append it to F̄
   end
end
```

Example 5. Starting with the canonical cover

$$F = \{C \to L, CT \to R, LT \to C, RT \to C\}$$

from examples 1 and 2, we use resolution:

$$\frac{RT \to C \quad C \to L}{RT \to L}, \frac{LT \to C \quad CT \to R}{LT \to R}$$

The newly found FDs $RT \to L$ and $LT \to R$ are already atomic, so we add them to $\overline{F}$. We then test whether new resolution steps have become possible:

$$\frac{CT \to R \quad RT \to L}{[CT \to L]}, \frac{C \to L \quad LT \to R}{[CT \to R]}$$

The FD $CT \to L$ gets LHS-minimized (we minimize the left hand side) to $C \to L$, which has already been found. The FD $CT \to R$ is already contained in $\overline{F}$ as well, so no further atomic FDs can be derived. We therefore get:

$$\overline{F} = F \cup \{RT \to L, LT \to R\}$$

Theorem 2. *The "linear resolution" algorithm computes $\overline{F}$ correctly.*

Proof: The algorithm computes all atomic FDs that can be derived using (1), which by theorem 1 are all. Instead of storing the potentially non-atomic FD $(XY \setminus A) \to B$ it stores the stronger result $U \to B$. Thus any derivation which requires $(XY \setminus A) \to B$ can be replaced by a derivation using $U \to B$ as initial FD instead.   ■

3   Faithful BCNF Decomposition

In the following we will use the atomic closure $\overline{F}$ to construct a faithful BCNF decomposition R', provided it exists. In this we take the same approach as was taken in [Os79], and refer the reader to proves given there.

The following lemma allows us to focus on making our decomposition faithful and in BCNF, and not worry about making it lossless.

Lemma 1. *[BD79] Every faithful decomposition of R containing a subschema which forms a key of R is lossless.*

Every minimal key is in BCNF, as the projection of F^+ on it contains only trivial FDs. Thus we can easily make a faithful BCNF decomposition R' lossless (without losing faithfulness or BCNF) by adding a minimal key as additional subschema if R' doesn't contain a key of R.

When synthesizing a decomposition from a set of FDs, we need to use a cover with the "right" FDs, so that the subschemas created are in BCNF.

Definition 2. *A FD $X \to Y \in F^+$ is called* critical *(w.r.t. F) if XY is not in BCNF (w.r.t. $F^+[XY]$). A cover G of F is called* critical *if it contains a critical FD. A FD or cover that is not critical is called* uncritical.

Theorem 3. *[Os79] The following are equivalent:*
(i) A schema (R, F) has a faithful, lossless decomposition into BCNF
(ii) F has an uncritical cover
(iii) F has an uncritical atomic cover

We may test whether a FD is critical by using $\overline{F}$ as follows:

Lemma 2. *[Os79] A FD $X \to A$ is critical w.r.t. F if and only if there exists a FD $Y \to B \in \overline{F}$ with $YB \subseteq XA$ and $XA \nsubseteq Y$.*

Guided by theorem 3, we try to find an uncritical canonical cover F'' of F as follows: for each $X \to A \in \overline{F}$ we check whether it is critical, and if so whether it is redundant. If it is redundant we discard it, otherwise we can be sure that there exists no uncritical atomic cover, and thus no FLBD. We can then create F'' from the remaining FDs by eliminating redundant ones.

We then apply the synthesis algorithms [BD79] using this cover, i.e. our decomposition will consist of those subschemas XA where $X \to A \in F''$, plus a minimal key if needed. Thus, even if no uncritical cover if found, we obtain a decomposition into 3NF [BD79].

Algorithm. "least critical cover synthesis"

```
compute F̄
F″ := F̄
for all  X → A ∈ F″ do
  for all Y → B ∈ F̄ do
    if YB ⊆ XA and XA ⊈ Ȳ then
      if  X → A ∈ (F″ \ {X → A})⁺ then
        remove X → A from F″
  end
end
for all  X → A ∈ F″ do
  if  X → A ∈ (F″ \ {X → A})⁺ then
    remove  X → A from F″
  else
    add schema XA to R
end
remove all schema Rᵢ ∈ R with Rᵢ ⊊ Rⱼ ∈ R
if R contains no key, add a minimal key
```

Theorem 4. *[Os79] Either the "least critical cover synthesis" algorithm computes a faithful, lossless BCNF decomposition $\mathcal{R}$, or no such decomposition exists.*

4   Improvements and Complexity Analysis

In this section we shall discuss possible improvements and implementation issues for the basic algorithms presented. Based on these improvements we present a brief complexity analysis. For this we shall use the following variables:

$$k = \text{number of attributes in } R$$
$$n = \text{number of FDs in } F$$
$$f = \text{number of FDs in } \overline{F}$$

4.1   Linear Resolution

We start with the algorithm for computing the atomic closure $\overline{F}$. By splitting up FDs by their right hand side attributes, the number of FDs in F can increase. To avoid this, we use the original set F and replace the singular resolution rule (1) by the generalized resolution rule

$$\frac{X \to Z \quad Y \to A}{X\left(Y \setminus \overline{X}\right) \to A} \quad Z \cap Y \neq \emptyset \tag{2}$$

Since all attributes in $\overline{X} \setminus X$ are extraneous in XY and thus could be removed when minimizing the left hand side, we don't lose any derivations. As with singular resolution, rule (2) only needs to be applied if $A \notin X$.

The set of all atomic FDs with right hand side A can be computed independently from other atomic FDs, as the linear resolution process only utilizes FDs in F and atomic ones with right hand side A. We may thus compute the minimal left hand sides for each attribute A individually, which allows us another optimization: For any FD $X \to Z \in F$ with $A \in \overline{X}$ we compute a minimal $U \subseteq X$ such that $U \to A$ is atomic (provided $A \notin X$, otherwise we discard $X \to Z$ completely). These FDs $U \to A$ are used as initial FDs for deriving $\overline{F}_A := \{X \to A \in \overline{F}\}$. The FD $X \to Z$ can then be removed from F for the purpose of linear resolution, since any application of (2) using $X \to Z$ with $A \in \overline{X}$ results in a FD $XY' \to A$ which can be LHS-minimized to a FD $U \to A$ already contained in the initial set.

Our optimized linear resolution algorithm is given below.

Algorithm. "linear resolution" revised

```
for each X → A ∈ F compute X̄
for each A ∈ RHS(F) :=   U   Z do
                       X→Z∈F
   F̄_A := ∅, F_A := ∅
   for each X → Z ∈ F do
     if A ∉ X̄ then
       add X → Z to F_A
     else if A ∉ X then
         find U ⊆ X with U → A  atomic
         if U → A ∉ F̄_A  then append it to F̄_A
   end
   for all Y → A ∈ F̄_A do
     for all X → Z ∈ F_A with Y ∩ Z ≠ ∅ do
       // derive X (Y \ X̄) → A by rule (2)
       find U ⊆ X (Y \ X̄) with U → A atomic
```

```
        if  U → A ∉ F̄_A  then append it to  F̄_A
      end
    end
  end
```
$$\overline{F} := \bigcup \overline{F}_A$$

A complexity analysis for the revised linear resolution algorithm is straight forward. Computing $\overline{X}$ for each $X \to A \in F$ can be performed in $O\left(kn^2\right)$, given that each $\overline{X}$ can be computed in $O\left(kn\right)$ using the "closure" algorithm [BB79]. The most time consuming step is the removal of extraneous attributes from $X\left(Y \setminus \overline{X}\right)$. Each test whether an attribute is extraneous requires one run of the "closure" algorithm. Thus finding a minimal U takes $O\left(k^2n\right)$ operations. Checking whether or not $U \to A \in \overline{F}_A$ can be done in $O\left(k\right)$ using an appropriate data structure to represent $\overline{F}_A$ (e.g. a binary tree with different branches indicating whether an attribute is contained in the left hand side of a FD). This is done at most $f \cdot n$ times (plus kn times for initialization), which leads to an overall complexity of $O\left(f \cdot k^2n^2\right)$.

4.2   Least Critical Cover Synthesis

While the improvements which we shall present in the following reduce the run-time of our "least critical cover synthesis" algorithm considerably, they do not improve its worst-case complexity behavior. We shall therefore begin with a brief complexity analysis.

Computation of $\overline{F}$ can be done in $O\left(f \cdot k^2n^2\right)$ as described earlier. Pre-computing $\overline{Y}$ for each $Y \to B \in \overline{F}$ can be done in $O\left(f \cdot kn\right)$ using the "closure" algorithm, so that the condition

$$Y \subseteq XA \text{ and } XA \nsubseteq \overline{Y}$$

can be tested in $O\left(k\right)$. The number of such tests is at most f^2, leading to a complexity of $O\left(f^2 \cdot k\right)$. The redundancy test

$$X \to A \in (F'' \setminus \{X \to A\})^+$$

can be performed in $O\left(f \cdot k\right)$ using the "closure" algorithm. At most f such tests are performed (both loops combined), which again gives us $O\left(f^2 \cdot k\right)$ as bound. Since $\mathcal{R}$ contains at most f schemas, removal of contained schemas from $\mathcal{R}$ can be performed in $O\left(f^2 \cdot k\right)$ as well. Checking whether $\mathcal{R}$ contains a key can be done in $O\left(f \cdot kn\right)$. A minimal key can be found in $O\left(k^2n\right)$ by starting with the trivial key R and testing for each attribute whether it can be removed while maintaining a key of R. Adding these complexities up leads to an overall bound of $O\left(f \cdot k^2n^2 + f^2 \cdot k\right)$.

We want to reduce the number of FDs for which we check for criticality and redundancy. Furthermore, we often do not start with a single relation schema, but rather with a decomposition from an earlier design. This decomposition is reflected by the form of the FDs given. It is therefore desirable to keep the cover F'' produced as close to the original cover F' as possible.

We do so by maintaining a cover F_R, which gets initialized with F'. We then check all FDs in F_R for being critical. If we find a critical one, we need to check whether it is redundant, and if so, substitute it for one or more non-critical ones from F''. These two tasks can be combined: when checking whether $X \to A \in F_R$ is redundant by computing the closure $\overline{X}$ (up to the point where A is added), we check any FD to be used for criticality (discarding FDs tested critical to avoid double-testing) and use only non-critical ones. If $X \to A$ turns out to be redundant, we replace it by the FDs used in computing the closure.

Algorithm. for substituting $X \to A \in F_R$

$X' := X, Subst := \emptyset$
while $A \notin X'$ do
  if $\exists Y \to B^2 \in F_R \setminus \{X \to A\}$ with $Y \subseteq X', B \notin X'$ then
   add B to X'
  else
    if $\exists Y \to B \in F''$ with $Y \subseteq X', B \notin X'$ then
     if $Y \to B$ critical then
      remove $Y \to B$ from F''
     else
      add B to X'
      add $Y \to B$ to $Subst$
    else
     return "$X \to A$ not redundant"
end
$F_R := (F_R \setminus \{X \to A\}) \cup Subst$
$F'' := F'' \setminus Subst$

Note that all critical $Y \to B \in F'' \setminus F_R$ we find are redundant since F_R is a cover for F, so we need not perform a costly check before removing them from F''. After substituting critical FDs from F_R, we may no longer have a non-redundant cover. Also, some of the sets XA for $X \to A \in F_R$ may contain additional atomic FDs. We can find those while checking for criticality. For $G \subseteq \overline{F}$ let $contained\,(G)$ denote the set of such additional atomic FDs, i.e.

$$contained\,(G) := \bigcup_{X \to A \in G} \overline{F}\,[XA]$$

Having computed $\overline{F}\,[XA]$ for each $X \to A \in F_R$, we now remove redundant FDs from F_R. When checking whether a FD $X \to A \in F_R$ is redundant, we check redundancy w.r.t. $contained\,(F_R \setminus \{X \to A\})$ rather than $F_R \setminus \{X \to A\}$. Having constructed $\mathcal{R}$, we can attempt to merge schemas to reduce their number. For this, we can use $\overline{F}$ to check whether the merged schema $R_i \cup R_j$ is in BCNF.

---

[2] The data structure used to speed up the "closure" algorithm as described in [BB79] can be used to perform this check quickly as well.

5   Other Applications and Related Work

The linear resolution algorithm can also be used to compute all minimal left hand sides X for any given right hand side Y, i.e. all minimal X with $X \to Y \in F^+$. For this, we simply start with some minimal left hand side $X \subseteq Y$ such that $X \to Y$.

Theorem 5. *Let F be any set of singular FDs over R and $Y \subseteq R$. Then any FD $X \to Y \in F^+$ with minimal left hand side X can be derived from $Y \to Y$ using linear resolution.*

Proof: Let X be minimal with $X \to Y \in F^+$. For every $A_i \in Y \setminus X$ there exists a minimal $X_i \subseteq X$ with $X_i \to A_i \in \overline{F}$. By theorem 1 every $X_i \to A_i \in \overline{F}$ has a linear resolution tree in F. Combining these resolution trees to substitute all $A_i \in Y \setminus X$ in $Y \to Y$ (possibly skipping some if the attribute A_i has been eliminated in an earlier step) yields a linear resolution of $X' \to Y$ for some $X' \subseteq \bigcup X_i \subseteq X$. Since X is minimal $X' = X$. ∎

In particular, linear resolution can be used to compute all minimal keys of a relation, and thus all prime attributes. This application was already found in [LO78]. The same author discusses an algorithm for finding a faithful BCNF decomposition in [Os79] similar to the basic version of our "least critical cover synthesis" algorithm, but does not provide efficient means for computing $\overline{F}$.

As well, the atomic closure $\overline{F}$ can be used to compute the atomic closure of the projection of F^+ over some subschema S, using the identity

$$\overline{F^+}[S] = \overline{F}[S]$$

This allows us to compute covers for the schemas produced in our BCNF decomposition. A similar approach was taken in [Go87], where the resolution rule is used in a non-linear fashion, but other optimizations are applied.

In [MC92] a sufficient but not necessary condition for the existence of a faithful BCNF decomposition is given, which can be checked in polynomial time. The condition actually ensures that *every* atomic FD is uncritical, and can be viewed as a way of identifying easy cases.

The subject of schema normalization has gained new interest in recent years with the advance of XML-databases [Sc05]. We note that the approach presented here can be generalized to complex-valued data models such as XML, although this is outside the scope of this paper.

6   Conclusion

We have shown how to compute the atomic closure of a set of FDs efficiently, and how to utilize it in finding a faithful BCNF decomposition. Our algorithm guarantees to find a faithful BCNF decomposition whenever one exists, and may likely generate better decompositions (for which fewer subschemas violate BCNF) than "standard" synthesis otherwise.

We have implemented the algorithms described, and run them on several thousand test schemas with up to 50 attributes and up to 100 FDs. In all cases a faithful BCNF decomposition was found (or determined that none exists) within seconds. This is in contrast to the algorithm given in [Os79], which computes $\overline{F}$ using brute-force and proves completely infeasible for schemas of that size.

From these test results and our complexity analysis we conclude that our approach is practical for reasonably large schemas (as long as the FD sets are not specifically designed to break the algorithm).

References

[LO78]   C. L. Luccesi & S.L. Osborn, Candidate Keys for Relations, Journal of Computer and System Sciences, Vol 17, No 2, pp. 270-279, 1978

[Os79]   Sylvia L. Osborn, Testing for Existance of a Covering Boyce-Codd Normal Form, Information Processing Letters, Vol 8, No 1, pp. 11-14, 1979

[BB79]   C. Beeri & P. A. Bernstein, Computational Problems Related to the Design of Normal Form Relational Schemas, ACM Transactions on Database Systems, Vol 4, No 1, pp. 30-59, 1979

[BD79]   J. Biskup, U. Dayal & P. A. Bernstein, Synthesizing Independent Databse Schemas in Proceedings of the 1979 ACM SIGMOD international conference on Management of data, pp. 143–151, 1979

[TF80]   D.-M. Tsou & P. C. Fischer, Decomposition of a Relation Scheme into Boyce-Codd Normal Form, in Proceedings of the ACM 1980 annual conference, pp. 411–417, 1980

[Go87]   Georg Gottlob, Computing Covers for Embedded Functional Dependencies, in Proceedings of the Sixth ACM SIGACT-SIGMOD-SIGART Symposium on Principles of Database Systems, pp. 58–69, 1987

[MR87]   H. Mannila & K.-J. Räihä, The Design of Relational Databases, Addison-Wesley, ISBN 0-201-56523-4, 1987

[MC92]   Mila E. Majster-Cederbaum, Ensuring the existence of a BCNF-decomposition that preserves functional dependencies in $O(N^2)$ time, in Information Processing Letters, Vol 43, No 2, pp. 95–100, 1992

[LL99]   M. Levene & G. Loizou, A Guided Tour of Relational Databases and Beyond, Springer, ISBN 1-85233-008-2, 1999

[Sc05]   Klaus-Dieter Schewe, Redundancy, Dependencies and Normal Forms for XML Databases, Proceedings of the Sixteenth Australasian Database Conference (ADC 2005), CRPIT, Vol 39, pp. 7–16, 2005

Instant Service Policy and Its Application to Deficit Round Robin

Jinoo Joung[1], Dongha Shin[1], Feifei Feng[2], and Hongkyu Jeong[2]

[1] Sangmyung University, Seoul, Korea
jjoung@smu.ac.kr
[2] Samsung Advanced Institute of Technology, Kiheung, Korea

Abstract. Many of scheduling algorithms that provide a predefined bandwidth to a traffic flow fall into a category of Latency-rate ($\mathcal{LR}$) server. A series of $\mathcal{LR}$ servers can be viewed as a virtual node with an $\mathcal{LR}$ server of the *latency* which is simply the summation of individual latencies of actual $\mathcal{LR}$ servers. Deficit Round Robin (DRR) is such an $\mathcal{LR}$ server and the simplest one to implement, so that it is adopted in many real systems. In this research we suggest a novel policy of Instant Service, which is applicable to any round-robin schedulers. We then apply this policy to DRR, make a variation called DRR with Instant Service (DRR-IS), and analyze it. We prove that the DRR-IS is still an $\mathcal{LR}$ server. We calculate its latency and investigate its fairness characteristics. We demonstrate the DRR-IS, compared with DRR, provides about 30% better latency while have the same complexity and the same fairness.

1 Introduction

The problem of guaranteeing a QoS (Quality of Service) within a packet switching network has been extensively studied and several solutions to this problem have been suggested. One of such suggestions is the IntServ (Integrated Services) in which the QoS is guaranteed by means of reserving, allocating and providing an amount of predefined resource to a data traffic unit, which often is called a flow or a session. Providing the allocated bandwidths, or service rates, or simply rates of an output link to multiple sharing flows plays a key role in this approach. A scheduling algorithm selects the next packet to transmit, and decides when it should be transmitted, on the basis of given performance metrics. Two well-established performance metrics of those algorithms are the delay and the fairness. The representative and pioneering solutions for scheduling algorithms that provide bandwidths to flows are Packet-level Generalized Processor Sharing (PGPS) [1] and Weighted Fair Queueing (WFQ) [2]. These early traffic scheduling algorithms and their variations, empowered by packet-level approximations of precise schedulers based on fluid-flow traffic model, are able to provide minimum delay and maximum fairness to each flows, while providing promised rates. This group of algorithms, called sorted priority scheduling algorithms however, generally need to sort packet deadlines and therefore suffer from the implementation complexity, which is $O(\log N)$ at best, while N is the number of active flows in a scheduler [3]. This sorting bottleneck makes practical implementations

S.-W. Cheng and C.K. Poon (Eds.): AAIM 2006, LNCS 4041, pp. 114–125, 2006.

of these algorithms problematic, and necessitates the design of simpler schemes. A simple round robin based algorithm, Deficit Round Robin (DRR) [4], and its variations [5, 6, 7] can maintain the crucial property of providing the allocated service rates, that of the PGPS and WFQ, without the complexity of sorted priority scheduling algorithms. In fact a simple weighted round-robin (WRR) scheduler would provide a rate to a flow if the size of packets is homogenous. When the packet size varies, however, as in the most existing networks, a flow with smaller packets gets disadvantage as the *rounds* go. Therefore redemptions for flows with shorter packets are necessary, and this is where the idea of registering *deficit* and redeeming later was introduced, and called DRR. The complexity of the basic DRR can be as low as $O(1)$. The DRR and its variations cannot match the performance of sorted priority algorithms, especially in terms of the delay bound. There are numerous upgrades to the basic DRR in this regards, but they all suffer from additional complexity, which one would like to avoid considering the original paradigm of adopting round robin based algorithms.

The DRR, with many other rate-providing servers, is proved to be a Latency-Rate server [10], or simply $\mathcal{LR}$ server. For a scheduling algorithm to belong to the $\mathcal{LR}$ server class, it is only required that the average rate of service offered by the server to an active flow, over any interval between the time Θ after the beginning of the busy period and the end of the busy period, is at least equal to its reserved rate. A busy period of a flow is a period of time during which the average arrival rate of the flow remains at or above its reserved rate. The parameter Θ is called the latency of the server. All the work-conserving servers that guarantee rates, including WFQ, PGPS, WRR, DRR, and many other schedulers exhibit this property and can therefore be modeled as $\mathcal{LR}$ servers. The behavior of an $\mathcal{LR}$ server is determined by two parameters, the latency and the allocated rate. The latency of an $\mathcal{LR}$ server may be considered as the worst-case delay seen by the first packet of the busy period of a flow. The latency of a particular scheduling algorithm may depend on its internal parameters, its transmission rate on the outgoing link, and the allocated rates of various flows. It was shown, however, that the maximum end-to-end delay experienced by a packet in a network of $\mathcal{LR}$ servers can be calculated from only the latencies of the individual servers on the path of the flow, and the traffic parameters of the flow that generated the packet. More specifically for a leaky-bucket constrained flow,

$$D_i \leq \frac{\sigma_i}{\rho_i} + \sum_{j=1}^{k} \Theta_i^{(S_j)}, \tag{1}$$

where D_i is the delay of flow i within a network, σ_i and ρ_i are the well known leaky bucket parameters, the burstiness and the average rate, respectively, and $\Theta_i^{(S_j)}$ is the latency of the server S_j.

In this research we suggest a novel concept of Instant Service, which is applicable to any round-robin scheduling algorithms. We then apply this concept to DRR, make a variation called DRR with Instant Service (DRR-IS), and analyze it. We prove that the DRR-IS is still an $\mathcal{LR}$ server and calculate its latency. The DRR-IS, compared with DRR, turns out to provide about 30% better latency while introducing no additional complexity at all. In the next section we describe the core algorithm and basic properties of DRR. We then introduce DRR-IS and analyze its performance, both in terms of delay and fairness, in section 3. In section 4 numerical results are shown from scenarios of real-time traffic services in residential network environments. DRR and other $\mathcal{LR}$ servers are compared.

2   Previous Works on Deficit Round Robin

In this section we describe the detail of DRR behavior, and its latency given by [10]. A DRR scheduler maintains a deficit counter per each flow, thus per each queue. A flow i is assigned with a quantum value ϕ_i, which represents a relative amount of service a flow will receive. A round is defined as a time interval during which all the active flows receive service opportunities, one per each flow. We will call this service opportunity a *turn* of a flow. At the start of the flow i's turn, the deficit value δ_i is incremented as much as the quantum value of the flow, ϕ_i. The size of the head packet of the flow i then is compared with the δ_i. If δ_i is larger or equal to the head packet size, then the head packet gets service and leave the queue. Whenever a packet is served, δ is decremented as much as the size of the served packet. The second head packet of the queue, which now becomes the head packet is then compared with the δ_i again. This process continues until the δ_i becomes smaller than the head packet. When this happens the next flow enters a turn and the packets within this flow will be served. Using this policy, the DRR can achieve $O(1)$ complexity, given that the ϕ_i is set to be greater than or equal to the maximum packet size of the flow i, for all i [9]. This is because otherwise a flow may not receive a service at all during a turn, and the amount of calculation required for serving a packet in a flow increases consequently. DRR can provide a lower bound of the service amount offered to a flow, in terms of number of turns. This important property of DRR is summarized in the following lemma.

Lemma 1. *Assume that flow i is continuously backlogged during $[t_1, t_2)$. Let k be the number of DRR turns given to i during the interval $[t_1, t_2)$. The service received by i during this period, $W_i(t_1, t_2)$, is given by*

$$W_i(t_1, t_2) \geq k\phi_i - \delta_i^k,\tag{2}$$

where δ_i^k is the deficit value of i, at the end of the kth round, counting from the first round within $[t_1, t_2)$.

Proof. See the proof of lemma 2 of [4].   □

In an accompanying research [10], DRR is proved to be an $\mathcal{LR}$ server. The latency of the DRR server is given as

$$\frac{3F - 2\phi_i}{r},\tag{3}$$

where F is defined as the frame size, which is the sum of all ϕ_i over i, and r is the output link capacity. Note that F does not represent the actual number of bytes served during a specific round, but the average number of bytes served in a round.

3 Analysis of DRR-IS

We assume a packet switch where a set of flows share a common output link. We denote with ρ_i the bandwidth, or the rate allocated to flow i. We assume that the switches are store-and-forward devices. We denote by *server* an output link controller of a switch, whose mission is to schedule packets destined for the output link and then transmit the scheduled packets accordingly. Let $A_i(\tau, t)$ denote the arrivals from flow i during the interval (τ, t) and $W_i(\tau, t)$ the amount of service received by flow i during the same interval. In a system based on the fluid-flow model both $A_i(\tau, t)$ and $W_i(\tau, t)$ are continuous functions of t. In the packet-by-packet model, however, we assume that $A_i(\tau, t)$ increases only when the last bit of a packet is received by the server; likewise $W_i(\tau, t)$ is increased only when the last bit of the packet in service leaves the server. We further denote that a flow i is *backlogged* when one or more packets of i are waiting for service. In other words, if $A_i(0, t) - W_i(0, t)$ is larger than zero then the flow i is backlogged at t. Therefore a *backlogged period* of flow i is any period of time during which packets belong to flow i are continuously queued in the server. When a packet from a previously inactive flow enters a server, the backlogged period of that flow starts and the flow is said to be now *activated*.

Every round-robin scheduler maintains an *active list*, with which information of active flows and the service order among them are updated. A newly activated flow is enqueued to the active list. When a flow becomes inactive, it is dequeued from the active list. Exactly where in this one dimensional linked list we should enqueue a flow is the key idea of this research. A round-robin scheduler including DRR has a head and a tail in the active list. When a flow is activated, this flow is enqueued after the current tail, so that it becomes a new tail. If this flow is activated during the head flow's turn, it has to wait almost a whole round's worth of time to receive a service. Instant Service policy lets an activated flow be served after the current flow's turn. The precise algorithm can be stated as the following:

1. Assume that a flow i is activated during another flow g's turn. The flow i gets service right after the flow g's turn is finished.
2. If more than one flow are activated during another flow's turn, the service order among these new flows may be determined arbitrarily.

Let us apply the IS policy to DRR and analyze it. Consider the following server with a set of flows V. The number of total flows is v. Among these v flows, n flows are backlogged and are initially being served according to the DRR algorithm. Let the set of these backlogged flows be N. Let the flow under our interest be i and assume i is activated during another flow g's service turn. g belongs to N. Further let the number of flows activated during the g's turn be m and let the set of these flows be M. According to the IS policy, the flows in M will receive service after the g's turn and the order among the flows in M will be determined arbitrarily, and therefore the flow i may be the last among them in the worst case.

Let us define the system $S(M, N)$ be the server in which a set of flows N is initially backlogged and a set of flows M is activated during an arbitrary flow

g's turn, where $g \in N$. In such a system $M \subset V$, $N \subset V$, and $M \cap N = \{\}$. Let us further define the following time instants.

T_0: the time flow i is activated (becomes backlogged).
t_k: the finish time of the kth round of the system.
τ_k: the finish time of flow i's turn within the kth round.

Figure 1 depicts the time instants, assuming that the flow i is the last one served among the flows in M.

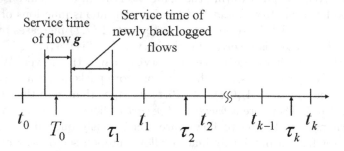

Fig. 1. Time instants of the system $S(M, N)$

3.1   Latency

Lemma 2. *For any system $S(M, N)$, $n(M) + n(N) < v$, there always exists a system $S(M', N')$, $M' \subset V$, $N' \subset V$, and $M' \cap N' = \{\}$, which satisfies the inequality,*

$$W_i^{S(M,N)}(T_0, t) \geq W_i^{S(M',N')}(T_0, t), M \neq M', N \neq N', \forall t > T_0. \qquad (4)$$

Proof. It is suffice for the proof to show a case in which the (4) is satisfied. Consider a system $S(M, N)$, where the sum of each number of elements, $n(M)$ plus $n(N)$, is less than the total number of flows, v. For such a system $S(M, N)$, there always exists a system $S(M^*, N)$, $M^* = M \cup \{i^*\}$, for i^* not belong to either M nor N. The flow under observation i is an element of both M and M^*. In the system $S(M^*, N)$, i^* and i become backlogged during the same flow's service time, therefore i^* can be served prior to i. In this case, $\tau_k^{S(M,N)} < \tau_k^{S(M^*,N)}$, $\forall k > 0$. If i^* receives service after i, then $\tau_k^{S(M,N)} = \tau_k^{S(M^*,N)}$. Therefore $\tau_k^{S(M,N)} \leq \tau_k^{S(M^*,N)}$ in any case. By the definition of τ_k,

$$W_i^{S(M,N)}(T_0, \tau_k^{S(M,N)}) = W_i^{S(M^*,N)}(T_0, \tau_k^{S(M^*,N)}), \forall k > 0. \qquad (5)$$

Moreover $W_i(T_0, t)$ is a monotonically increasing function, therefore

$$W_i^{S(M,N)}(T_0, t) \geq W_i^{S(M^*,N)}(T_0, t), \forall t > T_0, \qquad (6)$$

and the (4) holds. $\square$

Lemma 3. *For an arbitrary* N, $n(N) < v$, $N \subset V$,

$$W_i^{S(M,N)}(T_0, t) \geq W_i^{S(V-N,N)}(T_0, t), n(M) + n(N) < v, \forall t > T_0. \quad (7)$$

Proof. Lemma 3 is a direct consequence of lemma 2.   □

Lemma 3 implies that in a case where n flows are backlogged, the service received by i, W_i, is minimized when all the remaining $v - n$ flows, including i, are activated during another flow's single service turn.

Theorem 1.

$$W_i(T_0, t) \geq \max \left\{ 0, \rho_i \left(t - T_0 - \frac{2F + \phi_{\max} - \phi_i}{r} \right) \right\}, \quad (8)$$

where $\phi_{\max} = \max_{j \neq i, j \in V} \{\phi_j\}$.

Proof. Let us define the following sets of flows. Let A be the set of flows, which will receive service after the flow i in a round, among those which have been backlogged before i is activated. Similarly let B be the set of flows, which will receive service before the flow i in a round, among those which have been backlogged before i is activated. Further let C be the flows which are activated in the same flow's turn with i. In summary, $A \cup B = N$ and $C \cup \{i\} = M$. Let g be the flow during whose service time the flows in M are activated. The following inequalities hold.

$$\tau_1 - T_0 \leq \frac{1}{r} \left\{ \max_{j \in B}(\phi_j + \delta_j^0 - \delta_j^1) + \sum_{j \in C}(\phi_j - \delta_j^1) + (\phi_i - \delta_i^1) \right\}. \quad (9)$$

$$\tau_l - \tau_{l-1} \leq \frac{1}{r} \left\{ \sum_{j \in A}(\phi_j + \delta_j^{l-2} - \delta_j^{l-1}) + \sum_{j \in B}(\phi_j + \delta_j^{l-1} - \delta_j^l) \right.$$
$$\left. + \sum_{j \in C}(\phi_j + \delta_j^{l-1} - \delta_j^l) + (\phi_i + \delta_i^{l-1} - \delta_i^l) \right\}, \quad (10)$$

for l, $2 \leq l \leq k$. Summing the inequality (10) over l, from 2 to k, and the inequality (9) we get

$$\tau_k - T_0 \leq \frac{1}{r} \left\{ \max_{j \in B}(\phi_j + \delta_j^0 - \delta_j^1) + \underbrace{\sum \delta_j^0}_{A} + \underbrace{\sum \delta_j^1}_{B} + \underbrace{\sum \phi_j}_{C} + \phi_i \right.$$
$$\left. - (\underbrace{\sum \delta_j^{k-1}}_{A} + \underbrace{\sum \delta_j^k}_{B} + \underbrace{\sum \delta_j^k}_{C} + \delta_i^k) + (k-1)F \right\}. \quad (11)$$

$\max_B(\phi_j + \delta_j^0 - \delta_j^1) + \sum_B \delta_j^1 \leq \max_B \phi_j + \max_B(\delta_j^0 - \delta_j^1) + \sum_B \delta_j^1$, and without loss of generality, we can let $\max_{j \in B}(\delta_j^0 - \delta_j^1) = \delta_h^0 - \delta_h^1, h \in B$, and then $\max_B(\delta_j^0 - \delta_j^1) + \sum_B \delta_j^1 \leq \sum_B \phi_j$. Moreover $\sum_A \delta_j^0 \leq \sum_A \phi_j$. Therefore

$$\tau_k - T_0 \le \frac{1}{r}\{\max_{j \in B} \phi_j + \sum_{j \in \{A \cup B \cup C\}} \phi_j + \phi_i + (k-1)F - \delta\}$$

$$\le \frac{1}{r}\{\phi_{\max} + kF - \delta\}, \tag{12}$$

where $\phi_{\max} = \max_{j \ne i, j \in V}\{\phi_j\}$, and

$$\delta = \left(\sum_A \delta_j^{k-1} + \sum_B \delta_j^k + \sum_C \delta_j^k + \delta_i^k\right). \tag{13}$$

We get

$$k \ge \frac{1}{F}\{r(\tau_k - T_0) - \phi_{\max} + \delta\}. \tag{14}$$

Inequality (12) comes from that $\tau_k - T_0$ is maximized when $A \cup B \cup C \cup \{i\} = V$, because of lemma 3. Based on the inequality (14) and lemma 1, we obtain

$$W_i(T_0, \tau_k) = W_i(T_0, t_k) \ge \frac{1}{F}\{r(\tau_k - T_0) - \phi_{\max} + \delta\}\phi_i - \delta_i^k$$

$$= \rho_i(\tau_k - T_0) - \left(\frac{\phi_{\max} - \delta}{F}\right)\phi_i - \delta_i^k. \tag{15}$$

Note that $\rho_i/r = \phi_i/F$.

From here we investigate the lower bound of the amount of service received by flow i in an arbitrary time within the time interval $(\tau_{k-1}, \tau_k]$. For a given $W_i(T_0, \tau_k)$, $W_i(T_0, t)$ is minimized under the condition that the amount of data sent by flow i during kth turn is at its maximum, $2\phi_i - \delta_i^k$. Let's denote by t^* the time instant in the interval $(\tau_{k-1}, \tau_k]$, at which the service of the first packet of flow i is completed. Since the first packet size is less than or equal to ϕ_i, the total data served is $2\phi_i - \delta_i^k$, and by the definition of t^*, $t^* \le \tau_k - (\phi_i - \delta_i^k)/r$. Therefor, for any $t \le t^*$,

$$W_i(T_0, t) \ge \max\{0, \rho_i(\tau_k - T_0) - \left(\frac{\phi_{\max} - \delta}{F}\right)\phi_i - \delta_i^k - 2\phi_i + \delta_i^k\}$$

$$= \max\left\{0, \rho_i\left(\tau_k - T_0 - \frac{\phi_{\max} - \delta}{r} - \frac{2F}{r}\right)\right\}$$

$$\ge \max\left\{0, \rho_i\left(t - T_0 + \frac{\phi_i - \delta_i^k}{r} - \frac{\phi_{\max} - \delta}{r} - \frac{2F}{r}\right)\right\}$$

$$= \max\left\{0, \rho_i\left(t - T_0 - \frac{2F + \phi_{\max} - (\sum_A \delta_j^{k-1} + \sum_B \delta_j^k + \sum_C \delta_j^k) - \phi_i}{r}\right)\right\}$$

$$\ge \max\left\{0, \rho_i\left(t - T_0 - \frac{2F + \phi_{\max} - \phi_i}{r}\right)\right\}. \tag{16}$$

After t^*, flow i receives service with constant rate $r \ge \rho_i$ and therefore the above relation holds for the remaining time within the interval $(\tau_{k-1}, \tau_k]$ as well. Therefore the theorem holds for any time $t \ge T_0$.   □

From the definition of the $\mathcal{LR}$ servers and theorem 1, we conclude as the following.

Corollary 1. *Deficit Round Robin with Instant Service (DRR-IS) is an $\mathcal{LR}$ server and its latency is $(2F + \phi_{\max} - \phi_i)/r$.*

The service lower bound we calculated is in fact as tight as possible. It is sufficient to show an example that satisfies the above relation. Let us assume that connection i is activated at T_0, at which time $v-1$ flows, i.e. all the other flows are already backlogged. Moreover T_0 is exactly the start of flow g's turn. Right after the service completion of the flow g, according to DRR-IS policy, the flow i gets service. The size of the first packet of i is $\Delta\phi_i$, and the size of the second one is ϕ_i. In such a case, from the flow i, during the first round only the first packet is served, while only the second packet is served in the next round. From T_0 it takes to complete the first round $(\phi_g + \delta_g^0 - \delta_g^1 + \Delta\phi_i + \sum_{j \neq i, j \neq g}(\phi_j + \delta_j^0 - \delta_j^1))/r$. From the start of the second round to the service completion of the flow i in the second round, it takes $(\phi_g + \delta_g^1 - \delta_g^2 + \phi_i)/r$. Let us further assume the packets served during this time interval for all the other flows are of sizes as large as possible. That is to say $\delta_j^1 = 0, \forall j \neq i, \forall j \neq g, \delta_g^2 = 0$, and $\delta_j^0 = \phi_j - \Delta\phi_j, \forall j \neq i$. Further let $\max_{j \neq i} \phi_j = \phi_g$ and finally $\Delta\phi_i = \Delta\phi_j \ll F$, then from T_0 to the i's service completion in the second round it takes

$$(2\sum_{j=1}^{v}\phi_j + \phi_{\max} - \phi_i)/r, \tag{17}$$

which actually is the latency we obtained in the theorem 1. In other cases where some of flows including g are backlogged and all the other flows including i are activated during a single service time of g, the same results were obtained. We omit rigorous arguments because of the space limitation.

3.2 Fairness

It was suggested [3] the use of the difference in normalized service offered to any two flows as the measure of fairness for a scheduling algorithm. More precisely, an algorithm is considered close to fair if for any two flows i and j that are continuously backlogged in an interval of time (t_1, t_2),

$$\left| \frac{W_i^S(t_1, t_2)}{\rho_i} - \frac{W_j^S(t_1, t_2)}{\rho_j} \right| \leq \mathcal{F}^S, \tag{18}$$

where $W_i^S(t_1, t_2)$ is the service offered to flow i by a server S in the interval (t_1, t_2), and $\mathcal{F}^S$ is the *fairness measure* of server S. A server is perfectly fair if $\mathcal{F}^S$ is zero. The GPS is such a perfect server. Any packet-by-packet server, however, cannot be perfectly fair because packets must be served one at a time. It was proved that the $\mathcal{F}^S$ of DRR is $3F/r$, where F is the frame size [11]. We argue that DRR-IS has the same fairness measure as DRR.

Lemma 4. *For a DRR-IS server,*

$$\mathcal{F}^S = \frac{3F}{r}. \tag{19}$$

Proof. The proof takes a similar approach as the lemma 3.13 of [11], which is about the DRR's fairness. Let the flows i and j be backlogged since T_0. Consider an arbitrary time instant t, $t > T_0$. Without loss of generality, assume the flow i received more service than j, in $[T_0, t)$. Let k be the number of turns that i has experienced in $[T_0, t)$. For such t, the difference between the service offered to these two flows is at its maximum when i is served one more turn, the service offered to i is at its maximum, and the service offered to j is at its minimum. In other words, $W_i(T_0, t) \leq k\phi_i + \phi_i$ and $W_j(T_0, t) \geq (k-1)\phi_j - \phi_j$. The normalized service difference is therefore

$$\frac{W_i(T_0, t)}{\rho_i} - \frac{W_j(T_0, t)}{\rho_j} \leq \frac{3F}{r}, \tag{20}$$

since $\phi_i/F = \rho_i/r$. The lemma follows.   □

4   Numerical Results

We compare the latencies of well known $\mathcal{LR}$ servers, as well as the DRR-IS. We focus on a residential network environment, where the number of nodes, the number of hops to travel, and the number of flows are confined and predictable. Moreover in such networks the demand for real-time service is strong, especially for video and high quality audio applications. IEEE 802.3 Residential Ethernet Study Group [12] defines a bound for maximum end-to-end delay to be 2ms in a network of 7 hops for stringent audio applications[13]. This bound is obtained by considering human perception on interactive audio sessions, where many audio signals from various instruments and voices are generated, transmitted, processed and then heard by human. We assume the Fast Ethernet links are used across the network, therefore the link capacities are all 100Mbps. We compare the latencies of Packetized GPS, DRR, DRR-IS, and Aliquem DRR[8] in a single node, as the maximum packet (equivalently a Ethernet *frame* in a conventional expression) length varies. The Aliquem DRR is an implementation practice of a DRR with quantum size less than the maximum packet size, therefore can be considered as a generalized DRR. The maximum packet size varies from 84 to 1538 bytes, including the 12 bytes worth of inter-frame gap and 8 bytes of preamble. We consider the following two scenarios.

In the first scenario there are eight flows, demanding 10Mbps per each flow. Figure 2 depicts the latencies of $\mathcal{LR}$ servers. The quantum size of Aliquem DRR is set to 100 bytes, or to the maximum packet size when the maximum packet size is less than 100 bytes. The quantum size of DRR and DRR-IS is set to the maximum packet size. In the case where the maximum packet size is 1538 bytes, as in a normal Ethernet network, the latencies for PGPS, DRR, DRR-IS, and Aliquem DRR are 1.35ms, 2.71ms, 1.97ms, and 1.90ms, respectively. The above numerical results are for a single node, therefore the 2ms end-to-end delay requirement is never met. An immediate solution for this problem is to reduce the maximum packet size. With a reduced maximum packet size, 100 bytes for example, the latencies are 0.088ms, 0.176ms, 0.128ms, and 0.176ms,

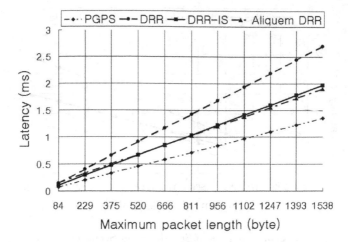

Fig. 2. Latencies of Schedulers in Scenario 1

respectively. The latency of DRR-IS is the smallest among DRR variations, since the Aliquem DRR can not have the advantage of the reduced quantum size. In this case the delay in a 7 hops network can be within the 2ms requirement, even with the leaky-bucket related delay element of the equation (1). For example, if the maximum burst size and the average rate of the leaky bucket are 100 bytes and 10Mbps, respectively, then the maximum end-to-end delay of seven DRR-IS servers' network with 100 bytes maximum packet size is 0.976ms, which is well below the 2ms requirement.

The second scenario is similar to the first one. There are two types of flows. There are thirty flows of the first type demanding 1Mbps per each flow. Five flows of the second type demanding 10Mbps each. Notice that the total offered load to an output link is 80%, which is the same as in the first scenario. Let us observe a flow from the first type, which demands 1Mbps. Figure 3 demonstrates the latencies experienced by an 1Mbps flow. The quantum size of this 1Mbps flow for Aliquem DRR is set to min(100byte, max. packet size) and for DRR and DRR-IS, it is set to the maximum packet size. The quantum sizes of 10Mbps flows should be ten times of these values. In this thiner flows scenario, the latencies are much longer. The smallest latency in this scenario is what the PGPS can provide with 83 bytes maximum packet, which is 0.682ms and far from satisfactory. This is due to the larger frame size because of the larger number of flows and the quantum size that is ten times larger than the maximum packet size for 10Mbps flows. There is a partial solution to this problem, called Stratified Round-robin, which is to hierarchically separate flows into several classes according to the demanding rates and then apply DRR only within a class, therefore eliminates any chance that a quantum size becomes much larger than the packet size [6]. This server, however, is not an $\mathcal{LR}$ server, therefore the exact maximum end to end delay bound cannot be calculated. The Instant Service policy can still be applied to Stratified RR, and an immediate performance improvement is expected.

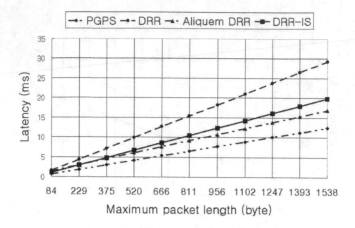

Fig. 3. Latencies of Schedulers in Scenario 2

5   Conclusion

The Instant Service (IS) policy is proved to perform well, and when applied to the basic DRR, it is shown to perform clearly better than the original by about 30% in terms of latency. In terms of fairness, the DRR and DRR-IS have the same performance. This is remarkable considering the minor modification on the enqueue process of the active list of a round robin scheduler. It can be interpreted that the IS policy improves the transient response of round-robin schedulers. While providing the reserved rates for equilibrium-state flows intact, the IS policy minimizes the penalty for flows in the transient state, near the activation time. This benefit can be easily appreciated for a flow which is activated and deactivated frequently. The IS policy may be applicable and advantageous to other schedulers than round-robin.

DRR-IS is proved to be still an $\mathcal{LR}$ server, therefore an end-to-end delay is obtainable by only summing individual latencies of each servers and the leaky-bucket related delay element. Moreover the IS policy is orthogonal to the other improvements such as the generalization of quantum size or the hierarchical round-robin. The IS policy can be applied to the basic DRR with one of these improvements at the same time.

References

1. Parekh A. and Gallager R. G.: A generalized processor sharing approach to flow control in integrated services networks: The single node case. IEEE/ACM Trans. Networking. vol. 1, no. 3 (1993) 344–357
2. Demers A., Keshav S., and Shenker S.: Analysis and Simulation of a Fair Queuing Algorithm. ACM SIGCOMM. (1989) 1–12
3. Golestani S.: A Self-clocked Fair Queueing Scheme for Broadband Applications. IEEE INFOCOM. (1994)

4. Shreedhar M. and Varghese G.: Efficient fair queueing using deficit round-robin. IEEE/ACM Trans. Networking. vol. 4, no. 3 (1996) 375–385
5. Guo C.: SRR, an O(1) Time Complexity Packet Scheduler for Flows in Multi-Service Packet Networks. ACM SIGCOMM. (2001)
6. Ramabhadran S. and Pasquale J.: Stratified Round Robin: A Low Complexity Packet Scheduler with Bandwidth Fairness and Bounded Delay. ACM SIGCOMM. (2003) 239–249
7. Yuan X. and Duan Z.: FRR: a Proportional and Worst-Case Fair Round Robin scheduler. IEEE INFOCOM. (2005)
8. Lenzini L., Mingozzi E., and Shea G.: Tradeoffs between low complexity, low latency, and fairness with deficit round-robin schedulers. IEEE/ACM Trans. Networking. vol. 12, no. 4. (2004) 681 –693
9. Kanhere S. S. and Sethu H.: On the latency bound of deficit round robin. Proceedings of the ICCCN. (2002)
10. Stiliadis D. and Varma A.: Latency-Rate servers: A general model for analysis of traffic scheduling algorithms. IEEE/ACM Trans. Networking. vol. 6, no. 5. (1998)
11. Stiliadis D.: Traffic Scheduling in Packet-Switched Networks: Analysis, Design and Implementation. Ph.D. Dissertation. U.C. Santa Cruz. (1996)
12. Residential Ethernet Study Group website, http://www.ieee802.org/3/re_study/index.html
13. Feng F. and Garner G. M.: Meeting Residential Ethernet Requirements: A Simulation Study. IEEE 802.3 Residential Ethernet Study Group. (2005)

A Compression-Boosting Transform for Two-Dimensional Data*

Qiaofeng Yang[1], Stefano Lonardi[1], and Avraham Melkman[2]

[1] Dept. of Computer Science & Engineering
University of California
Riverside, CA 92521
[2] Department of Computer Science
Ben-Gurion University of the Negev
Beer-Sheva, Israel 84105

Abstract. We introduce a novel invertible transform for two-dimensional data which has the objective of reordering the matrix so it will improve its (lossless) compression at later stages. The transform requires to solve a computationally hard problem for which a randomized algorithm is used. The inverse transform is fast and can be implemented in linear time in the size of the matrix. Preliminary experimental results show that the reordering improves the compressibility of digital images.

1 Introduction

Every day massive quantities of two-dimensional data are produced, stored and transmitted. Digital images are the most prominent type of data in this category. However, matrices over finite alphabets are used to represent all sorts of information, like graphs, database tables, geometric objects, etc. From the compression standpoint, two-dimensional data has to be treated differently than one-dimensional data. In order to obtain good compression of 2D data, one has to exploit the dependencies (or equivalently, expose the redundancies) both between the rows and the columns of the matrix.

Lossless compression algorithms are typically composed by a pipeline of independent stages, usually ending with a statistical encoder. For example, the celebrated `bzip2` employs a pipeline composed by the Burrows-Wheeler transform (BWT), a move-to-front encoder, and finally an Huffman compressor. Each step somewhat reorder the data so that redundancies get exposed and removed by the subsequent stages. The objective of the BWT is exactly that of elucidating the dependencies between adjacent symbols in the original text string.

In this paper we propose an invertible transform for two-dimensional data over an alphabet Σ. For simplicity, we assume $\Sigma = \{0, 1\}$. The extension to larger

---

* This project was supported in part by NSF CAREER IIS-0447773, and NSF DBI-0321756. AM was supported in part by the Paul Ivanier Center for Robotics Research and Production Management. A one-page abstract about this work appeared in the *Proceedings of Data Compression Conference*, Snowbird, Utah, 2005.

S.-W. Cheng and C.K. Poon (Eds.): AAIM 2006, LNCS 4041, pp. 126–137, 2006.

alphabets is immediate and is not pursued here. The goal of the transform is to "boost" the compression achieved by later stages. The transform is described by a simple recursive algorithm, which can be outlined as follows.

Given the matrix to be transformed, first search for the largest *columnwise-constant* (resp., *rowwise-constant*) submatrix, that is, a submatrix identified by a subset of rows and the columns (which are not necessarily contiguous) whose columns (resp., rows) are constant (i.e., either all 0 or 1). Reorder the rows and the columns such that the columnwise-constant (or rowwise-constant) submatrix is moved to the left-upper corner of the matrix. Recursively apply the transform on the rest of the matrix. Stop the recursion when the partition produces a matrix which is smaller than a predetermined threshold (see Figure 3 for an illustration of this process).

The intriguing question is whether this somewhat simple matrix transformation helps compression. Arguments can be made in favor or against. On one hand, each columnwise-constant (or rowwise-constant) submatrix can be represented compactly in a canonical form (first all the 0-columns, then all the 1-columns) by the list of its rows and columns. If a matrix can be decomposed into a small number of large constant submatrices, one would expect an improvement in compressibility. On the other hand, while the transform groups together portions of the matrix which are similar, the reordering can also break the local dependencies that exist in the original matrix between adjacent rows and columns. Breaking these dependencies could increase the entropy and have a negative impact on the compressibility.

The contribution of this paper is twofold. First, we present a novel invertible transform for 2D data. The design of the transform went through a series of refinements, and here we present the result of such process (Section 5). We also studied the computational cost of the transform, which turns out to be unbalanced. The inverse transform is extremely fast and simple, whereas the direct transform is very expensive. Our compression-boosting phase is therefore suitable to applications in which the data is compressed once and decompressed many times, like large repositories of digital images where images are stored and rarely modified.

The computational cost of the forward transform depends on the complexity of finding the largest columnwise-constant/rowwise-constant submatrix. In [1] we studied theoretically the general version of this problem. Although the problem turns out to be **NP**-hard, a relatively simple randomized algorithm that has good performance in practice was introduced in [1]. For completeness of presentation, we will briefly outline the algorithm in Section 4. The interested reader can refer to the original paper for more details.

Second, we study empirically the effects of the transform on the compressibility of two-dimensional data by comparing the performance of `gzip` and `bzip2` before and after the application of the transform on synthetic data and digital images. The preliminary results in Section 6 show that the transform boosts compression.

In closing, we want to point out that since our transform simply changes the representation of the data and it does not deal with the encoding problem (i.e., assigning bits to symbols), here we are not proposing a complete data compression software tool. Also, since we our transform is not optimized for digital images, the transform is not an image compression tool either. As said above, the primary use of our transform primary is as a preprocessing step to reorder the data so that the downstream compression with standard lossless encoder would be more efficient.

2   Related Works

Since we are not proposing a new compression method, we will not delve into the vast literature on lossless image compression. There are however, a few related works on the idea of reordering the columns and/or the rows of a matrix with the objective of reducing the storage space or the access time to the element of the matrix.

In [3, 4, 5], the main concern is to compress database tables by exploiting the dependencies between the columns. In [3], Buchsbaum *et al.* observe that partitioning the table into blocks of columns and compressing each block of columns individually could improve compression. The key problem is to find the optimal partition of columns. In the follow-up paper [4], the authors add the possibility of rearranging the columns. Their tool, called `pzip` outperforms `gzip` by a factor of two on database tables. Along the same line of research, the authors of [5] introduce the *k-transform* which captures the dependencies between $k + 1$ columns of a matrix. Although the problem of computing the k-transform for $k \geq 2$ is **NP**-hard, the proposed polynomial-time heuristic for the 2-transform performs remarkably well compared to `pzip`, `bzip2` and `gzip`.

The task of compressing boolean (sparse) matrices is also addressed in [6, 7, 8, 9, 10]. For example, in [9] the objective is to reorder the columns of a matrix such that the 1's in each row appear consecutively. Again, the problem of finding a reordering which minimizes the number of runs of 1's is **NP**-hard. This problem reduces to solving a traveling salesman problem for which the authors propose an heuristic algorithm. In [10] the objective is to find a reordering of both rows and columns of a boolean matrix so that the matrix can be broken into homogeneous rectangles and the description complexity involved in describing those rectangles (called *cross-association*) is minimized. The problem is defined in an information theoretical context and a two-stage heuristics algorithm is proposed.

3   Notations and Problem Definition

The input to the transform is a two-dimensional $n \times m$ matrix $X \in \{0, 1\}^{n \times m}$. The element (i, j) of X is denoted by $X_{[i,j]}$. A contiguous submatrix from row i_1 to row i_2 and from column j_1 to column j_2 is denoted by $X_{[i_1:i_2, j_1:j_2]}$.

A *row selection* of size k of X is defined as a subset of the rows $R = \{i_1, i_2, \ldots, i_k\}$, where $1 \leq i_s \leq n$ for all $1 \leq s \leq k$. Similarly, a *column selection* of size l of X is defined as a subset of the columns $C = \{j_1, j_2, \ldots, j_l\}$, where $1 \leq j_t \leq m$ for all $1 \leq t \leq l$. Given a row selection R, we say that a column j, $1 \leq j \leq m$, is *constant* with respect to R if the symbols in the j-th column of X restricted to the rows in R, are identical.

The submatrix $X_{[R,C]}$ *induced* by the pair (R, C) is defined as the matrix

$$X_{[R,C]} = \begin{vmatrix} X_{[i_1,j_1]} & X_{[i_1,j_2]} & \cdots & X_{[i_1,j_l]} \\ X_{[i_2,j_1]} & X_{[i_2,j_2]} & \cdots & X_{[i_2,j_l]} \\ \cdots & \cdots & \cdots & \cdots \\ X_{[i_k,j_1]} & X_{[i_k,j_2]} & \cdots & X_{[i_k,j_l]} \end{vmatrix}$$

A submatrix induced by a pair (R, C) is called *columnwise-constant* (resp., *rowwise-constant*) if all its columns (resp., rows) are constant. Hereafter, for brevity we will use the term *constant* submatrix to denote either a columnwise-constant or a rowwise-constant submatrix.

Example. Given the 6×6 matrix $X = \begin{vmatrix} 001011 \\ 101101 \\ 100110 \\ 101001 \\ 111101 \\ 110110 \end{vmatrix}$ over the alphabet $\Sigma = \{0, 1\}$, a selection $(R, C) = (\{2, 4, 5\}, \{1, 3, 5, 6\})$ results in the columnwise-constant submatrix $X_{[R,C]} = \begin{vmatrix} 1101 \\ 1101 \\ 1101 \end{vmatrix}$. $X_{[R,C]}$ is the largest area columnwise-constant submatrix in X.

The main computational problem is the following. Given a matrix $X \in \{0, 1\}^{n \times m}$, find a constant submatrix with the largest area. This problem is strongly related to the MAXIMUM EDGE BICLIQUE problem since a $n \times m$ binary matrix can also be interpreted as the adjacency matrix of a bipartite graph. The biclique problem requires to find the biclique which has the maximum number of edges which corresponds to the largest constant submatrix composed only of 1's. This problem was proved to be **NP**-hard in [11] by reduction to 3SAT. The weighted version of this problem was shown to be **NP**-hard by Dawande *et al.* [12]. A 2-approximation algorithm based on LP-relaxation was given in [13].

4   Finding the Largest Columnwise-Constant Submatrix

Given that the problem of finding the largest constant submatrix of 1's is **NP**-hard, it is unlikely that a polynomial time algorithm could be found. In [1] we introduced a randomized algorithm which is able to find the optimal solution with probability $1 - \epsilon$, where $0 < \epsilon < 1$. For completeness of presentation,

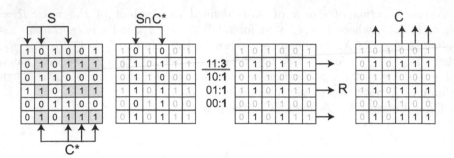

Fig. 1. An illustration of a recovery of a constant submatrix (shaded boxes), assuming $r^* = 3$

next we give a brief outline of the algorithm for the largest columnwise-constant submatrix. Rowwise-constant submatrices can be found along the same lines.

Recall that we are given a matrix $X \in \{0,1\}^{n \times m}$ and the objective of the algorithm is to discover a columnwise-constant submatrix $X_{(R^*,C^*)}$. Let us assume that the submatrix $X_{(R^*,C^*)}$ is maximal. To simplify the notation, let us call $r^* \equiv |R^*|$ and $c^* \equiv |C^*|$.

The key idea is the following. Observe that if we knew R^*, then C^* could be determined by selecting the constant columns with respect to R^*. If instead we knew C^*, then R^* could be obtained by taking the maximal set of rows which read the same symbol on the columns C^*. Unfortunately, neither R^* nor C^* is known. Our approach is to "sample" the matrix by randomly selecting subsets of columns (or rows), expecting that eventually one of the subsets will overlap with the solution (R^*, C^*).

In the following we describe how to retrieve the solution by sampling columns (one has also the choice to sample the rows). First, select a subset S of size k uniformly at random from the set of columns $\{1, 2, \ldots, m\}$. Assume for the time being that $S \cap C^* \neq \emptyset$. If we knew $S \cap C^*$, then (R^*, C^*) could be determined by the following three steps (1) select the string(s) w that appear exactly r^* times in the rows of $X_{[1:n, S \cap C^*]}$, (2) set R^* to be the set of rows in which w appears and (3) set C^* to be the set of constant columns corresponding to R^*. An example is illustrated in Figure 1.

The algorithm would work, but there are a few problems that need to be solved. First, the set $S \cap C^*$ could be empty. The solution is to try several different sets S, relying on the argument that the probability that $S \cap C^* \neq \emptyset$ *at least once* will approach one with more and more selections. The second problem is that we do not really know $S \cap C^*$. But, certainly $S \cap C^* \subseteq S$, so our approach is to check all possible subsets of S. The final problem is that we assumed that we knew r^*, but we do not. The solution is to introduce a *row threshold* parameter, called $\hat{r}$, that replaces r^*.

As it turns out, we need another parameter to avoid producing submatrices with small area which could potentially degrade the compressibility at later stages as discussed in Section 5. The *column threshold* parameter $\hat{c}$ is used to discard submatrices whose number of columns is smaller than $\hat{c}$. The algorithm

LARGEST_COLUMNWISE_CONSTANT_SUBMATRIX$(X, t, k, \hat{r}, \hat{c})$
INPUT: X is a $n \times m$ matrix over $\{0, 1\}$
       t is the number of iterations
       k is the selection size
       $\hat{r}, \hat{c}$ are the "thresholds" on the number of rows and columns, resp.

1   **repeat** t **times**
2   **select** randomly a subset S of columns such that $|S| = k$
3   **for** all subsets $U \subseteq S$ **do**
4           $D \leftarrow$ all strings composed of either 0 or 1 induced by $X_{[1:n, U]}$ that appear at least $\hat{r}$ times
5   **for** each string w in D
6               $V \leftarrow$ rows corresponding to w
7               $Z \leftarrow$ all constant columns corresponding to V
8   **if** $|Z| \geq \hat{c}$ **then save** (V, Z)
9   **return** the (V, Z) that maximizes $|V| \times |Z|$

Fig. 2. A sketch of the algorithm that discovers large columnwise-constant submatrices

considers all the submatrices which satisfy the user-defined row and column thresholds as candidates. Among all candidate submatrices, only the ones that maximize the total area are kept.

A sketch of the algorithm is shown in Figure 2. The algorithm depends on four key parameters, namely the selection size k, the row threshold $\hat{r}$, the column threshold $\hat{c}$, and the number of iterations t. A detailed discussion on how to choose each of these can be found in [1]. The worst case time complexity of the algorithm LARGEST_COLUMNWISE_CONSTANT_SUBMATRIX is $O\left(tk2^k(kn + nm)\right)$.

Because of the randomized nature of the approach, there is no guarantee that the algorithm will find the solution after a fixed number of iterations. We therefore need to choose t so that the probability that the algorithm will recover the solution in at least one of the t trials is $1 - \epsilon$, where $0 < \epsilon < 1$. In [1], we proved that the algorithm is able to find the maximal solution with probability $1 - \epsilon$ when the number of random selections t satisfies

$$t \geq \frac{\log \epsilon}{\log \left(\binom{m-c^*}{k} + \sum_{i=1}^{k} \left(1 - \left(1 - \frac{1}{|\Sigma|^i}\right)^{n-r^*}\right) \binom{c^*}{i}\binom{m-c^*}{k-i} \right) - \log \binom{m}{k}} \tag{1}$$

5   Forward Transform

As mentioned in the introduction, our strategy to boost the compressibility of two-dimensional data is to recursively decompose the input matrix based on the presence of large columnwise-constant or rowwise-constant submatrices found by the randomized search described above. The input to the recursive decomposition algorithm is the original matrix X along with user-defined thresholds ($\hat{r}$ and $\hat{c}$) and the number of iterations t. If one fixes ϵ, then the number of iterations t can be computed using equation (1).

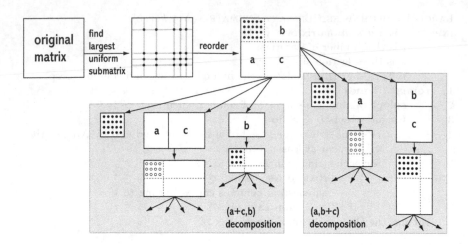

Fig. 3. Illustration of one step of the forward transform. Depending on the size of the constant submatrices in a+c, b, a, b+c either the decomposition $(a + c, b)$ or $(a, b + c)$ is chosen.

The recursive decomposition is carried out as follows. First, the procedure LARGEST_COLUMNWISE_CONSTANT_SUBMATRIX (and possibly also the procedure LARGEST_ROWWISE_CONSTANT_SUBMATRIX) is ran on X. If a constant submatrix is found, the rest of the matrix is partitioned into two submatrices depending on the size of the constant submatrices discovered at the next recursion level, as illustrated in Figure 3.

The decision whether to choose the partition (a+c, b) or the partition (a, b+c) depends on the size of the constant submatrices found in the resulting matrices a+c, b, b+c, and a. Let us call A_1, A_2, B_1, B_2 the areas of the constant submatrices found in a+c, b, a, b+c, respectively. Based on the values of A_1, A_2, B_1, and B_2 we studied three distinct criteria to determine the partition. The first is based on the condition $A_1 + A_2 > B_1 + B_2$ (hereafter called sum). The second and the third tests are $\max\{A_1, A_2\} > \max\{B_1, B_2\}$ (called max) and $A_1 > B_2$ (called indiv[1]). In all cases, if the test is true the algorithm chooses the partition $(a + c, b)$. Otherwise, the algorithm chooses the partition $(a, b + c)$. A discussion on how the test type affects the final compressed size is reported in Section 6.

Once the partition is determined, the randomized search is performed recursively on the newly formed matrices in the same manner. The recursion stops when the matrix becomes non-decomposable. We say that a matrix is *non-decomposable* if either it has less than $\hat{r}$ rows or less than $\hat{c}$ columns, or if the largest constant submatrix contained in it is smaller than $\hat{r} \times \hat{c}$.

The reason behind our choice of splitting in $(a + c, b)$ or $(a, b + c)$, instead of (a, b, c) is the following. Each time the algorithms partitions the matrix, we risk to split large constant submatrices that we could have potentially found later.

---

[1] Note that in this latter case we do not need to search in b and a.

The smaller is the number of matrices we split, the higher are the chances of finding large constant submatrices. Experimental results (not shown) confirmed our choice.

It should be noted that the user-defined thresholds ($\hat{r}$ and $\hat{c}$) play an important role in the transform. If the thresholds are too low, there is a danger of having a deep recursion tree and potentially finding a large number of tiny constant submatrices. If the thresholds are too high, there will be just a few constant submatrices. Both cases will have a negative impact on the compression. An experimental study regarding the choice of these thresholds is reported in Section 6.

6   Implementation, Experiments and Results

We now describe how the transformed data is represented. Clearly, each constant submatrix can be represented very succinctly. The column indices of columnwise-constant submatrices are reordered so that each row reads 00...0011...11. Thus, each constant submatrix can be represented by the list of rows and column indices, and where the transition from 0 to 1 takes place. Non-decomposable submatrices are saved contiguously in row-major order. The content of non-decomposable matrices is saved in a file called string.

Row and column indices of constant and non-decomposable submatrices are saved in another file called index. For each set of row and column indices, the first index is saved as it is, while the rest is saved as differences between adjacent indices. The length file is used to record the number of rows and the number of columns for constant and non-decomposable submatrices, along with a binary flag to indicate whether the submatrix is constant or non-decomposable.

The information contained in the files string, index and length allows one to invert the transform and reconstruct the original matrix. The inverse transform is simple and extremely fast. Basically, the matrix is reconstructed element by element in the order of the indices stored in index. The inverse transform was

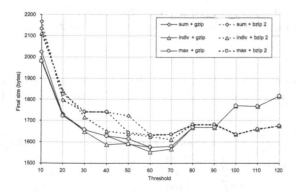

Fig. 4. The performance of the algorithm on the image **bird** for different strategies (sum, max, indiv) in choosing how to partition the matrix

implemented and tested to make sure that we had all the information necessary to recover the original matrix. The time complexity of the inverse transform is linear in the size of input.

In order to determine whether the transform improves compression, we compared the size of the file obtained by compressing the original matrix against the overall size of the files string, index and length compressed with the same program. We employed two popular lossless compression algorithms, namely gzip and bzip2.

We tested the three criteria (sum, max, indiv) discussed in Section 5 on several images and simulated data. The result on the image bird is reported in Figure 4 for different choices of the thresholds. In the majority of our experiments, the strategy indiv appeared to be the best. Therefore, all experimental tests that follow employ the indiv test.

6.1   Simulations on Synthetic Data

We generated several datasets, each composed of four random matrices of size 256×256 over a binary alphabet. In each of the four matrices we embedded one, two, three, and four columnwise-constant submatrices of size 64×64, respectively. The position and the content of each embedded submatrix were randomly chosen under the condition that the submatrices did not overlap with each other.

For each matrix we compared the compression size obtained with gzip and bzip2 before and after the transform. The performance of the transform was measured on several datasets. Table 1 shows the results averaged on all datasets. The results are very stable with respect to the choices of $\hat{r}$ and $\hat{c}$. Any choice in the range 10 to 60 produces almost identical results. The number of iterations t was set to $10,000$.

The goal of these simulations was twofold. First, it allowed us to test the ability of our randomized search to recover the embedded submatrices. Failures in recovering all the submatrices are typically due to the recursive partitioning. If the partitioning process happens to split an embedded submatrix, there is no hope of recovering it as a single piece. This observation is behind the idea of partitioning into $(a+c, b)$ or $(a, b+c)$ instead of partitioning into. (a, b, c). That may avoid splitting a potentially large constant submatrix that lies in $a + c$ or $b + c$.

Table 1. Results on 256×256 synthetic data. File matrix$_i$ contains i embedded columnwise-constant submatrices of size 64×64. Parameters: $\hat{r} = 10, \hat{c} = 10, t = 10,000$.

filename	gzip	transform+gzip	bzip2	transform+bzip2
matrix$_1$	11,121	**10,041**	11,014	**10,197**
matrix$_2$	11,111	**9,536**	11,051	**9,712**
matrix$_3$	11,094	**8,989**	10,951	**9,194**
matrix$_4$	11,061	**8,395**	10,919	**8,530**

Table 2. Comparing the compressibility of 256 × 256 binary images before and after the transform. Threshold parameters $\hat{r} = 60, \hat{c} = 60$. Iterations $t = 10,000$.

filename	size	gzip	transform+gzip	bzip2	transform+bzip2
bird	65,792	1,978	**1521**	1,778	**1581**
camera	65,792	4,330	**3693**	3,839	**3664**
lena	65,792	3,450	**3186**	3,026	**2930**
peppers	65,792	3,186	**2941**	2,757	**2671**
tulips	65,792	5,133	**4695**	4,483	**4329**

Second, this synthetic data is arguably the most favorable type of data for our transform. The "background" of the matrix is random, and therefore there are very few dependencies between rows and columns. The large majority of the dependencies are the ones created by the embedding of the constant submatrices. In some sense, this data represents the best case scenario. This is shown by a considerable improvement in the file size after the transform is applied.

6.2 Experiments on Digital Binary Images

In order to determine whether the transform can boost the compressibility of general data, we tested the transform on five 256 × 256 images downloaded from the Internet (namely bird, camera, lena, peppers, and tulips, all of which are commonly used in the data compression community). Each 8 bpp greyscale image was converted to binary by setting to black each pixel whose brightness was below 128 and to white each pixel above 128. Table 2 shows the compression results before and after the transform (for $\hat{r} = 60, \hat{c} = 60, t = 10,000$). In all the images we tested, the transform improves the lossless compression downstream.

In these experiments, we considered only columnwise-constant matrices. We tested an implementation of the transform that also searches for rowwise-constant matrices, but it did not boost the compressibility further.

Although a comparison of the results in Table 2 against a specialized lossless image tool, say JBIG, would appear appropriate, it is not. Our transform is (1) general purposed (i.e., not optimized for digital images), and (2) not a complete compression tool. As said in the introduction, we do not deal with the encoding problem (we are in fact relying on gzip and bzip2), nor we are necessarily bound to process digital images. If we were to compare the results of the table against JBIG, the performance of gzip/bzip2 would also part of the equation, and these tools have not been designed specifically to compress digital images.

Next, we tested how sensitive is the transform to the choice of the parameters $\hat{r}$ and $\hat{c}$, and to the number of iteration t. We selected the image bird, and we ran the transform on different parameter choices. We computed the total number and the average area of the columnwise-constant submatrices found (Figure 5), for several choices of $\hat{r} = \hat{c}$ and for two values for t. We also recorded the total proportion of the matrix which was covered by columnwise-constant submatrices

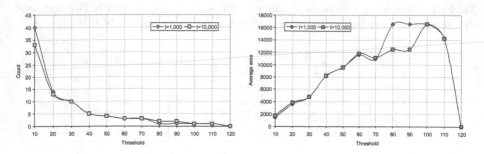

Fig. 5. LEFT: Number of columnwise-constant submatrices in image `bird`. RIGHT: Average area of columnwise-constant submatrices in image `bird`.

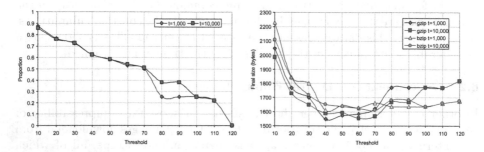

Fig. 6. LEFT: Proportion of columnwise-constant submatrices in image `bird`. RIGHT: Comparing the final compression size for the image `bird` for different choices of the threshold parameters.

(the rest is non-decomposable), and the final size of the files after compression (Figure 6). Observe that when the thresholds are low, the proportion of the matrix covered by columnwise-constant submatrices is quite high. However with low thresholds, the transform finds a large number of columnwise-constant submatrices which average area is low (Figure 5), which in turn results in large file sizes for `index` and `length`. Compared to the file `string`, files `index` and `length` are considerably harder to compress. Therefore, the consequence of choosing thresholds too low is poor compression boosting. Good compression relies on finding a balance between the gain of representing a portion of the matrix a single bit and the cost of adding the extra information necessary to reconstruct the original matrix.

The optimal value of the thresholds $\hat{r} = \hat{c}$ for the image `bird` is around 40, but other values in the range 40 to 70 achieve very similar results. We carried out the same analysis on other 256×256 images, and the same general considerations apply. With respect to the final compression, in most cases the larger is the number of iterations t, the better is the compression.

References

1. Lonardi, S., Szpankowski, W., Yang, Q.: Finding biclusters by random projections. In: Proceedings of Symposium on Combinatorial Pattern Matching (CPM'04). Volume 3109 of LNCS., Istanbul, Turkey, Springer (2004) 74–88
2. Storer, J.A., Helfgott, H.: Lossless image compression by block matching. Comput. J. **40** (1997) 137–145
3. Buchsbaum, A.L., Caldwell, D.F., Church, K.W., Fowler, G.S., Muthukrishnan, S.: Engineering the compression of massive tables: an experimental approach. In: Proceedings of the ACM-SIAM Annual Symposium on Discrete Algorithms, San Francisco, CA (2000) 213–222
4. Buchsbaum, A.L., Fowler, G.S., Giancarlo, R.: Improving table compression with combinatorial optimization. In: Proceedings of the ACM-SIAM Annual Symposium on Discrete Algorithms, San Francisco, CA (2002) 175–184
5. Vo, B.D., Vo, K.P.: Using column dependency to compress tables. In Storer, J.A., Cohn, M., eds.: Data Compression Conference, Snowbird, Utah, IEEE Computer Society Press, TCC (2004) 92–101
6. Galli, N., Seybold, B., Simon, K.: Compression of sparse matrices: Achieving almost minimal table sizes. In: Proceedings of Conference on Algorithms and Experiments (ALEX98), Trento, Italy (1998) 27–33
7. Bell, T., McKenzie, B.: Compression of sparse matrices by arithmetic coding. In Storer, J.A., Cohn, M., eds.: Data Compression Conference, Snowbird, Utah, IEEE Computer Society Press, TCC (1998) 23–32
8. McKenzie, B., Bell, T.: Compression of sparse matrices by blocked Rice coding. IEEE Trans. Inf. Theory **47** (2001) 1223 – 1230
9. Johnson, D.S., Krishnan, S., Chhugani, J., Kumar, S., Venkatasubramanian, S.: Compressing large boolean matrices using reordering techniques. In: To appear in Proceedings of International Conference on Very Large Data Bases (VLDB 2004), Toronto, Canada (2004)
10. Chakrabarti, D., Papadimitriou, S., Modha, D., Faloutsos, C.: Fully automatic cross-assocations. In: Proceedings of the Eighth ACM SIGKDD International Conference on Knowledge Discovery and Data Mining (KDD-04), ACM Press (2004) 89–98
11. Peeters, R.: The maximum-edge biclique problem is NP-complete. Technical Report 789, Tilberg University: Faculty of Economics and Business Adminstration (2000)
12. Dawande, M., Keskinocak, P., Swaminathan, J.M., Tayur, S.: On bipartite and multipartite clique problems. Journal of Algorithms **41** (2001) 388–403
13. Hochbaum, D.S.: Approximating clique and biclique problems. Journal of Algorithms **29** (1998) 174–200

Non-metric Multicommodity and Multilevel Facility Location

Rudolf Fleischer*, Jian Li, Shijun Tian, and Hong Zhu**

Department of Computer Science and Engineering
Shanghai Key Laboratory of Intelligent Information Processing
Fudan University
Shanghai 200433, China
{rudolf, lijian83, tiansj, hzhu}@fudan.edu.cn

Abstract. We give logarithmic approximation algorithms for the non-metric uncapacitated multicommodity and multilevel facility location problems. The former algorithms are optimal up to a constant factor, the latter algorithm is far away from the lower bound, but it is the first algorithm to solve the general multilevel problem. To solve the multicommodity problem, we also define a new problem, the friendly tour operator problem, which we approximate by a greedy algorithm.

1 Introduction

The facility location problem and many variants have been studied extensively in both the operation research and computer science niterature [11, 13, 17]. In the basic *uncapacitated facility location problem (UFL)* we are given a set $\mathcal{C}$ of n *clients* and a set $\mathcal{F}$ of m *facilities*. Each facility $f \in \mathcal{F}$ has an opening cost f^0, and connecting a client $c \in \mathcal{C}$ to f costs c^f. These costs can be arbitrary real numbers, although they will be positive in most applications. In this paper, all facility location problems will be uncapacitated, so we will henceforth omit 'uncapacitated' when speaking about facility location problems.

We may consider the sets $\mathcal{C}$ and $\mathcal{F}$ as the two sides of a bipartite graph. Consider a set E of edges (or *links*) between $\mathcal{C}$ and $\mathcal{F}$. Let F_E be the subset of facilities incident to at least one edge. If (c, f) is an edge in E, then we say c can *satisfy* its demand, and f *satisfies* the demand of c. E is a feasible solution if every client in $\mathcal{C}$ can satisfy its demand (i.e., every client is incident to at least one edge). The cost of E is defined as

$$cost(E) = \sum_{f \in F_E} f^0 + \sum_{(c,f) \in E} c^f \, ,$$

* The work described in this paper was partially supported by the National Natural Science Fund China (grants #60496321, #60373021, and #60573025) and the Shanghai Science and Technology Development Fund (grant #03JC14014).
** The order of authors follows the international standard of alphabetic order of the last name. In China, where first-authorship is the only important aspect of a publication, the order of authors should be Jian Li, Shijun Tian, Rudolf Fleischer, and Hong Zhu.

S.-W. Cheng and C.K. Poon (Eds.): AAIM 2006, LNCS 4041, pp. 138–148, 2006.

where the first sum is the *startup cost* of F_E and the second sum is the *connection cost* (or *link cost*). UFL is the problem of finding a solution of minimum cost.

In *metric* UFL, the link costs obey the triangle inequality. In particular, in *geometric* UFL the clients and facilities are points in the plane (or more general, in $\mathbb{R}^d$).

In this paper we will discuss four variants of non-metric UFL, two variants of the multicommodity facility location problem, the multilevel facility location problem, and the multilevel concentrator location problem. These problems are defined in Section 2. In Section 3 we state our new results and review previous related work. In Section 4 we give our new asymptotically optimal approximation algorithms for the two multicommodity facility location variants. In Section 5, we then give approximation algorithms for the multilevel facility location problem and for the multilevel concentrator location problem. We end the paper with some remarks and open problems in Section 6.

2   The Models

2.1   Multicommodity Facility Location

The *multicommodity facility location problem (*MCFL*)* generalizes UFL by introducing a set S of k different *commodities* (or *services*). In UFL, we only have a single commodity. Each client c *demands* one unit of each commodity in a subset $S_c \subseteq S$, whereas each facility f can only offer a subset $S_f \subseteq S$ of commodities (but arbitrary many units of each type). A collection of links E is a feasible solution if each client can satisfy its demand for each of its commodities. In a more general setting, we might also consider a weighted version of MCFL where clients have a certain non-negative demand for each commodity and facilities have only limited capacity for each demand.

Note that the link costs do not scale with the number of commodities served by the link. Once established, a link can be used to satisfy the demands for several commodities without additional cost. MCFL is a natural model, for example, for planning the locations of network switches (for a computer network in a large building, or telephone switchboards in a city) where we want to minimize the setup cost plus the cost of connecting each client to a switch.

We could generalize MCFL by charging an independent link cost for the commodities, i.e., if a client satisfies his demands for several commodities from one facility, it must pay the link cost for each commodity. However, this problem can be reduced to UFL by splitting each client into several clients at the same location, one for each commodity.

Another generalization of UFL is the *facility location with service installation cost problem (*FLSC*)* [14]. If facility f satisfies the demand for commodity s of some client, it must pay a one-time *installation cost* f^s for this commodity. Note that now a feasible solution must specify the links E and for each facility f the set $D_f \subseteq S_f$ of commodities provided by f, and each client must be able to satisfy its demands from some facilities that provide the commodities and have paid the respecitve startup costs. The startup cost of f is then

$$f^0 + \sum_{s \in D_f} f^s \,.$$

Ravi and Sinha called these cost functions *linear* [12].

2.2   Multilevel Facility Location

Let $k \geq 1$ be some integer. In the *k-level facility location problem (k-UFL)* we consider a $(k+1)$-layer graph, where the first layer $\mathcal{C} = \mathcal{F}_0$ is the set of clients, and the next k layers $\mathcal{F}_1, \ldots, \mathcal{F}_k$ are sets of facilities. Each edge (link) has a cost. We are interested in paths connecting a client in layer $\mathcal{F}_0$ with some facility in the last layer $\mathcal{F}_k$. The cost of such a path, call a *link path*, is the sum of its individual link costs. A feasible solution is a set E of link paths such that each node in $\mathcal{F}_0$ is incident to at least one link path. Intuitively, each client has a demand for a commodity available at any facility in layer k, which then must be routed to the client via facilities at the intermediate $k-1$ layers. Note that in this model an edge can incur multiple cost if it is used in several link paths.

If an edge only incurs cost once even if it is shared by several link paths, we are dealing with the *k-level concentrator location problem (k-LCLP)*. Here, each client must be satisfied by a facility in $\mathcal{F}_1$, each facility in $\mathcal{F}_1$ must be satisfied by a facility in $\mathcal{F}_2$, etc. Formally, our goal is to choose subsets $\emptyset \neq V_t \subseteq \mathcal{F}_t$, for $1 \leq t \leq k$, such that

$$\sum_{j \in D} \min_{k \in V_1} c_{jk} + \sum_{t=1}^{k-1} \sum_{j \in V_t} \min_{i \in V_{t+1}} c_{ji} + \sum_{t=1}^{k} \sum_{i_t \in V_t} f_{i_t}$$

is minimized.

3   Background and New Results

3.1   Multicommodity Facility Location

Facility location problems are usually NP-hard, and approximation algorithms for many variants have been studied [11, 13, 17]. The multicommodity facility location problem, however, has only been studied recently. Ravi and Sinha [12] gave a first $O(\log |\mathcal{S}|)$-approximation algorithm for metric UFL when each client can only demand a single commodity. The result generalizes to the case of clients demanding several commodities, but if they satisfy them over the same link, the link cost will also be charged several times (so this model is different from MCFL). Their result is based on an IP formulation of the problem that can be approximated by rounding fractional LP solutions.

Shmoys *et al.* [14] gave a primal-dual 6-approximation algorithm for FLSC under the assumption that facilities can be ordered by increasing installation costs, with the same order for all commodities.

We present in this paper the first approximation algorithms for non-metric MCFL and FLSC. They are purely combinatorial, not based on IP-approximations.

For MCFL we give an H_h-approximation, where h is the total number of commodities demanded by all the clients, i.e., $h = \sum_{c \in \mathcal{S}} |S_c|$. Since $h \leq nk$, this is an $O(\log(nk))$-approximation. For FLSC we give a $(3H_h)$-approximation, which is also an $O(\log(nk))$-approximation. If all facilities f have startup-cost $f^0 = 0$, the approximation ratio is only $2H_h$. We also show that our approximation ratios are asymptotically optimal. This follows easily from the non-approximability lower bound bound for the set cover problem by Feige [5].

Both algorithms are based on the well-known greedy *minimum weight set cover (SC)* approximation algorithm by Chvátal [4]. This algorithm iteratively picks the set for which the ratio of weight over newly covered elements is minimized, giving a SC approximation of ratio of H_d, where d is the number of elements in the set to be covered. As Hochbaum observed [8], the same algorithm, with the same approximation factor, can be applied to other problems as long as they can be reduced to SC and as long as it is possible to compute in every step in polynomial time the subset (or its equivalent structure) minimizing the relative weight of the newly covered elements. For UFL, this condition is fulfilled, with $\mathcal{C}$ the set to be covered, so there is an H_n-approximation for UFL [8].

We will see in Section 4 that we can also easily use the SC approximation for MCFL. However, for FLSC computing the minimum relative weight set in every step is rather difficult. Since we cannot easily compute an optimal set, we use a 3-approximation, which is the reason for the factor of 3 in the $3H_n$-approximation ratio of the FLSC algorithm. The 3-approximation is the solution of a new problem we define, the *friendly tour operator problem (FTO)*. Such a quasi-greedy approach has been used before, see for example [7].

3.2 Multilevel Facility Location

Multilevel facility location has a long history in operations research [1, 3, 9, 16]. Of course, 1-UFL is nothing but UFL. Shmoys *et al.* [15] gave the first constant factor approximation algorithm for the metric case, and the current best known result is a 1.52-approximation by Mahdian *et al.* [10]. Guha and Khuller showed that it is unlikely to be approximated within a factor of 1.463 [6].

Shmoys *et al.* [15] extended their filtering and rounding technique for metric 1-UFL to metric 2-UFL, resulting in a 3.16-approximation algorithm. Later, Aardal *et al.* [2] showed that metric k-UFL can be approximated in polynomial time by a factor of 3 for any positive integer k using a linear programming relaxation. For small values of k, better approximation algorithms are known: 1.77 for $k = 2$, 2.51 for $k = 3$, and 2.81 for $k = 4$ [18]. For non-metric 2-UFL, Zhang gave an $O(\ln n)$-approximation [18].

In this paper, we present the first approximation algorithm for general non-metric k-UFL. The approximation ratio of our algorithm is $O(\ln^k n)$. The algorithm is defined inductively, starting with the classical $O(\ln n)$-approximation for 1-UFL. In the inductive step, we again make use of the greedy SC approximation technique. With a very similar algorithm, we can also solve the k-level concentrator location problem where we just have a hierarchy of k levels of facilities.

4 Multicommodity Facility Location

4.1 Set Cover and Facility Location

We first quickly review the relationship between UFL and SC, since this is at the heart of all our algorithms. We follow loosely the exposition by Vygen [17, Section 3.1].

In the set cover problem we are given a finite set U, a family $\mathcal{X}$ of subsets of U which together cover U, and non-negative weights $c(V)$ on the sets $V \in \mathcal{X}$. The task is to find a subset $\mathcal{Y} \subseteq \mathcal{X}$ covering U of minimum total weight. SC is a special case of UFL: let the elements in U be the clients, the subsets in $\mathcal{X}$ the facilities, the weight of a set $V \in \mathcal{X}$ the startup cost of the facility, and let the link cost of client $c \in U$ to facility $V \in \mathcal{X}$ be zero if $c \in V$ and infinity if $c \notin V$. Now, every solution to UFL corresponds to a set cover of the same cost, and vice versa.

Conversely, UFL can be considered a special case of set cover. For an instance of UFL, define a *star* to be a pair (f, C) with $f \in \mathcal{F}$ and $C \subseteq \mathcal{C}$ (meaning that we link all the clients in C to facility f). The *cost* of this star is

$$f^0 + \sum_{c \in C} c^f ,$$

and its *effectiveness* is

$$\frac{f^0 + \sum_{c \in C} c^f}{|C|} ,$$

i.e., the relative cost per client in the star. Then we can define a SC instance by choosing $\mathcal{C}$ as the set U and all possible subsets of $\mathcal{C}$ as $\mathcal{X}$, where $C \subseteq \mathcal{C}$ has cost equal to the minimum cost of a star (f, C), minimized over all $f \in \mathcal{F}$. Now, an optimal solution to SC corresponds to an optimal solution to UFL of the same cost, and vice versa.

Chvátal's greedy SC approximation algorithm iteratively picks a set for which the ratio of weight over newly covered elements is minimized [4]. If we apply this algorithm to UFL, we must in every step pick the most effective star. Although there are exponentially many stars, we do not need to compute them all. Instead, we can find the most effective star among the stars (f, C_k^f), where f is an arbitrary facility, and C_k^f denotes the first k clients in a linear order with nondecreasing link cost to f, for $k = \{1, \ldots, n\}$. Having identified the most effective star, we then open the facility and henceforth disregard all clients in this star. We refer to this algorithm as the *standard star algorithm*.

4.2 Approximating MCFL

In MCFL, each client can demand several commodities. If the client decides to satisfy its demand for one commodity from a facility, then it can, without additional cost, satisfy all demands that the facility provides from that facility. This means, if we pick a star in the standard star algorithm, the facility should satisfy all unsatisfied demands of the clients in the star. Thus, we should change the definition of effectiveness of a star (f, C) to

$$\frac{f^0 + \sum_{c \in C} c^f}{\sum_{c \in C} |S_f \cap S_c|} .$$

In the definition of C_k^f we now order the clients in linear order with nondecreasing link cost divided by number of demands that could maximally be satisfied by f, i.e., we sort them by nondecreasing

$$\frac{c^f}{|S_f \cap S_c|} .$$

Theorem 1. *The modified standard star algorithm gives an H_h-approximation, where h is the total number of commodities demanded by all the clients.*

Proof. We have to show that the modified linear order of clients in the definition of C_k^f guarantees that we indeed find a most effective star. This proof is straightforward and omitted in this extended abstract.   □

4.3   The Friendly Tour Operator Problem

To solve FLSC, we must define a new problem that we need as a subroutine, the *friendly tour operator problem* (FTO). Consider a tour operator who would like to organize a tour for tourists. Each tour t incurs a fixed cost t^0 (maybe the profit of the tour operator). There is a certain finite set $\mathcal{A}$ of actions that can be arbitrarily combined in a tour. Let A_t denote the set of actions offered in tour t. Each action a incurs a cost t^a (maybe an entrance fee). There is also a set T of tourists. Each tourist x demands to participate in some set A_x of actions. He will only join the tour t if $A_x \subseteq A_t$. The total cost of t will be

$$t^0 + \sum_{a \in A_t} t^a ,$$

which is equally shared by all participants. The goal of the friendly tour operator is not to maximize his profit, but to offer a tour of minimum cost for the participants.

We could model the problem as a hypergraph problem, where the nodes are the actions and the hyperedges are the tourists. Then the problem generalizes the densest subgraph problem which is NP-hard. So we cannot solve FTO optimally in polynomial time. But we can find a good approximation to the best tour by a simple greedy algorithm, Approx-FTO.

Starting with all actions, in each step we first compute the average cost of the current action set and then discard that action (and all tourists demanding it) that maximizes the quotient of the cost of the action and the number of tourists demanding the action (i.e., intuitively we discard an action if it has high cost and is not high in demand). In the sequence of action sets computed, we then choose the one with lowest average cost.

Theorem 2. *Let d be the maximum number of actions any tourist demands. Then, Approx-FTO achieves an approximation factor of d if $t^0 = 0$ and a factor of $d + 1$ if $t^0 \geq 0$.*

Proof. Let $A^\star$ be an optimal set of actions and OPT be the value of the optimal solution. Let $T^\star$ be the number of tourists participating in the optimal tour, and let $D_b^\star$ denote how many of them are demanding action $b \in A^\star$.

Let a be the first action in $A^\star$ deleted by `Approx-FTO`. Right before this happens, let A be the current set of actions, T be the number of remaining tourists, $cost$ be the current average cost, and for any $b \in A$ let D_b denote the number of tourists demanding action b. Clearly, $D_b^\star \leq D_b$ for all b. Therefore, $\frac{t^a}{D_a} \leq OPT$, because otherwise $A^\star - \{a\}$ would be a better solution than $A^\star$.

We choose a in the next step because $\frac{t^a}{D_a} \geq \frac{t^b}{D_b}$, for all $b \in A$. Since each tourist can demand at most d actions, we have $\sum_{b \in A} D_b \leq d \cdot T$. Putting all together, we obtain

$$cost \leq \frac{t^0 + \sum_{b \in A} t^b}{T} \leq \frac{t^0}{T^\star} + \frac{d \cdot \sum_{b \in A} t^b}{\sum_{b \in A} D_b} \leq OPT + d \cdot \frac{t^a}{D_a} \leq (d+1) \cdot OPT \, .$$

If $t^0 = 0$, the first OPT term vanishes and we get a d-approximation.   □

As the following example shows, our analysis of `Approx-FTO` is tight if $d = 2$. In this case, we can model `FTO` as a graph problem with actions as nodes and edges as tourists. We assume startup cost $t^0 = 0$. Consider the graph G which is the union of $K_{n,n}$ and S_{2n}, where $K_{n,n}$ is the complete bipartite graph with node partitions U and V, where $|U| = |V| = n$, and S_{2n} is a star with $2n + 1$ nodes, namely a center node v and $2n$ leaves. Each node in U has cost $1 + \epsilon$, where $\epsilon > 0$ is sufficiently small. The cost of v is 2, while the leaves all have cost $\frac{1}{n}$. The nodes in V have cost zero. The optimal solution is in this case the $K_{n,n}$, with minimum average cost $\frac{1+\epsilon}{n}$. But `Approx-FTO` will first delete a node in $K_{n,n}$ (which has maximum ratio $\frac{1+\epsilon}{n}$) and eventually find S_{2n} as the solution with average cost $\frac{2 + 2n \cdot \frac{1}{n}}{2n} = \frac{2}{n}$.

4.4   Approximating FLSC

In `FLSC`, each facility has some additional startup cost for providing a commodity. Therefore, it may now happen that a client satisfies one demand from one facility but a second demand from another facility although the first facility could also satisfy the second demand (but its startup cost for this demand is too high).

We must redefine cost and effectiveness of a star, and even stars itself. Consider a facility f at some step of the algorithm. If it had been used before, its startup cost is now zero. If some of its commodities are already in use from earlier clients, their startup costs are also zero. A star is now a triple (f, C, S), where S is a subset of commodities provided by the star. We may assume that S always includes all commodities that are already in use at the facility (they can now be used for free by other clients). The cost of the star is then defined as

$$f^0 + \sum_{s \in S} f^s + \sum_{c \in C} c^f \, ,$$

and its *effectiveness* is

$$\frac{f^0 + \sum_{s \in S} f^s + \sum_{c \in C} c^f}{\sum_{c \in C} |S \cap S_c|}.$$

After choosing a most effective star, we only discard the demands of the clients that have been satisfied (a client can be discarded when all its demands are satisfied).

The problem is how to find a most effective star in polynomial time. There does not seem to be a natural linear order of clients in the definition of C_k^f that guarantees that we indeed find the most effective star among the C_k^f. Since we cannot find the best star, we approximate it. Note that to compute a most effective star we only have to solve an FTO for each facility f and then choose the cheapest of all of them. To be more precise, for fixed f, the FTO uses $t^0 = f^0$. There are $n + k$ actions, one for each commodity and one for each link from f to a client. The cost of an action is the corresponding cost in FLSC. For each client $c \in C$ and unsatisfied comodity $s \in S_c$, there is a tourist demanding the two actions c and the link from f to c.

Theorem 3. *The modified standard star algorithm using* Approx-FTO *as a subroutine to approximate a most efficient star gives a* $2H_h$-*approximation, where h is the total number of commodities demanded by all the clients, if $f^0 = 0$ for all $f \in \mathcal{F}$, and a $3H_h$-approximation in the general case.*

Proof. The theorem follows from the standard star algorithm together with the approximation of the most effective star given in Theorem 2. $\qquad\square$

4.5   Lower Bounds

FLSC is clearly a generalization of MCFL, so any lower bound for the approximation factor of MCFL is also a lower bound for FLSC.

Theorem 4. *There is no polynomial approximation algorithm for* MCFL *and* FLSC *with an approximation factor of* $(1 - \epsilon) \cdot max\{\ln n, \ln k\}$, *for any $\epsilon > 0$, where n is the number of clients and k is the number of commodities.*

Proof. We give two reductions from SC. Let $|U| = n$. Recall that there is no polynomial time approximation algorithm for SC with an approximation factor of $(1 - \epsilon) \cdot \ln n$, for any $\epsilon > 0$, unless $NP \subseteq DTIME[n^{O(\log \log n)}]$ [5].

The reduction given in Subsection 4.1, where we have a single commodity and clients correspond to elements in U, gives a lower bound of $\ln n$.

In the second reduction, let each commodity correspond to a unique element in U. There is only one client demanding all commodities. For each subset in $\mathcal{X}$, there is a facility with startup cost 1 providing the corresponding commodities. All connection costs are zero. Now any set cover corresponds to a MCFL solution of the same cost. Thus, we cannot approximate MCFL with a factor better than $\ln k$. $\qquad\square$

5 k-Level Facility Location

We must define a more general version of k-UFL, k-UFL$_\ell$, which has an additional input parameter ℓ. In this problem, we can first choose s subset of ℓ clients which is then optimally served by some set of facilities. Note that k-UFL$_m$ is just the original k-UFL.

We define our approximation algorithm for k-UFL inductively. First, we give an $O(\ln \ell)$-approximation algorithm for 1-UFL$_\ell$. Then we show how to lift an $O(\ln^{k-1} \ell)$-approximation for $(k-1)$-UFL$_\ell$ up to an $O(\ln^k \ell)$-approximation for k-UFL$_\ell$.

5.1   Approximating 1-UFL$_\ell$

The $\ln \ell$-approximation algorithm for 1-UFL$_\ell$ is very similar to the greedy algorithm for 1-UFL. When we compute the most effective star for facility f, we only consider sets C_k^f for $k = 1, \ldots, \ell$, and we stop when we have satisfied ℓ clients.

Theorem 5. *The modified standard star algorithm computes a $\ln \ell$-approximation for 1-UFL$_\ell$, for any $1 \leq \ell \leq m$.*   □

5.2   Approximating k-UFL$_\ell$

Suppose we have an approximation algorithm APPROX-$(k-1)$-UFL$_\ell$ for $(k-1)$-UFL$_\ell$ for every $1 \leq \ell \leq n$. Then we can construct an algorithm APPROX-k-UFL$_\ell$ for k-UFL$_\ell$ as follows.

Consider a fixed facility $f \in \mathcal{F}_k$. We construct an instance for $(k-1)$-UFL$_\ell$ as follows. The set of clients remains unchanged, also the set of facility levels $\mathcal{F}_1, \ldots, \mathcal{F}_{k-1}$. What changes is the connection cost between $\mathcal{F}_{k-2}$ and $\mathcal{F}_{k-1}$. We increase the cost of each original edge (u, v) between the two levels by the cost of the original edge (v, f). Intuitively, we are extending the last edge on a path from a client to a node in level $k-1$ by the edge leading to f in level k.

In the standard star algorithm, we would now compute, for each facility, the best way to connect it with $1, 2, 3, \ldots$ clients, and then choose the cheapest star. Here we cannot easily compute these values. Instead, we again approximate them.

Let $cost(f, j)$ be the cost of an approximation computed by APPROX-$(k-1)$-UFL$_j$, for $1 \leq j \leq \ell$. We compute all these values for all f and j and determine the smallest one. This tells us which facility f in level k to choose. We choose all the facilities and connections computed in the corresponding approximation of the $(k-1)$-level problem, and we connect f to all facilities chosen on level $k-1$.

Theorem 6. *If APPROX-$(k-1)$-UFL$_\ell$ can achieve an approximation factor of $O(\ln^{k-1} \ell)$, for all $1 \leq \ell \leq n$, then APPROX-k-UFL$_\ell$ computes an $O(\ln^k \ell)$-approximation.*   □

Theorem 7. *There exists a $\ln^k n$-approximation algorithm for k-UFL.*   □

5.3   The k-Level Concentrator Location Problem

It is not hard to modify our algorithm for k-UFL to approximate k-LCLP. The only change is in the inductive step when we change the connection costs of edges between layers $k - 2$ and k. Instead, we now increase the startup costs of facilities on layer $k - 1$ by the cost of the edge to facility f on layer k.

Theorem 8. *There exists a* $\ln^k n$-*approximation for* k-LCLP. $\square$

6   Conclusions

We presented the first logarithmic approximation algorithms for the non-metric multicommodity facility location problem. Note that in our model the connection costs do not scale with the number of commodities that use a connection. This actually generalizes the case where connection costs scale. For FLSC, our algorithms have an additional constant factor of 2 or 3, which may not be necessary for an optimal approximation algorithm.

We also presented the first poly-logarithmic approximation algorithm for the non-metric k-level facility location problem. We conjecture that this problem admits a logarithmic approximation for any $k \geq 1$.

Acknowledgements

We would like to thank the reviewers for reading our hastily prepared last-minute submission, for which we apologize. Their comments helped us to improve the presentation of our results.

References

1. K. Aardal. *On the solution of one and two-level capacitated facility location problems by the cutting plane approach.* Ph.D. thesis, Université Catholique de Louvain, Louvain-la-Neuve, Belgium, 1992.
2. K. Aardal, F. A. Chudak, and D. B. Shmoys. A 3-approximation algorithm for the k-level uncapacitated facility location problem. *Information Processing Letters*, 72(5-6):161–167, 1999.
3. K. Aardal, M. Labbé, J. Leung, and M. Queyranne. On the two-level uncapacitated facility location problem. *INFORMS Journal on Computing*, 8:289–301, 1996.
4. V. Chvátal. A greedy heuristic for the set cover problem. *Mathematics of Operations Research*, 4:233–235, 1979.
5. U. Feige. A threshold of $\ln n$ for approximating set-cover. *Journal of the ACM*, 45(4):634–652, 1998.
6. S. Guha and S. Khuller. Greedy strikes back: Improved facility location algorithms. *Journal of Algorithms*, 31:228–248, 1999.
7. S. Guha and S. Khuller. Improved methods for approximating node weighted Steiner trees and connected dominating sets. *Information and Computation*, 150:57–74, 1999.

8. D. S. Hochbaum. Heuritics for the fixed cost median problem. *Mathematical Programming*, 22(2):148–162, 1982.
9. L. Kaufmann, M. vanden Eede, and P. Hansen. A plant and warehouse location problem. *Operational Research Quaterly*, 28:547–557, 1977.
10. M. Mahdian, Y. Ye, and J. Zhang. A 1.52-approximation algorithm for the uncapacitated facility location problem. In *Proceedings of the 5th International Workshop on Approximation Algorithms for Combinatorial Optimization (APPROX'02)*. Springer Lecture Notes in Computer Science 2462, pages 127–137, 2002.
11. P. Mirchandani and R. Francis, editors. *Discrete Location Theory*. John Wiley & Sons, Inc., Chichester, 1990.
12. R. Ravi and A. Sinha. Multicommodity facility location. In *Proceedings of the 15th ACM-SIAM Symposium on Discrete Algorithms (SODA'04)*, pages 342–349, 2004.
13. D. B. Shmoys. Approximation algorithms for facility location problems. In *Proceedings of the 3rd International Workshop on Approximation Algorithms for Combinatorial Optimization (APPROX'00)*. Springer Lecture Notes in Computer Science 1913, pages 27–33, 2000.
14. D. B. Shmoys, C. Swamy, and R. Levi. Facility location with service installation costs. In *Proceedings of the 15th ACM-SIAM Symposium on Discrete Algorithms (SODA'04)*, pages 1088–1097, 2004.
15. D. B. Shmoys, E. Tardos, and K. I. Aardal. Approximation algorithms for facility location problems. In *Proceedings of the 29th ACM Symposium on the Theory of Computation (STOC'97)*, pages 265–274, 1997.
16. D. Tcha and B. Lee. A branch-and-bound algorithm for the multi-level uncapacitated location problem. *European Journal on Operations Research*, 18:35–43, 1984.
17. J. Vygen. Approximation Algorithms for Facility Location Problems (Lecture Notes). Technical Report Report No. 05950-OR, Research Institute for Discrete Mathematics, University of Bonn, 2005.
18. J. Zhang. Approximating the two-level facility location problem via a quasi-greedy approach. In *Proceedings of the 15th ACM-SIAM Symposium on Discrete Algorithms (SODA'04)*, pages 808–817, 2004.

Sublinear Time Width-Bounded Separators and Their Application to the Protein Side-Chain Packing Problem⋆

Bin Fu[1,2] and Zhixiang Chen[3]

[1] Dept. of Computer Science, University of New Orleans, LA 70148, USA
[2] Research Institute for Children, 200 Henry Clay Avenue
New Orleans, LA 70118
fu@cs.uno.edu
[3] Dept. of Computer Science, University of Texas - Pan American
TX 78539, USA
chen@cs.panam.edu

Abstract. Given $d > 2$ and a set of n grid points Q in $\Re^d$, we design a randomized algorithm that finds a w-wide separator, which is determined by a hyper-plane, in $O(n^{\frac{2}{d}} \log n)$ sublinear time such that Q has at most $(\frac{d}{d+1} + o(1))n$ points one either side of the hyper-plane, and at most $c_d w n^{\frac{d-1}{d}}$ points within $\frac{w}{2}$ distance to the hyper-plane, where c_d is a constant for fixed d. In particular, $c_3 = 1.209$. To our best knowledge, this is the first sublinear time algorithm for finding geometric separators. Our 3D separator is applied to derive an algorithm for the protein side-chain packing problem, which improves and simplifies the previous algorithm of Xu [26].

1 Introduction

The work in this paper aims for efficient identification of width-bounded separators for a given set of points in the d-dimensional Euclidean space and their applications to intractable practical problems. Intuitively, a width-bounded separator utilizes a simple structured hyper-plane to divide the set into two *"balanced"* subsets, while at the same time maintaining a *"low density"* of the set within a given distance to the hyper-plane. This new notion of separators was initially introduced by Fu in [11], and it was shown that these separators are very suitable in solving a number of distance-bounded geometric problems such as the protein folding problem in the HP model in [10] and some other intractable problems in [11, 6].

The main contributions of this paper are summarized as follows:

In section 5, we present an $O(n^{\frac{2}{d}} \log n)$ sublinear time randomized algorithm for finding a with-bounded separator in the dimensioinal Euclidean space $\Re^d$

⋆ This research is supported by Louisiana Board of Regents fund under contract number LEQSF(2004-07)-RD-A-35, and in part by NSF Grant CNS-0521585.

for $d > 2$. To our best knowledge, this is the first sublinear time algorithm for finding geometric separators. For many other geometric problems, a higher dimension brings higher computational complexity. However, it is interesting to notice that the exponent of our algorithm's computational complexity is reversely propotional to the dimension of the space.

In section 6, we exhibit an application of our sublinear time separator to the protein side-chain packing problem. One of the most fundamental problems in the molecular biology is to predict a protein's 3D structure when given its 1D amino-acid sequence. Although much effort has been made for decades, this problem remains unsolved. An important component of the general protein structure prediction problem is the protein side-chain packing problem. It determines the side-chain positions onto the fixed backbone [23]. This problem has been proved to be NP-complete [1]. Recently, a $r_{ave}^{O(n^{\frac{2}{3}} \log n)}$ time algorithm was shown by Xu [26], where r_{ave} is the average number of side-chain rotamers in a protein. We apply width-bounded separators to the protein side-chain packing problem. The length of side-chain of each amino acid is small compared to the size of one protein. Two side-chains in a protein molecular do not interact with each other if their distance is slightly larger than the sum of their lengths according to models used in (e.g. [4, 5, 26]). Using our width-bounded separators, we obtain an algorithm with computational time $r_{\max}^{O(n^{\frac{2}{3}})}$, where $r_{\max}$ is the maximal number of side-chain rotamers among a protein. Since the number of rotamers is usually small, we assume both r_{ave} and r_{max} are constants, hence our new algorithm has a better complexity bound.

2   The Related Work

There have been extentive efforts on finding separators due to their critical roles in many issues of algorithm design and analysis. Because of space limit we cannot give a comprehensive review of the related work but list some representative results in this area. Lipton and Tarjan [16] proved that every n vertex planar graph has at most $\sqrt{8n}$ vertices whose removal separates the graph into two disconnected parts of size at most $\frac{2}{3}n$. Their $\frac{2}{3}$-separator has been improved by a series of papers [7, 12, 2, 8] with the best record $1.97\sqrt{n}$ by Djidjev and Venkatesan [8]. Spielman and Teng [25] showed a $\frac{3}{4}$-separator with size $1.82\sqrt{n}$ for planar graphs. Separators for more general graphs were derived in [13, 3, 22]. A planar graph can be induced by a set of non-overlapping discs on the plane such that every vertex corresponds to a disc center and each edge corresponds to a tangent relationship between two discs. The separator developed by Miller, Teng and Vavasis [17] is a generalization of planar graph separators to the d-dimensional Euclidean space. Some $O(\sqrt{k \cdot n})$ size separators for k-thick systems and the related algorithms were derived in [18, 19, 17, 24].

The study of width-bounded separators were initiated by Fu in [11] and has yielded successful applications in [10, 6]. Our width-bounded geometric separator has some interesting advantages over previous geometric separators such as the popular geometric separator by Miller, Teng and Vavasis [17]. First, the width-bounded separator has a simple linear structure as the separator is determined by a hyper-plane and a width parameter w, but Miller *et al.*'s separator is a sphere, which can be also found in linear time [9]. The linear structure is very crucial for us in deriving sublinear time algorithm in this paper. Second, the width-bounded separator has a smaller constant in its size upper bound factor than other separators. The constant factor was not clearly given in Miller *et al.*'s separator. Furthermore, their separator only has a balance condition bounded by $\frac{d+1}{d+2}n$ due to their transformation to a higher dimension, while The balance condition of the width-bounded separator is bounded by $\frac{d}{d+1}n$. Third, the width-bounded separator can be used to deal with an arbitrary set of points via using a set of grid points and weights to characterize the distribution of points from the input set.

3   Notations, Definitions, and Width-Bounded Separators

For any finite set A, $|A|$ denotes the number of elements in A. Let $\Re$ be the set of all real numbers. For two points p_1, p_2 in the d-dimensional Euclidean space $\Re^d$, $\mathrm{dist}(p_1, p_2)$ is the Euclidean distance between p_1 and p_2. For a set $A \subseteq \Re^d$, $\mathrm{dist}(p_1, A) = \min_{q \in A} \mathrm{dist}(p_1, q)$. The *diameter* of any $P \subseteq \Re^d$ is $\max_{p_1, p_2 \in P} \mathrm{dist}(p_1, p_2)$. For $a > 0$ and a set A of points in $\Re^d$, if the distance between every two points in A is at least a, then A is called a-*separated*. For $\epsilon > 0$ and a set Q of points in $\Re^d$, an ϵ-*sketch* of Q is another set P of points in $\Re^d$ such that each point in Q has a distance $\leq \epsilon$ to some point in P. We say P is a sketch of Q if P is an ϵ-sketch of Q for some constant $\epsilon > 0$ (that does not necessarily depend on the size of Q). A sketch set is usually a 1-separated set such as a grid point set. A weight function $w : P \to [0, \infty)$ is often used to measure the density of Q near each point in P. Let $f : \Re^d \to \Re$ be a smooth function. Its *surface* is the set $L(f) = \{v \in \Re^d | f(v) = 0\}$. A *hyper-plane* in $\Re^d$ through a fixed point $p_0 \in \Re^d$ is defined by the equation $(p - p_0) \cdot v = 0$, where v is a normal vector of the plane and " $\cdot$ " is the usual vector inner product. A hyper-plane in $\Re^d$ is determined by $L(f)$ for some linear function $f : \Re^d \to \Re$.

Definition 1. *Given any $Q \subseteq \Re^d$ with a sketch $P \subseteq \Re^d$, a constant $a > 0$, and a weight function $w : P \to [0, \infty)$, an a-wide-separator is determined by the surface $L(f)$ for some linear function $f : \Re^d \to \Re$. The separator has two measurements for its quality of separation: (1)* $\mathrm{balance}(L(f), Q) = \frac{\max(|Q_1|, |Q_2|)}{|Q|}$*, where $Q_1 = \{q \in Q | f(q) < 0\}$ and $Q_2 = \{q \in Q | f(q) > 0\}$; and (2)* $\mathrm{density}(L(f), P, \frac{a}{2}, w)$*, where in general* $\mathrm{density}(A, P, x, w) = \sum_{p \in P, \mathrm{dist}(p, A) \leq x} w(p)$ *for any $A \subseteq \Re^d$ and $x > 0$. When f is fixed or no confusion arises, we use* $\mathrm{balance}(L, Q)$ *and* $\mathrm{density}(L, P, \frac{a}{2}, w)$ *to stand for* $\mathrm{balance}(L(f), Q)$ *and* $\mathrm{density}(L(f), P, \frac{a}{2}, w)$*, respectively.*

Definition 2. *A (b, c)-partition of $\Re^d$ divides the space into a disjoint union of regions $P_1, P_2, \ldots$, such that each P_i, called a regular region, has a volume of b*

and a diameter $\leq c$. A (b,c)-regular point set A is a set of points in $\Re^d$ with a (b,c)-partition $P_1, P_2, \ldots$, such that each P_i contains at most one point from A. For two regions A and B, if $A \subseteq B$ $(A \cap B \neq \emptyset)$, we say B contains (intersects resp.) A.

Let $B_d(r, o)$ be the d-dimensional ball of radius r at center o. Its volume is $V_d(r) = \frac{2^{(d+1)/2}\pi^{(d-1)/2}}{1 \cdot 3 \cdots (d-2) \cdot d} r^d$ if d is odd, or $\frac{2^{d/2}\pi^{d/2}}{2 \cdot 4 \cdots (d-2) \cdot d} r^d$ otherwise. Let $V_d(r) = v_d \cdot r^d$, where v_d is a constant for the fixed dimension d. In particular, $v_1 = 2, v_2 = \pi$ and $v_3 = \frac{4\pi}{3}$. We will use the following well-known fact that can be easily derived from Helly Theorem (see [21]).

Lemma 1. *For an n-element set P in the d-dimensional space $\Re^d$, there is a point q with the property that any half-space that does not contain q, covers at most $\frac{d}{d+1}n$ elements of P. (Such a point q is called a centerpoint of P.)*

Definition 3. *Let $a > 0$, p and o be two points in $\Re^d$. Define $Pr_d(a, p_0, p)$ to be the probability that the point p has $\leq a$ perpendicular distance to a random hyper-plane L through the point p_0. Define function $f_{a,p,o}(L) = 1$ if p has a distance $\leq a$ to the hyper-plane L through o, or 0 otherwise. The expectation of function $f_{a,p,o}(L)$ is $E(f_{a,p,o}(L)) = Pr_d(a, o, p)$. Assume $P = \{p_1, p_2, \ldots, p_n\}$ is a set of n points in $\Re^d$ and each p_i has weight $w(p_i) \geq 0$. Define function $F_{a,P,o}(L) = \sum_{p \in P} w(p) f_{a,p,o}(L)$.*

We give an upper bound for the expectation $E(F_{a,P,o}(L))$ for $F_{a,P,o}(L)$ in the lemma below.

Lemma 2. *[11] Let $d \geq 2$. Let o be a point in $\Re^d$, $a, b, c > 0$ be constants and $\epsilon, \delta > 0$ be small constants. Assume that $P_1, P_2, \ldots$, form a (b,c)-partition for $\Re^d$, and the weights $w_1 > \cdots > w_k > 0$ satisfy $k \cdot \max_{i=1}^{k}\{w_i\} = O(n^\epsilon)$. Let P be a set of n weighted (b,c)-regular points in a d-dimensional plane with $w(p) \in \{w_1, \ldots, w_k\}$ for each $p \in P$. Let n_j be the number of points $p \in P$ with $w(p) = w_j$ for $j = 1, \ldots, k$. We have $E(F_{a,P,o}(L)) \leq (k_d \cdot (\frac{1}{b})^{\frac{1}{d}} + \delta) \cdot a \cdot \sum_{j=1}^{k} w_j \cdot n_j^{\frac{d-1}{d}} + O(n^{\frac{d-2}{d}+\epsilon})$, where $k_d = \frac{d \cdot h_d}{d-1} \cdot v_d^{\frac{1}{d}}$ with $h_d = \frac{2(d-1)v_{d-1}}{d \cdot v_d}$. In particular, $k_2 = \frac{4}{\sqrt{\pi}}$ and $k_3 = \frac{3}{2}\left(\frac{4\pi}{3}\right)^{\frac{1}{3}}$.*

Definition 4. *Let $a_1, \ldots, a_d > 0$ be positive constants. A $(a_1, \ldots, a_d)$-grid regular partition divides $\Re^d$ into a disjoint union of $a_1 \times \cdots \times a_d$ rectangular regions. A $(a_1, \ldots, a_d)$-grid (regular) point is a corner point of a rectangular region. Under certain translation and rotation, each $(a_1, \ldots, a_d)$-grid regular point is represented as $(a_1 t_1, \ldots, a_d t_d)$ for some integers $t_1, \ldots, t_d$. For a point $p = (x_1, \ldots, x_d) \in \Re^d$, if $x_1, \ldots, x_d$ are all integers, then p is simply called a grid point (it is a $(1, \ldots, 1)$-grid regular point). For each point q and a hyper-plane L in $\Re^d$, define $sd(q, L)$ to be the signed distance from q to L, which is $sd(q, L) = (q - q_0) \cdot v_L$, where q_0 is a point on L, and v_L is the normal vector of the plane L with the first nonzero coordinate to be positive.*

For a hyper-plane L in $\Re^d$, if L is through a point q_0 and has the normal vector v, then it has linear equation $(u - q_0) \cdot v = 0$. If $q \in \Re^d$ and $l_q = sd(q, L)$, then the hyper-plane L' through q and parallel to L has equation $(u - (q_0 + l_q v)) \cdot v = 0$. We use $L(l_q)$ to represent such a hyper-plane L'.

For an interval $I \subseteq R$, $\|I\|$ is the length of I. For example, $\|[a, b)\| = b - a$. We often use $Pr(E)$ to represent the probability of an event E. For a real number x, $\lfloor x \rfloor$ is the largest integer $y \leq x$, and $\lceil x \rceil$ are the least integer $z \geq x$. For an interval $[a, b] \subseteq R$, define $center([a, b])$ to be $\frac{a+b}{2}$.

Lemma 3. *Let P be a finite set of points in $\Re^d$ and q_0 be a fixed point in $\Re^d$. Then for a random hyper-plane L through q_0, $Pr(sd(p_1, L) = sd(p_2, L)$ for $p_1, p_2 \in P$ with $p_1 \neq p_2) = 0$.*

Proof. A random hyper-plane L through a fixed point q_0 can be characterized by the equation $(q - q_0) \cdot v_L = 0$, where v_L is the normal vector of L. Each unit vector can be considered as a point of the surface of the unit ball $B_d(1, o)$, where $o = (0, \ldots, 0)$ is the origin point. The surface area size of $B_d(r, o)$ is equal to $\frac{dV_d(r)}{d_r} = dv_d r^{d-1}$. The surface area of $B_d(r, o)$ is of dimension $d - 1$.

For two fixed points p_1 and p_2, if $sd(p_1, L) = sd(p_2, L)$, then $(p_1 - q_0) \cdot v_L = (p_2 - q_0) \cdot v_L$. It implies that $(p_1 - p_4) \cdot v_L = 0$. Consider the sub-area on the surface of $B(1, o)$: $\{v | (p_1 - p_2) \cdot v = 0$ and $v \cdot v = 1\}$, which is the intersection between a plane $(p_1 - p_2) \cdot v = 0$ and $B_d(1, o)$, and is of dimension $d - 2$. It is easy to see that it has area size 0 in the d-dimensional space. The lemma follows since the union of a finite number of areas of area size 0 still has 0 area size. $\square$

4   An Overview of Our Techniques

Given any set Q of points in $\Re^d$ with a sketch P, the idea of our techniques for finding an a-width-bound separator is to transform the problem from the d-dimensional space to the 1-dimensional space. By Lemma 1 and Lemma 2, we can see the existence of a hyper-plane that satisfies both the balance and the density conditions. Lemma 2 gives an upper bound on the expectation of $F_{a,P,o}(L)$. By Markov's inequality, $Pr(F_{a,P,o}(L) > (1 + \alpha)E(F_{a,P,o}(L))) \leq \frac{1}{1+\alpha}$. Thus, a random hyper-plane L has probability $\geq 1 - \frac{1}{1+\alpha} = \frac{\alpha}{1+\alpha}$ that $F_{a,P,o}(L) \leq (1 + \alpha)E(F_{a,P,o}(L))$. The chance is amplified if we repeat the random selection of the hyper-plane L multiple times.

Let $n_P = |P|$ and $n_Q = |Q|$. After a hyper-plane L is fixed, we try to find another hyper-plane L' that is parallel to L. We want L' to guarantee the desired balance and density conditions. To do so, we compute signed distances for all the points in Q and P to the hyper-plane L. Those signed distances are all different for the points in Q and, respectively, for the points in P (by Lemma 3). These signed distances are all in the 1-dimensional real axis, and finding L' can be done via finding a "right position" among these distances, hence this transforms the problem from the d-dimensional space into to the 1-dimensional space as follows: Find the interval b $[D_{1,d+1}, D_{d,d+1}]$ such that both the left side $(-\infty, D_{1,d+1})$ and the right side $(D_{d,d+1}, +\infty)$ have roughly $\frac{n_Q}{d+1}$ signed distances from Q to L.

So, every hyper-plane L' (parallel to L) with a signed distance in $[D_{1,d+1}, D_{d,d+1}]$ to L guarantees the balance condition. For an interval I, we compute its weight as the sum of the weights of the points of P with their signed distances in I. We then look for an interval $[x - a, x + a]$ that has $x \in [D_{1,d+1}, D_{d,d+1}]$ and the smallest weight. Finally, we let L' be a hyper-plane with a signed distance x to L. The balance boundaries $D_{1,d+1}$ and $D_{d,d+1}$ can be detected by sampling a small number of points from Q. Using the Chernoff bound, we have a high probability that there is a small fraction difference from the exact boundaries. Similarly, the desired interval can be also detected by sampling a small number of points from P.

5   The Sublinear Time Randomized Algorithm

We use the following well-known Chernoff bound (see [20] for a proof) and simplied version in Lemma 4. The proofs of many lemmas are omitted in the conference version of this paper and will be included in the full version of this paper.

Theorem 1. [20] Let $X_1, \cdots, X_n$ be n independent random $0, 1$ variables, where X_i takes 1 with probability p_i. Let $X = \sum_{i=1}^{n} X_i$, and $\mu = E[X]$. Then for any $\delta > 0$, (1) $Pr(X < (1-\delta)\mu) < e^{-\frac{1}{2}\mu\delta^2}$, and (2)$Pr(X > (1+\delta)\mu) < \left[\frac{e^\delta}{(1+\delta)^{(1+\delta)}}\right]^\mu$.

Lemma 4. Let $X_1, \cdots, X_n$ be n independent random $0, 1$ variables, where X_i takes 1 with probability p. Let $X = \sum_{i=1}^{n} X_i$. Then for any $\frac{1}{3} > \epsilon > 0$, (1) $Pr(X < pn - \epsilon n) < e^{-\frac{1}{2}n\epsilon^2}$, and (2)$Pr(X > pn + \epsilon n) < e^{-\frac{1}{3}n\epsilon^2}$.

Theorem 2. Let $d \geq 2$ be the fixed dimension number and v be a positive parameter. Let $a, b, c > 0$ be constants and $\delta, s_1, s_2 > 0$ be small constants. Let Q be another set of n_Q points in $\Re^d$, and P be a set of n_P (b, c)-regular points, which form a sketch for Q. Let $w_1 > w_2 \cdots > w_k > 0$ be positive weights with $k \cdot w_1 = O(n_P^{s_1})$, $\frac{w_1}{w_k} = o(n_P^{\frac{1}{d}})$, $\frac{k}{w_k} = O(n_P^{s_2})$, and w be a mapping from P to $\{w_1, \cdots, w_k\}$. There exists an $O(v^2 \cdot (n_P^{\frac{d}{d+1}+2(s_1+s_2)} \cdot \log n_P + \log n_Q))$ time randomized algorithm to find a hyper plane M with probability $\geq 1 - \frac{1}{2^v}$ such that (1) each half space has $\leq (\frac{d}{d+1} + \delta)n_Q$ points from Q, and (2) $\sum_{p \in P}$ and $\text{dist}(p,M) \leq a$ $w(p) \leq \left(k_d \cdot b^{\frac{-1}{d}} + \delta\right) \cdot a \cdot \sum_{j=1}^{k} w_j n_j^{\frac{d-1}{d}} + O(n_P^{\frac{d-2}{d}+s_1})$ for all large n_P, where $n_j \geq 1$ is the number of points $p \in P$ with $w(p) = w_j$ $(j = 1, \cdots, k)$.

Proof. We use two phases to find the separator hyper-plane. The first phase determines the orientation of the hyper-plane by selecting a random hyper-plane, and finds the region of the separator hyper-plane for a balanced partition. The second phase finds the position of the separator plane with a small sum of weights for the points of the set P close to it. Without loss of generality, we assume that $0 < \delta < 1$. Since $n_j \geq 1(j = 1, \cdots, k)$, we have $k \leq n_P$. Let $b = \prod_{i=1}^{d} a_i$. Select constant $c_0 > 0$ and let $\delta_1 = c_0\delta$ so that $(k_d \cdot b^{\frac{-1}{d}} + 3\delta_1)(1+\delta_1)^2 \leq (k_d \cdot b^{\frac{-1}{d}} + \frac{\delta}{2})$. Let $a_1 = a(1 + \delta_1)$ and $\alpha = \delta_1$. Let c_1 be a constant such that

$$k \cdot w_1 \le c_1 n_P^{s_1} \text{ and } \frac{k}{w_k} \le c_1 n_P^{s_2}. \tag{1}$$

Let o be the center point from Lemma 1 (our algorithm does not need to find such a center point o, but will use its existence). By Lemma 2, $E(F_{a_1,P,o}) \le (k_d \cdot b^{\frac{-1}{d}} + \delta_1) \cdot a_1 \cdot \sum_{j=1}^{k} w_j n_j^{\frac{d-1}{d}} + O(n_P^{\frac{d-2}{d}+s_1})$. By Markov inequality, $Pr(F_{a_1,P,o}(L) \ge (1+\alpha)E(F_{a_1,P,o})) \le \frac{1}{1+\alpha}$. This tells us that a random hyper-plane L has the probability at least $1 - \frac{1}{1+\alpha}$ such that there exists a separator hyper-plane L' (it may be through o) that satisfies the conditions of the theorem and is parallel to L. We assign the values to some parameters:

$$r = c_4 v, \text{where } c_4 \text{ is a constant to be fixed later} \tag{2}$$

$$\delta_2 = \frac{\delta_1 \cdot a}{c_1} \tag{3}$$

$$\epsilon = \frac{\delta_2}{3c_1 n_P^{\frac{1}{d}+s_1+s_2}} \tag{4}$$

$$\epsilon_0 = \frac{\delta}{6} \tag{5}$$

$$\epsilon_1 = 5\epsilon_0 \tag{6}$$

$$m_1 = \frac{3(\ln 100 + r + \log n_Q)}{\epsilon_0^2} \tag{7}$$

$$m_2 = \frac{(\ln 100 + 2 \log n_P + r)}{\epsilon^2} \tag{8}$$

Phase 1 of the algorithm: The input of our algorithm is $P, Q, n_Q = |Q|$, and $n_P = |P|$. Each input point $p \in P$ has the format $< (x_1, \cdots, x_d), w(p) >$, where $p = (x_1, \cdots, x_d)$ and $w(p)$ is the weight of p. The algorithm starts with the following steps: Select a fixed point $o^* \in \Re^d$ and a random plane L through o^* (random hyper-plane can be selected via selecting a random normal vector). Select m_1 random points $q_1, \cdots, q_{m_1}$ from Q and let $Q' =< q_1, \cdots, q_{m_1} >$ represent the list of these points (one point may appear multiple times). For each $q_j \in Q'$, compute its signed distance $d_{q_i} = sd(q_i, L)$ to L. Find the $\lfloor (\frac{1}{d+1} - \epsilon_1)m_1 \rfloor$-th least point $D_{1,d+1}^* = sd(q_1^*, L)$ for $d_{q_1}, \cdots, d_{q_{m_1}}$. Find the $\lceil (\frac{d}{d+1} + \epsilon_1)m_1 \rceil$-th least point $D_{d,d+1}^* = sd(q_2^*, L)$ for $d_{q_1}, \cdots, d_{q_{m_1}}$. Select m_2 random points $p_1, \cdots, p_{m_2}$ from P and let $P' =< p_1, \cdots, p_{m_2} >$ represent the list of these points. For each $p_i \in P'$, compute $d_{p_i} = sd(p_i, L)$. It is well-known that finding the i-th element from a list takes linear steps. The computation above takes $O(m_1 + m_2)$ steps. In the rest of the algorithm, we locate the position of the separator hyper-plane by finding its signed distance to L. Its position will be at the center of an interval of size $2a$. In the rest of the proof, we treat both P and Q as lists of points from $\Re^d$. Each point appears only at most once on both P and Q. Let $t_d = k_d \cdot b^{\frac{-1}{d}} + \delta$. For $q \in \Re^d$ and $A \subseteq \Re^d$, define $Pr(A, L, \leftarrow q) = \frac{|\{q' | q' \in A \text{ and } sd(q',L) \le sd(q,L)\}|}{|A|}$. For a list of points $B =< x_1, \cdots, x_m >$ from $\Re^d$ and a point $q \in \Re^d$, define $X_{B,L,q}(i) = 1$ if $sd(q_i, L) \le sd(q, L)$, or 0 otherwise. We also define $Y(B, L, q) = \sum_{i=1}^{m} X_{B,L,q}(i)$.

Lemma 5. *It has probability* $\geq 1 - \frac{e^{-r}}{50}$ *such that* $Pr(Q, L, \leftarrow q_1^*) \in [\frac{1}{d+1} - \delta, \frac{1}{d+1} - \frac{\delta}{6}]$ *and* $Pr(Q, L, \leftarrow q_2^*) \in [\frac{d}{d+1} + \frac{\delta}{6}, \frac{d}{d+1} + \delta]$.

Phase 2 of the algorithm: In this phase, we will find a position of L' (parallel to L) with the signed distance to L in the range $[D_{1,d+1}^*, D_{d,d+1}^*]$. Lemma 5 guarantees (with high probability) that each position in the interval $[D_{1,d+1}^*, D_{d,d+1}^*]$ gives a balance partition. We look for the position that has the small sum of weights for the points of P close to L'.

For a list $A =< x_1, \cdots, x_m >$, $|A| = m$ is denoted to be the *length of* A and $x \in A$ means that x is one of the elements in A ($x = x_i$ for some $1 \leq i \leq m$). For a real subset $J \subseteq R$ and a list A of finite points in $\Re^d$, define

$$Pr_*(A, L, J, w_j) = \frac{|\{|p|p \in A \text{ and } w(p) = w_j \text{ and } sd(p, L) \in J\}|}{|A|},$$

and $Z(A, L, J, w_j) = \sum_{p \in A} X_{L,p,J,w_j}^*$, where $X_{L,p,J,w_j}^* = 1$ if $sd(p, L) \in J$ and $w(p) = w_j$, or 0 otherwise. We also define $W(A, L, J) = \sum_{p \in A \text{ and } sd(p,L) \in J} w(p)$. By the definitions, It is easy to see that

$$W(A, L, J) = \sum_{j=1}^{k} w_j Z(A, L, J, w_j) = \sum_{j=1}^{k} w_j Pr_*(A, L, J, w_j)|A|. \qquad (9)$$

Since $\sum_{j=1}^{k} n_j = n_P$, we have that $n_j \geq \frac{n_P}{k}$ for some $1 \leq j \leq k$. By (1), we have that $n_P^{s_2} \geq \frac{k}{c_1 w_k} \geq \frac{k^{\frac{d-1}{d}}}{c_1 w_j}$. This implies that $n_P^{\frac{d-1}{d} - s_2} \leq c_1 w_j (\frac{n_P}{k})^{\frac{d-1}{d}} \leq c_1 w_j n_j^{\frac{d-1}{d}}$ for some $1 \leq j \leq k$. By (3), for some $1 \leq j \leq k$,

$$\delta_2 \cdot n_P^{\frac{d-1}{d} - s_2} \leq \delta_1 \cdot a \cdot w_j n_j^{\frac{d-1}{d}}. \qquad (10)$$

Lemma 6. *Let* $f \leq n_P$ *be an integer and* $H_1, H_2, \cdots, H_f \subseteq \Re$ *be* f *real intervals. It has probability* $\geq 1 - \frac{1}{100} e^{-r}$ *such that* $W(P, L, H_i) \in [W(P', L, H_i)\frac{n_P}{m_2} - \delta_2 n_P^{\frac{d-1}{d} - s_2}, W(P', L, H_i)\frac{n_P}{m_2} + \delta_2 n_P^{\frac{d-1}{d} - s_2}])$ *for* $i \leq f$.

Case 1. $|D_{1,d+1}^* - D_{d,d+1}^*| \geq 3a n_P^{\frac{2}{d}}$. Partition $[D_{1,d+1}^*, D_{d,d+1}^*]$ into disjoint intervals $[l_1, l_2), [l_2, l_3), \cdots, [l_{u-1}, l_u), [l_u, l_{u+1}]$ such that each $l_{i+1} - l_i (i = 1, \cdots, u)$ is equal to $\frac{|D_{1,d+1}^* - D_{d,d+1}^*|}{g_1(n_P)} \geq 3a$, where $g_1(n_P) = u = n_P^{\frac{2}{d}}$. Let $J_i = [l_i, l_{i+1})$ if $i < u$, and $J_u = [l_u, l_{u+1}]$. Compute $W(P', L, J_i)$ for $i = 1, \cdots, u$, which takes $O(m_2 + g_1(n_P)) = O(m_2)$ steps. The algorithm selects $J = J_{i_0}$ that has the least $W(P', L, J_{i_0})$ and let $L' = L(center(J_{i_0}))$, which takes $O(g_1(n_P)) = O(m_2)$ steps. Assume that J_{i_1} is the interval with the least $W(P, L, J_{i_1})$.

Lemma 7. *It has probability* $\geq 1 - \frac{1}{50} e^{-r}$ *such that* $W(P, L, J_{i_0}) \leq \left(k_d \cdot b^{\frac{-1}{d}} + \delta\right) \cdot a \cdot \sum_{j=1}^{k} w_j \cdot n_j^{\frac{d-1}{d}}$.

Case 2. $|D^*_{1,d+1} - D^*_{d,d+1}| < 3an^{\frac{2}{d}}_P$. Let J^* be interval such that $center(J^*) \in [D^*_{1,d+1}, D^*_{d,d+1}]$ and $|J^*| = 2a_1 = 2a(1 + \delta_1)$ and $W(P, L, J^*)$ is the least.

Subcase 2.1. $|D^*_{1,d+1} - D^*_{d,d+1}| \le \delta_1 a$. Let $J = [D^*_{1,d+1} - a, D^*_{1,d+1} + a]$ and let $L' = L(D^*_{1,d+1})$ (In other words, $L' = L(center(J))$). Clearly, $J \subseteq J^*$ and $W(P, L, J) \le W(P, L, J^*)$.

Subcase 2.2. $\delta_1 a < |D^*_{1,d+1} - D^*_{d,d+1}| < 3an^{\frac{2}{d}}_P$. Let $g_2(n_P)$ be the least integer $v \ge 2$ such that $\frac{|D^*_{d,d+1} - D^*_{1,d+1}| + 2a}{v} \le \frac{\delta_1 a}{3}$. Since $v \ge 2$ and $\frac{|D^*_{d,d+1} - D^*_{1,d+1}| + 2a}{v-1} > \frac{\delta_1 a}{3}$, we have $\frac{|D^*_{d,d+1} - D^*_{1,d+1}| + 2a}{v} = \frac{v-1}{v} \frac{|D^*_{d,d+1} - D^*_{1,d+1}| + 2a}{v-1} > \frac{v-1}{v} \frac{\delta_1 a}{3} \ge \frac{\delta_1 a}{6}$. Therefore, $v \le \frac{|D^*_{d,d+1} - D^*_{1,d+1}| + 2a}{\frac{\delta_1 a}{6}} \le \frac{3an^{\frac{2}{d}}_P + 2a}{\frac{\delta_1 a}{6}} = \frac{6(3n^{\frac{2}{d}}_P + 2)}{\delta_1} = O(n^{\frac{2}{d}}_P)$. Let $s = \frac{|D^*_{d,d+1} - D^*_{1,d+1}| + 2a}{g_2(n_P)} \in [\frac{\delta_1 a}{6}, \frac{\delta_1 a}{3}]$. Partition $[D^*_{1,d+1} - a, D^*_{d,d+1} + a]$ into the union of $g_2(n_P)$ disjoint intervals of size s: $[r_1, r_2) \cup [r_2, r_3) \cup \cdots \cup [r_{v-1}, r_v) \cup [r_v, r_{v+1}]$, where $v = g_2(n_P)$ and $r_{i+1} = r_i + s$ for $i = 1, \cdots, v$. Let $I_i = [r_i, r_{i+1})$ for $i = 1, \cdots, v - 1$ and $I_v = [r_v, r_{v+1}]$. Let $J^*_i = I_i \cup I_{i+1} \cdots \cup I_{i+h-1}$ for $i = 1, \ldots, v - h + 1$, where h is an integer with $2a < h \cdot s < 2a + 2s$. The algorithm selects the interval $J = J^*_{i_2}$ that has the least $W(P', L, J^*_{i_2})$. Finally, the algorithm outputs $L' = L(center(J))$ for the separator hyper-plane. We analyze the algorithm for the case 2.

Lemma 8. *Assume that J is the interval output from the case 2 (either subcase 2.1 or subcase 2.2). It has probability $\ge 1 - \frac{1}{100}e^{-r}$ such that $W(P, L, J) \le W(P, L, J^*) + 2\delta_1 \cdot aw_j n^{\frac{d-1}{d}}_j$ for some $j \le k$.*

For a list A of finite points in $\Re^d$ and a hyper-plane M_1, define $F_1(M_1, a, A) = \sum_{p_i \in A}$ and $dist(p_i, M_1) \le a$ $w(p_i)$. If M_1 and M_2 are two hyper-planes with signed distance $d_{M_1, M_2} = sd(p, M_1)$ for some point p in the M_2, then $F_1(M_2, a, A) = W(A, M_1, J)$, where J is the interval $[d_{M_1, M_2}, -a, d_{M_1, M_2} + a]$. The the hyper-plane $L(center(J^*_{i_2}))$ output by the algorithm has that $F_1(L(center(J^*_{i_2})), a, P') \le F_1(L(center(J^*)), a_1, P') + 2\delta_1 \cdot aw_j n^{\frac{d-1}{d}}_j$ for some $j \le k$. See the section **??** for the algorithm description in the Appendix.

Time and accuracy of the algorithm: After the hyper-plane L is selected in phase one, by Lemma 5 we have the probability at least $1 - e^{-r}$ that both $Pr(Q, L, \leftarrow q^*_1) \in [\frac{1}{d+1} - \delta, \frac{1}{d+1} - \frac{\delta}{6}]$ and $Pr(Q, L, \leftarrow q^*_2) \in [\frac{d}{d+1} + \frac{\delta}{6}, \frac{d}{d+1} + \delta]$. This means every L' (parallel to L) with the signed distance in the interval $[D^*_{1,d+1}, D^*_{d,d+1}]$, it has at most $(\frac{d}{d+1} + \delta)n_Q$ points of Q in each of the half spaces. In phase 2, we have probability at least $1 - e^{-r}$ to output the separator L' such that $F_1(L', a, P) \le \left(k_d \cdot b^{\frac{-1}{d}} + \delta\right) \cdot a \cdot \sum_{j=1}^k w_j \cdot n^{\frac{d-1}{d}}_j$ (case 1 of phase 1, see Lemma 7) or $F_1(L', a, P)) \le F_1(L(J^*), a_1, P) + 2\delta_2 w_j n^{\frac{d-1}{d}}_j$ (case 2 of phase 2, see Lemma 8), where J^* is the interval of length $2a_1$ with the least $F_1(L(J^*), a_1, P)$ and center between $D^*_{1,d+1}$ and $D^*_{d,d+1}$.

Assume that L is a fixed hyper-plane and L^* is a another hyper-plane that is parallel to L and $F_1(L^*, a_1, P)$ is the least. By Lemma 7 and Lemma 8, it has probability $\geq (1 - e^{-r})^2$ such that we can get another L' (parallel to L) such that $F_1(L', a, P) \leq F_1(L^*, a_1, P) + 2\delta_1 w_j n_j^{\frac{d-1}{d}}$ for some $j \leq k$ or $F_1(L', a, P) \leq \left(k_d \cdot b^{\frac{-1}{d}} + \delta\right) \cdot a \cdot \sum_{j=1}^k w_j \cdot n_j^{\frac{d-1}{d}}$. The number of points in Q in each side of L' is $\leq (\frac{d}{d+1} + \delta)n_Q$.

We have probability at most $\frac{1}{1+\alpha}$ that $F_{a_1,P,o}(L) \geq (1 + \alpha)E(F_{a_1,P,o})$. If the algorithm repeats z times, let $L_1, \cdots, L_z$ be the random hyper planes selected for L. With probability $\geq (1 - (\frac{1}{1+\alpha})^z)$, one of those L_is has another hyper-plane L_i^* such that L_i^* is parallel to L_i and has $F_{a_1,P,o}(L_i^*) \leq (1 + \alpha)E(F_{a_1,P,o})$. Therefore, we have probability at least $(1 - (\frac{1}{\alpha+1})^z)(1 - e^{-r})^{2z}$ to find out such a L' with $F_1(L', a, P) \leq (1 + \alpha)E(F_{a_1,P,o}) + 2\delta_1 w_j n_j^{\frac{d-1}{d}}$ for some $j \leq k$ or $F_1(L', a, P) \leq \left(k_d \cdot b^{\frac{-1}{d}} + \delta\right) \cdot a \cdot \sum_{j=1}^k w_j \cdot n_j^{\frac{d-1}{d}}$. Thus, $F_1(L', a, P) \leq \left(k_d \cdot b^{\frac{-1}{d}} + \delta\right) \cdot a \cdot \sum_{j=1}^k w_j n_j^{\frac{d-1}{d}} + O(n_P^{\frac{d-2}{d}+s_1})$.

Now we give a bound for the probability. Let $z = \frac{2r}{\ln(1+\alpha)}$. Then $1 - (\frac{1}{1+\alpha})^z > 1 - e^{-r}$. Therefore, $(1 - (\frac{1}{1+\alpha})^z)(1 - e^{-r})^{2z} > (1 - e^{-r})^{2z+1} > 1 - (2z+1)e^{-r} > 1 - \frac{1}{2^v}$, where we let $r = c_4 v$ for some constant c_4 large enough.

The phase 1 of the algorithm takes $O(m_1 + m_2)$ steps. The case 1 of phase 2 takes $O(m_2)$ steps. The case 2 of phase 2 takes $O(m_2)$ steps. Totally, it takes $O(z(m_1+m_2)) = O(v \cdot (n_P^{\frac{2}{d}+2(s_1+s_2)} \cdot (\log n_P+v)+v \log n_Q)) = O(v^2 \cdot (n_P^{\frac{2}{d}+2(s_1+s_2)} \cdot \log n_P + \log n_Q))$ steps. $\square$

Corollary 1. *Let $d \geq 2$ be the dimension number and the parameter $v > 0$. Let $a > 0$ be a constant and $\delta > 0$ be a small constant. There exists a randomized $O(v^2 n^{\frac{2}{d}} \log n)$ time such that given a set Q of n grid points in $\Re^d$, the algorithm finds a hyper-plane L with probability at least $1 - \frac{1}{2^v}$ such that each side of L has at most $(\frac{d}{d+1} + \delta)n$ points of Q, and the number of points of Q with distance $\leq a$ to L is $\leq (k_d + \delta)an^{\frac{d-1}{d}}$.*

6   An Application to Protein Side-Chain Packing Problem

We follow the description of Xu [26] for the model of protein side chain packing. The side-chain prediction problem can be formulated as follows. We use a reside interaction graph $G = (V, E)$ to represent a protein resides and their interactions. Each vertex in V represents a residue of the protein. For each reside $i \in V$, $D(i)$ is the set of all possible rotamers of side chain i. There is an interaction edge $(i, j) \in E$ if and only if there are $l \in D(i)$ and $k \in D(j)$ such that there exist an atom in the rotamer l conflicts with another atom in the rotamer k. Two atoms conflict each other iff their distance is less than the sum of their radii. For each two rotamers $l \in D(i)$ and $k \in D(j)$ ($i \neq j$), there is an associated score $P_{i,j}(l, k)$ if residue i interacts with residue j. For each rotamer $l \in D(i)$, there is a score $S_i(l)$, which characterizes the interaction energy between l and the backbone of the protein. The prediction problem is to give $A(i) \in D(i)$

to residues $i \in V$ so that the following energy value is minimized. $E(G) = \sum_{i \in V} S_i(A(i)) + \sum_{i \neq j, (I,j) \in E} P_{i,j}(A(i), A(j))$.

For more detailed description about the protein side chain packing, see (e.g. [23, 4, 26, 5]). Let d_u^* be distance such that there is no interaction between two resides if their distance is $\geq d_u^*$. Let d_l^* be the minimal distance between two amino acids. Both d_u^* and d_l^* are constants.

Theorem 3. *There exists a $r_{max}^{O(n^{\frac{2}{3}})}$-time algorithm to find the optimal solution for the protein side chain packing problem, where r_{max} is the maximal number of rotamers of one amino acid. In other words, $r_{max} = \max_i |D(i)|$.*

Proof. Our algorithm is based on the divide and conquer method. Let $d_0 = d_l^* \frac{\sqrt{2}}{2}$ be the unit distance. Since $d_l^* = \sqrt{2}d_0$, we consider that the minimal distance between two amino acids is $d_l = \sqrt{2}$ and the minimal distance for the interaction between two side chains is $d_u = \frac{d_u^*}{d_0}$. For a grid point $p = (x, y, z)$ $(x, y, z$ are integers), define $cube(p) = \{(u, v, w) \in \Re^3 | x - \frac{1}{2} \leq u < x + \frac{1}{2}$ and $y - \frac{1}{2} \leq v < y + \frac{1}{2}$ and $z - \frac{1}{2} \leq w < z + \frac{1}{2}\}$. The 3D space $\Re^3$ is partitioned into many cubes: $\Re^3 = cube(p_0) \cup cube(p_1) \cup \cdots$. For different grid points $p \neq p'$, $cube(p) \cap cube(p') = \emptyset$. Each amino acid is represented by the position of its C_α. Therefore, no two amino acids can stay at the same $cube(p)$ for any grid point p. Let P be the set of all grid points p such that $cube(p)$ contains the C_α for an amino acid.

Let $w = d_u + 2\sqrt{2}$. By Corollary 1, there exists a w-wide separator L plane such that each side has at most $(\frac{3}{4} + \delta)n$ contain amino acid, and the number of grid points (with amino acids in its cube) is bounded by $1.209wn^{\frac{2}{3}}$, where $\delta > 0$ is an arbitrary small constant. The w-wide separator partitions the problem into P_1, S and P_2, where S is the separator area. Clearly, a side chain whose amino acid C_α is in $cube(p)$ with $p \in P_1$ does not interact another side chain in P_2 because of the w-wide separator between P_1 and P_2.

The number of ways to arrange the side chains in the separator area S is bounded by $r_{max}^{1.209wn^{\frac{2}{3}}}$. We only need $O(n)$ time for computing the separator. We assume that $r_{max} \geq 2$ (otherwise, it is trivial). Let $T(n)$ is the computational time for the protein side chain packing problem with n resides. Solving each sub-problem $P_i(i = 1, 2)$ takes $T((\frac{3}{4} + \delta)n)$ steps. We have the recursive $T(n) \leq 2(r_{max}^{1.209wn^{\frac{2}{3}}} + O(n))T((\frac{3}{4} + \delta)n)$. This gives that $T(n) = r_{max}^{O(n^{\frac{2}{3}})}$. $\square$

References

1. T. Akutsu, NP-hardness results for protein side-chain packing, In S. Miyano and T. Takagi, editors, Genome Informatics 8, 1997, pp. 180-186.
2. N. Alon, P. Seymour, and R. Thomas, Planar Separator, SIAM J. Discr. Math. 7,2(1990) 184-193.
3. N. Alon, P. Seymour, and R. Thomas, A separator theorem for graphs with an excluded minor and its applications, STOC'90, pp. 293-299.

4. A. A. Canutescu, A. A. Shelenkov, and R. L. Dunbrack Jr., A graph-theory algorithm for rapid protein side-chain prediction, b Protein science, 12: 2003, pp. 2001-2014.

5. B. Chazelle, C. Kingsford, and M. Singh, A semidefinite programming approach to side-chain positioning with new rounding strategies, INFORMS Journal on Computing, 2004, pp. 86-94.

6. Z. Chen, B. Fu, Y. Tang, and B. Zhu, A PTAS for a DISC covering problem using width-bounded separator, COCOON'05, 2005.

7. H. N. Djidjev, On the problem of partitioning planar graphs. SIAM journal on discrete mathematics, 3(2) June, 1982, pp. 229-240.

8. H. N. Djidjev and S. M. Venkatesan, Reduced constants for simple cycle graph separation, Acta informatica, 34(1997), pp. 231-234.

9. D. Eppstein, G. L. Miller, S. Teng, A Deterministic Linear Time Algorithm for Geometric Separators and its Applications, SOCG'93, pp. 99-108.

10. B. Fu and W. Wang, A $2^{O(n^{1-1/d}\log n)}$-time algorithm for d-dimensional protein folding in the HP-model, ICALP'04, pp. 630-644.

11. B. Fu, Theory and application of width bounded geometric separator. Full draft is in ECCC 2005, TR05-13. Extended abstract is in STACS'06, pp. 277-288.

12. H. Gazit, An improved algorithm for separating a planar graph, manuscript, USC, 1986.

13. J. R. Gilbert, J. P. Hutchinson, and R. E. Tarjan, A separation theorem for graphs of bounded genus, Journal of algorithm, (5)1984, pp. 391-407.

14. S. Jadhar and A. Mukhopadhyay, Computing a center of a finite planar set of points in linear time, SOCG'93, pp. 83-90.

15. D. Lichtenstein, Planar formula and their uses, SIAM journal on computing, 11,2(1982), pp. 329-343.

16. R. J. Lipton and R. Tarjan, A separator theorem for planar graph, SIAM Journal on Applied Mathematics, 36(1979) 177-189.

17. G. L. Miller, S.-H. Teng, and S. A. Vavasis, An unified geometric approach to graph separators, FOCS'91, pp. 538-547.

18. G. L. Miller and W. Thurston, Separators in two and three dimensions, STOC'90, pp. 300-309.

19. G. L. Miller and S. A. Vavasis, Density graphs and separators, SODA'91, pp. 331-336.

20. R. Motwani and P. Raghavan, Randomized algorithms, Cambridge University Press, 1995.

21. J. Pach and P. K. Agarwal, Combinatorial geometry, Wiley-Interscience Publication, 1995.

22. S. Plotkin, S. Rao, and W. D. Smith, Shallow excluded minors and improved graph decomposition, SODA'90, pp. 462-470.

23. J. W. Ponter, and F. M. Richards, Tertiary templates for proteins: use of packing criteria and the enumeration of allowed sequences for different structural classes. J. molecular biology, 193, 1987, pp. 775-791.

24. W. D. Smith and N. C. Wormald, Application of geometric separator theorems, FOCS'98, pp. 232-243.

25. D. A. Spielman and S. H. Teng, Disk packings and planar separators, SOCG'96, pp.349-358.

26. J. Xu, Rapid protein side-chain packing via tree decomposition, RECOMB,05, pp. 408-422.

Polygonal Curve Approximation Using Grid Points with Application to a Triangular Mesh Generation with Small Number of Different Edge Lengths

Shin-ichi Tanigawa* and Naoki Katoh

Department of Architecture and Architectural Engineering, Kyoto University
Kyotodaigaku-Katsura, Nishikyo, Kyoto 615-8540, Japan
{is.tanigawa, naoki}@archi.kyoto-u.ac.jp

Abstract. For a given x-monotone polygonal curve each of whose edge lengths is between $\underline{l}$ and $2\underline{l}$, we consider the problem of approximating it by another x-monotone polygonal curve using points of a square grid so that there exists a small number of different edge lengths and every edge length is between $\underline{l}$ and $\beta\underline{l}$, where β is a given parameter satisfying $1 \le \beta \le 2$. Our first algorithm computes an approximate polygonal curve using fixed square grid points in $O((n/\alpha^4)\log(n/\alpha))$ time. Based on this, our second algorithm finds an approximate polygonal curve as well as an optimal grid placement simultaneously in $O((n^3/\alpha^{12})\log^2(n/\alpha))$ time, where α is a parameter that controls the closeness of approximation. Based on the approximate polygonal curve, we shall give an algorithm for finding a uniform triangular mesh for an x-monotone polygon with a constant number of different edge lengths.

1 Introduction

In this paper, we consider the following problem: given an x-monotone polygonal curve P each of whose edge lengths is between $\underline{l}$ and $2\underline{l}$, where $\underline{l}$ is the predetermined standard edge length, the problem is to approximate P by another x-monotone polygonal curve using points of a given square grid. This problem is motivated by the problem of finding a triangular mesh with a constant number of different edge lengths [18]. Since it seems to be difficult in general to find a triangulation for a given polygon such that the number of different edge lengths is constant, we considered in our previous paper [18] the following problem:

Input: An x-monotone polygon P each of whose edge lengths is between $\underline{l}$ and $2\underline{l}$ and a parameter β with $1 \le \beta \le 2$.
Output: An x-monotone polygon Q which approximates P appropriately, and triangulation of Q such that (i) boundary edge lengths of Q are between $\underline{l}$ and

---

* Supported by JSPS Grant-in-Aid for Scientific Research on priority areas of New Horizons in Computing.

S.-W. Cheng and C.K. Poon (Eds.): AAIM 2006, LNCS 4041, pp. 161–172, 2006.

$\beta \underline{l}$, (ii) the edges and the angles of the triangles are *uniform* and (iii) the number of different edge lengths is constant, where *uniformity* is measured by ratio of the maximum edge length to minimum one and by minimum angle.

In [18], we have presented an algorithm for this problem which can be regarded as an extension of the results of [14, 19]. The idea of the algorithm in [18] is that the location of every vertex of Q and every Steiner point for the triangulation is determined so that it coincides with a point of square grid of width $\alpha \underline{l}$, where $0 < \alpha < 1$. This obtained a triangular mesh with $3\pi/8\alpha^2 + o(1/\alpha^2)$ different edge lengths in $O(n/\alpha^4(1/\alpha^2 + \log(n/\alpha)))$ time. A parameter α controls the closeness of approximation of Q as well as the number of different edge lengths. In this paper, we will study the problem for finding an optimal placement of the square grid and an approximate polygon simultaneously which has not been studied before. We only consider *translation* of the grid in finding an optimal placement.

Motivating application. In the field of architecture, the surface of large-span structure such as dome structure is represented by triangulation consisting of the bars and joints called *triangular truss*. From practical viewpoint, there are constraints on structural strength and construction cost. As to the strength of the structure, the lengths of members and angles between consecutive members incident to a joint are critical properties that determine the performance. On the other hand, a cost to realize the structure heavily depends on the number of different member lengths. From this standpoint, we are concerned with how to realize an architectural structure with a limited number of different elements.

Historical Perspective. The *polygonal curve approximation*, especially *curve simplification*, has various applications for cartography, computer graphics, and geographic information system, and has been extensively studied. In particular, Imai and Iri [13] proposed two different approaches to this problem. The first one is *min-# problem*: given an error ε, find an approximation curve within ε with minimum number of vertices. The second one is *min-ε problem*: given an integer m, find an approximation curve consisting of at most m vertices with minimum error. A number of algorithms have been developed to solve the above problems under various constraints and error criteria (see e.g. [2, 4, 6, 7, 10, 12, 13, 17]). In this paper, we will consider the error criterion mostly used in the literature of curve approximations, which is called *tolerance zone criteria* [4, 6, 7, 12, 13, 17], or also called *Hausdorff error measure* in [2]. It is assumed in most of papers that original vertices are used for vertices of the approximation curve. In our problem, we do not make such assumption since we consider the problem of approximating a polygonal curve by using points of a square grid. Several algorithms for the min-# and the min-ε problems under *uniform* metric, namely L_∞ metric, whose vertices need not be a subset of vertices of the original curve are developed in [9, 11, 12]. For L_1 and L_2 metric, Aronov et al. [5] gave fully polynomial-time algorithm under *min-sum* criteria, however there is no polynomial time algorithm for min-ε problem under L_2 metric to our knowledge.

Our results. In this paper, we will propose two algorithms for polygonal curve approximation using grid points. The first one gives an optimal polygonal curve approximation using fixed grid points in $O((n/\alpha^4)\log(n/\alpha))$ time (Section 2). Using this algorithm as a subroutine, we propose an algorithm for finding both an optimal polygonal curve approximation and an optimal placement of the square grid simultaneously in $O((n^3/\alpha^{12})\log^2(n/\alpha))$ time (Section 3).

For the easiness of exposition, we assume that $1/\alpha$ is an integer throughout the paper.

2   Approximation of a Polygonal Curve Using a Fixed Grid

Let P be a given piecewise linear curve in the plane which is defined as a sequence of vertices $\langle p_1, \ldots, p_n \rangle$, such that any two consecutive vertices p_i and p_{i+1} are connected by the line segment $p_i p_{i+1}$. First, we introduce the function which measures the error between the original curve and the approximate one with respect to the previous function used in many curve simplification algorithm mentioned in the introduction. The basic approach adopted therein assumes that the vertices $\langle p_{i_1}, \ldots, p_{i_m} \rangle$ of Q is a subset of the vertices of P with $i_1 = 1$ and $i_m = n$, and the error of an approximate curve is defined as *Hausdorff error measure* with L_h metric, (e.g. $h = 1, 2$ or ∞). In this paper, we consider L_2 metric, $d : \mathbb{R}^2 \times \mathbb{R}^2 \to \mathbb{R}$. For a vertex x and a set A, let $d(x, A)$ be the distance between x and A, i.e. $d(x, A) = \min_{y \in A} d(x, y)$. The Hausdorff error under L_2 metric between a line segment $p_j p_k$ and P is defined as $d_H(p_j p_k, P) = \max_{j \le l \le k} d(p_l, p_j p_k)$. Thus, the error between P and Q under Hausdorff error measure is defined as $d_H(Q, P) = \max_{1 \le j \le m-1} d_H(p_{i_j} p_{i_{j+1}}, P)$. Notice that in our problem the vertices of an approximate polygon are not a subset of the vertices of P. Let $Q = \langle q_1, \ldots, q_m \rangle$ be an approximate polygonal curve of P. Define a mapping $\phi : \mathbb{R}^2 \to \mathbb{R}^2$ such that, for a vertex q_i, $\phi(q_i)$ is a point (not necessarily a vertex) of P which is nearest from q_i, i.e. $d(q_i, \phi(q_i)) = \min_{k=1, \ldots, n-1} d(q_i, p_k p_{k+1})$. Then, we define the Hausdorff error between the segment $q_i q_{i+1}$ and P as

$$d_H(q_i q_{i+1}, P) = \begin{cases} \max\{d(q_i, \phi(q_i)), d(q_{i+1}, \phi(q_{i+1})), \max\limits_{j+1 \le l \le k}\{d(p_l, q_i q_{i+1})\}\} & \text{if } j < k \\ \max\{d(q_i, \phi(q_i)), d(q_{i+1}, \phi(q_{i+1}))\} & \text{if } j = k, \end{cases}$$

(1)

where j and k are indices such as $\phi(q_i) \in p_j p_{j+1}$ and $\phi(q_{i+1}) \in p_k p_{k+1}$. Then,

$$d_H(Q, P) = \max_{1 \le i \le m-1} d_H(q_i q_{i+1}, P).$$

(2)

We now give the rigorous definition of our problem:

Input: An x-monotone polygonal curve P with vertices $V_P = \langle p_1, \ldots, p_n \rangle = \langle (x_1, y_1), \ldots, (x_n, y_n) \rangle$ and $x_1 \le \cdots \le x_n$ such that each edge length is between $\underline{l}$ and $2\underline{l}$, a square grid G of width $\alpha \underline{l}$ with $0 < \alpha < 1$ and a parameter β with $1 \le \beta \le 2$.
Output: An x-monotone polygonal curve $Q = \langle q_1, \ldots, q_m \rangle$ with minimum Hausdorff error such that each q_i coincides with a grid point, and each edge length is between $\underline{l}$ and $\beta \underline{l}$.

The algorithm consists of the following steps. In the first step we enumerate *candidate grid points* which are defined as $C = \{c \in V_G \mid d(c, P) \leq \delta\}$, where V_G is a set of vertices of G and δ will be determined later. Next we construct a network containing geometric information, by which the problem reduces to finding an optimal path on the network.

Candidate grid points. For computational efficiency, it is necessary to enumerate candidate grid points. From the following lemma, we set δ to $\sqrt{2}l$.

Lemma 1. *Let Q be the approximate polygon which minimizes Hausdorff error. Then, $d_H(Q, P)$ is less than $\sqrt{2}l$.*

For each $p_j p_{j+1}$ of P, there are $O(1/\alpha^2)$ candidate grid points in $\{x \in V_G \mid d(x, p_j p_{j+1}) \leq \sqrt{2}l\}$, because every $p_j p_{j+1}$ is assumed to be less than $2l$. Then, we have that the total number of candidate grid points is $O(n/\alpha^2)$.

Constructing a network. We introduce a directed network $\mathcal{N} = (V, E)$, where $V = C$ and $E = \{(c_1, c_2) \mid c_1, c_2 \in C, x\text{-coordinate of } c_2 \text{ is greater than or equal to that of } c_1 \text{and } l \leq d(c_1, c_2) \leq \beta l\}$. The edge (c_1, c_2) is assigned a weight $d_H(c_1 c_2, P)$ which is the error between $c_1 c_2$ and P. Notice that if the segment $c_1 c_2$ is vertical, (c_1, c_2) is bidirected.

Let ε^* be the error of the optimal solution. Since there is at least one vertex of the optimal solution in a circle with radius ε^* centered at p_1, we add an auxiliary node n_s to V as a starting node and edges from n_s to the nodes associated with the candidate points in $\{c \in C \mid d(c, p_1) \leq \sqrt{2}l\}$. Similarly, we add an auxiliary point n_t as a terminal node as well as edges from the candidate points in $\{c \in C \mid d(c, p_n) \leq \sqrt{2}l\}$ to n_t. The arcs added have no weights.

Finding an optimal path. Sorting all the edge weights, we perform a binary search on the edge weights. When the binary search focuses on the edge weight w, we check by a depth-first search whether a path from n_s to n_t exists in the network which consists of the edges whose weights are at most w. The total number of nodes in $\mathcal{N}$ is $O(n/\alpha^2)$. Number of edges incident to one node are $O(1/\alpha^2)$. Therefore, the total number of edges in $\mathcal{N}$ is $O(n/\alpha^4)$. Therefore, the total time to get an optimal solution is $O((n/\alpha^4)\log(n/\alpha))$.

3   Approximation of a Polygonal Curve with an Optimal Grid Layout

The grid layout affects the value of the error between the original polygonal curve P and an optimal approximate curve Q^*. The method that we stated in the previous section produces a different solution Q^* depending on a position of grid. Thus, in this section we will be concerned with finding a grid layout such that $d_H(Q^*, P)$ is minimum. We call such grid layout *an optimal grid layout*. The difficulty of this problem lies in that (i) candidate grid points move depending on the choice of the square grid position and (ii) the nearest edge of P from a candidate point as well as the weight $d_H(c_1 c_2, P)$ of an edge (c_1, c_2) also changes depending on the grid. In order to overcome the difficulty (i), we enlarge the set of candidate grid points appropriately. For the second difficulty (ii), we do as follows. First, instead of considering optimization problem, we consider a *decision problem* $\mathcal{D}(\varepsilon)$. Given an error bound $\varepsilon > 0$, $\mathcal{D}(\varepsilon)$ asks whether there exists an approximate polygonal curve Q and a translation vector t in $\mathbb{R}^2$ such that $d_H(Q \oplus t, P) \le \varepsilon$, where $A \oplus t = \{a + t \mid a \in A\}$ denotes the *Minkowski sum* between a set A and a vector t. Now, note that t can be restricted to be in $[-\frac{\alpha l}{2}, \frac{\alpha l}{2}] \times [-\frac{\alpha l}{2}, \frac{\alpha l}{2}]$ because we consider the grid of width αl. We use the simple notation R to stand for $[-\frac{\alpha l}{2}, \frac{\alpha l}{2}] \times [-\frac{\alpha l}{2}, \frac{\alpha l}{2}]$. In order to solve $\mathcal{D}(\varepsilon)$, R is decomposed into regions such that in each region f the set E of candidate edges is classified into two classes E_1 and E_2 such that $d_H(e \oplus t, P) \le \varepsilon$ holds for any $e \in E_1$ and for any $t \in f$ while $d_H(e \oplus t, P) > \varepsilon$ holds for any $e \in E_2$. If $e \in E_1$ in f, it is called *available* in f. Once such decomposition is obtained, we can solve $\mathcal{D}(\varepsilon)$ easily: we can test whether there exists a path from a starting node n_s to a terminal node n_t using only available edges in each face of R. If there exists such a path for some region in R, then the answer of $\mathcal{D}(\varepsilon)$ is "yes", otherwise "no". Such decomposition of R can be defined as follows. For every candidate edge $c_i c_j$, we compute the curves defining the region of $d_H(c_i c_j \oplus t, P) \le \varepsilon$ based on Eq.(1). Union of such curves over all candidate edges gives the desired decomposition of R.

Candidate grid points and the associated network. Define $C(t) = \{c \in V_G \oplus t \mid d(c, P) \le \sqrt{2}l\}$ which is a set of candidate grid points shown by Lemma 1. From Lemma 1, we have at least one approximate polygonal curve using the points of $C(t)$. Then, redefining $C = \{c \in V_G \mid d(c, P) \le \sqrt{2}(1 + \alpha)l\}$ as a set of candidate grid points, we can easily see that $C(t) \subset C$ holds for any $t \in R$.

In a manner similar to the way explained in the previous section, if available edges are given, we can construct the associated network $\mathcal{N} = (V, E)$ although the edge weights are not given because t is not fixed.

Decomposition of the space of t. Consider a line segment $c_1 c_2$ which is associated with an edge $(c_1, c_2) \in E$. Translating it by a vector t results in a line segment $c_1 c_2 \oplus t$. We now consider the curves defining the region of $d_H(c_1 c_2 \oplus t, P) \le \varepsilon$ in R.

First, let us define some notations. For two sets A and B, the *Minkowski sum* between A and B is defined as $A \oplus B = \{a + b \mid a \in A, b \in B\}$. Let B^ε denote a disk of radius ε around the origin. For a segment $p_i p_{i+1}$ of P, let $p_i p_{i+1} \oplus B^\varepsilon$ is called *racetrack* which consists of a rectangle of width 2ε with two semicircles of radius ε attached to its sides. Let P^ε be a union of the racetracks of all edges of P, i.e. $P^\varepsilon = \bigcup_{i=1}^{n-1}(p_i p_{i+1} \oplus B^\varepsilon)$.

Suppose $\phi(c_1 + t)$ and $\phi(c_2 + t)$ belong to $p_j p_{j+1}$ and $p_k p_{k+1}$ respectively. Then, from the edge weight definition in (1), $d_H(c_1 c_2 \oplus t, P) \le \varepsilon$ is equivalent to

$$d(c_1 + t, \phi(c_1 + t)) \le \varepsilon \text{ and } d(c_2 + t, \phi(c_2 + t)) \le \varepsilon \tag{3}$$

and if $j < k$

$$d(p_l, c_1 c_2 \oplus t) \le \varepsilon \text{ for } l = j + 1, j + 2, \ldots, k. \tag{4}$$

First, we consider the inequalities (3). The inequality $d(c_1 + t, \phi(c_1 + t)) \le \varepsilon$ implies that the vertex $c_1 + t$ is inside of P^ε. Thus, it is rewritten as $t \in P^\varepsilon \oplus (-c_1)$. Similarly, the inequality $d(c_2 + t, \phi(c_2 + t)) \le \varepsilon$ is rewritten as $t \in P^\varepsilon \oplus (-c_2)$. Then, the inequality (3) is defined by the curves $d(c_1 + t, \phi(c_1 + t)) = \varepsilon$ and $d(c_2 + t, \phi(c_2 + t)) = \varepsilon$ which form the boundaries of $(P^\varepsilon \oplus (-c_1)) \cap R$ and $(P^\varepsilon \oplus (-c_2)) \cap R$ in R. See Fig.1(a) for example.

Next, let us consider the inequality (4). Note that the indices j and k change depending on t. The region in R such that j and k do not change is determined by the *Voronoi diagram*, Vor(P), of the segments of P each of whose *Voronoi edges* is either line segment or paraboloid arc. Then the Voronoi diagram Vor($P \oplus (-c_1)) \cap R$ decomposes the space of t into the regions in each of which the segment of P nearest to $c_1 + t$ is determined. So we decompose R into the faces of Vor($P \oplus (-c_1)) \cap R$. Similarly Vor($P \oplus (-c_2)) \cap R$ decomposes R into regions in each of which the segment of P nearest to $c_2 + t$ is determined. Thus, the decomposition of R is also defined by union of Voronoi edges of Vor($P \oplus (-c_1)) \cap R$ and Vor($P \oplus (-c_2)) \cap R$. In each face f of such decomposition, the closest segment of P to $c_1 + t$ as well as the closest one to $c_2 + t$ are fixed, (see Fig.1(b)). Let $p_j p_{j+1}$ and $p_k p_{k+1}$ be the nearest segments from $c_1 + t$ and $c_2 + t$ for $t \in f$, respectively, and let $B^\varepsilon(p_l)$ be a closed disk of radius ε centered at p_l. The inequality $d(p_l, c_1 c_2 \oplus t) \le \varepsilon$ implies $(c_1 c_2 \oplus t) \cap B^\varepsilon(p_l) \ne \emptyset$, which says that there exists a point $x \in c_1 c_2$ such that $t + x \in B^\varepsilon(p_l)$. This is equivalent to $t \in B^\varepsilon(p_l) \oplus (-x)$. Then, $d(p_l, c_1 c_2 \oplus t) \le \varepsilon$ if and only if $t \in B^\varepsilon(p_l) \oplus (-c_1 c_2)$. Define $I^\varepsilon_{p_l}(c_1, c_2) := B^\varepsilon(p_l) \oplus (-c_1 c_2)$, which is also a racetrack whose boundary is defined by two line segments and two semicircles (see Fig.1(c)).

In summary, the desired decomposition in each of whose faces a set of available edges do not change is defined by a collection of

(i)   the boundary of $(P^\varepsilon \oplus (-c_i)) \cap R$ for all $c_i \in C$,

(ii) Voronoi edges of Vor($P \oplus (-c_i)) \cap R$ for all $c_i \in C$ and

(iii) the boundary of $I^\varepsilon_{p_l}(c_i, c_j) \cap R$ for all $p_l \in V_P$ and for all $(c_i, c_j) \in E$.

Decision problem $\mathcal{D}(\varepsilon)$. The outline of our algorithm for solving $\mathcal{D}(\varepsilon)$ is as follows:

1. Calculate the curves (i), (ii) and (iii).
2. Compute the arrangement $\mathcal{A}$ defined by a collection of all these curves.
3. Sweep the arrangement $\mathcal{A}$. Every time we visit a new face f, we construct a network $\mathcal{N}$ consisting of only available edges, and check whether a path from n_s to n_t exists by depth-first search. The algorithm returns "yes", if such a path exists until sweeping is finished.

Now, let us show how we efficiently implement Steps 3. The arrangement $\mathcal{A}$ is defined by the collection of the curves (i), (ii) and (iii). Therefore, there exist three kinds of events as follows while sweeping the arrangement:

(a) The sweep line crosses a segment of $(P^\varepsilon \oplus (-c_i)) \cap R$, which informs that c_i gets inside of P^ε or gets out of P^ε.

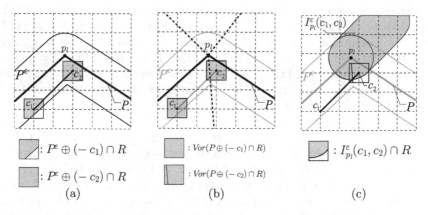

$$\text{\rotatebox{90}{▨}}: P^\varepsilon \oplus (-c_1) \cap R \qquad \text{▨}: Vor(P \oplus (-c_1) \cap R) \qquad \text{▨}: I^\varepsilon_{p_l}(c_1, c_2) \cap R$$

$$\text{▨}: P^\varepsilon \oplus (-c_2) \cap R \qquad \text{▧}: Vor(P \oplus (-c_2) \cap R)$$

(a)   (b)   (c)

Fig. 1. Example of the decomposition of the space of t. (a) Shaded areas represent $P^\varepsilon \oplus (-c_1) \cap R$ and $P^\varepsilon \oplus (-c_2) \cap R$. (b) Dashed line segments are Voronoi edges of $Vor(P)$. $Vor(P \oplus (-c_2)) \cap R$ decomposes R into two regions. (c) Shaded region represents $I^\varepsilon_{p_l}(c_1, c_2)$.

(b) The sweep line crosses a segment of $Vor(P \oplus (-c_i)) \cap R$, which informs the change of the nearest segment of c_i among the edges of P.

(c) The sweep line crosses a segment of $I^\varepsilon_{p_l}(c_i, c_j) \cap R$, which informs that either $d(p_l, c_i c_j) \leq \varepsilon$ or $d(p_l, c_i c_j) > \varepsilon$ newly holds.

We now describe four data structures which are updated corresponding to above events. (1) Array PL for candidate points $c_i \in C$ is such that PL[i]$= 1$ if $c_i \in P^\varepsilon$, PL[i]$= 0$ otherwise. (2) Array NL for candidate points is such that NL[i]$= k$ represents that the nearest segment from c_i is $p_k p_{k+1}$. (3) Three-dimensional array IA for candidate edge $(c_i, c_j) \in E$ and $p_l \in V_P$ is such that IA[i][j][l]$= 1$ if $d(p_l, c_i c_j) \leq \varepsilon$ holds, IA[i][j][l]$= 0$ otherwise. (4) Two-dimensional array AL for candidate edge $(c_i, c_j) \in E$ is such that AL[i][j]$= 1$ if the edge (c_i, c_j) is available, AL[i][j]$= 0$ otherwise. Then, let us show how they are updated according to the events (a), (b) and (c) while sweeping the arrangement.

Lemma 2. *The time to update the data structures* PL, NL, IA *and* AL *is* $O(n/\alpha^2)$ *for each event* (a), (b) *and* (c).

Proof. Event (a): Two subcases are possible: (a1) c_i gets inside of P^ε, and (a2) it gets outside of P^ε. In case (a1), PL[i] is updated from 0 to 1. For each $(c_i, c_j) \in E$ (or $(c_j, c_i) \in E$) incident to c_i AL[i][j] (or AL[j][i]) may change. Updating AL[i][j] is done as follows: compute indices k_i, k_j of the nearest segments of c_i and c_j by NL[i] and NL[j], and then set AL[i][j]$= 1$ if PL[j]$= 1$ and IA[i][j][l]$= 1$ for all $k_i + 1 \leq l \leq k_j$. Checking IA[i][j][l] for all $k_i + 1 \leq l \leq k_j$ takes $O(n)$ time. Since there are $O(1/\alpha^2)$ edges incident to c_i in the network, the update of the date structures for (a1) takes $O(n/\alpha^2)$ time. In case (a2), PL[i] is updated from 1 to 0 and AL[i][j] (or AL[j][i]) is set to 0 for all edges $(c_i, c_j) \in E$ $((c_j, c_i) \in E)$. This can be done in $O(1/\alpha^2)$ time.

Event (b): Suppose the nearest segment of c_i changes to $p_{k_i}p_{k_i+1}$. NL[i] is updated to k_i. Then, consider c_ic_j or c_jc_i incident to c_i. Updating AL[i][j] is done by computing index k_j of the nearest segment of c_j by NL[j], and then setting AL[i][j]= 1 if PL[i]= 1, PL[i]= 1 and IA[i][j][l]= 1 for all $k_i + 1 \leq l \leq k_j$, setting AL[i][j]= 0 otherwise. Checking IA[i][j][l] for all $k_i + 1 \leq l \leq k_j$ takes $O(n)$ time. Thus, updating of the date structures for (b) can takes $O(n/\alpha^2)$ time.

Event (c): Two subcases are possible, i.e. (c1) $d(p_l, c_ic_j)$ becomes less than or equal to ε, and (c2) it becomes larger than ε. In case (c1), IA[i][j][l] is changed to 1. Compute indices k_i, k_j of the nearest segments of c_i and c_j by NL[i] and NL[j]. Update AL[i][j] to 1 if PL[i]= 1, PL[j]= 1 and IA[i][j][l]= 1 for all $k_i + 1 \leq l \leq k_j$. Checking IA[i][j][l] for all $k_i + 1 \leq l \leq k_j$ takes $O(n)$ time in worst case. In case (c2), IA[i][j][l] is changed to 0, and accordingly update AL[i][j] to 0. Thus, updating of the date structures for (c) takes $O(n)$ time. $\square$

Finally, to analyse the complexity of $\mathcal{A}$, we prove the following lemma.

Lemma 3. *The arrangement $\mathcal{A}$ is defined by $O(n/\alpha^4)$ line segments, circular arcs and paraboloid arcs. Then, the complexity of the arrangement $\mathcal{A}$ is $O(n^2/\alpha^8)$.*

Proof. First, let us define some notations. For a region A, define $N(A)$ to be the total number of segments bounding the region A. For a point x, let $R(x)$ be a closed square region centered at x with width αl. Then, the total number of segments of $\mathcal{A}$ is $\sum_{c_i \in C} N((P^\varepsilon \oplus (-c_i)) \cap R) + \sum_{c_i \in C} N(\text{Vor}(P \oplus (-c_i)) \cap R) + \sum_{(c_i,c_j) \in E} \sum_{p_l \in V_P} N(I_{p_l}^\varepsilon(c_i, c_j) \cap R)$. (i) Let us consider the first term. We have $\sum_{c_i \in C} N((P^\varepsilon \oplus (-c_i)) \cap R) = \sum_{c_i \in C} N(P^\varepsilon \cap R(c_i))$. This is at most the number of intersection points between the squares of G and the boundary of P^ε. Since all of edge lengths of P are given less than $2l$, the boundary of each racetrack intersects $O(1/\alpha)$ squares of G. Hence, $\sum_{c_i \in C} N(P^\varepsilon \cap R(c_i)) = O(n/\alpha)$.

(ii) Next let us consider the second term. We have $\sum_{c_i \in C} N(\text{Vor}(P \oplus (-c_i)) \cap R) = \sum_{c_i \in C} N(\text{Vor}(P) \cap R(c_i))$. This is at most the number of intersection points between the squares cantered at $c_i \in C$ and Voronoi edges of $\text{Vor}(P)$. Let us consider the region $P^{\delta'}$, where $\delta' = \sqrt{2}(1 + 2\alpha)l$. Note that $\bigcup_{c_i \in C} R(c_i) \subset P^{\delta'}$. Considering the Voronoi edges which are inside $P^{\delta'}$, we can show that each of such Voronoi edges is included in one racetrack of $P^{\delta'}$. Since diameter of the racetrack is at most $2l + 2\delta'$, each Voronoi edge appeared inside $P^{\delta'}$ has the length less than $2l + 2\delta'$. Therefore, the number of intersections between one Voronoi edge and squares $R(c_i)$ for all $c_i \in C$ is $O(1/\alpha)$. Since the total number of Voronoi edges of $\text{Vor}(P)$ is $O(n)$ (see [15]), we have $\sum_{c_i \in C} N(\text{Vor}(P \oplus (-c_i)) \cap R) = O(n/\alpha)$.

(iii) Finally, consider the third term. For $p_l \in V_P$ and $c_i, c_j \in C$, $I_{p_l}^\varepsilon(c_i, c_j)$ represents the space of t such that the line segment $c_ic_j \oplus t$ intersects a circle $C_{p_l}^\varepsilon$ centered at p_l with radius ε. Let us consider how many c_ic_j with $(c_i, c_j) \in E$ intersect a circle $C_{p_l}^\varepsilon$. The number of endpoints of the segments which intersect $C_{p_l}^\varepsilon$ is $O(1/\alpha^2)$. Then, the total number of segments which intersects $C_{p_l}^\varepsilon$ is

$O(1/\alpha^4)$. Hence, $\sum_{(c_i,c_j)\in E}\sum_{p_l\in V_P} N(I_{p_l}^\varepsilon(c_i,c_j)) = O(n/\alpha^4)$. This proves the theorem. □

Let us analyse the time complexity of $\mathcal{D}(\varepsilon)$. Since there are $O(n/\alpha^4)$ segments from Lemma 3, $\mathcal{A}$ can be computed in time $O((n^2/\alpha^8)\log(n/\alpha))$, (see [8]). In each face of $\mathcal{A}$, the algorithm computes a network consisting of available edges and check whether a path from n_s to n_t exists by depth-first search. For two adjacent faces, the update of the network can be done in $O(n/\alpha^2)$ time from Lemma 2. For each face, depth-first search can be computed in time $O(n/\alpha^4)$ because the number of edges in the network is $O(n/\alpha^4)$. In summary, we have proved the following theorem.

Theorem 1. *Given a parameter ε, the running time to compute the decision problem $\mathcal{D}(\varepsilon)$ is $O((n^3/\alpha^{12})\log(n/\alpha))$.*

Minimization problem. We solve the minimization problem by applying a *parametric search* on the parameter ε. See [16] for the definition and description of the parametric search technique, and see [1,3] for the details of how it is applied in geometric optimisation problems. Now, we have a sequential decision problem $\mathcal{D}(\varepsilon)$, and clearly $\mathcal{D}(\varepsilon)$ is *monotone* in ε, meaning that if $\mathcal{D}(\varepsilon_0)$ answers "yes" for some ε_0, then $\mathcal{D}(\varepsilon)$ also answer "yes" for all $\varepsilon > \varepsilon_0$. We wish to find the smallest value ε^* such that $\mathcal{D}(\varepsilon^*)$ answers "yes". Megiddo's idea is to run a "generic" version of the decision problem on the unknown ε^*. The generic algorithm does not have to solve the same problem as the concrete decision algorithm. For us, we consider the following generic algorithm:

- Compute the curves (i), (ii) and (iii) on ε^*.
- Compute all intersections among the curves (i), (ii) and (iii).

Let T_s be the running time of the sequential algorithm for the decision problem. This generic algorithm is consisting of the intersection queries among line segments, circular arcs and paraboloid arcs, each of which depends on the sign of a polynomial. From Lemma 3, there are $O(n^2/\alpha^8)$ such pairs of segments, then generic algorithm can be done in $T_p = O(1)$ steps using $\mathcal{P} = O(n^2/\alpha^8)$ processors, each for a pair of segments. Each of the intersection queries can be answered by examining the sign of at most quadratic polynomial in ε, and such comparison has at most two critical values. In each step of the parametric search, to resolve the intersection queries on ε^*, we call the decision problem on these critical values. It is known that, for each step in the parallel generic algorithm, we take $O(\mathcal{P}+T_s\log\mathcal{P})$ time. Then, the total time to obtain an optimal solution is $O(T_p(\mathcal{P} + T_s\log\mathcal{P})) = O((n^3/\alpha^{12})\log^2(n/\alpha))$.

Theorem 2. *Given an x-monotone polygonal curve P with n vertices, one can solve the curve approximation problem in time $O((n^3/\alpha^{12})\log^2(n/\alpha))$ with minimum Hausdorff error such that each vertex of the approximate curve coincides with a point of a square grid of width $\alpha\underline{l}$, where $0 < \alpha < 1$.*

4   Polygon Approximation

Suppose that we are given an x-monotone polygon P which consists of two x-monotone curves, one is an upper x-monotone curve $\langle p_1, \ldots, p_m \rangle$ and the other is a lower x-monotone curve $\langle p_m, \ldots, p_{n+1} = p_1 \rangle$. In this section, we will study the problem for approximating P by another x-monotone polygon Q using grid points. In the same manner as in polygonal curve approximation, we consider the decision problem $\mathcal{D}(\varepsilon)$ and the directed network $\mathcal{N} = (V, E)$ consisting of available edges. Let us explain how to construct $\mathcal{N}$. First, we calculate the candidate grid points C_{up} and C_{low} for the upper and lower curves, and construct two directed networks $\mathcal{N}_{up}$ and $\mathcal{N}_{low}$ corresponding to C_{up} and C_{low}, respectively. Direction of an edge is defined as was explained in Section 2. Finally, we combine $\mathcal{N}_{up}$ and $\mathcal{N}_{low}$ as follows. Let V_1, V_2, V_3 and V_4 be sets of candidate points in $\{c \in C_{up} \mid d(c, p_1) \le \sqrt{2}(1 + \alpha)\underline{l}\}$, $\{c \in C_{up} \mid d(c, p_m) \le \sqrt{2}(1 + \alpha)\underline{l}\}$, $\{c \in C_{low} \mid d(c, p_m) \le \sqrt{2}(1 + \alpha)\underline{l}\}$ and $\{c \in C_{low} \mid d(c, p_1) \le \sqrt{2}(1 + \alpha)\underline{l}\}$, respectively. We connect a directed edge from each node $c_2 \in V_2$ to the node $c_3 \in V_3$ with weight 0 if both c_2 and c_3 represent the same vertex. The resulting combined network is the desired $\mathcal{N}$. Although $V_1 = V_4$, the nodes in V_1 and V_4 are regarded as distinct nodes. To construct the network consisting of available edges, we decompose the space of the translation vector t as in the previous section, which is defined by a collection of curves, (i), (ii) and (iii). After $\mathcal{N}$ is constructed, for each node c_1 in V_1, we test whether there exists a path from c_1 to $c_4 \in V_4$ where c_4 is the same point as c_1. Since there are $O(1/\alpha^2)$ elements in V_1, the time to solve the decision problem is $O((1/\alpha^2) \cdot (n^3/\alpha^{12}) \log(n/\alpha))$. The minimization problem can be done in a manner similar to the one explained in Section 3. Thus, we can derive the following:

Theorem 3. *Given an x-monotone polygon P with n vertices, one can solve the polygon approximation problem in time $O((n^3/\alpha^{14}) \log^2(n/\alpha))$ with minimum Hausdorff error such that each vertex of the approximate curve coincides with a point of a square grid of width $\alpha \underline{l}$, where $0 < \alpha < 1$.*

5   Experimental Results

We have implemented the algorithm for polygon approximation with a fixed grid and triangular mesh generation with a constant number of different edge lengths which has presented in [18]. To show how a grid layout affects the solutions, we have performed the polygon approximation with several grid layouts. Given initial x-monotone polygon in Fig. 2(a) and setting parameters such that $\underline{l} = 50$, $\alpha = 0.5$ and $\beta = 2.0$, we shall show the results of polygon approximations for the best grid layout and the worst one in Fig. 2(b) and (c), respectively. Fig. 2(d) is the result of the triangulation for the polygon approximation with the best grid layout. The numerical results are given in Table 1. It is observed from the table that the total number of different edge lengths is very small.

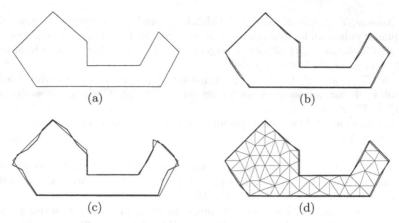

Fig. 2. (a) Initial polygon. (b) Approximate polygon with the best grid layout. (c) Approximate polygon with the worst grid layout. (d) Triangular mesh with 7 kinds of different edge lengths for the polygon in (c). ($\underline{l} = 50, \alpha = 0.5$ and $\beta = 2.0$).

Table 1. The experimental results with best grid layout and worst one. ($\underline{l} = 50, \alpha = 0.5$ and $\beta = 2.0$)

	♯ of lengths	Max length	Min length	Hausdorff error
The best layout	7	100.0	50.0	6.7
The worst layout	7	100.0	50.0	21.0

6   Conclusion and Future Works

In this paper, we considered the polygonal curve approximation problem using square grid points. We presented a developed algorithm for finding an approximate polygonal curve as well as an optimal grid layout simultaneously in $O((n^3/\alpha^{12}) \log^2(n/\alpha))$ time.

Considering the application to architecture, the following problems are left for future works: (i) Extension to curved surface and (ii) Characterization of curved surfaces realized by a constant number of different edge lengths.

References

1. P. K. Agarwal, B. Aronov, M. Sharir, and S. Suri. Selecting distances in the plane. *Algorithmica*, 9(5):495–514, 1993.
2. P. K. Agarwal, S. Har-Peled, N. H. Mustafa, and Y. Wang. Near-linear time approximation algorithms for curve simplification. In *Proc. of the 10th Ann. European Symp. Alg. (ESA)*, volume 2461 of *LNCS*, pages 29–41. Springer-Verlag, 2002.
3. P. K. Agarwal, M. Sharir, and S. Toledo. Applications of parametric searching in geometric optimization.
4. P. K. Agarwal and K. R. Varadarajan. Efficient algorithms for approximating polygonal chains. *Discrete & Comp, Geom.*, 23(2):273–291, 2000.

5. B. Aronov, T. Asano, N. Katoh, K. Mehlhorn, and T. Tokuyama. Polyline fitting of planar points under min-sum criteria. In *Proc. of 15th Symp. on Alg. and Comp. (ISAAC)*, volume 3341 of *LNCS*, pages 77–88. Springer, 2004. (also to appear in Int. J. of Comput. Geom. & Appl.).

6. W. S. Chan and F. Chin. Approximation of polygonal curves with minimum number of line segments or minimum error. *Int. J. Comput. Geometry Appl.*, 6(1):59–77, 1996.

7. D. Z. Chen and O. Daescu. Space-efficient algorithms for approximating polygonal curves in two-dimensional space. *Int. J. Comput. Geometry Appl.*, 13(2):95–111, 2003.

8. H. Edelsbrunner, L. J. Guibas, J. Pach, R. Pollack, R. Seidel, and M. Sharir. Arrangements of curves in the plane - topology, combinatorics and algorithms. *Theor. Comput. Sci.*, 92(2):319–336, 1992.

9. M. T. Goodrich. Efficient piecewise-linear function approximation using the uniform metric. *Discrete & Comput. Geom.*, 14(4):445–462, 1995.

10. J. Gudmundsson, G. Narasimhan, and M. H. M. Smid. Distance-preserving approximations of polygonal paths. In *Proc. of the 23rd Conf. Foundations of Software Tech. and Theor. Comp. Science*, volume 2914 of *LNCS*, pages 217–228. Springer-Verlag, 2003.

11. L. J. Guibas, J. Hershberger, J. S. B. Mitchell, and J. Snoeyink. Approximating polygons and subdivisions with minimum link paths. *Int. J. Comput. Geometory Appl.*, 3(4):383–415, 1993.

12. H. Imai and M. Iri. An optimal algorithm for approximating a piecewise linear function. *Info. Proc. Letters*, 9(3):159–162, 1986.

13. H. Imai and M. Iri. Polygonal approximations of a curve - formulatons and algorithms. In G. T. Toussaint, editor, *Comp. Morphology*, pages 71–86. North-Holland, Amsterdam, 1988.

14. N. Katoh, M. Ohsaki, and Y. Xu. A uniform triangle mesh generation of curved surfaces. In *Proc. of the Japan Conf. Discrete and Comput. Geom.*, volume 3742 of *LNCS*. Springer-Verlag, 2004.

15. D. T. Lee and I. Robert L. Drysdale. Generalization of voronoi diagrams in the plane. *SIAM J. Comput.*, 10(1):73–87, 1981.

16. N. Megiddo. Applying parallel computation algorithms in the design of serial algorithms. *J. ACM*, 30(4):852–865, 1983.

17. A. Melkman and J. O'Rourke. On polygonal chain approximation. In G. T. Toussaint, editor, *Comp. Morphology*, pages 87–95. North-Holland, Amsterdam, 1988.

18. S. Tanigawa and N. Katoh. Finding a triangular mesh with a constant number of different edge lengths. In *Proc. of the 17th Canad. Conf. on Comput. Geom.*, 2005. http://cccg.cs.uwindsor.ca/papers/41.pdf.

19. Y. F. Xu, W. Dai, N. Katoh, and M. Ohsaki. Triangulating a convex polygon with small number of non-standard bars. In *Proc. of the 11th Int. Comput. Comb. Conf. (COCOON)*, volume 3595 of *LNCS*, pages 481–489. Springer-Verlag, 2005.

Distributions of Points and Large Convex Hulls of k Points

Hanno Lefmann

Fakultät für Informatik, TU Chemnitz, D-09107 Chemnitz, Germany
lefmann@informatik.tu-chemnitz.de

Abstract. We consider a variant of Heilbronn's triangle problem by asking for fixed integers $d, k \geq 2$ and any integer $n \geq k$ for a distribution of n points in the d-dimensional unit cube $[0, 1]^d$ such that the minimum volume of the convex hull of k points among these n points is as large as possible. We show that there exists a configuration of n points in $[0, 1]^d$, such that, simultaneously for $j = 2, \ldots, k$, the volume of the convex hull of any j points among these n points is $\Omega(1/n^{(j-1)/(1+|d-j+1|)})$. Moreover, for fixed $k \geq d+1$ we provide a deterministic polynomial time algorithm, which finds for any integer $n \geq k$ a configuration of n points in $[0, 1]^d$, which achieves, simultaneously for $j = d+1, \ldots, k$, the lower bound $\Omega(1/n^{(j-1)/(1+|d-j+1|)})$ on the minimum volume of the convex hull of any j among the n points.

1 Introduction

For integers $n \geq 3$, Heilbronn's problem asks for the supremum $\Delta_2(n)$ of the minimum area of a triangle formed by three of n points over all distributions of n points in the unit square $[0, 1]^2$. It has been observed by Erdős, see [16], that $\Delta_2(n) = \Omega(1/n^2)$, which can be seen by considering for primes n the points $P_k = 1/n \cdot (k \bmod n, k^2 \bmod n)$, $k = 0, 1, \ldots, n-1$. Komlós, Pintz and Szemerédi [10] improved this lower bound to the currently known best lower bound $\Delta_2(n) = \Omega(\log n/n^2)$, see [4] for a deterministic polynomial time algorithm achieving this lower bound. Upper bounds were given in a series of papers by Roth [16, 17, 18, 19] and Schmidt [20], and the currently known best upper bound is due to Komlós, Pintz and Szemerédi [9], who proved that $\Delta_2(n) = O(2^{c\sqrt{\log n}}/n^{8/7})$ for some constant $c > 0$. We remark that for n points, which are chosen uniformly at random in $[0, 1]^2$, the expected value of the minimum area of a triangle is $\Theta(1/n^3)$, as was shown recently by Jiang, Li and Vitany [8].

A variant of Heilbronn's problem in dimension $d \geq 2$, which has been considered by Barequet, asks for the supremum $\Delta_{d+1,d}(n)$ – over all distributions of n points in the d-dimensional unit cube $[0, 1]^d$ – of the minimum volume of a $(d + 1)$-point simplex among n points. Barequet showed in [2] the lower bound $\Delta_{d+1,d}(n) = \Omega(1/n^d)$ for fixed $d \geq 2$, see [3] for an on-line version for dimensions $d = 3, 4$. His lower bound was improved in [11] to $\Delta_{d+1,d}(n) = \Omega(\log n/n^d)$, and in [15] for dimension $d = 3$ a deterministic polynomial time algorithm was given, which achieves $\Delta_{4,3}(n) = \Omega(\log n/n^3)$. Recently, Brass [5] improved the upper bound $\Delta_{d+1,d}(n) = O(1/n)$ to $\Delta_{d+1,d}(n) = O(1/n^{(2d+1)/(2d)})$ for odd $d \geq 3$.

S.-W. Cheng and C.K. Poon (Eds.): AAIM 2006, LNCS 4041, pp. 173–184, 2006.

Here we consider the following generalization of Heilbronn's problem: for fixed integers $d, k \geq 2$ and any integer $n \geq k$ find n points in the d-dimensional unit cube $[0, 1]^d$, such that the minimum volume of the convex hull of any k points among these n points is as large as possible. Let the corresponding supremum values – over all distributions of n points in $[0, 1]^d$ – on the minimum volumes of the convex hull of k points among n points be denoted by $\Delta_{k,d}(n)$.

This problem has been investigated also by Chazelle, who considered it in connection with lower bounds on the query complexity of range searching problems. He proved in [7] that for any fixed dimension $d \geq 2$ there exists a constant $c > 0$ such that a random set of n points in the unit cube $[0, 1]^d$ satisfies with probability greater than $1 - 1/n$, that the volume of the convex hull of any $k \geq \log n$ points is $\Omega(k/n)$, indeed it holds $\Delta_{k,d}(n) = \Theta(k/n)$ for $\log n \leq k \leq n$ for fixed $d \geq 2$. An extension of the range of k might also improve his lower bounds on the query complexity, see [7].

Here we consider the case of fixed values k and d. Areas of triangles arising from n points in $[0, 1]^d$ have been investigated in [12], where for fixed dimension $d \geq 2$ it has been shown that $\Delta_{3,d}(n) = \Omega((\log n)^{1/(d-1)}/n^{2/(d-1)})$ and $\Delta_{3,d}(n) = O(1/n^{2/d})$. Moreover, for fixed $k \leq d + 1$ it has been proved recently in [14] that $\Delta_{k,d}(n) = \Omega((\log n)^{1/(d-k+2)}/n^{(k-1)/(d-k+2)})$. For the special case of dimension $d = 2$ and arbitrary $k \geq 3$ it was shown in [13] that $\Delta_{k,2}(n) = \Omega((\log n)^{1/(k-1)}/n^{(k-1)/(k-2)})$.

Here we prove the following lower bounds, in particular for $k > d$.

Theorem 1. *Let $d, k \geq 2$ be fixed integers.*

(i) *Then, for any integer $n \geq k$ there exists a configuration of n points in the unit cube $[0, 1]^d$, such that, simultaneously for $j = 2, \ldots, k$, the volume of the convex hull of any j points among these n points is*

$$\Omega(1/n^{(j-1)/(1+|d-j+1|)}). \tag{1}$$

(ii) *Moreover, for fixed $k \geq d + 1$ there is a deterministic polynomial time algorithm, which finds for any integer $n \geq k$ a configuration of n points in $[0, 1]^d$, which, simultaneously for $j = d+1, \ldots, k$, achieves the lower bound $\Omega(1/n^{(j-1)/(1+|d-j+1|)})$ on the volume of the convex hull of any j among the n points in $[0, 1]^d$.*

Our arguments remain valid if d and k are functions of n, but then the lower bound (1) will depend on d and j. Notice that for fixed integers $d, j \geq 2$, Theorem 1 yields $\Delta_{j,d} = \Omega(1/n^{(j-1)/(1+|d-j+1|)})$. Concerning upper bounds, for fixed integers $d, j \geq 2$ a partition of $[0, 1]^d$ into d-dimensional subcubes each of volume $\Theta(n^{-1/j})$, yields $\Delta_{j,d}(n) = O(1/n^{(j-1)/d})$ for $j \leq d + 1$ and $\Delta_{j,d}(n) = O(1/n)$ for $j \geq d + 1$. Moreover, for even integers j, $2 \leq j \leq d + 1$, the upper bound can be improved to $\Delta_{j,d}(n) = O(1/n^{(j-1)/d+(j-2)/(2d(d-1))})$, see [14].

Somewhat surprisingly, achieving by a deterministic polynomial time algorithm for the same n points in $[0, 1]^d$ the lower bound $\Delta_{j,d}(n) = \Omega(1/n^{(j-1)/(1+|d-j+1|)})$, simultaneously for $j = 2, \ldots, k$, where $d, k \geq 2$ are fixed integers, causes so far some difficulties w.r.t. the lower dimensional simplices, i.e., for $4 \leq j \leq d$.

2   Lower Bounds

Let dist (P_i, P_j) be the *Euclidean distance* between the points $P_i, P_j \in [0,1]^d$. A *simplex* given by the points $P_1, \ldots, P_j \in [0,1]^d$, $2 \leq j \leq d+1$, is the set of all points $P_1 + \sum_{i=2}^{j} \lambda_i \cdot (P_i - P_1)$ with $\sum_{i=2}^{j} \lambda_i \leq 1$ and $\lambda_2, \ldots, \lambda_j \geq 0$. The $((j-1)$-dimensional) *volume of a simplex* given by j points $P_1, \ldots, P_j \in [0,1]^d$, $2 \leq j \leq d+1$, is defined by vol $(P_1, \ldots, P_j) := 1/(j-1)! \cdot \prod_{i=2}^{j}$ dist $(P_i; \langle P_1, \ldots, P_{i-1} \rangle)$, where dist $(P_i; \langle P_1, \ldots, P_{i-1} \rangle)$ is the Euclidean distance of the point P_i from the affine real space $\langle P_1, \ldots, P_{i-1} \rangle$ generated by the vectors $P_2^\top - P_1^\top, \ldots, P_{i-1}^\top - P_1^\top$ attached at P_1. For j points $P_1, \ldots, P_j \in [0,1]^d$, $j \geq d+1$, let vol $(P_1, \ldots, P_j)$ be the (d-dimensional) volume of the convex hull of the points $P_1, \ldots, P_j$.

First we prove part (i) of Theorem 1.

Proof. Let $d, k \geq 2$ be fixed integers. For arbitrary integers $n \geq k$, we select uniformly at random and independently of each other $N := k \cdot n$ points $P_1, P_2, \ldots, P_N$ from the unit cube $[0,1]^d$.

Set $v_j := \beta_j / n^{\gamma_j}$ for constants $\beta_j, \gamma_j > 0$, $j = 2, \ldots, k$, which will be fixed later. Let $V := \{P_1, P_2, \ldots, P_N\}$ be the random set of chosen points in $[0,1]^d$. For $j = 2, \ldots, k$, let $\mathcal{E}_j$ be the set of all j-element subsets $\{P_{i_1}, \ldots, P_{i_j}\} \in [V]^j$ of points in V such that vol $(P_{i_1}, \ldots, P_{i_j}) \leq v_j$. We estimate the expected numbers $E(|\mathcal{E}_j|)$ of j-element sets in $\mathcal{E}_j$, $j = 2, \ldots, k$, and we show that for a suitable choice of the parameters $v_2, \ldots, v_k$ all numbers $E(|\mathcal{E}_j|)$ are not too big, i.e., $E(|\mathcal{E}_2|) + \cdots + E(|\mathcal{E}_k|) \leq (k-1) \cdot n$. Thus, there exists a choice of N points $P_1, P_2, \ldots, P_N \in [0,1]^d$ such that $|\mathcal{E}_2| + \cdots + |\mathcal{E}_k| \leq (k-1) \cdot n$. Then, for $j = 2, \ldots, k$, we delete one point from each j-element set of points in $\mathcal{E}_j$. The remaining points yield at least n points such that the volume of the convex hull of any j points of these at least n points is at least v_j.

Lemma 1. *Let $d, k \geq 2$ be fixed integers. For $j = 2, \ldots, k$, there exist constants $c_{j,d} > 0$ such that for every real $v_j > 0$ it is*

$$E(|\mathcal{E}_j|) \leq c_{j,d} \cdot N^j \cdot v_j^{1+|d-j+1|}. \tag{2}$$

Proof. For reals $v_j > 0$ and random points $P_1, \ldots, P_j \in [0,1]^d$ we give an upper bound on the probability Prob (vol $(P_1, \ldots, P_j) \leq v_j$). We assume that the points $P_1, \ldots, P_j$ are numbered such that for $2 \leq g \leq h \leq j$ and $g \leq d+1$ it is

$$\text{dist } (P_g; \langle P_1, \ldots, P_{g-1} \rangle) \geq \text{dist } (P_h; \langle P_1, \ldots, P_{g-1} \rangle). \tag{3}$$

The point P_1 can be anywhere in $[0,1]^d$. Given the point P_1, the probability, that the point $P_2 \in [0,1]^d$ has from P_1 a Euclidean distance within the infinitesimal range $[r_1, r_1 + dr_1]$, is at most the difference of the volumes of the d-dimensional balls with center P_1 and with radii $(r_1 + dr_1)$ and r_1, respectively, hence

$$\text{Prob } (r_1 \leq \text{ dist } (P_1, P_2) \leq r_1 + dr_1) \leq d \cdot C_d \cdot r_1^{d-1} dr_1,$$

where C_d denotes the volume of the d-dimensional unit ball in $\mathbb{R}^d$.

Given the points P_1 and P_2 with dist $(P_1, P_2) = r_1$, the probability that the Euclidean distance of the point $P_3 \in [0,1]^d$ from the affine line $\langle P_1, P_2 \rangle$ is within the infinitesimal range $[r_2, r_2 + dr_2]$ is at most the difference of the volumes of two cylinders centered at the line $\langle P_1, P_2 \rangle$ with radii $r_2 + dr_2$ and r_2, respectively, and, by assumption (3), with height $2 \cdot r_1 = 2 \cdot$ dist (P_1, P_2), thus

$$\text{Prob } (r_2 \leq \text{ dist } (P_3; \langle P_1, P_2 \rangle) \leq r_2 + dr_2) \leq 2 \cdot r_1 \cdot (d-1) \cdot C_{d-1} \cdot r_2^{d-2} dr_2 \,.$$

In general, let the points $P_1, \ldots, P_g$, $g < j$ and $g < d + 1$, be given with dist $(P_x; \langle P_1, \ldots, P_{x-1} \rangle) = r_{x-1}$ for $x = 2, \ldots, g$. For $g \leq j - 2$ and $g \leq d - 1$, by (3) the projection of the point P_{g+1} onto the affine space $\langle P_1, \ldots, P_g \rangle$ is contained in a $(g-1)$-dimensional box with volume $2^{g-1} \cdot r_1 \cdots r_{g-1}$, hence

$$\text{Prob } (r_g \leq \text{ dist } (P_{g+1}; \langle P_1, \ldots, P_g \rangle) \leq r_g + dr_g)$$
$$\leq 2^{g-1} \cdot r_1 \cdots r_{g-1} \cdot (d-g+1) \cdot C_{d-g+1} \cdot r_g^{d-g} dr_g \,. \tag{4}$$

For $g = j - 1 < d$, to satisfy vol $(P_1, \ldots, P_j) \leq v_j$, we must have $1/(j-1)! \cdot \prod_{i=2}^{j} \text{dist } (P_i; \langle P_1, \ldots, P_{i-1} \rangle) \leq v_j$. By (3) the projection of the point P_j onto the affine space $\langle P_1, \ldots, P_{j-1} \rangle$ is contained in a $(j-2)$-dimensional box with volume $2^{j-2} \cdot r_1 \cdots r_{j-2}$, and the point P_j has Euclidean distance at most $((j-1)! \cdot v_j)/(r_1 \cdots r_{j-2})$ from the affine space $\langle P_1, \ldots, P_{j-1} \rangle$, which happens with probability at most

$$2^{j-2} \cdot r_1 \cdots r_{j-2} \cdot C_{d-j+2} \cdot \left(\frac{(j-1)! \cdot v_j}{r_1 \cdots r_{j-2}} \right)^{d-j+2} \,. \tag{5}$$

For $d \leq g \leq j - 1$, the projection of the point P_{g+1} onto the affine space $\langle P_1, \ldots, P_d \rangle$ is contained in a $(d-1)$-dimensional box with volume at most $2^{d-1} \cdot r_1 \cdots r_{d-1}$. Since vol $(P_1, \ldots, P_d, P_{g+1}) \leq v_j$ by monotonicity, the point P_{g+1} has Euclidean distance at most $(d! \cdot v_j)/(r_1 \cdots r_{d-1})$ from the affine space $\langle P_1, \ldots, P_d \rangle$, which happens with probability at most

$$2^{d-1} \cdot r_1 \cdots r_{d-1} \cdot \frac{2 \cdot d! \cdot v_j}{r_1 \cdots r_{d-1}} = d! \cdot 2^d \cdot v_j \,. \tag{6}$$

Thus, for $j \leq d$ with (4) and (5) and some constants $c_{j,d}^*, c_{j,d}^{**} > 0$, we obtain

$$\text{Prob } (\text{vol } (P_1, \ldots, P_j) \leq v_j)$$

$$\leq \int_{r_{j-2}=0}^{\sqrt{d}} \cdots \int_{r_1=0}^{\sqrt{d}} 2^{j-2} \cdot \frac{C_{d-j+2} \cdot ((j-1)!)^{d-j+2} \cdot v_j^{d-j+2}}{(r_1 \cdots r_{j-2})^{d-j+1}} \,.$$

$$\cdot \prod_{g=1}^{j-2} \left(2^{g-1} \cdot r_1 \cdots r_{g-1} \cdot (d-g+1) \cdot C_{d-g+1} \cdot r_g^{d-g} \right) dr_{j-2} \ldots dr_1$$

$$\leq c_{j,d}^{**} \cdot v_j^{d-j+2} \cdot \int_{r_{j-2}=0}^{\sqrt{d}} \cdots \int_{r_1=0}^{\sqrt{d}} \prod_{g=1}^{j-2} \left(r_g^{2j-2g-3} \right) dr_{j-2} \ldots dr_1$$

$$\leq c_{j,d}^* \cdot v_j^{d-j+2} \qquad \text{as } 2 \cdot j - 2 \cdot g - 3 \geq 1$$

$$= c_{j,d}^* \cdot v_j^{1+|d-j+1|} \qquad \text{as } j \leq d. \tag{7}$$

Moreover, for $j = d+1, \ldots, k$, by (4) and (6) for constants $c_{j,d}^*, c_{j,d}^{**} > 0$ we infer

$$\text{Prob}\,(\text{vol}\,(P_1, \ldots, P_j) \leq v_j)$$

$$\leq \int_{r_{d-1}=0}^{\sqrt{d}} \cdots \int_{r_1=0}^{\sqrt{d}} (d! \cdot 2^d \cdot v_j)^{j-d} \cdot$$

$$\cdot \prod_{g=1}^{d-1} \left(2^{g-1} \cdot r_1 \cdots r_{g-1} \cdot (d-g+1) \cdot C_{d-g+1} \cdot r_g^{d-g} \right) dr_{d-1} \ldots dr_1$$

$$\leq c_{j,d}^{**} \cdot v_j^{j-d} \cdot \int_{r_{d-1}=0}^{\sqrt{d}} \cdots \int_{r_1=0}^{\sqrt{d}} \prod_{g=1}^{d-1} \left(r_g^{2d-2g-1} \right) dr_{d-1} \ldots dr_1$$

$$\leq c_{j,d}^* \cdot v_j^{j-d} \qquad \text{as } 2 \cdot d - 2 \cdot g - 1 \geq 1$$

$$= c_{j,d}^* \cdot v_j^{1+|d-j+1|} \qquad \text{as } j \geq d+1. \tag{8}$$

By (7) and (8) we have $\text{Prob}\,(\text{vol}\,(P_1, \ldots, P_j) \leq v_j) \leq c_{j,d}^* \cdot v_j^{1+|d-j+1|}$ for constants $c_{j,d}^* > 0$, $j = 2, \ldots, k$. Since there are $\binom{N}{j}$ choices for j out of the N random points $P_1, \ldots, P_N \in [0,1]^d$, inequality (2) follows. $\qquad\square$

By (2) and Markov's inequality there exist $N = k \cdot n$ points $P_1, \ldots, P_N$ in the unit cube $[0,1]^d$ such that for $j = 2, \ldots, k$:

$$|\mathcal{E}_j| \leq k \cdot c_{j,d} \cdot N^j \cdot v_j^{1+|d-j+1|}. \tag{9}$$

Lemma 2. *Let $d, k \geq 2$ be fixed integers. Then, for every β_j, γ_j with $0 < \beta_j \leq 1/(c_{j,d} \cdot k^{j+1})^{1/(1+|d-j+1|)}$ and $\gamma_j \geq (j-1)/(1+|d-j+1|)$, $j = 2, \ldots, k$, it is*

$$|\mathcal{E}_j| \leq N/k. \tag{10}$$

Proof. For $j = 2, \ldots, k$, by (9) and using $v_j = \beta_j / n^{\gamma_j}$ we infer

$$|\mathcal{E}_j| \leq N/k$$

$$\Longleftarrow k \cdot c_{j,d} \cdot N^j \cdot v_j^{1+|d-j+1|} \leq N/k$$

$$\Longleftrightarrow k^{j+1} \cdot c_{j,d} \cdot \beta_j^{1+|d-j+1|} \cdot n^{j-1-\gamma_j(1+|d-j+1|)} \leq 1,$$

which holds for $j - 1 \leq \gamma_j \cdot (1 + |d-j+1|)$ and $k^{j+1} \cdot c_{j,d} \cdot \beta_j^{1+|d-j+1|} \leq 1$. $\qquad\square$

Fix $\gamma_j := (j-1)/(1+|d-j+1|)$ and $\beta_j := 1/(c_{j,d} \cdot k^{j+1})^{1/(1+|d-j+1|)}$, $j = 2, \ldots, k$. By Lemma 2 we have $|\mathcal{E}_2| + \cdots + |\mathcal{E}_k| \leq ((k-1)/k) \cdot N$. For $j = 2, \ldots, k$, we discard one point from each j-element set in $\mathcal{E}_j$. Then, the set $I \subseteq V$ of remaining points contains a subset of size $N/k = n$. These n points in $[0,1]^d$ satisfy, simultaneously for $j = 2, \ldots, k$, that the volume of the convex hull of each j of these n points is bigger than $v_j = \beta_j / n^{(j-1)/(1+|d-j+1|)}$, which finishes the proof of part (i) and (1) in Theorem 1. $\qquad\square$

3 A Deterministic Algorithm

Here we derandomize the probabilistic arguments from Section 2 to show Theorem 1, part (ii). Throughout this section, let $k \geq d+1$. Let $B^d(T)$ denote the d-dimensional ball with radius T around the origin. Then $B^d(T) \cap \mathbb{Z}^d$ is the set of all points $P \in \mathbb{Z}^d$, which have Euclidean distance at most T from the origin. To provide a deterministic polynomial time algorithm which, for any integer $n > 0$, finds a configuration of n points in $[0,1]^d$, such that the volume of the convex hull of small sets of points is large, we discretize the unit cube $[0,1]^d$ by considering, for T large enough, but bounded from above by a polynomial in n, all points in $B^d(T) \cap \mathbb{Z}^d$. This set $B^d(T) \cap \mathbb{Z}^d$ will be rescaled later by the factor T^d. However, with this discretization we have to take care of degenerate sets of points, where a set $\{P_1, \ldots, P_j\} \subset [0,1]^d$ with $j \geq d+1$ is called *degenerate*, if all points $P_1, \ldots, P_j$ are contained in a $(d-1)$-dimensional affine subspace of $\mathbb{R}^d$, otherwise $\{P_1, \ldots, P_j\}$ is called *non-degenerate*.

Set $v_j := \beta_j \cdot T^d / n^{(j-1)/(j-d)}$ for suitable constants $\beta_j > 0$, $j = d+1, \ldots, k$, which will be fixed later. We construct for $j = d+1, \ldots, k$ two types of j-element edges. For points $P_{i_1}, \ldots, P_{i_j} \in B^d(T) \cap \mathbb{Z}^d$, let $\{P_{i_1}, \ldots, P_{i_j}\} \in \mathcal{E}_j$ if and only if $\mathrm{vol}\,(P_{i_1}, \ldots, P_{i_j}) \leq v_j$ and $\{P_{i_1}, \ldots, P_{i_j}\}$ is not contained in a $(d-1)$-dimensional affine subspace of $\mathbb{R}^d$, i.e., the set $\{P_{i_1}, \ldots, P_{i_j}\}$ is non-degenerate. Moreover, let $\{P_{i_1}, \ldots, P_{i_j}\} \in \mathcal{E}_j^0$ if and only if $\{P_{i_1}, \ldots, P_{i_j}\}$ is contained in a $(d-1)$-dimensional affine subspace of $\mathbb{R}^d$.

To give upper bounds on these numbers $|\mathcal{E}_j|$ and $|\mathcal{E}_j^0|$ of j-element sets, $j = d+1, \ldots, k$, we use *lattices* in $\mathbb{Z}^d$.

A *lattice* L in $\mathbb{Z}^d$ is a subset of $\mathbb{Z}^d$, which is generated by all integral linear combinations of some linearly independent vectors $b_1, \ldots, b_m \in \mathbb{Z}^d$, hence $L = \mathbb{Z}b_1^\top + \cdots + \mathbb{Z}b_m^\top$. The parameter $m = \mathrm{rank}(L)$ is called the *rank* of the lattice L, and the set $\mathcal{B} = \{b_1, \ldots, b_m\}$ is called a *basis* of L. The set $F_\mathcal{B} := \{\sum_{i=1}^m \alpha_i \cdot b_i \mid 0 \leq \alpha_i \leq 1, i = 1, \ldots, m\} \subseteq \mathbb{R}^d$ is called the *fundamental parallelepiped* $F_\mathcal{B}$ of $\mathcal{B}$, its *volume* is $\mathrm{vol}(F_\mathcal{B}) := (\det(G(\mathcal{B})^\top \cdot G(\mathcal{B})))^{1/2}$, where $G(\mathcal{B}) := (b_1, \ldots, b_m)_{d \times m}$ is the $d \times m$ *generator matrix* of $\mathcal{B}$ (up to the ordering of the vectors). If $\mathcal{B}$ and $\mathcal{B}'$ are two bases of a lattice L in $\mathbb{Z}^d$, then the volumes of the fundamental parallelepipeds are equal, i.e., $\mathrm{vol}(F_\mathcal{B}) = \mathrm{vol}(F_{\mathcal{B}'})$, see [6].

For integers $a_1, \ldots, a_n \in \mathbb{Z}$, which are not all equal to 0, let $\gcd(a_1, \ldots, a_n)$ denote the *greatest common divisor* of $a_1, \ldots, a_n$. For vectors $a = (a_1, \ldots, a_d)^\top \in \mathbb{R}^d$ and $b = (b_1, \ldots, b_d)^\top \in \mathbb{R}^d$ let $\langle a, b \rangle := \sum_{i=1}^d a_i \cdot b_i$ be the standard scalar product. The length of a vector $a \in \mathbb{R}^d$ is defined by $\|a\| := \sqrt{\langle a, a \rangle}$. For a lattice L in $\mathbb{Z}^d$ let $\mathrm{span}(L)$ be the linear space over the reals, which is generated by the vectors in L. For a subset $S = \{P_1, \ldots, P_k\} \subset \mathbb{R}^d$ of points the *rank* of S is the dimension of the linear space over the reals, which is generated by the vectors $P_2^\top - P_1^\top, \ldots, P_k^\top - P_1^\top$.

A vector $a = (a_1, \ldots, a_d)^\top \in \mathbb{Z}^d \setminus \{0^d\}$ is called *primitive*, if $\gcd(a_1, \ldots, a_d) = 1$ and $a_j > 0$ with $j = \min\{i \mid a_i \neq 0\}$. A lattice L in $\mathbb{Z}^d$ is called *m-maximal*, if $\mathrm{rank}(L) = m$ and no other lattice $L' \neq L$ in $\mathbb{Z}^d$ with $\mathrm{rank}(L') = m$ contains L as a proper subset. There is a one-to-one correspondence between *m-maximal* lattices in $\mathbb{Z}^d$ and primitive vectors $a = (a_1, \ldots, a_d)^\top \in \mathbb{Z}^d \setminus \{0^d\}$:

(i) For each lattice L in $\mathbb{Z}^d$ with $\operatorname{rank}(L) = d - 1 \geq 1$ there is exactly one primitive vector $a_L = (a_1, \ldots, a_d)^\top \in \mathbb{Z}^d \setminus \{0^d\}$ with $\langle a_L, x^\top \rangle = 0$ for every $x \in L$. This vector $a_L \in \mathbb{Z}^d \setminus \{0^d\}$ is called the *primitive normal vector* of the lattice L.

(ii) For each lattice L' in $\mathbb{Z}^d$ with $\operatorname{rank}(L') = d - 1$ there is exactly one $(d - 1)$-maximal lattice L in $\mathbb{Z}^d$ with $L' \subseteq L$.

(iii) There exists a bijection between the set of all $(d - 1)$-maximal lattices L in $\mathbb{Z}^d$ and the set of all primitive vectors a_L in $\mathbb{Z}^d$.

For a $(d - 1)$-maximal lattice L in $\mathbb{Z}^d$, a *residue class* of L is a set L' of the form $L' = x + L$ with $x \in \mathbb{Z}^d$.

The proofs of Lemmas 3 – 6 concerning lattices can be found in [15].

Lemma 3 ([15]). *Let L be a $(d-1)$-maximal lattice in $\mathbb{Z}^d$ with primitive normal vector $a_L \in \mathbb{Z}^d$ and with basis $\mathcal{B}$.*

(i) *There exists a point $v \in \mathbb{Z}^k \setminus L$ such that $\mathbb{Z}^d$ can be partitioned into the residue classes $s \cdot v + L$, $s \in \mathbb{Z}$, and, for each point $x \in L$, it is $\operatorname{dist}(s \cdot v + x, \operatorname{span}(L)) = |s| / \|a_L\|$.*

(ii) *The volume of the fundamental parallelepiped $F_\mathcal{B}$ fulfills $\operatorname{vol}(F_\mathcal{B}) = \|a_L\|$.*

Lemma 4 ([15]). *Let $d \in \mathbb{N}$ be fixed. Let $S \subseteq B^d(T) \cap \mathbb{Z}^d$ be a set of points with $\operatorname{rank}(S) \leq d - 1$. Then there exists a $(d - 1)$-maximal lattice L of $\mathbb{Z}^d$ such that S is contained in some residue class $L' = v + L$ of L for some $v \in \mathbb{Z}^d$, and L has a basis $b_1, \ldots, b_{d-1} \in \mathbb{Z}^d$ with $\max_{i=1,\ldots,d-1} \|b_i\| = O(T)$.*

The next lemma is crucial in our considerations to estimate the numbers $|\mathcal{E}_j|$ and $|\mathcal{E}_j^0|$ of j-element sets, $j = d + 1, \ldots, k$.

Lemma 5 ([15]). *Let $d \in \mathbb{N}$ be fixed. Let L be a $(d - 1)$-maximal lattice of $\mathbb{Z}^d$ with primitive normal vector $a_L \in \mathbb{Z}^d$, and let $\mathcal{B} = \{b_1, \ldots, b_{d-1}\}$ be a basis of L with $\max_{i=1,\ldots,d-1} \|b_i\| = O(T)$. Then the following hold:*

(i) *The primitive normal vector a_L satisfies $\|a_L\| = O(T^{d-1})$.*

(ii) *For every residue class L' of L it is $|L' \cap B^d(T)| = O(T^{d-1} / \|a_L\|)$.*

For integers $g, l \in \mathbb{N}$ let $r_g(l)$ be the number of representations $x_1^2 + \cdots + x_g^2 = l$ with $x_1, \ldots, x_g \in \mathbb{Z}$.

Lemma 6 ([15]). *Let $g, r \in \mathbb{N}$ be fixed integers. Then, for all integers $m \in \mathbb{N}$:*

$$\sum_{l=1}^{m} \frac{r_g(l)}{l^r} = \begin{cases} O\left(m^{g/2-r}\right) & \text{if } g/2 - r > 0 \\ O\left(\log m\right) & \text{if } g/2 - r = 0 \\ O(1) & \text{if } g/2 - r < 0. \end{cases}$$

Lemma 7. *Let $d, k \geq 2$ be fixed integers with $k \geq d + 1$. For $j = d + 1, \ldots, k$, there exist constants $c_{j,0} > 0$, such that the numbers $|\mathcal{E}_j^0|$ of j-element degenerate sets of points in $B^d(T) \cap \mathbb{Z}^d$ satisfy*

$$|\mathcal{E}_j^0| \leq c_{j,0} \cdot T^{(d-1)j+1} \cdot \log T. \tag{11}$$

Proof. By Lemma 4, each degenerate j-element subset of points in $B^d(T) \cap \mathbb{Z}^d$ is contained in a residue class L' of some $(d-1)$-maximal lattice L in $\mathbb{Z}^d$, and L has a basis $b_1, \ldots, b_{d-1} \in \mathbb{Z}^d$ with $\|b_i\| = O(T)$, $i = 1, \ldots, d-1$. By Lemma 5(i), it suffices to consider all $(d-1)$-maximal lattices L with primitive normal vectors $a_L \in \mathbb{Z}^d$ of length $\|a_L\| = O(T^{d-1})$.

Having fixed a $(d-1)$-maximal lattice L in $\mathbb{Z}^d$, which is determined by its primitive normal vector $a_L \in \mathbb{Z}^d$, by Lemma 3(i), there are $O(T \cdot \|a_L\|)$ residue classes L' of the lattice L with $L' \cap B^d(T) \neq \emptyset$. By Lemma 5(ii), each set $L' \cap B^d(T)$ contains $O(T^{d-1}/\|a_L\|)$ points. From each set $L' \cap B^d(T)$ we can select j points in $\binom{O(T^{d-1}/\|a_L\|)}{j}$ ways to obtain a degenerate set of j points. This implies

$$|\mathcal{E}_j^0| = O\left(\sum_{a \in \mathbb{Z}^d, \|a\| = O(T^{d-1})} T \cdot \|a\| \cdot \binom{T^{d-1}/\|a\|}{j} \right)$$

$$= O\left(T^{(d-1)j+1} \cdot \sum_{a \in \mathbb{Z}^d, \|a\| = O(T^{d-1})} \frac{1}{\|a\|^{j-1}} \right)$$

$$= O\left(T^{(d-1)j+1} \cdot \sum_{l=1}^{O(T^{2d-2})} \frac{r_d(l)}{l^{(j-1)/2}} \right) = O\left(T^{(d-1)j+1} \cdot \log T \right),$$

since, by Lemma 6, we have $\sum_{l=1}^m r_d(l)/l^{(j-1)/2} = O(\log m)$ for $j = d+1$ and $\sum_{l=1}^m r_d(l)/l^{(j-1)/2} = O(1)$ for $j = d+2, \ldots, k$. $\qquad\square$

Lemma 8. *Let $d, k \geq 2$ be fixed integers with $k \geq d+1$. For $j = d+1, \ldots, k$, there exist constants $c_j > 0$, such that the numbers $|\mathcal{E}_j|$ of j-element non-degenerate sets of points in $B^d(T) \cap \mathbb{Z}^d$ with the volume of their convex hull at most v_j, fulfill*

$$|\mathcal{E}_j| \leq c_j \cdot T^{d^2} \cdot v_j^{j-d}. \tag{12}$$

Proof. For $j = d+1, \ldots, k$, consider j points $P_1, \ldots, P_j \in B^d(T) \cap \mathbb{Z}^d$ with vol $(P_1, \ldots, P_j) \leq v_j$, where $\{P_1, \ldots, P_j\}$ is non-degenerate. Let these points be numbered such that for $2 \leq g \leq h \leq j$ and $g \leq d+1$ it is

$$\text{dist } (P_g; \langle P_1, \ldots, P_{g-1} \rangle) \geq \text{dist } (P_h; \langle P_1, \ldots, P_{g-1} \rangle). \tag{13}$$

By Lemma 4, the points $P_1, \ldots, P_d \in B^d(T) \cap \mathbb{Z}^d$ are contained in a residue class L' of some $(d-1)$-maximal lattice L in $\mathbb{Z}^d$ with primitive normal vector $a_L \in \mathbb{Z}^d$, where L has a basis $b_1, \ldots, b_{d-1} \in \mathbb{Z}^d$ with $\|b_i\| = O(T)$ for $i = 1, \ldots, d-1$. By Lemma 5(i), it suffices to consider all $(d-1)$-maximal lattices L with primitive vectors $a_L \in \mathbb{Z}^d$ of length $\|a_L\| = O(T^{d-1})$.

We fix a $(d-1)$-maximal lattice L in $\mathbb{Z}^d$, which is determined by its primitive normal vector $a_L \in \mathbb{Z}^d$. By Lemma 3(i), there are $O(T \cdot \|a_L\|)$ residue classes L' of L with $L' \cap B^d(T) \neq \emptyset$. By Lemma 5(ii), from each set $L' \cap B^d(T)$ we can select d points $P_1, \ldots, P_d$ in $\binom{O(T^{d-1}/\|a_L\|)}{d}$ ways. By (13) we infer for the $(d-1)$-dimensional volume vol $(P_1, \ldots, P_d) > 0$, as otherwise $\{P_1, \ldots, P_j\}$ is degenerate. Also by (13) the projection of each point $P_i \in B^d(T) \cap \mathbb{Z}^d$, $i = d+1, \ldots, j$, onto the residue class L' is contained in a $(d-1)$-dimensional box of volume $2^{d-1} \cdot (d-1)! \cdot \text{vol} (P_1, \ldots, P_d)$, which, by Lemma 3(ii), contains at most

$$2^{d-1} \cdot (d-1)! \cdot 2^{d-1} \cdot \text{vol} (P_1, \ldots, P_d)/\|a_L\| \tag{14}$$

points of L', since $P_1, \ldots, P_d \in L'$. With vol $(P_1, \ldots, P_d, P_i) \leq v_j$ it follows that dist $(P_i, \langle P_1, \ldots, P_d \rangle) \leq d \cdot v_j/\text{vol} (P_1, \ldots, P_d)$, and, by Lemma 3(i), each point $P_i \in B^d(T) \cap \mathbb{Z}^d$, $i = d+1, \ldots, j$, is contained in one of at most

$$\|a_L\| \cdot d \cdot v_j/\text{vol} (P_1, \ldots, P_d) \tag{15}$$

residue classes L'' of L. By (14) in each residue class L'' we can choose at most $(d-1)! \cdot 2^{2d-2} \cdot \text{vol} (P_1, \ldots, P_d)/\|a_L\|$ points $P_i \in B^d(T) \cap \mathbb{Z}^d$, hence with (15) each point P_i, $i = d+1, \ldots, j$, can be chosen in at most $d! \cdot 2^{2d-2} \cdot v_j$ ways. Applying this to each point $P_{d+1}, \ldots, P_j \in B^d \cap \mathbb{Z}^d$, we infer the upper bound

$$|\mathcal{E}_j| = O \left(\sum_{a \in \mathbb{Z}^d, \|a\|=O(T^{d-1})} T \cdot \|a\| \cdot \binom{T^{d-1}/\|a\|}{d} \cdot v_j^{j-d} \right)$$

$$= O \left(T^{d^2-d+1} \cdot v_j^{j-d} \cdot \sum_{a \in \mathbb{Z}^d, \|a\|=O(T^{d-1})} \frac{1}{\|a\|^{d-1}} \right)$$

$$= O \left(T^{d^2-d+1} \cdot v_j^{j-d} \cdot \sum_{l=1}^{O(T^{2d-2})} \frac{r_d(l)}{l^{(d-1)/2}} \right) = O(T^{d^2} \cdot v_j^{j-d}),$$

since, by Lemma 6, we have $\sum_{l=1}^{m} r_d(l)/l^{(d-1)/2} = O(m^{1/2})$. $\square$

For fixed integers $d, j, k \geq 2$ the sets $\mathcal{E}_j$ and $\mathcal{E}_j^0$, can easily be constructed in time polynomial in T. Namely, by considering every j-element subset $S \subset B^d(T) \cap \mathbb{Z}^d$ of points, we determine all degenerate sets of j points in $B^d(T) \cap \mathbb{Z}^d$ and all non-degenerate sets of j points in $B^d(T) \cap \mathbb{Z}^d$ with volume of their convex hulls at most v_j in time $O(T^{dj})$, since there are $\binom{O(T^d)}{j}$ j-element subsets in $B^d(T) \cap \mathbb{Z}^d$.

Let $|B^d(T) \cap \mathbb{Z}^d| = C_d' \cdot T^d$, where $C_d' > 0$ is a constant. We enumerate the points in $B^d(T) \cap \mathbb{Z}^d$ by $P_1, \ldots, P_{C_d' \cdot T^d}$. To each point P_i associate a parameter $p_i \in [0, 1]$, $i = 1, \ldots, C_d' \cdot T^d$, and define a potential function $F(p_1, \ldots, p_{C_d' \cdot T^d})$:

$$F(p_1, \ldots, p_{C_d' \cdot T^d}) := 2^{p C_d' T^d/2} \cdot \prod_{i=1}^{C_d' T^d} \left(1 - \frac{p_i}{2} \right) +$$

$$+ \sum_{j=d+1}^{k} \frac{\sum_{\{i_1, \ldots i_j\} \in \mathcal{E}_j} p_{i_1} \cdots p_{i_j}}{2 \cdot k \cdot p^j \cdot c_j \cdot T^{d^2} \cdot v_j^{j-d}} + \sum_{j=d+1}^{k} \frac{\sum_{\{i_1, \ldots, i_j\} \in \mathcal{E}_j^0} p_{i_1} \cdots p_{i_j}}{2 \cdot k \cdot p^j \cdot c_{j,0} \cdot T^{(d-1)j+1} \cdot \log T}.$$

With the initialisation $p_1 := \cdots := p_{C'_d \cdot T^d} := p = (2 \cdot k \cdot n)/(C'_d \cdot T^d) \leq 1$, i.e., say $T^d = \omega(n)$, we infer by Lemmas 7 and 8 that $F(p, \ldots, p) < (2/e)^{p C'_d T^d /2} + (2k - 2d)/(2k)$, which is less than 1 for $p \cdot C'_d \cdot T^d \geq 7 \cdot \ln k$. Using the linearity of $F(p_1, \ldots, p_{C'_d \cdot T^d})$ in each p_i, we minimize $F(p_1, \ldots, p_{C'_d \cdot T^d})$ by choosing one after the other $p_i := 0$ or $p_i := 1$ for $i = 1, \ldots, C'_d \cdot T^d$, and finally we obtain $F(p_1, \ldots, p_{C'_d \cdot T^d}) < 1$. With $V^* = \{P_i \in B^d(T) \cap \mathbb{Z}^d \mid p_i = 1\}$ this yields a subset $V^* \subseteq B^d(T) \cap \mathbb{Z}^d$ of points and subsets $\mathcal{E}_j^{0*} := [V^*]^j \cap \mathcal{E}_j^0$ and $\mathcal{E}_j^* := [V^*]^j \cap \mathcal{E}_j$ of j-element sets, $j = d+1, \ldots, k$, such that

$$|V^*| \geq p \cdot C'_d \cdot T^d /2 \tag{16}$$

$$|\mathcal{E}_j^*| \leq 2 \cdot k \cdot p^j \cdot c_j \cdot T^{d^2} \cdot v_j^{j-d} \tag{17}$$

$$|\mathcal{E}_j^{0*}| \leq 2 \cdot k \cdot p^j \cdot c_{j,0} \cdot T^{(d-1)j+1} \cdot \log T . \tag{18}$$

By choice of the parameters v_j, $j = d+1, \ldots, k$, the running time of this derandomization is $O(T^d + \sum_{j=d+1}^{k}(|\mathcal{E}_j| + |\mathcal{E}_j^0|)) = O(T^{dk})$, which is polynomial in T for fixed integers $d, k \geq 2$.

Lemma 9. *For $j = d+1, \ldots, k$, and $0 < \beta_j \leq (C_d'^{j}/(2^{j+2} \cdot k^{j+1} \cdot c_{j,d}))^{1/(j-d)}$, it is*

$$|\mathcal{E}_j^*| \leq |V^*|/(2 \cdot k) .$$

Proof. By (16) and (17) with $v_j := \beta_j \cdot T^d/n^{\frac{j-1}{j-d}}$, and $p = (2 \cdot k \cdot n)/(C'_d \cdot T^d)$, and with $\beta_j > 0$ it is

$$|\mathcal{E}_j^*| \leq |V^*|/(2 \cdot k)$$

$$\Longleftarrow 2 \cdot k \cdot p^j \cdot c_j \cdot T^{d^2} \cdot v_j^{j-d} \leq p \cdot C'_d \cdot T^d/(4 \cdot k)$$

$$\Longleftrightarrow 8 \cdot k^2 \cdot \left(\frac{2 \cdot k \cdot n}{C'_d \cdot T^d}\right)^{j-1} \cdot c_j \cdot T^{d^2-d} \cdot \left(\frac{\beta_j \cdot T^d}{n^{\frac{j-1}{j-d}}}\right)^{j-d} \leq C'_d$$

$$\Longleftrightarrow 2^{j+2} \cdot k^{j+1} \cdot c_j \cdot \beta_j^{j-d} \leq C_d'^{j} ,$$

which holds for $\beta_j^{j-d} \leq C_d'^{j}/(2^{j+2} \cdot k^{j+1} \cdot c_j)$, $j = d+1, \ldots, k$. $\square$

Lemma 10. *For $j = d+1, \ldots, k$, and $T/(\log T)^{1/(j-1)} = \omega(n)$, it is*

$$|\mathcal{E}_j^{0*}| \leq |V^*|/(2 \cdot k) .$$

Proof. By (16) and (18), with $p = (2 \cdot k \cdot n)/(C'_d \cdot T^d)$, $j = d+1, \ldots, k$, we infer

$$|\mathcal{E}_j^{0*}| \leq |V^*|/(2 \cdot k)$$

$$\Longleftarrow 2 \cdot k \cdot p^j \cdot c_{j,0} \cdot T^{(d-1)j+1} \cdot \log T \leq p \cdot C'_d \cdot T^d/(4 \cdot k)$$

$$\Longleftrightarrow 8 \cdot k^2 \cdot \left(\frac{2 \cdot k \cdot n}{C'_d \cdot T^d}\right)^{j-1} \cdot c_{j,0} \cdot T^{(d-1)j-d+1} \cdot \log T \leq C'_d$$

$$\Longleftrightarrow 2^{j+2} \cdot k^{j+1} \cdot c_{j,0} \cdot \frac{n^{j-1}}{T^{j-1}} \cdot \log T \leq C_d'^{j} ,$$

which holds for $T/(\log T)^{1/(j-1)} = \omega(n)$. $\square$

With $T := n \cdot \log n$ and $\beta_j := (C_d'^j/(2^{j+2} \cdot k^{j+1} \cdot c_j)^{1/(j-d)}$, $j = d+1, \ldots, k$, the assumptions of Lemmas 9 and 10 are fulfilled. By deleting in time $O(|V^*| + \sum_{j=d+1}^{k}(|\mathcal{E}_j^*| + |\mathcal{E}_j^{0*}|))O(T^{kd})$ one point from each j-element set in $\mathcal{E}_j^*$ and $\mathcal{E}_j^{0*}$, $j = d+1, \ldots, k$, the remaining points yield a subset $V^{**} \subseteq V^*$ of size at least $|V^*|/k \geq p \cdot C_d' \cdot T^d/(2 \cdot k) = n$. Then these at least n points in $B^d(T) \cap \mathbb{Z}^d$ satisfy that the volume of the convex hull of any j of these points, $j = d+1, \ldots, k$, is at least v_j, i.e., $\Omega(T^d/n^{(j-1)/(j-d)})$. After rescaling by the factor T^d, we have at least n points in the unit cube $[0,1]^d$ such that the volume of the convex hull of any j of these points is $\Omega(1/n^{(j-1)/(j-d)})$, $j = d+1, \ldots, k$. Altogether the running time of this deterministic algorithm is $O((n \cdot \log n)^{dk})$ for fixed $d, k \geq 2$, hence polynomial in n, which finishes the proof of Theorem 1, part (ii).

4   Concluding Remarks

Our arguments yield a deterministic polynomial time algorithm for obtaining a distribution of n points in $[0,1]^d$, which, for fixed integers $j \geq d+1$, shows $\Delta_{j,d}(n) = \Omega(1/n^{(j-1)/(1+|d-j+1|)})$. With the results from [14], i.e., using a result of Ajtai, Komlós, Pintz, Spencer and Szemerédi [1] on uncrowded hypergraphs we can improve Theorem 1 slightly (Details are omitted here), namely for fixed integers $d, k \geq 3$ and a fixed integer j_0 with $3 \leq j_0 \leq d+1$ one can find in polynomial time a configuration of n points in $[0,1]^d$, such that, simultaneously for $j = 2, \ldots, k$ but $j \neq j_0$, the volume of the convex hull of j points among these n points is at least $\Omega(1/n^{(j-1)/(1+|d-j+1|)})$ and $\Delta_{j_0,d}(n) = \Omega((\log n)^{1/(d-j_0+2)}/n^{(j_0-1)/(d-j_0+2)})$. It would be interesting to get such an improvement by a logarithmic factor for the same n points in $[0,1]^d$, simultaneously for $3 \leq j \leq k$, for fixed d, k.

Moreover, improvements of the existing upper bounds, which were given in the introduction, are desirable. Also investigations of this problem for non-constant values of k might be of interest in view of the results of Chazelle [7].

References

1. M. Ajtai, J. Komlós, J. Pintz, J. Spencer and E. Szemerédi, *Extremal Uncrowded Hypergraphs*, Journal of Combinatorial Theory Ser. A, 32, 1982, 321–335.
2. G. Barequet, *A Lower Bound for Heilbronn's Triangle Problem in d Dimensions*, SIAM Journal on Discrete Mathematics 14, 2001, 230–236.
3. G. Barequet, *The On-Line Heilbronn's Triangle Problem in Three and Four Dimensions*, Proceedings '8rd Annual International Computing and Combinatorics Conference COCOON'02', LNCS 2387, Springer, 2002, 360–369.
4. C. Bertram-Kretzberg, T. Hofmeister and H. Lefmann, *An Algorithm for Heilbronn's Problem*, SIAM Journal on Computing 30, 2000, 383–390.
5. P. Brass, *An Upper Bound for the d-Dimensional Heilbronn Triangle Problem*, SIAM Journal on Discrete Mathematics 19, 192–195, 2005.
6. J. W. S. Cassels, *An Introduction to the Geometry of Numbers*, Vol. 99, Springer-Verlag, New York, 1971.

7. B. CHAZELLE, *Lower Bounds on the Complexity of Polytope Range Searching*, Journal of the American Mathematical Society 2, 637–666, 1989.
8. T. Jiang, M. Li and P. Vitany, *The Average Case Area of Heilbronn-type Triangles*, Random Structures & Algorithms 20, 2002, 206–219.
9. J. Komlós, J. Pintz and E. Szemerédi, *On Heilbronn's Triangle Problem*, Journal of the London Mathematical Society, 24, 1981, 385–396.
10. J. Komlós, J. Pintz and E. Szemerédi, *A Lower Bound for Heilbronn's Problem*, Journal of the London Mathematical Society, 25, 1982, 13–24.
11. H. Lefmann, *On Heilbronn's Problem in Higher Dimension*, Combinatorica 23, 2003, 669–680.
12. H. Lefmann, *Large Triangles in the d-Dimensional Unit-Cube*, Proceedings 10th Annual International Conference Computing and Combinatorics COCOON'04, eds. K.-Y. Chwa and J. I. Munro, LNCS 3106, Springer, 2004, 43–52.
13. H. Lefmann, *Distributions of Points in the Unit-Square and Large k-Gons*, Proceedings ACM-SIAM Syposium on Discrete Algorithms, SODA'05, ACM und SIAM, 241–250, 2005.
14. H. Lefmann, *Large Simplices in the d-Dimensional Unit-Cube (Extended Abstract)*, Proceedings 11th Annual International Conference Computing and Combinatorics COCOON'05, ed. L. Wang, LNCS 3595, Springer, 2005, 514–523.
15. H. Lefmann and N. Schmitt, *A Deterministic Polynomial Time Algorithm for Heilbronn's Problem in Three Dimensions*, SIAM Journal on Computing 31, 2002, 1926–1947.
16. K. F. Roth, *On a Problem of Heilbronn*, Journal of the London Mathematical Society 26, 1951, 198–204.
17. K. F. Roth, *On a Problem of Heilbronn, II, and III*, Proc. of the London Mathematical Society (3), 25, 1972, 193–212, and 543–549.
18. K. F. Roth, *Estimation of the Area of the Smallest Triangle Obtained by Selecting Three out of n Points in a Disc of Unit Area*, Proc. of Symposia in Pure Mathematics, 24, 1973, AMS, Providence, 251–262.
19. K. F. Roth, *Developments in Heilbronn's Triangle Problem*, Advances in Mathematics, 22, 1976, 364–385.
20. W. M. Schmidt, *On a Problem of Heilbronn*, Journal of the London Mathematical Society (2), 4, 1972, 545–550.

Throwing Stones Inside Simple Polygons*

Otfried Cheong[1], Hazel Everett[2], Hyo-Sil Kim[1],
Sylvain Lazard[2], and René Schott[2]

[1] Division of Computer Science, KAIST, Daejeon, South Korea
{otfried, hyosil}@tclab.kaist.ac.kr
[2] LORIA & IECN – INRIA Lorraine, Universities Nancy 1 & 2, Nancy, France
Firstname.Name@loria.fr

Abstract. Given two sets A and B of m non-intersecting line segments in the plane, we show how to compute in $O(m \log m)$ time a data structure that uses $O(m)$ space and allows to answer the following query in $O(\log m)$ time: Given a parabola $\gamma : y = ax^2 + bx + c$, does γ separate A and B? This structure can be used to build a data structure that stores a simple polygon and allows ray-shooting queries along parabolic trajectories with vertical main axis. For a polygon with complexity n, we can answer such "stone throwing" queries in $O(\log^2 n)$ time, using $O(n \log n)$ space and $O(n \log^2 n)$ preprocessing time. This matches the best known bound for circular ray shooting in simple polygons.

1 Introduction

Ray shooting is a fundamental problem in computational geometry. We are given a set of geometric objects in $\mathbb{R}^d$ (usually $d = 2$ or 3) that we wish to pre-process and store in such a way that we can quickly answer queries of the form: given a query ray (a half-infinite line segment), determine the first object hit (intersected) by the ray. Ray shooting arises in computer graphics, in visualization, and in other geometric problems such as collision detection and motion planning.

In some applications, ray-shooting queries need to be performed along rays that are not straight. Motion planning for car-like robots, for instance, makes use of ray-shooting queries along circular arcs. In this paper, we consider another natural ray-shooting query: which object will be hit first by a flying stone that moves under the influence of gravity along a parabolic trajectory?

We are aware of only one previous result that addresses ray shooting along parabolic trajectories: Sharir and Shaul [9] very recently gave a near-linear size data structure for triangles in 3D with sublinear query time.

We concentrate here on ray shooting inside a simple polygon of complexity n. For straight rays, this problem has been solved by Hershberger and Suri [6],

---

* This research was supported by the French-Korean Science and Technology Amicable Relationships program (STAR).

S.-W. Cheng and C.K. Poon (Eds.): AAIM 2006, LNCS 4041, pp. 185–193, 2006.

who gave a data structure that requires linear space and answers queries in time $O(\log n)$. For circular rays, Agarwal and Sharir [1] gave a data structure achieving $O(\log^4 n)$ query time with $O(n \log^3 n)$ space. This was improved to $O(\log^2 n)$ query time with $O(n \log n)$ space by Cheng et al. [4] using a novel hierarchical decomposition of simple polygons.

We make use of Cheng et al.'s hierarchical decomposition and of their framework for ray shooting. This framework guides the search for the answer to a ray-shooting query inside a simple polygon. All that remains to be done to implement parabolic ray-shooting queries is to provide a data structure that stores two sets A and B of line segments and allows queries of the form: given a parabola γ, decide whether A lies entirely *above* γ, and whether B lies entirely *below* γ. (Note that since we are interested in trajectories under the influence of gravity, our parabolas are concave and have a vertical main axis. In other words, they can be expressed in the form $y = ax^2 + bx + c$, with $a < 0$.) We will call this a *parabola separation query*.

Our result is a data structure for parabola separation queries that stores m segments in space $O(m \log m)$ and has query time $O(\log m)$. Plugging this data structure into Cheng et al.'s framework [4] results in a data structure that stores a simple polygon P of complexity n in space $O(n \log n)$ and allows to answer ray-shooting queries along parabolic arcs originating inside P in query time $O(\log^2 n)$. These bounds equal the best known bounds for circular ray shooting. We omit a detailed description of the application of this framework; the result follows from Lemma 5 in Cheng et al. [4].

Separation queries are of interest indepently of their application to ray shooting. Let A and B be sets of planar line segments. A line ℓ is a *strong separator* of A and B if all segments of A lie in one closed half-plane defined by ℓ, and all segments of B lie in the other closed half-plane defined by ℓ. Note that segments from both sets are allowed to lie on ℓ.

Given A and B, a strong separator ℓ can be found in linear time by solving a two-dimensional linear program (it suffices to ensure that the endpoints of A and B are separated). The query version of this problem is to preprocess A and B into a data structure that allows us to determine quickly whether a given line is a strong separator. This can be done by computing the feasible region of the linear program, and preprocessing it for point location.

Our result answers the analogous question when ℓ is a parabola, albeit only for the case of parabolas with vertical main axis.

Parabola separation queries consist of two independent queries: (a) Determine whether A lies above γ; and (b) determine whether B lies below γ. In Section 2 we give a solution for part (b) that is very similar to the solution for lines mentioned above: We simply compute the space of all feasible parabolas, and preprocess it for point location. Our solution for (a) in Section 3 is much more complicated. This is due to the fact that it does not suffice to test the parabola against the endpoints of the segments. We describe a solution based on *Abstract Voronoi diagrams* as defined by Klein [7].

2   Does B Lie Below the Parabola?

For a parabola γ given by its equation $\gamma : y = ax^2 + bx + c$ where $a < 0$, let γ^- denote the closed region lying below the parabola, that is $\gamma^- := \{(x, y) \in \mathbb{R}^2 \mid y \leqslant ax^2 + bx + c\}$.

We are given a set B of m line segments, which we wish to preprocess and store in a data structure such that we can answer the following query: Given a parabola γ, does B lie entirely in γ^-?

Since γ^- is convex, a segment pq lies in γ^- if and only if both p and q lie in γ^-. It therefore suffices to test whether the set S of the $2m$ endpoints of segments in B lies in γ^-.

We represent the parabola $\gamma : y = ax^2 + bx + c$ as the point $(a, b, c) \in \mathbb{R}^3$. Each point $p_i = (x_i, y_i)$ defines a linear constraint in this space: $p_i \in \gamma^-$ if and only if $y_i \leqslant ax_i^2 + bx_i + c$. Since all $2m$ constraints can be written in the form $c \geqslant y_i - x_i^2 a - x_i b$, the set of parabolas γ with $B \subset \gamma^-$ is the region in (a, b, c)-space above these $2m$ planes.

We can now solve our problem as follows: We compute the upper envelope of these $2m$ planes in time $O(m \log m)$, project it onto the (a, b)-plane, and preprocess it for planar point location, using $O(m \log m)$ preprocessing time and $O(m)$ space [5]. For each face of the subdivision, we store the point p_i defining the plane supporting the corresponding facet of the upper envelope. To answer a query for a parabola $\gamma : y = ax^2 + bx + c$, we locate the point (a, b) in our subdivision in time $O(\log m)$, and determine the corresponding point $p_i \in S$. We then have $B \subset \gamma^-$ if and only if $y_i \leqslant ax_i^2 + bx_i + c$, which we can test in constant time.

Theorem 1. *Given a set B of m line segments, we can build in time $O(m \log m)$ a data structure of size $O(m)$ that allows us to answer in $O(\log m)$ time queries of the form: Given a parabola $\gamma : y = ax^2 + bx + c$, where $a < 0$, does $B \subset \gamma^-$?*

3   Does A Lie Above the Parabola?

We will make use of an entirely different parametrization of parabolas in this section. Recall that any parabola can be expressed as the locus of points equidistant from a point (the *focus*) and a line (the *directrix*). Since our parabola γ has vertical main axis, its directrix is a horizontal line $y = k$, and its focus w lies below the directrix.

We will express a parabola γ using the two parameters $k \in \mathbb{R}$ and $w \in \mathbb{R}^2$, such that $\gamma = \{p \in \mathbb{R}^2 \mid ||wp|| = |k - y_p|\}$, where $p = (x_p, y_p)$. Since the focus lies below the parabola, the closed region γ^+ lying above the parabola γ is then $\gamma^+ := \{p \in \mathbb{R}^2 \mid ||wp|| \geqslant |k - y_p|\}$.

We are given a set A of m non-intersecting line segments,[1] which we wish to preprocess and store in a data structure such that we can answer the following query: Given a parabola γ, does A lie entirely in γ^+? The answer to this question

---

[1] Segments are allowed to share endpoints.

does not change if we replace A by its lower envelope. We start by computing this lower envelope, in time $O(m \log m)$, so that in the following we can assume that any vertical line intersects only one segment, or perhaps the right endpoint of one segment and the left endpoint of another segment. (There are no vertical segments, as they have been replaced by a point.) In other words, in the following we assume that A is x-*monotone*.

We first observe that for a point $p \in \mathbb{R}^2$, we can rewrite the condition $p \in \gamma^+$ as follows:

$$p \in \gamma^+ \Leftrightarrow ||wp|| \geqslant |k - y_p| \Leftrightarrow ||wp|| \geqslant k - y_p \Leftrightarrow ||wp|| + y_p \geqslant k.$$

Here we made use of the fact that if $k - y_p$ is negative, then p lies above the directrix, and therefore in γ^+.

Let us now define a pseudo-distance function $d(u, p)$ with additive weight (for the point p) as follows:

$$d(u, p) := ||up|| + y_p.$$

From the above we find that $p \in \gamma^+$ if and only if $d(u, p) \geqslant k$.

Consider now a segment $s \in A$. Since s is compact and $d(u, p)$ is continuous, the set $\{d(u, p) \mid p \in s\}$ attains its minimum, and so we can define $d(u, s) := \min_{p \in s} d(u, p)$. Now we observe that $s \subset \gamma^+$ if and only if for all $p \in s$ we have $d(w, p) \geqslant k$, which is equivalent to $d(w, s) \geqslant k$. Similarly, $A \subset \gamma^+$ if and only if for all $s \in A$ we have $d(w, s) \geqslant k$, which is equivalent to $\min_{s \in A} d(w, s) \geqslant k$. It follows that we can decide whether $A \subset \gamma^+$ if we are able to compute $\min_{s \in A} d(u, s)$ for a given query point $u \in \mathbb{R}^2$ (namely the focus w of the parabola).

We have now reduced the problem to the well-known *post-office problem* (but using a somewhat unusual pseudo-distance function): We want to store our set A of line segments in such a way that we can quickly find the element $s \in A$ nearest to a given query point u. Our solution to this problem will be completely analogous to the standard solution of the Euclidean post-office problem: We will compute the Voronoi diagram of the set A (under our distance function d) and preprocess it for point-location. It remains to show that the Voronoi diagram has linear complexity, and can be computed in $O(m \log m)$ time (point-location with linear storage and $O(\log m)$ query time can then be done using standard techniques).

In general, the Voronoi diagram of segments where each segment carries an additive weight is not a well-behaved Voronoi diagram: Voronoi regions can be disconnected, and the diagram can have quadratic complexity. On first sight, our problem looks even harder, as our distance function is more general: our weights vary along each segment site. Nevertheless, we will be able to show that our Voronoi diagram is well behaved, making use of the special structure of our problem: First, our set A is x-monotone. Second, our additive weights are rather special—the weight is identical to the y-coordinate of the point on the site.

We start by giving a geometric interpretation of our distance function $d(u, p)$. We observe that the Voronoi diagram is invariant under translations, so we

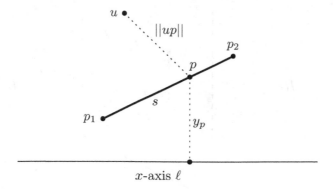

Fig. 1. Geometric interpretation of $d(u, p)$

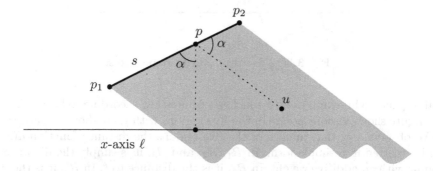

Fig. 2. Fermat's law enforces equal angles at p

can and will assume from now on that the set A lies entirely above the x-axis (denoted ℓ). For $p \in s$, $s \in A$, the weight y_p of p is then the distance of p from ℓ. It follows that for $u \subset \mathbb{R}^2$, the distance $d(u, p)$ is the length of the shortest path from u to ℓ passing through p, see Fig. 1.

Assume now that $p \in s$ is the point realizing $d(u, s)$, that is $d(u, p) = d(u, s)$. That means that p is the point in s that minimizes the length of the path $up\ell$.

If u lies vertically above s, then clearly the shortest possible path from u to ℓ through s is a vertical segment, and so p lies vertically below u. On the other hand, if u lies below the supporting line of s, then the path $up\ell$ enters and leaves s from below. If p is an interior point of s, then this path must be locally optimal, and so Fermat's law implies that it enters and leaves s under equal angles α, see Fig. 2. The angle α is fixed by the slope of s, and so this situation arises whenever u lies in the shaded region. For all other $u \in \mathbb{R}^2$, the shortest path from u to ℓ through s uses an endpoint of s.

We can therefore partition the plane into five convex regions $R_1, \ldots, R_5$ as in Fig. 3. In regions R_1 and R_2, the shortest path to ℓ goes through endpoint p_1, in region R_3 it goes through endpoint p_2. For $u \in R_4$, the shortest path to ℓ is a ver-

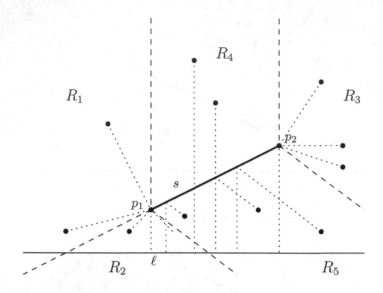

Fig. 3. The five regions with respect to s

tical segment, while for $u \in R_5$ it touches s from below according to Fermat's law. The figure shows various points in the five regions with their shortest path to ℓ.

We observe now that in each of the five regions the distance function $d(x, s)$ can be written in a simple form. In R_1, R_2, and R_3, it is simply the distance to a point with an additive weight. In R_4, it is the distance to ℓ. In R_5, it is the distance to ℓ', the mirror image of ℓ when reflected around the supporting line of s.

We now define the Voronoi region $VR(s, A)$ of a segment $s \in A$ as

$$VR(s, A) := \{u \in \mathbb{R}^2 \mid \forall s' \in A \text{ with } s' \neq s \text{ we have } d(u, s) < d(u, s')\}. \quad (1)$$

Lemma 1. *Let A be an x-monotone set of line segments, and let $s \in A$. Then we have:*

- *All interior points p of s, and all points u vertically above such an interior point $p \in s$ lie in $VR(s, A)$.*
- *Let $u \in VR(s, A)$, and let $p \in s$ be the point with $d(u, p) = d(u, s)$. Then the segment up lies entirely in $VR(s, A)$.*
- *$VR(s, A)$ is path-connected.*

Proof. (i) Let p be an interior point of s, and let u either be identical to p or lie vertically above p. Then the vertical segment from u to ℓ intersects s, and its length is $d(u, s)$. Since A is x-monotone and p is an interior point of s, this segment does not intersect any other segment $s' \in A$. It follows that any path from u to ℓ through another segment $s' \neq s$ cannot be straight, and is therefore longer than $d(u, s)$.

(ii) Let v be a point on the segment up, and assume $v \notin VR(s, A)$. This implies that there is a point $q \in s'$, $s' \neq s$, such that $d(v, q) \leqslant d(v, s) \leqslant d(v, p)$. Then the length of the path $uvq\ell$ is at most $d(u,p)$, which implies $d(u,q) \leqslant d(u,p)$, and $u \notin VR(s,A)$, a contradiction.

(iii) Follows from (i) and (ii): Let u and v be in $VR(s,A)$, and let $p, q \in s$ be their nearest points on s. Then the path $upqv$ lies in $VR(s,A)$.   □

Consider now two segments $s, s' \in A$. We are interested in their bisector, that is, the set

$$J(s, s') := \{u \in \mathbb{R}^2 \mid d(u, s) = d(u, s')\}. \tag{2}$$

Let us also define the regions "dominated" by s and s':

$$D(s, s') := \{u \in \mathbb{R}^2 \mid d(u, s) < d(u, s')\}, \tag{3}$$

$$D(s', s) := \{u \in \mathbb{R}^2 \mid d(u, s') < d(u, s)\}. \tag{4}$$

The regions $D(s, s')$, $J(s, s')$, and $D(s', s)$ form a disjoint partition of $\mathbb{R}^2$.

Lemma 2. *Let A be an x-monotone set of line segments, and let $s, s' \in A$. Then the bisector $J(s, s')$ is an infinite simple curve consisting of at most 25 conic arcs. $J(s, s')$ partitions $\mathbb{R}^2$ into the two unbounded connected regions $D(s, s')$ and $D(s', s)$.*

Proof. We partition the plane into the five convex regions R_i for s, and similarly into the five region R_i' for s'. The intersection of each pair $R_i \cap R_j'$ is a convex polygon R_{ij}. Each R_{ij} either belongs entirely to $D(s, s')$ or $D(s', s)$, or intersects both of them. In the latter case, $R_{ij} \cap J(s, s')$ is the intersection of R_{ij} with either a line, a parabola, or a hyperbola. It follows that $J(s, s')$ is the union of at most 25 conic arcs.[2]

Applying Lemma 1 to the set $\{s, s'\}$ implies that $D(s, s') = VR(s, \{s, s'\})$ is path-connected and unbounded, and the same holds for $D(s', s)$. It follows that $J(s, s')$ is a single simple infinite curve.   □

We now have all the necessary ingredients to prove that our Voronoi diagram is an Abstract Voronoi diagram as defined by Klein [7]. We start by recalling Klein's definition of Abstract Voronoi diagrams: We are given a set A of (abstract) objects with a total order $\prec$. For any pair $s, s' \in A$ with $s \neq s'$, let $D(s, s')$ be either empty or an open unbounded subset of the plane, and let $J(s, s')$ be the boundary of $D(s, s')$. $J(s, s')$ is called the *bisecting curve* of s and s'. The following conditions must hold:

(i) $J(s, s') = J(s', s)$, and the regions $D(s, s')$, $J(s, s')$ and $D(s', s)$ form a partition of $\mathbb{R}^2$ (into three disjoint sets).

(ii) If $\emptyset \neq D(s, s') \neq \mathbb{R}^2$ then $J(s, s')$ is homeomorphic to the open interval $(0, 1)$.

(iii) Any two bisecting curves intersect in a finite number of connected components.

---

[2] Clearly, the bound 25 is far too pessimistic. For instance, R_4 and R_4' can never intersect. We leave it to the reader to determine the exact number of pieces.

Define now $R(s, s')$ as $D(s, s') \cup J(s, s')$ if $s \prec s'$, and as $D(s, s')$ otherwise. The *extended Voronoi region* $EVR(s, A)$ of s is the intersection of all regions $R(s, s')$ for $s \in A$, $s' \neq s$, and the *Voronoi region* $VR(s, A)$ of s is the interior of $EVR(s, A)$. For any non-empty subset $A' \subset A$, the Voronoi regions must satisfy the following two conditions:

(iv) For all $s \in A'$ with $EVR(s, A') \neq \emptyset$: $VR(s, A') \neq \emptyset$ and both $EVR(s, A')$ and $VR(s, A')$ are path-connected.

(v) $\mathbb{R}^2 = \bigcup_{s \in A'} EVR(s, A')$.

Lemma 3. *Let A be an x-monotone set of line segments. Then $\{VR(s, A) \mid s \in A\}$ is an Abstract Voronoi diagram.*

Proof. We define $D(s, s')$ as in (3). Lemma 2 implies that $D(s, s')$ is an open unbounded subset of $\mathbb{R}^2$, that $J(s, s')$ as defined in (2) is indeed its boundary, and that conditions (i), (ii), and (iii) hold.

It is easy to see that $VR(s, A')$ as defined in (1) is the interior of $EVR(s, A')$, and that $EVR(s, A')$ is a subset of the closure of $VR(s, A')$. Lemma 1 therefore implies condition (iv), and condition (v) holds trivially. □

We can now state the main result of this section.

Theorem 2. *Given a set A of m non-intersecting line segments, we can build in time $O(m \log m)$ a data structure of size $O(m)$ that allows us to answer in $O(\log m)$ time queries of the form: Given a parabola $\gamma : y = ax^2 + bx + c$, where $a < 0$, does $A \subset \gamma^+$?*

Proof. As mentioned, we first replace A by its lower envelope in time $O(m \log m)$ to obtain a set of m segments that is x-monotone. We then compute in expected time $O(m \log m)$ the Voronoi diagram of A under our distance function, using the randomized incremental algorithm by Klein et al. [8]. The diagram is a planar subdivision of complexity $O(m)$, and can be preprocessed in time $O(m \log m)$ using space $O(m)$ to answer point-location queries in time $O(\log m)$.

To answer a query, we locate the focus w of the parabola in the Voronoi diagram. This tells us a segment $s \in A$ such that $d(w, s) = \min_{s' \in A} d(w, s')$. It suffices to test whether s lies in γ^+ to finish the query. □

4   Conclusions

We gave a data structure for parabola separation queries based on an Abstract Voronoi diagram. Similar diagrams had been studied by Ahn et al. [2] and by Bae and Chwa [3].

It remains an interesting open problem to find an efficient solution for ray shooting along general parabolic arcs, that is, where the direction of the directrix is not known in advance. It would also be interesting to perform ray shooting along general conic. The eccentricity would become another parameter of the query.

References

1. P. Agarwal and M. Sharir. Circle shooting in a simple polygon. *J. Algorithms*, 14:69–87, 1993.
2. H.-K. Ahn, O. Cheong, and R. van Oostrum. Casting a polyhedron with directional uncertainty. *Computational Geometry: Theory and Applications*, 26:129–141, 2003.
3. S.-W. Bae and K.-Y. Chwa. Voronoi diagrams with a transportation network on the Euclidean plane. In *Proc. of ISAAC*, volume 3341 of *LNCS*, pages 101–112, 2004.
4. S.-W. Cheng, O. Cheong, H. Everett, and R. van Oostrum. Hierarchical decompositions and circular ray shooting in simple polygons. *Discrete Comput. Geom.*, 32:401–415, 2004.
5. M. de Berg, M. van Kreveld, M. Overmars, and O. Schwarzkopf. *Computational Geometry: Algorithms and Applications*. Springer-Verlag, Berlin, Germany, 2nd edition, 2000.
6. J. Hershberger and S. Suri. A pedestrian approach to ray shooting: Shoot a ray, take a walk. *J. Algorithms*, 18:403–431, 1995.
7. R. Klein. *Concrete and Abstract Voronoi Diagrams*, volume 400 of *Lecture Notes Comput. Sci.* Springer-Verlag, 1989.
8. R. Klein, K. Mehlhorn, and S. Meiser. Randomized incremental construction of abstract Voronoi diagrams. *Computational Geometry*, 3:157–184, 1993.
9. M. Sharir and H. Shaul. Ray shooting and stone throwing with near-linear storage. *Computational Geometry*, 30:239–252, 2005.

Some Basics on Tolerances[*]

Boris Goldengorin[2,3], Gerold Jäger[1], and Paul Molitor[1]

[1] University of Halle-Wittenberg, Computer Science Institute,
D-06099 Halle (Saale), Germany
[2] Faculty of Economic Sciences, University of Groningen,
9700 AV Groningen, The Netherlands
[3] Department of Applied Mathematics,
Khmelnitsky National University, Ukraine

Abstract. In this paper we deal with sensitivity analysis of combinatorial optimization problems and its fundamental term, the tolerance. For three classes of objective functions $(\Sigma, \Pi, \text{MAX})$ we give some basic properties on upper and lower tolerances. We show that the upper tolerance of an element is well defined, how to compute the upper tolerance of an element, and give equivalent formulations when the upper tolerance is $+\infty$ or > 0. Analogous results are given for the lower tolerance and some results on the relationship between lower and upper tolerances are given.

Keywords: Sensitivity analysis, upper tolerance, lower tolerance.

1 Introduction

After an optimal solution to a combinatorial optimization problem has been determined, a natural next step is to apply *sensitivity analysis* (see Sotskov et al. [22]), sometimes also referred to as *post-optimality analysis* or *what-if analysis* (see e.g., Greenberg [11]). Sensitivity analysis is also a well-established topic in linear programming (see Gal [5]) and mixed integer programming (see Greenberg [11]). The purpose of sensitivity analysis is to determine how the optimality of the given optimal solution depends on the input data. There are several reasons for performing sensitivity analysis. In many cases the data used are inexact or uncertain. In such cases sensitivity analysis is necessary to determine the credibility of the optimal solution and conclusions based on that solution. Another reason for performing sensitivity analysis is that sometimes rather significant considerations have not been built into the model due to the difficulty of formulating them. Having solved the simplified model, the decision maker wants to know how well the optimal solution fits in with the other considerations.

The most interesting topic of sensitivity analysis is the special case when the value of a single element in the optimal solution is subject to change. The goal of

---

[*] This work was supported by the Deutsche Forschungsgemeinschaft DFG under Grant SI 657/5.

S.-W. Cheng and C.K. Poon (Eds.): AAIM 2006, LNCS 4041, pp. 194–206, 2006.

such perturbations is to determine the *tolerances* being defined as the maximum changes of a given individual cost (weight, distance, time etc.) preserving the optimality of the given optimal solution. The first successful implicit application of upper tolerances for improving the Transportation Simplex Algorithm is appeared in the so called Vogel's Approximation Method (see Reinfeld and Vogel [19]) and has been used for a straightforward enumeration of the k-best solutions for some positive integer k (see e.g., Murty [16] and Van der Poort et al. [25]) as well as a base of the MAX-REGRET heuristic for solving the three-index assignment problem (see Balas and Saltzman [1]). The values of upper tolerances have been applied for improving the computational efficiency of heuristics and branch-and-bound algorithms for solving different classes of NP-hard problems (for example of the traveling salesman problem (TSP) see Goldengorin and Jäger [6], Goldengorin et al. [9], Turkensteen et al. [24]). Also for the TSP, Helsgaun [13] improved the Lin-Kernighan heuristic by using the lower tolerances to the minimum 1-tree with great success.

Computational issues of tolerances to the minimum spanning tree problem and TSP are addressed in Chin and Hock [3], Gordeev et al. [10], Gusfield [12], Kravchenko et al. [14], Libura [15], Ramaswamy and Chakravarti [17], Shier and Witzgall [20], Sotskov [21], Tarjan [23]. Recently, Volgenant ([27]) has suggested an $O(n^3)$ algorithm for computing the upper and lower tolerances for all arcs in the Assignment Problem. Ramaswamy et al. have reviewed the sensitivity analysis problem for the maximum capacity path problem (see [18] and references within) and suggested an elegant reduction of the sensitivity analysis problem for the shortest path and maximum capacity path problems in an undirected network to the minimum cost interval problem. For an extensive account on computational issues of upper and lower tolerances in the context of sensitivity analysis in combinatorial optimization, see among others, Gal [4], Gal and Greenberg [5], Goldengorin and Sierksma [8] and Greenberg [11].

The purpose of this paper is to give an overview over the terms of upper and lower tolerances for the three most natural types $\sum, \prod, \text{MAX}$ of objective functions. To our best knowledge we have not found any publications treating the sensitivity analysis problem for a general class of combinatorial optimization problems with different types of objective functions. The paper is the first which deals with tolerances in an exact, general and comprehensive way, so that discrepancies of previous descriptions can be avoided, e.g. all of above mentioned papers have used but not indicated an important assumption that the set of feasible solutions to a combinatorial optimization problem under consideration is independent of the cost (objective) function. Furthermore, this coherent consideration leads to new results about tolerances.

The paper is organized as follows. In section 2 we define a combinatorial minimization problem and give all notations which are necessary for the terms of upper and lower tolerances. In section 3 we define the upper tolerance and give characteristics of it. Especially, we show that the upper tolerance is well defined with respect to the problem instance, i.e., that the upper tolerance of an element with respect to an optimal solution S^* of a problem instance $\mathcal{P}$ doesn't depend on $S^\star$ but only on $\mathcal{P}$ itself. Furthermore we show how to characterize elements with upper tolerance $+\infty$ or > 0 and how the upper tolerance can be computed. In section 4 we show similar relations for the lower tolerance. In section 5 we give relationships between lower and upper tolerances which mostly are direct conclusions from the sections 3 and 4. Our main result for objective functions of type $\sum$ is that under certain conditions the minimum value of upper tolerance equals the minimum value of lower tolerance and the maximum value of upper tolerance equals the maximum value of lower tolerance. Similar results for objective functions of type $\prod$, MAX do not hold. We summarize our paper in section 6 and propose directions for future research. For the non-trivial proofs of the statements we refer to the full version of the paper ([7]).

2   Combinatorial Minimization Problems

A *combinatorial minimization problem* $\mathcal{P}$ is given by a tuple $(\mathcal{E}, D, c, f_c)$ with

- $\mathcal{E}$ is a finite ground set of elements.
- $D \subseteq 2^{\mathcal{E}}$ is the set of the feasible solutions.
- $c : \mathcal{E} \to \mathbf{R}$ is the function which assigns costs to each single element of $\mathcal{E}$.
- $f_c : 2^{\mathcal{E}} \to \mathbf{R}$ is the objective (cost) function which depends on function c and assigns costs to each subset of $\mathcal{E}$.

A subset $S^\star \subseteq \mathcal{E}$ is called an *optimal solution* of $\mathcal{P}$, if $S^\star$ is a feasible solution and the costs $f_c(S^\star)$ of $S^\star$ are minimal[1], i.e., $S^\star \in D$ and $f_c(S^\star) = \min\{f_c(S); S \in D\}$. We denote the set of optimal solutions by $D^\star$. There are some particular monotone cost functions which often occur in practice:

- [**Type $\sum$**] The cost function $f_c : 2^{\mathcal{E}} \to \mathbf{R}$ is of type $\sum$, if for each $S \in 2^{\mathcal{E}}$: $f_c(S) = \sum_{\overline{e} \in S} c(\overline{e})$ holds.
- [**Type $\prod$**] The cost function $f_c : 2^{\mathcal{E}} \to \mathbf{R}$ is of type $\prod$, if for each $S \in 2^{\mathcal{E}}$: $f_c(S) = \prod_{\overline{e} \in S} c(\overline{e})$ and for each $e \in \mathcal{E}$: $c(e) > 0$ holds.
- [**Type MAX**] The cost function $f_c : 2^{\mathcal{E}} \to \mathbf{R}$ is of type MAX[2], if for each $S \in 2^{\mathcal{E}}$: $f_c(S) = \max\{c(\overline{e}); \overline{e} \in S\}$ holds.

These three objective functions are *monotone*, i.e., the costs of a subset of $\mathcal{E}$ don't become cheaper if the costs of a single element of $\mathcal{E}$ are increased.

In the remainder of the paper, we only consider combinatorial minimization problems $\mathcal{P} = (\mathcal{E}, D, c, f_c)$ which fulfill the following three conditions.

Condition 1. *The set D of the feasible solutions of $\mathcal{P}$ is independent of function c.*

Condition 2.   *The cost function $f_c : 2^{\mathcal{E}} \to \mathbf{R}$ is either of type $\sum$, type $\prod$, or type MAX.*

---

[1] Analogous considerations can be made if the costs have to be maximized, i.e., for combinatorial maximization problems.

[2] Such a cost function is also called *bottleneck function*.

Condition 3. *There is at least one optimal solution of $\mathcal{P}$, i.e., $D^\star \neq \emptyset$.*

Note that the Traveling Salesman Problem (TSP), Minimum Spanning Tree (MST), and many other combinatorial minimization problems fulfill these three conditions (see Bang-Jensen and Gutin [2]).

Given a combinatorial minimization problem $\mathcal{P} = (\mathcal{E}, D, c, f_c)$, we obtain a new combinatorial minimization problem if we increase the costs of a single element $e \in \mathcal{E}$ by some constant $\alpha \in \mathbf{R}$. We will denote the new problem by $\mathcal{P}_{\alpha,e} = (\mathcal{E}, D, c_{\alpha,e}, f_{c_{\alpha,e}})$, which is formally defined by $c_{\alpha,e}(\bar{e}) = \begin{cases} c(\bar{e}) & \text{, if } \bar{e} \neq e \\ c(\bar{e}) + \alpha & \text{, if } \bar{e} = e \end{cases}$
for each $\bar{e} \in \mathcal{E}$ and $f_{c_{\alpha,e}}$ is of the same type as f_c. Further define $\mathcal{P}_{-\infty,e} = \lim_{\alpha \to -\infty} \mathcal{P}_{\alpha,e}$ and $\mathcal{P}_{+\infty,e} = \lim_{\alpha \to +\infty} \mathcal{P}_{\alpha,e}$.

We need some more notations with respect to a combinatorial minimization problem $\mathcal{P}$. Let e be a single element of $\mathcal{E}$.

- $f_c(\mathcal{P})$ denotes the costs of an optimal solution $S^\star$ of $\mathcal{P}$.
- For $M \subseteq D$, $f_c(M)$ denotes the costs of the best solution included in M. The costs $f_c(S)$ of either infeasible or empty set S are defined as $+\infty$. Obviously, for each $M \subseteq D$: $f_c(\mathcal{P}) \leq f_c(M)$ holds.
- $D_-(e)$ denotes the set of the feasible solutions of D each of which doesn't contain the element $e \in \mathcal{E}$, i.e., $D_-(e) = \{ S \in D; e \notin S \}$.
  Analogously, $D_+(e)$ denotes the set of the feasible solutions D each of which contains the element $e \in \mathcal{E}$, i.e., $D_+(e) = \{ S \in D; e \in S \}$.
- $D_-^\star(e)$ denotes the set of the best feasible solutions of D each of which doesn't contain the element $e \in \mathcal{E}$, i.e.,

$$D_-^\star(e) = \{ S \in D; e \notin S \text{ and } (\forall S' \in D)(e \notin S' \Rightarrow f_c(S) \leq f_c(S')) \}$$

The elements of $D_-^\star(e)$ are called $S_-^\star(e)$.
Analogously, $D_+^\star(e)$ denotes the set of the best feasible solutions D each of which contains the element $e \in \mathcal{E}$, i.e.,

$$D_+^\star(e) = \{ S \in D; e \in S \text{ and } (\forall S' \in D)(e \in S' \Rightarrow f_c(S) \leq f_c(S')) \}$$

The elements of $D_+^\star(e)$ are called $S_+^\star(e)$.

3   Upper Tolerances

Let $\mathcal{P} = (\mathcal{E}, D, c, f_c)$ be a combinatorial minimization problem which fulfills Conditions 1, 2, and 3. Consider an optimal solution $S^\star$ of $\mathcal{P}$ and fix it.

For a single element e of this optimal solution $S^\star$, let the *upper tolerance* $u_{S^\star}(e)$ of element e with respect to $S^\star$ be the supremum by which the costs of e can be increased such that $S^\star$ remains an optimal solution, provided that the costs of all other elements $\bar{e} \in \mathcal{E} \setminus \{e\}$ remain unchanged, i.e., for each $e \in S^\star$ the upper tolerance is defined as follows:

$$u_{S^\star}(e) := \sup\{\alpha \in \mathbf{R}; \ S^\star \text{ is an optimal solution of } \mathcal{P}_{\alpha,e}\}$$

Because of the monotonicity of the cost function it holds:

$$u_{S^*}(e) := \inf\{\alpha \in \mathbf{R}; \; S^* \text{ is } not \text{ an optimal solution of } \mathcal{P}_{\alpha,e}\}$$

As S^* is an optimal solution of $\mathcal{P}_{0,e}$, which is $\mathcal{P}$, the upper tolerance $u_{S^*}(e)$ is either a non-negative number or $+\infty$. Because of Condition 2, for each $e \in S^*$ with $u_{S^*}(e) < +\infty$, it holds:

$$u_{S^*}(e) = \max\{\alpha \in \mathbf{R}; \; S^* \text{is an optimal solution of } \mathcal{P}_{\alpha,e}\}$$

Theorem 1. *Let S^* be an optimal solution of $\mathcal{P}$ with $e \in S^*$. e is contained in every feasible solution of $\mathcal{P}$ if and only if $u_{S^*}(e) = +\infty$, i.e., $e \in \bigcap_{S \in D} S \iff u_{S^*}(e) = +\infty$.*

Theorem 2. *The upper tolerance of an element doesn't depend on a particular optimal solution of $\mathcal{P}$, i.e.,*

$$(\forall S_1, S_2 \in D^*) \; (\forall e \in S_1 \cap S_2) \qquad u_{S_1}(e) = u_{S_2}(e)$$

Thus, if a single element $e \in \mathcal{E}$ is contained in at least one optimal solution S^* of $\mathcal{P}$, the upper tolerance of e does not depend on that particular optimal solution S^* but only on problem $\mathcal{P}$ itself. Hence, we can refer to the upper tolerance of e with respect to an optimal solution S^* as upper tolerance of e with respect to $\mathcal{P}$, $u_{\mathcal{P}}(e)$.

Note that the upper tolerance of an element e which is not contained in any optimal solution is not defined. For these elements $e \in \mathcal{E}$, we set $u_{\mathcal{P}}(e) :=$ UNDEFINED.

Theorem 3. *If $e \in \mathcal{E}$ with $u_{\mathcal{P}}(e) \notin \{\text{UNDEFINED}, +\infty\}$, then for all $\epsilon > 0$ the element e is not contained in any optimal solution of $\mathcal{P}_{u_{\mathcal{P}}(e)+\epsilon,e}$.*

Theorem 3 states that, for all $e \in \mathcal{E}$ with $u_{\mathcal{P}}(e) \neq$ UNDEFINED and $u_{\mathcal{P}}(e) \neq +\infty$, increasing the costs of e by $u_{\mathcal{P}}(e) + \epsilon$ for $\epsilon > 0$ makes the element uninteresting for optimal solutions.

Theorem 4. *For each single element $e \in \mathcal{E}$ which is contained in at least one optimal solution S^* of $\mathcal{P}$, the upper tolerance of e is given by*

- $u_{\mathcal{P}}(e) = f_c(D_-^*(e)) - f_c(\mathcal{P})$, *if the cost function is of type $\sum$*
- $u_{\mathcal{P}}(e) = \frac{f_c(D_-^*(e)) - f_c(\mathcal{P})}{f_c(\mathcal{P})} \cdot c(e)$, *if the cost function is of type $\prod$*
- $u_{\mathcal{P}}(e) = f_c(D_-^*(e)) - c(e)$, *if the cost function is of type MAX*

Theorem 5. *For each single element $e \in \mathcal{E}$ it holds for a cost function of type $\sum$, $\prod$ and MAX: $f_c(D_-^*(e)) = f_{c+\infty,e}(\mathcal{P})$.*

Theorem 4 and Theorem 5 tell us how to compute the upper tolerance of a single element $e \in \mathcal{E}$ with respect to $\mathcal{P}$. We observe (see also Ramaswamy and Chakravarti [17], Van Hoesel and Wagelmans [26]):

Corollary 1. *The upper tolerance of one element $e \in \mathcal{E}$ can be computed by solving two different instances of $\mathcal{P}$ for a cost function of type $\sum, \prod$ and solving one instance of $\mathcal{P}$ for a cost function of type MAX, i.e., the computation of the upper tolerance has the same complexity as $\mathcal{P}$ itself.*

Theorem 6. *If the cost function is either of type $\sum$ or $\prod$, then a single element e in at least one optimal solution is contained in every optimal solution if and only if its upper tolerance is greater than 0, i.e., $e \in \bigcap_{S^* \in D^*} S^* \iff u_{\mathcal{P}}(e) > 0$ or equivalently $\bigcap_{S^* \in D^*} S^* = \{e;\ u_{\mathcal{P}}(e) > 0\}$.*

Theorem 6 characterizes those elements which are contained in every optimal solution. We only have to know the upper tolerance of an element. Unfortunately, this property doesn't hold for a cost function of type MAX.

Remark 1. In general, for a cost function of type MAX only the direction "$\Rightarrow$" of Theorem 6 holds, but not the direction "$\Leftarrow$".

Corollary 2. *Let the cost function be either of type $\sum$ or of type $\prod$. There is only one optimal solution of $\mathcal{P}$ if and only if the upper tolerance $u_{\mathcal{P}}(e) > 0$ for all e with $u_{\mathcal{P}}(e) \neq$ UNDEFINED.*

Remark 2. Note that Condition 1 is crucial for all these properties, in particular for Theorem 4.

4 Lower Tolerances

Now, let S^* be an optimal solution of $\mathcal{P}$ which doesn't contain the element $e \in \mathcal{E}$. Analogously to the considerations which we have made with respect to the upper tolerance, we can ask for the supremum by which the costs of element e can be decreased such that S^* remains an optimal solution, provided that the costs of all other elements remain unchanged. More formally, we define for all $e \in \mathcal{E} \setminus S^*$:

$$l_{S^*}(e) := \sup\{\alpha \in \mathbb{R};\ f_{c-\alpha,e} \text{ is monot. and } S^* \text{ is an optimal solution of } \mathcal{P}_{-\alpha,e}\}$$

Because of the monotonicity of the cost function it holds:

$$l_{S^*}(e) := \inf\{\alpha \in \mathbb{R};\ f_{c-\alpha,e} \text{ is monot. and } S^* \text{ is } not \text{ an optimal solution of } \mathcal{P}_{-\alpha,e}\}$$

Note that if the cost function of the combinatorial minimization problem is of type $\prod$, the costs of the elements have to be greater than zero to guarantee monotonicity. In the following, let $\delta_{max}(e)$ be defined as

$$\delta_{max}(e) := \begin{cases} +\infty \text{ , if } f_c \text{ is either of type } \sum \text{ or of type MAX} \\ \\ c(e) \text{ , if } f_c \text{ is of type } \prod \end{cases}$$

$\delta_{max}(e)$ is the supremum by which element e can be decreased such that the cost function remains either of type $\sum, \prod$, or MAX.

As $S^\star$ is an optimal solution of $\mathcal{P}_{-0,e}$ which is $\mathcal{P}$, the lower tolerance $l_{S^\star}(e)$ is either a non-negative number or $+\infty$ if $e \notin S^\star$. More exactly, it holds for each $e \in \mathcal{E} \setminus S^\star$:

$$0 \le l_{S^\star}(e) \le \delta_{max}(e)$$

Because of Condition 2, for each $e \in \mathcal{E} \setminus S^\star$ and each $l_{S^\star}(e) < \delta_{max}(e)$, it holds:

$$l_{S^\star}(e) = \max\{\alpha \in \mathbf{R};\ f_{c-\alpha,e} \text{ is monot. and } S^\star \text{ is an optimal solution of } \mathcal{P}_{-\alpha,e}\}$$

Theorem 7. *Let the cost function be of type $\sum$ or $\prod$ and let $S^\star$ be an optimal solution of $\mathcal{P}$. Then, an element e isn't contained in a feasible solution if and only if $l_{S^\star}(e) = \delta_{max}(e)$, i.e., $e \in \mathcal{E} \setminus \bigcup_{S \in D} S \iff l_{S^\star}(e) = \delta_{max}(e)$.*

Remark 3. In general, for a cost function of type MAX only the direction "$\Rightarrow$" of Theorem 7 holds, but not the direction "$\Leftarrow$".

Remark 3 partly puts lower tolerances with respect to a cost function of type MAX in question. It states that the lower tolerance of an element can be very large, namely $+\infty$, although this element can be included in a feasible solution. It can be shown that the element can be included in an optimal solution. This contradicts the intuition that an element with large lower tolerance is not a "good" element and should not be included in solutions by heuristics.

Theorem 8. *The lower tolerance of an element doesn't depend on a particular optimal solution of $\mathcal{P}$, i.e., $(\forall S_1, S_2 \in D^\star)(\forall e \notin S_1 \cup S_2):\ l_{S_1}(e) = l_{S_2}(e)$.*

Thus, if there is at least one optimal solution $S^\star$ of $\mathcal{P}$ which doesn't contain element e, the lower tolerance of e doesn't depend on that particular optimal solution but only on problem $\mathcal{P}$ itself. As for upper tolerances, we can refer to the lower tolerance of e with respect to an optimal solution $S^\star$ as lower tolerance of e with respect to $\mathcal{P}$, $l_{\mathcal{P}}(e)$.

The lower tolerance of an element e which is contained in every optimal solution is not defined, yet. For these elements e, we set $l_{\mathcal{P}}(e) :=$ UNDEFINED.

Theorem 9. *If $e \in \mathcal{E}$ is a single element with $l_{\mathcal{P}}(e) \notin \{\text{UNDEFINED}, \delta_{max}(e)\}$, then element e is contained in every optimal solution of $\mathcal{P}_{-(l_{\mathcal{P}}(e)+\epsilon),e}$ for all $0 < \epsilon < \delta_{max}(e) - l_{\mathcal{P}}(e)$.*

Theorem 9 states that if we decrease the costs of e by more than $l_{\mathcal{P}}(e)$, then an optimal solution will contain element e, provided that $l_{\mathcal{P}}(e)$ is neither UNDEFINED nor $\delta_{max}(e)$.

Let for a single element $e \in \mathcal{E}$ and a cost function of type MAX

$$g(e) := \begin{cases} \min_{S \in D_+(e)} \max_{a \in S \setminus \{e\}}\{c(a)\} & \text{, if } D_+(e) \ne \emptyset \\ +\infty & \text{, if } D_+(e) = \emptyset \end{cases}$$

Obviously, it holds:

$$f_{c-\infty,e}(\mathcal{P}) = \min\{g(e), f_c(D_-^\star(e))\}$$

Theorem 10. *For each single element* $e \in \mathcal{E}$ *it holds*

- $f_c(D_+^\star(e)) = \lim_{K \to +\infty}(f_{c-K,e}(\mathcal{P}) + K)$, *if the cost function is of type* $\sum$
- $f_c(D_+^\star(e)) = \lim_{K \to c(e)^-}\left(\frac{f_{c-K,e}(\mathcal{P})}{c(e)-K} \cdot c(e)\right)$, *if the cost function is of type* $\prod$
- $f_c(D_+^\star(e)) = \max\{g(e), c(e)\}$, *if the cost function is of type MAX*

Theorem 11. *For each single element* $e \in \mathcal{E}$ *with* $l_\mathcal{P}(e) \neq$ UNDEFINED, *the lower tolerance of* e *with respect to* $\mathcal{P}$ *is given by*

- $l_\mathcal{P}(e) = f_c(D_+^\star(e)) - f_c(\mathcal{P})$, *if the cost function is of type* $\sum$
- $l_\mathcal{P}(e) = \frac{f_c(D_+^\star(e))-f_c(\mathcal{P})}{f_c(D_+^\star(e))} \cdot c(e)$, *if the cost function is of type* $\prod$
- $l_\mathcal{P}(e) = \begin{cases} c(e) - f_c(\mathcal{P}) \text{, } & \text{if } g(e) < f_c(\mathcal{P}) \\ +\infty & \text{, } \text{otherwise} \end{cases}$, *if the cost function is of type MAX*

Theorem 10 and Theorem 11 tell us how to compute the lower tolerance of a single element $e \in \mathcal{E}$ with respect to $\mathcal{P}$. We observe

Corollary 3. *The lower tolerance of a single element* $e \in \mathcal{E}$ *can be computed by solving two different instances of* $\mathcal{P}$ *for a cost function of type* $\sum, \prod$ *and solving one instance of* $\mathcal{P}$ *for a cost function of type MAX, i.e., the computation of the lower tolerance has the same complexity as* $\mathcal{P}$ *itself.*

Theorem 12. *If the cost function is either of type* $\sum$ *or* $\prod$, *then a single element* $e \in \mathcal{E}$ *isn't contained in any optimal solution if and only if its lower tolerance is greater than 0, i.e.,* $e \notin \bigcup_{S^\star \in D^\star} S^\star \iff l_\mathcal{P}(e) > 0$ *or equivalently* $\mathcal{E} \setminus \bigcup_{S^\star \in D^\star} S^\star = \{e;\ l_\mathcal{P}(e) > 0\}$.

Theorem 12 characterizes those elements which are never included in an optimal solution.

Remark 4. In general, for a cost function of type MAX only the direction "$\Rightarrow$" of Theorem 12 holds, but not the direction "$\Leftarrow$".

5   Relationship Between Lower and Upper Tolerances

The following properties hold for each cost function f_c either of type $\sum$ or $\prod$.

Corollary 4. *Let the cost function be either of type* $\sum$ *or of type* $\prod$. *For all* $e \in \mathcal{E}$, *the equivalence* $l_\mathcal{P}(e) =$ UNDEFINED $\iff u_\mathcal{P}(e) > 0$ *holds.*

Proof. The statement follows from Theorem 6 and the definition of lower tolerance. $\square$

Corollary 5. *Let the cost function be either of type* $\sum$ *or of type* $\prod$. *For all* $e \in \mathcal{E}$, *the equivalence* $u_\mathcal{P}(e) =$ UNDEFINED $\iff l_\mathcal{P}(e) > 0$ *holds.*

Proof. The statement follows from Theorem 12 and the definition of upper tolerance. $\square$

Corollary 6. *Let the cost function be either of type $\sum$ or of type $\prod$. For each $e \in \mathcal{E}$ which is contained in at least one optimal solution of $\mathcal{P}$ but not in all, i.e., $e \in \cup_{S^\star \in D^\star} S^\star$ and $e \notin \cap_{S^\star \in D^\star} S^\star$, the equation $u_\mathcal{P}(e) = l_\mathcal{P}(e) = 0$ holds.*

Proof. Both the upper tolerance and the lower tolerance of e are defined. $u_\mathcal{P}(e) = 0$ holds because of Theorem 6. $l_\mathcal{P}(e) = 0$ holds because of Theorem 12. □

Actually, there are much more close interrelations between lower and upper tolerances.

Let $u_{\mathcal{P},min} = \min\{u_\mathcal{P}(e); e \in \mathcal{E} \text{ and } u_\mathcal{P}(e) \neq \text{UNDEFINED}\}$ and $l_{\mathcal{P},min} = \min\{l_\mathcal{P}(e); e \in \mathcal{E} \text{ and } l_\mathcal{P}(e) \neq \text{UNDEFINED}\}$ be the smallest upper and lower tolerance with respect to $\mathcal{P}$. Furthermore, let $\Delta_{\mathcal{P},min}$ be defined as $\Delta_{\mathcal{P},min} = \min\{\delta_{max}(e); e \in \mathcal{E}\}$.

Corollary 7. *Let the cost function be either of type $\sum$ or of type $\prod$. Provided that there are at least two different optimal solutions, i.e., $|D^\star| \geq 2$, the equation $u_{\mathcal{P},min} = l_{\mathcal{P},min} = 0$ holds.*

Proof. As there are at least two optimal solutions S_1 and S_2, there is an element e_1 with $e_1 \in S_1 \setminus S_2$ or $e_1 \in S_2 \setminus S_1$. Thus, $e_1 \in \cup_{S^\star \in D^\star} S^\star$ and $e_1 \notin \cap_{S^\star \in D^\star} S^\star$. By Corollary 6, these two properties of e_1 implies $u_\mathcal{P}(e_1) = 0$ and $l_\mathcal{P}(e_1) = 0$. Thus $u_{\mathcal{P},min} = l_{\mathcal{P},min} = 0$ holds. □

Much more interesting is the case that there is only one optimal solution. Here, both the minimal upper tolerance and the minimal lower tolerance are greater than 0. Nevertheless, they are equal. First, we analyze the special case that there is only one feasible solution of $\mathcal{P}$.

Lemma 1. *Let the cost function be either of type $\sum$ or of type $\prod$. If the set D of the feasible solutions of $\mathcal{P}$ consists of only one element, say S, i.e., $|D| = 1$, then $u_{\mathcal{P},min} = +\infty$ and*

- $l_{\mathcal{P},min} = +\infty$,   *if $S = \mathcal{E}$*
- $l_{\mathcal{P},min} = \Delta_{\mathcal{P},min}$,   *if $S = \emptyset$*
- $l_{\mathcal{P},min} \geq \Delta_{\mathcal{P},min}$,   *if $S \neq \mathcal{E}$ and $S \neq \emptyset$*

Remark 5. Note that for the set of the feasible solutions D we have $D \neq \emptyset$ (Condition 3), but nevertheless it might hold: $\emptyset \in D$.

Corollary 8. *Let the cost function be of type $\sum$. If the set D consists of only one element, i.e., $|D| = 1$, then $u_{\mathcal{P},min} = l_{\mathcal{P},min} = +\infty$ holds.*

Proof. The corollary is implied by Lemma 1 as $\Delta_{\mathcal{P},min} = +\infty$ for a cost function of type $\sum$. □

Lemma 2. *Let the cost function be of type $\sum$. Provided that no feasible solution is a subset of another feasible solution and there are at least two different feasible solutions but only one optimal solution, i.e., $|D \geq 2|$ and $|D^\star| = 1$, then the equation $u_{\mathcal{P},min} = l_{\mathcal{P},min}$ holds. In particular, $0 < l_{\mathcal{P},min} \neq +\infty$ and $0 < u_{\mathcal{P},min} \neq +\infty$.*

Theorem 13. *Let the cost function be of type* $\sum$. *Provided that no feasible solution is a subset of another feasible solution, then the equation* $u_{\mathcal{P},min} = l_{\mathcal{P},min}$ *holds.*

Proof. The statement is implied by Corollary 7, Corollary 8, and Lemma 2.   □

Remark 6. If we relax the condition that no feasible solution is a subset of another feasible solution, then Theorem 13 doesn't hold.

Remark 7. In general, Theorem 13 doesn't hold for a cost function of type $\prod$.

Remark 8. In general, Theorem 13 doesn't hold for a cost function of type MAX.

Corollary 9. *Let the cost function be of type* $\sum$. *Provided that no feasible solution is a subset of another feasible solution, there is only one optimal solution of* $\mathcal{P}$ *if and only if the lower tolerance* $l_{\mathcal{P}}(e) > 0$ *for all e with* $l_{\mathcal{P}}(e) \neq$ UNDEFINED.

Proof. The statement follows from Corollary 2, Theorem 13 and the definition of $u_{\mathcal{P},min}$ and $l_{\mathcal{P},min}$.   □

Finally, we consider the largest upper and lower tolerance with respect to $\mathcal{P}$, $u_{\mathcal{P},max} = \max\{\, u_{\mathcal{P}}(e); \; e \in \mathcal{E}$ and $u_{\mathcal{P}}(e) \neq$ UNDEFINED $\}$ and $l_{\mathcal{P},max} = \max\{\, l_{\mathcal{P}}(e);$ $e \in \mathcal{E}$ and $l_{\mathcal{P}}(e) \neq$ UNDEFINED $\}$. We define $G := \{e \in \bigcup_{S^* \in D^*} S^*; u_{\mathcal{P}}(e) = u_{\mathcal{P},max}\}$ and $H := \{e \in \mathcal{E} \setminus \bigcap_{S^* \in D^*} S^*; l_{\mathcal{P}}(e) = l_{\mathcal{P},max}\}$.

We call the set of feasible solutions D *connected*, if D satisfies

a) $\left(\bigcup_{e \in \bigcup_{S^* \in D^*} S^*} \bigcup_{S^*_-(e) \in D^*_-(e)} S^*_-(e) \right) \cap H \neq \emptyset$

b) $\left(\bigcup_{e \in \mathcal{E} \setminus \bigcap_{S^* \in D^*} S^*} \bigcup_{S^*_+(e) \in D^*_+(e)} (\mathcal{E} \setminus S^*_+(e)) \right) \cap G \neq \emptyset$

It is easy to see that conditions a) and b) are equivalent to the conditions a') and b'):

a') $\exists e \in \bigcup_{S^* \in D^*} S^* \; \exists S^*_-(e) \in D^*_-(e) : \; S^*_-(e) \cap H \neq \emptyset$

b') $\exists e \in \mathcal{E} \setminus \bigcap_{S^* \in D^*} S^* \; \exists S^*_+(e) \in D^*_+(e) : \; (\mathcal{E} \setminus S^*_+(e)) \cap G \neq \emptyset$

Theorem 14. *Let the cost function be of type* $\sum$. *If the set of the feasible solutions D is connected, then the equation* $u_{\mathcal{P},max} = l_{\mathcal{P},max}$ *holds.*

We illustrate the conditions a) and b) and Theorem 14 by the following combinatorial minimization problem $\mathcal{P} = (\mathcal{E}, D, c, f_c)$:

- $\mathcal{E} = \{v, x, y, z\}$ with $c(v) = 1$, $c(x) = 2$, $c(y) = 4$, and $c(z) = 8$
- $D = \{\{v, x\}, \{y, z\}\}$
- f_c is a cost function of type $\sum$

The only optimal solution is $\{v, x\}$. It holds $u_{\mathcal{P}}(v) = 9$ and $u_{\mathcal{P}}(x) = 9$ which implies $u_{\mathcal{P},max} = 9$ and $l_{\mathcal{P}}(y) = 9$ and $l_{\mathcal{P}}(z) = 9$ which implies $l_{\mathcal{P},max} = 9$. Therefore $u_{\mathcal{P},max} = l_{\mathcal{P},max}$. Furthermore it holds $G = \{v, x\}$, $H = \{y, z\}$, $D^*_-(v) = \{\{y, z\}\}$, $D^*_-(x) = \{\{y, z\}\}$, $D^*_+(y) = \{\{y, z\}\}$, and $D^*_+(z) = \{\{y, z\}\}$. As condition a') and condition b') hold, D is connected.

Remark 9. The condition that the set of the feasible solutions D is connected is only a sufficient, but not a necessary condition for $u_{\mathcal{P},max} = l_{\mathcal{P},max}$, i.e., there is a combinatorial minimization problem, where $u_{\mathcal{P},max} = l_{\mathcal{P},max}$, although D is not connected,

Remark 10. In general, Theorem 14 doesn't hold for a cost function of type $\prod$.

Remark 11. In general, Theorem 14 doesn't hold for a cost function of type MAX.

6   Summary and Future Research Directions

In this paper we have rigorously defined and studied the properties of upper and lower tolerances for a general class of combinatorial optimization problems with three types of objective functions, namely with types $\sum$, $\prod$, and MAX. Theorems 2 and 8 indicate that the upper and lower tolerances do not depend on a particular optimal solution under the condition that the set of the feasible solutions is independent on the costs of ground elements.

For problems with the objective functions of types $\sum$ and $\prod$ Theorem 6 implies that the upper tolerances can be considered as an *invariant* characterizing the structure of the set of all optimal solutions as follows. If all upper tolerances are positive (see Corollary 2), then the set of optimal solutions contains a unique optimal solution. If some upper tolerances are positive and others are zeros, then the set of optimal solutions contains at least two optimal solutions such that the cardinality of their intersection is equal to the number of positive upper tolerances. If all upper tolerances are zeros, then the set of optimal solutions contains at least two optimal solutions such that the cardinality of their intersection is equal to zero, i.e., there is no common element in all optimal solutions. Similar conclusions can be made from Theorem 12 and Corollary 9 if we replace each optimal solution by its complement to the ground set.

One of the major problems, when solving NP-hard problems by means of the branch-and-bound approach, is the choice of the branching element which keeps the search tree as small as possible. Using tolerances we are able to ease this choice. Namely, if there is an element from the optimal solution of the current relaxed NP-hard problem (we assume that this optimal solution is a non-feasible solution to the original NP-hard problem) with a positive upper tolerance, then this element is in all optimal solutions of the current relaxed NP-hard problem. Hence, branching on this element means that we enter a common part in all possible search trees emanating from each particular optimal solution of the current relaxed NP-hard problem. Therefore, branching on an element with a positive upper tolerance is not only necessary for finding a feasible solution to the original NP-hard problem but also is a best possible choice. An interesting direction of research is to develop tolerance based b-n-b type algorithms for different NP-hard problems with the objective functions of types $\sum$ and $\prod$.

Many modern heuristics for finding high quality solutions to a NP-hard problem delete high cost elements and save the low cost ones from a relaxed NP-hard problem. A drawback of this strategy is that in terms of either high or low cost elements the structure of all optimal solutions to a relaxed NP-hard problem cannot be described. A tolerance of an element is the cost of excluding or including that element from the solution at hand. Hence, another direction of research is to develop tolerance based heuristics for different NP-hard problems with the objective functions of types $\sum$ and $\prod$.

Acknowledgement

This article is dedicated to the former project leader, Prof. Dr. Jop Sibeyn, who is missed since a snow-hike in spring 2005. He was involved in the application of the DFG project SI 657/5 and has contributed to the results presented here by lively and inspiring discussions.

References

1. Balas, E., Saltzman, M.J.: An algorithm for the three-index assignment problem. Oper. Res. **39** (1991) 150–161.
2. Bang-Jensen, J. Gutin, G.: Digraphs: Theory, Algorithms and Applications. Springer-Verlag London (2002).
3. Chin, F., Hock, D.: Algorithms for Updating Minimal Spanning Trees. J. Comput. System Sci. **16** (1978) 333–344.
4. Gal, T.: Sensitivity Analysis, Parametric Programming, and Related Topics: Degeneracy, Multicriteria Decision Making, Redundancy. W. de Gruyter, Berlin and New York (1995).
5. Gal, T., Greenberg, H.J. (eds.): Advances in Sensitivity Analysis and Parametric Programming, Internat. Ser. Oper. Res. Management Sci. **6**. Kluwer Academic Publishers, Boston (1997).
6. Goldengorin, B., Jäger, G.: How To Make a Greedy Heuristic for the Asymmetric Traveling Salesman Competitive. SOM Research Report 05A11, University of Groningen, The Netherlands, 2005 (http://som.eldoc.ub.rug.nl/reports/themeA/2005/05A11/).
7. Goldengorin, B., Jäger, G., Molitor, P.: Some Basics on Tolerances. SOM Research Report 05A13, University of Groningen, The Netherlands, 2005 (http://som.eldoc.ub.rug.nl/reports/themeA/2005/05A13/).
8. Goldengorin, B., Sierksma, G:. Combinatorial optimization tolerances calculated in linear time. SOM Research Report 03A30, University of Groningen, The Netherlands, 2003 (http://som.eldoc.ub.rug.nl/reports/themeA/2003/03A30/).
9. Goldengorin, B., Sierksma, G., Turkensteen, M.: Tolerance Based Algorithms for the ATSP. Graph-Theoretic Concepts in Computer Science. 30th International Workshop, WG 2004, Bad Honnef, Germany, June 21-23, 2004, Hromkovic, J., Nagl, M., Westfechtel, B. (eds.). Lecture Notes in Comput. Sci. **3353** (2004) 222–234.
10. Gordeev, E.N., Leontev, V.K., Sigal, I.K.: Computational algorithms for finding the radius of stability in selection problems. USSR Comput. Math. Math. Phys. **23** (1983) 973–979.

11. Greenberg, H.J.: An annotated bibliography for post-solution analysis in mixed integer and combinatorial optimization. In: Woodruff, D.L. (ed.), Advances in Computational and Stochastic Optimization, Logic Programming, and Heuristic Search. Kluwer Academic Publishers (1998) 97–148.
12. Gusfield, D.: A note on arc tolerances in sparse minimum-path and network flow problems. Networks **13** (1983) 191–196.
13. Helsgaun, K.: An effective implementation of the Lin-Kernighan traveling salesman heuristic. European J. Oper. Res. **126** (2000) 106–130.
14. Kravchenko, S.A., Sotskov, Y.N., Werner, F.: Optimal schedules with infinitely large stability radius. Optimization **33** (1995) 271–280.
15. Libura, M.: Sensitivity analysis for minimum hamiltonian path and traveling salesman problems. Discrete Appl. Math. **30** (1991) 197–211.
16. Murty, K.G.: An algorithm for ranking all the assignments in order of increasing cost. Oper. Res. **16** (1968) 682–687.
17. Ramaswamy, R., Chakravarti, N.: Complexity of determining exact tolerances for min-sum and min-max combinatorial optimization problems. Working Paper WPS-247/95, Indian Institute of Management, Calcutta, India, (1995) 34.
18. Ramaswamy, R., Orlin, J.B., Chakravarti, N.: Sensitivity analysis for shortest path problems and maximum capacity path problems in undirected graphs. Math. Program., Ser. A, **102** (2005) 355–369.
19. Reinfeld, N.V., Vogel, W.R.: Mathematical Programming. Prentice-Hall, Englewood Cliffs, N.J. (1958).
20. Shier, D.R., Witzgall, C.: Arc tolerances in minimum-path and network flow problems. Networks **10** (1980) 277–291.
21. Sotskov, Y.N.: The stability of the approximate boolean minimization of a linear form. USSR Comput. Math. Math. Phys. **33** (1993) 699–707.
22. Sotskov, Y.N., Leontev, V.K., Gordeev, E.N.: Some concepts of stability analysis in combinatorial optimization. Discrete Appl. Math. **58** (1995) 169–190.
23. Tarjan, R.E.: Sensitivity Analysis of Minimum Spanning Trees and Shortest Path Trees. Inform. Process. Lett. **14**(1) (1982) 30–33.
24. Turkensteen, M., Ghosh, D., Goldengorin, B., Sierksma, G.: Tolerance-Based Branch and Bound Algorithms. A EURO conference for young OR researches and practitioners, ORP3 2005, 6 – 10 September 2005, Valencia, Spain. Proceedings Edited by C. Maroto et al., ESMAP, S.L. (2005) 171–182.
25. Van der Poort, E.S., Libura, M., Sierksma, G., Van der Veen, J.A.A.: Solving the k-best traveling salesman problem. Comput. Oper. Res. **26** (1999) 409–425.
26. Van Hoesel, S., Wagelmans, A.: On the complexity of postoptimality analysis of 0/1 programs. Discrete Appl. Math. **91** (1999) 251–263.
27. Volgenant, A.: An addendum on sensitivity analysis of the optimal assignment. European J. Oper. Res. **169** (2006) 338–339.

Note on a Class of Admission Control Policies for the Stochastic Knapsack Problem

Adriana F. Gabor[1] and Jan-Kees C.W. van Ommeren[2]

[1] EURANDOM and Faculty of Mathematics and Computer Science,
TUE, P.O.Box 513, 5600 MB, Eindhoven, The Netherlands
a.f.gabor@tue.nl

[2] Faculty of Electrical Engineering, Mathematics and Computer Science, University
of Twente, PO Box 217, 7500 AE Enschede, The Netherlands
j.c.w.vanommeren@ewi.utwente.nl

Abstract. In this note we discuss a class of exponential penalty function policies recently proposed by Iyengar and Sigman for controlling a stochastic knapsack. These policies are based on the optimal solution of some related deterministic linear programs. By finding explicitly their optimal solution, we reinterpret the exponential penalty function policies and show that they belong to the class of threshold policies. This explains their good practical behavior, facilitates the comparison with the thinning policy, simplifies considerably their analysis and improves the bounds previously proposed.

1 Introduction

Recently, Iyengar and Sigman [1] proposed an exponential penalty function policy for controlling a loss network. A loss network is a network of resources, each with a known capacity. Requests for using the capacity are divided into classes, corresponding to arrival rates, service duration, resource requirements, and the profit they will bring for the network. There is no waiting room, so at every arrival of a request, it must be decided whether to accept the request or not. An admitted request occupies the allocated resource for the service duration and releases all the resources when it leaves the network. The objective is to design an admission policy that optimizes an appropriate performance measure of the revenue.

A major part of [1] is dedicated to the stochastic knapsack, which is a loss network with only one resource. For a review on other policies proposed for controlling the stochastic knapsack, see [5].

In this note we will focus on the exponential penalty function policy proposed in [1] for controlling a stochastic knapsack. This policy is based on the solution to a linear program. By solving this LP explicitly, we will show that for the stochastic knapsack, this policy reduces to a threshold policy. From the optimal solution of this LP we will derive an index, called the "threshold" index, which will divide the classes of different indices into two groups: one that will always be rejected, and one that will be accepted if there is enough capacity to accommodate them. The requests belonging to the class with the threshold index, are

S.-W. Cheng and C.K. Poon (Eds.): AAIM 2006, LNCS 4041, pp. 207–219, 2006.
© Springer-Verlag Berlin Heidelberg 2006

accepted only if they satisfy an extra condition, given by the penalty function. By interpreting the exponential penalty function policy as a threshold policy, we are then able to improve the bounds on the expected reward rate obtained in [1] and to compare the exponential policy with the thinning policy proposed in [3].

This note closely follows [1] and is organized as follows. In Section 2 we present in detail the stochastic knapsack problem. With the exception of the last section, we will present the analysis for exponential service times. In Section 3 we discuss bounds for the expected reward rate achieved by admission policies in a stochastic knapsack. We start by discussing the upper bound proposed in [1] on the expected reward rate achieved by a policy and tighten it. Then we focus on the exponential penalty function policy and show that it is a threshold policy. This will lead to improved lower bounds for the expected reward rate. We continue by discussing the bounds in the "steady-state" regime. In Section 4 we will compare the exponential penalty function policy and the thinning policy (for the stochastic knapsack). In Section 5 we generalize the results presented in the previous sections to service times with a general distribution. We conclude with some remarks on the exponential penalty function policy for the stochastic knapsack and discuss why the results presented in this note are not easily generalized to loss networks.

2   Admission Control in the Stochastic Knapsack

The stochastic knapsack problem is a special case of a loss network problem and can be formulated as follows. There is a knapsack (network) of capacity $b \in R_+$. Requests for using the network belong to m independent Poisson arrival classes. Class i requests have an arrival rate λ_i and a service duration S_i which is exponentially distributed with rate μ_i (with the exception of Section 5), i.e., $S_i \sim exp(\mu_i)$. The requests in class i need a capacity b_i and pay r_i per unit time during their service duration. There is no waiting room in the system, therefore, each arriving request must either be accepted to the system and assigned a capacity allocation or rejected. When an accepted request departs after service completion, it releases all the allocated resources simultaneously. For simplicity, we will assume that the system is initially empty (all results easily generalize to the case when the system is not initially empty).

Let $T_{(i,n)}$, $i = 1, ..., m, n \geq 1$ denote the arrival epoch of the nth class i request. Since all admission decisions are made at arrival epochs, a feasible admission control policy π can be described as a collection of random variables $\pi = \{\pi_{(i,n)}, i = 1, ..., m, n \geq 1\}$, with $\pi_{(i,n)} = 0$ denoting that request of type i arriving at the epoch $T_{(i,n)}$ is rejected and $\pi_{(i,n)} = 1$ denoting that the request is accepted.

Let $x_i^\pi(t)$ be the number of class i requests in the system at time t under policy π. Define $x^\pi(t) = (x_1(t), ..., x_m(t))$. A request class i can be accepted only if there is sufficient capacity to accommodate it, that is

$$\sum_{i'=1}^{m} b_{i'} x_{i'}(t) + b_i \leq b.$$

The system controller is permitted to reject requests even if there is sufficient capacity.

The instantaneous reward rate $R^\pi(t)$ under policy π at time t is given by

$$R^\pi(t) = \sum_{i=1}^{m} r_i x_i^\pi(t).$$

The objective of the controller is to choose a policy π that maximizes a certain function of the reward rate process $\{R^\pi(t), t \geq 0\}$. Common performance measures for finite time horizon problems are either the expected total reward $E[\int_0^T R^\pi(s)ds]$ or the expected discounted reward $E[\int_0^T e^{-\beta s} R^\pi(s)ds]$, with $\beta > 0$; for infinite horizon problems, appropriate measures are either the discounted reward $E[\int_0^\infty e^{-\beta s} R^\pi(s)ds]$, $\beta > 0$ or the long-run average reward limit $\lim_{T\to\infty} \frac{1}{T}E[\int_0^T R^\pi(s)ds]$.

In [1] the authors construct feasible policies that perform well both in the transient period and in steady state. They first establish an upper bound $R^*(t)$ on the achievable expected reward rate $E[R^\pi(t)]$ and then construct a feasible policy $\bar\pi$ with expected reward rate $\mathbf{E}[R^\pi(t)] \simeq R^*(t)$. Thus, the policy $\bar\pi$ satisfies

$$E[\int_0^T e^{-\beta s} R^\pi(s)ds] \leq \int_0^T e^{-\beta s} R^*(s)ds \simeq E[\int_0^T e^{-\beta s} R^{\bar\pi}(s)ds],$$

for $\beta > 0$, which means that $\bar\pi$ is approximately optimal for any finite time horizon, and

$$\lim_{T\to\infty} \frac{1}{T}E[\int_0^T R^\pi(s)ds] \leq \lim_{T\to\infty} \frac{1}{T}\int_0^T R^*(s)ds \simeq \lim_{T\to\infty} \frac{1}{T}E[\int_0^T R^{\bar\pi}(s)ds],$$

that is, $\bar\pi$ is approximately optimal in steady state as well.

In the next sections we will discuss the admission policy $\bar\pi$ proposed in [1]. We will prove that it is a threshold policy, i.e., only classes of a certain index are admitted to the network. This will also lead to improved bounds and an analytical comparison with the thinning policy proposed by Kelly.

3   Control Policies for the Stochastic Knapsack

3.1   Upper Bound on the Achievable Reward Rate

In this section we discuss the upper bound on the achievable reward at time t proposed in [1] and show a simple way of calculating it.

Let π denote any feasible control policy for the single resource model. Let $x_i^\pi(t)$ denote the number of class i requests at time t. Since $(x_i^\pi(t))_{i=\overline{1,m}}$ is feasible,

$$\sum_{i=1}^{m} b_i E[x_i^\pi(t)] \leq b.$$

Clearly, $E[x_i^\pi(t)] \leq E[q_i(t)]$, where $q_i(t)$ is the number of class i requests at time t in a corresponding $M/M/\infty$ system. Since the system is initially empty, $E[q_i(t)] = \rho_i(1 - e^{-\mu_i t})$, where $\rho_i = \frac{\lambda_i}{\mu_i}$ (see e.g. [6] page 75). Hence,

$$\alpha = (\frac{E[x_1^\pi(t)]}{\rho_1}, ..., \frac{E[x_m^\pi(t)]}{\rho_m})$$

is feasible for the following linear program:

$$maximize \qquad \sum_{i=1}^{m} r_i \rho_i \alpha_i$$

$$P(t) \qquad s.t. \sum_{i=1}^{m} b_i \rho_i \alpha_i \leq b,$$

$$0 \leq \alpha_i \leq 1 - e^{-\mu_i t}.$$

Let $\alpha^*(t)$ denote an optimal solution of the linear program P(t) and let $R^*(t)$ denote its optimal value. Clearly,

$$E[R^\pi(t)] = \sum_{i=1}^{m} r_i E[x_i^\pi(t)] \leq R^*(t).$$

In [1] the authors find an upper bound on $E[R^\pi(t)]$ by finding an upper bound on $R^*(t)$. Next we show how the exact value of $R^*(t)$ can be directly calculated.

Note that the problem P(t) is a continuous knapsack problem (see e.g. [4]). Thus, an optimal solution can be found as follows. Suppose from now on that the classes are indexed in decreasing order of the profit to capacity ratio, i.e.,

$$\frac{r_1}{b_1} \geq ... \geq \frac{r_m}{b_m}.$$

Let $k^*(t)$ be the index with the following property:

$$\sum_{i=1}^{k^*(t)-1} b_i \rho_i (1 - e^{-\mu_i t}) \leq b \text{ and } \sum_{i=1}^{k^*(t)} b_i \rho_i (1 - e^{-\mu_i t}) > b. \qquad (1)$$

Then, the optimal solution of P(t) is given by:

$$\alpha_i^*(t) = \begin{cases} 1 - e^{-\mu_i t}, \text{ for } i < k^*(t) - 1 \\ \frac{b - \sum_{i=1}^{k^*(t)} b_i \rho_i (1 - e^{-\mu_i t})}{b_{k^*(t)} \rho_{k^*(t)} (1 - e^{-\mu_{k^*(t)} t})}, \text{ for } i = k^*(t) \\ 0, \text{ for } i > k^*(t). \end{cases} \qquad (2)$$

Hence, we have obtained the following upper bound.

Theorem 1. *The reward rate $R^\pi(t)$ of any feasible policy π satisfies*

$$E[R^\pi(t)] \leq R^*(t) = \sum_{i=1}^{k^*(t)} r_i \rho_i \alpha_i^*(t),$$

where $R^*(t)$ is the optimal value of (P) and $\alpha_i^*(t)$ is given by (2).

3.2   The Exponential Penalty Function Policy

In this section we describe the penalty function policy proposed in [1] and show that it is a threshold policy. This leads to improved lower bounds for the expected reward obtained by the penalty function policy and facilitates the comparison with the thinning policy proposed by Kelly [3].

Next we introduce two linear programs which play an essential role in describing and analyzing the penalty policy.

Define the "steady state" version of $P(t)$ as

$$maximize \qquad \sum_{i=1}^{m} r_i \rho_i \alpha_i$$

$$P \qquad s.t. \sum_{i=1}^{m} b_i \rho_i \alpha_i \le b,$$

$$0 \le \alpha_i \le 1.$$

Since P is a continuous knapsack problem, it's optimal solution α^* has the following structure:

$$\alpha_i^* = \begin{cases} 1, & \text{for } i < k^* \\ \frac{b - \sum_{i=1}^{k^*} b_i \rho_i}{b_{k^*} \rho_{k^*}}, & \text{for } i = k^* \\ 0, & \text{for } i > k^*, \end{cases} \qquad (3)$$

where k^* is the index for which

$$\sum_{i=1}^{k^*-1} b_i \rho_i \le b \text{ and } \sum_{i=1}^{k^*} b_i \rho_i > b.$$

Consider the following perturbation of the program P.

$$maximize \qquad \sum_{i=1}^{m} r_i \rho_i \alpha_i$$

$$P_\epsilon \qquad s.t. \sum_{i=1}^{m} b_i \rho_i \alpha_i \le \frac{b}{1 + 4\epsilon},$$

$$0 \le \alpha_i \le 1.$$

The optimal solution α^ϵ of P_ϵ is : $\alpha_i^\epsilon = 1$, for $i \le k^\epsilon$, $\alpha_{k^\epsilon}^\epsilon \in (0,1)$ and $\alpha_i^\epsilon = 0$, for $i \ge k^\epsilon$, where k^ϵ is the index for which

$$\sum_{i=1}^{k^\epsilon-1} b_i \rho_i \le \frac{b}{1 + 4\epsilon} \text{ and } \sum_{i=1}^{k^\epsilon} b_i \rho_i > \frac{b}{1 + 4\epsilon}.$$

Denote by R^*, respectively R_ϵ^*, the optimal value of P, respectively P_ϵ. By comparing the feasibility regions and the optimal solutions of the problems $P(t)$, P and P_ϵ, we obtain the following relationships among them.

Lemma 1. a) $k_\epsilon \leq k^* \leq k^*(t)$
   b) $R_\epsilon^* \leq R^* \leq R^*(t)$.

In our analysis, we will also make use of the dual problems D, respectively D_ϵ, of P, respectively P_ϵ:

$$minimize \ ub + \sum_{i=1}^{m} v_i \qquad\qquad minimize \quad u\frac{b}{1+4\epsilon} + \sum_{i=1}^{m} v_i$$

$$D \quad s.t. \quad v_i + b_i\rho_i u \geq r_i\rho_i, i = 1, ..., m \quad D_\epsilon \quad s.t. \quad v_i + b_i\rho_i u \geq r_i\rho_i, i = 1, ..., m$$
$$\mathbf{v} \geq 0, \mathbf{u} \geq 0 \qquad\qquad\qquad \mathbf{v} \geq 0, \mathbf{u} \geq 0$$

The next lemma will prove useful in the analysis of the exponential penalty policy.

Lemma 2. If (u^*, v^*) is optimal solution for both D and D_ϵ, then $\frac{r_{k^*}}{b_{k^*}} = \frac{r_{k\epsilon}}{b_{k\epsilon}}$.

Proof. From the complementary slackness conditions follows that the optimal solutions (u^*, v^*) of D and (u_ϵ, v_ϵ) of D_ϵ are equal to:

$$u^* = \frac{r_{k^*}}{b_{k^*}}$$

$$v^* = \begin{cases} (r_i - \frac{r_{k^*}}{b_{k^*}}b_i)\rho_i, \text{ for } i = 1, ..., k^* \\ 0, \text{ for } i \geq k^* + 1 \end{cases}$$

and

$$u_{k\epsilon} = \frac{r_{k\epsilon}}{b_{k\epsilon}}$$

$$v_{k\epsilon} = \begin{cases} (r_i - \frac{r_{k\epsilon}}{b_{k\epsilon}}b_i)\rho_i, \text{ for } i = 1, ..., k^\epsilon \\ 0, \text{ for } i \geq k^\epsilon + 1. \end{cases}$$

Hence, for (u^*, v^*) to be optimal for D_ϵ, it is necessary that $\frac{r_{k^*}}{b_{k^*}} = \frac{r_{k\epsilon}}{b_{k\epsilon}}$.

The penalty function policy $\bar{\pi}$ proposed in [1] can be described as follows. The classes of requests that may be accepted by the penalty function policy are restricted to the ones with $\alpha_i^* \neq 0$, where α^* is the optimal solution of P. Hence, only the classes of index at most k^* are considered.

"Construct" an "augmented network" as follows. Additional to the initial system, called system 0, consider a fictitious infinite capacity system, called system 1. The state of the augmented network (formed by system 0 and system 1 together) at time t is $s(t) = (x(t), y(t)) \in Z^{2m}$, where $x_i(t), i \in \{1, ..., m\}$ denotes the number of class i requests in system 0 at time t and $y_i(t), i \in \{1, ..., m\}$ denotes the number of class i requests being served in system 1 at time t. System 0 is initially empty, and system 1 is initialized with $y_i^0(0^-) = (1 - \alpha_i^\epsilon)\rho_i$, for $i = 1, k_\epsilon$. Note that $y_i^0(0^-) = 0$, for $i < k^\epsilon$.

For each class i, define the following penalty function $\Psi_i(s)$:

$$\Psi_i(s(t)) = exp(\beta \frac{b_i x_i(t)}{c_i^0}) + exp(\beta \frac{b_i y_i(t)}{c_i^1}).$$

An incoming request of type i is accepted in server 0 if it fits into the knapsack and the following condition holds

$$\frac{\partial \Psi_i(s(t))}{\partial x_i} \leq \frac{\partial \Psi_i(s(t))}{\partial y_i}, \tag{4}$$

otherwise it is sent to server 1, where it stays its service time and then leaves the network. The constants c_i^0, c_i^1 and β are defined as

$$c_i^0 = (1 + 4\epsilon)\alpha_i^\epsilon b_i \rho_i \text{ and } c_i^1 = (1 + 4\epsilon)(1 - \alpha_i^\epsilon) b_i \rho_i, \tag{5}$$

$$\beta \leq \epsilon \min\{\frac{c_i^0}{b_i}, \frac{c_i^1}{b_i} : c_i^0 \neq 0 \text{ and } c_i^1 \neq 0\} \tag{6}$$

where α^ϵ is the optimal solution of (P_ϵ).

Remark 1. Condition (4) is equivalent with:

$$\frac{x_i(t)}{c_i^0} \leq \frac{y_i(t)}{c_i^1} + \frac{1}{\beta b_i} log(\frac{c_i^0}{c_i^1}).$$

Interpretation of the penalty policy. Based on the exact expression of α^ϵ, we are now able to reinterpret the penalty policy $\bar{\pi}$ as follows:

ACCEPT all the requests of classes $i < k^\epsilon$ that fit into the knapsack,
REJECT all the requests of classes $i > k^\epsilon$
ACCEPT the requests of class k^ϵ if

$$\frac{x_{k^\epsilon}(t)}{\alpha_{k^\epsilon}^\epsilon} \leq \frac{y_{k^\epsilon}(t)}{1 - \alpha_{k^\epsilon}^\epsilon} + \frac{(1 + 4\epsilon)\rho_{k^\epsilon}}{\beta} log(\frac{\alpha_{k^\epsilon}^\epsilon}{1 - \alpha_{k^\epsilon}^\epsilon})$$

The rejected requests are sent to system 1, where they remain for the duration of their service time.

Remark 2. Since all requests of class $i, i < k^\epsilon$, are accepted as long as there is capacity, we conclude that the exponential penalty policy proposed in [1] is a threshold policy with the threshhold index k^ϵ.

3.3   On a Lower Bound on the Expected Reward Achieved by $\bar{\pi}$

In this section we will show how the analysis of the exponential penalty function policy presented in [1] can be simplified and improved by interpreting the policy as a threshold policy. We will first summarize the results obtained in [1].

Let $\xi_i(t)$, respectively $\eta_i(t)$ be the number of class i requests in system 1 at time t that were rejected by the penalty function, respectively by the capacity constraints.

Clearly, for each $i \leq m$,

$$E[x_i(t)] = E[q_i(t)] + E[y_i^0(t)] - E[y_i(t)]$$
$$\geq E[q_i(t)] + E[y_i^0(t)] - (E[\xi_i(t)] + E[\eta_i(t)]). \tag{7}$$

Hence, one way to obtain lower bounds for $E[x_i(t)]$, is to obtain upperbounds for $E[\xi_i(t)]$, respectively $E[\eta_i(t)]$.

These upper bounds are obtained by comparison with $\tilde{x}_i(t)$, respectively $\tilde{y}_i(t)$, the number of requests of type i present at time t in system 0, respectively system 1, in the network if the capacity was infinite.

Between $x_i(t)$, $\tilde{x}_i(t)$, $\xi_i(t)$ and $\eta_i(t)$, the following relationships exist (see [1] for the proofs):

Lemma 3. *a)For each $i \leq m$, $x_i(t) \overset{d}{\leq} \tilde{x}_i(t)$ and $\tilde{y}_i(t) \overset{d}{\leq} y_i(t)$, where $X \overset{d}{\leq} Y$ denotes the fact that, for all $u \geq 0$, $P(X \geq u) \leq P(Y \geq u)$.*

b)For each $i \leq m$, $E[e^{\frac{\beta b_i \tilde{x}_i(t)}{b}}] \leq \left(2e^{(1-\frac{\epsilon}{2})\beta}\right)^{\frac{c_i^0}{b}}$ and $E[\tilde{y}_i(t)] \leq (1+\varsigma)(1-\alpha_i^\epsilon)\rho_i$, where $\varsigma = \left(\frac{log(2)}{\beta} + 1 - \frac{\epsilon}{2}\right)(1 + 4\epsilon) - 1$.

c) For each $i \leq m$, $E[\xi_i(t)] \leq E[\tilde{y}_i(t)]$.

d) For each $i \leq m$, $E[\eta_i(t)] \leq 2\rho_i e^{-\frac{\epsilon}{2}(\beta-4)}(1 - e^{-\mu_i t})$.

Substituting the bounds obtained in Lemma 3 in formula (7), one can lower bound the expected reward achieved by policy $\bar{\pi}$:

Theorem 2. *For $\epsilon < \frac{1}{4}$,*

$$E[\bar{R}(t)] \geq \max\{\sum_{i=1}^m r_i\rho_i(1 - e^{-\mu_i t})(\alpha_i^\epsilon - 2e^{-\frac{\epsilon}{2}(\beta-4)}) - \varsigma(1 - \alpha_i^\epsilon)), 0\}, \tag{8}$$

where $\varsigma = \left(\frac{log(2)}{\beta} + 1 - \frac{\epsilon}{2}\right)(1 + 4\epsilon) - 1$ and c_i^0, c_i^1, β are given by (5) and (6).

We proceed now with the tightening of the bound in Theorem 2.

First, remark that for $i > k_\epsilon$, $x_i(t) = 0$, hence these types of requests will not bring any profit. Therefore, in the remainder of this note, we will omit from the analysis the classes of index higher then k^ϵ. Moreover, the definition of $\tilde{x}_i(t)$, together with the fact that k^ϵ is the threshold index, implies that for $i < k^\epsilon$, $\tilde{x}_i(t) \overset{d}{=} q_i(t)$. Hence, for these classes $E[e^{\frac{\beta b_i \tilde{x}_i(t)}{b}}]$ can be obtained exactly, namely:

$$E[e^{\frac{\beta b_i \tilde{x}_i(t)}{b}}] = E[e^{\frac{\beta b_i q_i(t)}{b}}] = e^{\rho_i(1-e^{-\mu_i t})(e^{\frac{\beta b_i}{b}}-1)} \leq e^{\rho_i(\epsilon+1)(1-e^{-\mu_i t})\frac{\beta b_i}{b}}, \tag{9}$$

where for the last inequality we have used the fact that for $x \in (0,1)$, $e^x \leq x + x^2$ and that $\frac{\beta b_i}{b} \leq \epsilon$.

Also, for $i < k^\epsilon$, $\tilde{y}_i(t) \overset{d}{=} \xi_i(t) \overset{d}{=} 0$.
Consider now $E[\eta_i(t)]$. For $i \leq k^\epsilon$,

$$E[\eta_i(t)] = \int_0^t \lambda_i P(\sum_{i=1}^{k^\epsilon} b_i x_i(u) \geq b - b_i) e^{-\mu_i(t-u)} du \tag{10}$$

$$\leq \int_0^t \lambda_i P(\sum_{i=1}^{k^\epsilon} b_i \tilde{x}_i(u) \geq b - b_i) e^{-\mu_i(t-u)} du \tag{11}$$

$$\leq e^{-\beta(1-\frac{b_i}{b})} \int_0^t \lambda_i E[e^{\sum_{i=1}^{k^\epsilon} \frac{\beta b_i}{b} \tilde{x}_i(u)}] e^{-\mu_i(t-u)} du \tag{12}$$

$$= e^{-\beta(1-\frac{b_i}{b})} \int_0^t \lambda_i \prod_{i=1}^{k^\epsilon} E[e^{\frac{\beta b_i}{b} \tilde{x}_i(u)}] e^{-\mu_i(t-u)} du, \tag{13}$$

where in (11) we have used Lemma 3 a), in (12) we have used Markov's inequality and in (13) we have used the independency of the $\tilde{x}_i$'s.

By substituting in (13) the expression for $E[e^{\frac{\beta b_i \tilde{x}_i(t)}{b}}]$ obtained in (9) for indices $i < k^\epsilon$ and the bound given in Lemma 3 b) for $i = k^\epsilon$, we obtain that:

$$E[\eta_i(t)] \leq 2^{\frac{c_{k^\epsilon}^0}{b}} e^{-\frac{\epsilon}{2}(\beta-4)} (1 - e^{-\mu_i t}). \tag{14}$$

Finally, by combining (7), the bound in Lemma 3 b) and c) for $i = k^\epsilon$ (for $i \neq k^\epsilon$, $\xi_i(t) = 0$), and (14), we improve the lower bounds on the expected number of requests of each type in the network at time t and on the expected reward achieved by policy $\bar{\pi}$ as follows.

Theorem 3. *a) For $i < k^\epsilon$,*

$$E[x_i(t)] \geq \rho_i(1 - e^{-\mu_i t}) max\{1 - 2^{\frac{c_{k^\epsilon}^0}{b}} e^{-\frac{\epsilon}{2}(\beta-4)}, 0\}.$$

For $i = k^\epsilon$,

$$E[x_{k^\epsilon}(t)] \geq \rho_{k^\epsilon} max\{(1 - e^{-\mu_{k^\epsilon} t})(\alpha_{k^\epsilon} - 2^{\frac{c_{k^\epsilon}^0}{b}} e^{-\frac{\epsilon}{2}(\beta-4)}) - \varsigma(1 - \alpha_{k^\epsilon}^\epsilon), 0\}.$$

b) For $\epsilon < \frac{1}{4}$, the average return $E[\bar{R}(t)]$ obtained by policy $\bar{\pi}$ can be bounded from below as follows:

$$E[\bar{R}(t)] \geq \sum_{i=1}^{k^\epsilon-1} r_i \rho_i(1 - e^{-\mu_i t}) max\{1 - 2^{\frac{c_{k^\epsilon}^0}{b}} e^{-\frac{\epsilon}{2}(\beta-4)}, 0\} +$$

$$+ r_{k^\epsilon} \rho_{k^\epsilon} max\{(1 - e^{-\mu_{k^\epsilon} t})(\alpha_{k^\epsilon}^\epsilon - 2^{\frac{c_{k^\epsilon}^0}{b}} e^{-\frac{\epsilon}{2}(\beta-4)}) - \varsigma(1 - \alpha_{k^\epsilon}^\epsilon), 0\}. \tag{15}$$

Remark 3. From the comparison of the upper bound on the achievable reward $R^*(t)$ and the lower bound given in Theorem 3, we conclude that if k^ϵ is close to $k^*(t)$, and if $\beta >> 1$, the quality of the bounds is very good. However, if ϵ is chosen such that $\frac{\epsilon^2}{2} \leq \frac{b_{k^\epsilon}}{b}$, then it can be proven that $1 < 2^{\frac{c_{k^\epsilon}^0}{b}} e^{-\frac{\epsilon}{2}(\beta-4)}$, which implies that the lower bound given in the previous theorem is 0.

3.4   Bounds of the Exponential Penalty Policy in a Limiting Regime

In this section we will discuss the behaviour of policy $\bar{\pi}$ when $t \to \infty$ and the influence of the choice of ϵ on the policy in this regime.

Denote by $L(t)$ the lower bound in Theorem 3. Clearly, the following relation holds:

$$
\lim_{t \to \infty} \frac{L(t)}{R^*} = \max\{1 - 2^{\frac{c^0_{k^\epsilon}}{b}} e^{-\frac{\epsilon}{2}(\beta - 4)}, 0\}(1 - \frac{\mathbf{I}_{\{k^\epsilon < k^*\}} \sum_{i=k^\epsilon}^{k^*-1} r_i \rho_i + r_{k^*} \rho_{k^*} \alpha_{k^*}}{\sum_{i=1}^{k^*-1} r_i \rho_i + r_{k^*} \rho_{k^*} \alpha_{k^*}})
$$

$$
- \max\{(1 - e^{-\mu_{k^\epsilon} t})(\alpha^\epsilon_{k^\epsilon} - 2^{\frac{c^0_{k^\epsilon}}{b}} e^{-\frac{\epsilon}{2}(\beta - 4)})
$$

$$
- \varsigma(1 - \alpha^\epsilon_{k^\epsilon}), 0\} \frac{r_{k^\epsilon} \rho_{k^\epsilon}}{\sum_{i=1}^{k^*-1} r_i \rho_i + r_{k^*} \rho_{k^*} \alpha_{k^*}}, \tag{16}
$$

where $\mathbf{I}_{\{k^\epsilon < k^*\}} = 1$ if $k^\epsilon < k^*$ and 0 otherwise.

Note that the classes that cause the bound in (16) to deviate from 1 are the ones that are admitted in the knapsack problem P but are not admitted in the perturbed knapsack problem P_ϵ. It is then intuitive that by restricting the number of such classes, the bound improves. This is exactly what happens by choosing e.g. ϵ such that $\epsilon < \max\{\epsilon_0, \frac{1}{4}\}$, with $\epsilon_0 = \max\{\epsilon :$ D and D_ϵ have the same optimal solution$\}$, as in [1], Corollary 1. From Lemma 1 follows that if $\epsilon < \epsilon_0$, for each k such that $k^\epsilon \le k \le k^*$, $\frac{r_k}{b_k} = \frac{r_{k^*}}{b_{k^*}}$. If for each $k \ne k^*$, $\frac{r_k}{b_k} \ne \frac{r_{k^*}}{b_{k^*}}$, then the classes admitted into the knapsack in problem P and P_ϵ coincide ($k^\epsilon = k^*$). The only difference is that in P_ϵ, a lower fraction of class k^* is admitted.

4   On the Penalty Function Policy and the Thinning Policy

The thinning policy was proposed by Kelly in [3]. In [1], the authors compare experimentally the exponential penalty function policy with the thinning policy and conclude that the first policy performs better in the transient period and the second in steady state. In this section we will see that by interpreting both policies as threshold policies, one can explain to a certain extent their behaviour.

The thinning policy, which we will denote by $\tilde{\pi}$, is based on α^*, the optimal solution of the "steady state program" P. It accepts a request of type i with probability α^*_i if it fits into the knapsack and if it does not fit, it rejects it. Based on the exact calculation of α^*, we conclude that the thinning policy can be described as follows:

ACCEPT a request of type $i < k^*$ if it fits into the knapsack,
REJECT all requests of types $i > k^*$,
ACCEPT a request of type k^* with probability α_{k^*}.

Note that the definitions of the problems P and P_ϵ imply that $k^\epsilon < k^*$. Hence, the exponential penalty policy and the thinning policy treat the classes $i < k^\epsilon$ and $i > k^*$ in the same way. The only difference between the two policies consists in the way they treat the classes $k^\epsilon \le i \le k^*$. The superior behavior of the exponential penalty policy on the thinning policy in transient period, observed experimentally in [1], may be due to the fact that by rejecting "some less profitable" classes, i.e., the classes of index $k^\epsilon < i \le k^*$, there will be more space in the knapsack for "the more profitable" ones.

5   General Service

In this subsection we assume that the service duration S_i has a general distribution with mean $\frac{1}{\mu_i}$, $i = 1, ..., m$. Let g_i denote the density and G_i denote the cumulative distribution function (CDF) of the service duration $i = 1, ..., m$.

Since the LP's P and D only depend on the mean service time, they will remain the same. The program $P(t)$ changes as follows. For the number of users $q_i(t)$ in service at time t in an $M/G/\infty$ system, it is known that $E[q_i(t)] = \rho_i(1 - G_i^e(t))$, where $G_i^e(t)$ is the tail of the equilibrium CDF of the class i service time distribution (see e.g. [6]). Thus, the only change in $P(t)$ is that the tail $e^{-\mu_i t}$ is replaced by $G_i^e(t)$.

Denote this new LP by $\tilde{P}(t)$, by $\tilde{\alpha}$ his optimal solution and by $\tilde{R}(t)$ the optimal value of $\tilde{P}(t)$. Again, $\tilde{P}(t)$ is a continuous knapsack problem, so the optimal solution is $0 - 1$, but for at most one class. Let $\tilde{k}(t)$ be the index of this class.

Theorem 1 can be easily generalized for the case where the service times have a genral distribution.

Theorem 4. *For general service times, the reward rate $R^\pi(t)$ of any feasible policy π satisfies*

$$E[R^\pi(t)] \le \tilde{R}(t) = \sum_{i=1}^{\tilde{k}(t)-1} r_i \rho_i + r_{\tilde{k}(t)} \rho_{\tilde{k}(t)} \tilde{\alpha}_{\tilde{k}(t)}.$$

Consider next the exponential penalty policy $\bar{\pi}$. For finding similar lower bounds to the one in Theorem 3, in [1] extra assumptions on G_i are introduced. Let g_i^t and G_i^t be the density and the CDF of the remaining service time of a class i request conditioned on that it has been in service for time t units. Then, the tail

$$\bar{G}_i^t(s) = 1 - G_i^t(s) = \frac{G_i^e(t+s) - G_i^e(s)}{G_i^e(t)}$$

and

$$g_i^t(s) = -\frac{d\bar{G}_i^t(s)}{ds} = \frac{g_i^e(s) - g_i^e(t+s)}{G_i^e(t)}.$$

Assumption 1. *The function $g_i^t(s)$ is a decreasing function of t for all $i = 1, ..., m$, i.e., $g_i^t(0) \ge \lim_{tu \to \infty} g_i^u(0) = g_i^e(0) = \mu_i$, for all $i = 1, ..., m$.*

Note that, since for classes $i < k^\epsilon$ one can obtain better bounds by estimating the number of users of class i accepted in the knapsack with the number of users in service at time t in an $M/G/\infty$ queue, the assumption above is not necessary only for the class k^ϵ (see also Remark 9). However, unless the class k^ϵ is fixed from before (e.g. equal to k^*), we cannot renounce at the assumption above for all classes $i < k^\epsilon$. Since $\epsilon < \frac{1}{4}$, we can though assume general service times for the classes $i < k_{\frac{1}{4}}$ (the classes accepted into the knapsack when the total capacity is $\frac{b}{2}$). Also, since the classes of index $i, i > k^*$ are never admitted into the knapsack, we can assume general service time for them as well.

Under Assumption 1 for the classes $k_{\frac{1}{4}} < i < k^*$, Theorem 3 has the following equivalent.

Theorem 5. *For $\epsilon < \frac{1}{4}$, the average return $E[\bar{R}(t)]$ obtained by policy $\bar{\pi}$ can be bounded from below as follows:*

$$E[\bar{R}(t)] \geq \sum_{i=1}^{k^\epsilon-1} r_i \rho_i \max\{1 - \bar{G}_i^e(t) - 2^{-\frac{c_{k^\epsilon}^0}{b}} e^{-\frac{\epsilon}{2}(\beta-4)}(1 - e^{-\mu_i t}), 0\}+$$

$$+ r_{k^\epsilon} \rho_{k^\epsilon} \max\{(1 - \bar{G}_{k^\epsilon}(t) + \varsigma)\alpha_{k^\epsilon}^\epsilon - 2^{-\frac{c_{k^\epsilon}^0}{b}} e^{-\frac{\epsilon}{2}(\beta-4)}(1 - e^{-\mu_{k^\epsilon}})$$
$$+ \bar{G}_{k^\epsilon}(t) - \bar{G}_{k^\epsilon}^e(t) - \varsigma, 0\}.$$

5.1 Concluding Remarks

In this note we have shown, based on the optimal solution of some continuous knapsack problems, that the exponential penalty function policy proposed in [1] for controlling loss networks reduces to a threshold policy in the case of the stochastic knapsack. Thus, all requests up to a certain index (the "threshold" index) are accepted if there is enough space in the knapsack. Only for accepting the requests of the class with the threshold index one makes use of the penalty function. As a consequence, the question whether the exponentiality of the penalty functions is necessary is reduced to one single class, namely the class with the "threshold" index. Furthermore, we were able to improve the bounds proposed in [1] and to compare the exponential penalty policy with the thinning policy proposed in [3].

In the last section of [1], the authors generalize the penalty approach to control loss networks and to problems in which the constraints in the LP characterizing the "steady state" define a general polytope. Since the optimal solution of this LP's is not as structured as the optimal solution of continuous knapsack problems, the simplified analysis and the improved bounds presented in this note do not extend to the general case.

References

1. Iyengar, G. and Sigman K. (2004). Exponential penalty function control of loss networks, *The Annals of Applied Probability*, 2004, Vol. 14, No. 4, 1698-1740.
2. Halfin, S. and Whitt, W. (1981) Heavy-traffic limits for queues with many exponential servers. *Operations research* 29, 567-588.

3. Kelly, F.P. (1991). Loss Networks, *Annals of Applied Probability* 1, 319-378.
4. Martello, S. and Toth, P. (1990) Knapsack Problems: Algorithms and Computer Implementations. Wiley, Chichester, West Sussex, England.
5. Ross, K.W. (1995) *Multiservice Loss Models for Broadband Telecommunication Networks*, Springer-Verlag.
6. Wolff, R.W. (1989) *Stochastic Modeling and the Theory of Queues*. Prentice-Hall, Englewood Cliffs, NJ.

Inverse Bottleneck Optimization Problems on Networks

Xiucui Guan[1] and Jianzhong Zhang[2,*]

[1] Department of Mathematics, Southeast University, Nanjing, 210096, P.R. China
[2] Department of System Engineering and Engineering Management, Chinese
University of Hong Kong, Hong Kong, P. R. China
mazhang@cityu.edu.hk

Abstract. The bottleneck optimization problem is to find a feasible
solution that minimizes the bottleneck cost. In this paper, we consider
the inverse bottleneck optimization problems with bound constraints on
modification under weighted l_1 norm, weighted sum-Hamming distance
and weighted bottleneck-Hamming distance. That is, given a feasible
solution F^*, we aim to modify the cost function under some measure
such that F^* becomes an optimal solution to the bottleneck optimization problem. We show that the inverse problem under weighted l_1 norm
and weighted sum-Hamming distance can be reduced to $O(m)$ minimum cut problems, while the inverse problem under weighted bottleneck-
Hamming distance can be reduced to $O(\log m)$ cut feasibility problems,
where $m = |E|$.

1 Introduction

Inverse optimization problems have always been the focus of extensive research
recently, which span a wide variety of applications and vary from traffic planning to high speed communication, to computerized tomography, to isotonic
regression, to conjoint analysis in marketing, etc [4, 2]. Given a candidate solution x^* for an optimization problem, the inverse optimization problem is to
perturb the model parameters by a minimum cost so that x^* is optimal for
the perturbed problem. The generally used measures of the modification cost
are (weighted) l_1, l_2, l_∞ norms and (weighted) Hamming distance. Specially, the
(weighted) Hamming distance has generated wide interest in [2,5,6,9].

We consider the Bottleneck Optimization Problem (BOP) on a network $G =
(V, E, c)$, where c is a cost function defined on the edge set E. Let F be a set
of edges that have some required properties. For example, F might be a path
between a pair of nodes, an assignment of some of the nodes to others, or a
spanning tree of G. We call F a feasible solution and let $\mathcal{F}$ be the set of all
feasible solutions [3]. The problem (BOP) is to find a feasible solution that
minimizes the bottleneck cost of F, which is mathematically formulated below.

$$(BOP) \quad \min_{F \in \mathcal{F}} \max_{e \in F} c(e)$$

---

* Corresponding author.

S.-W. Cheng and C.K. Poon (Eds.): AAIM 2006, LNCS 4041, pp. 220–230, 2006.
© Springer-Verlag Berlin Heidelberg 2006

Although many inverse optimization problems have been well studied (see the survey [4]), only few results on the inverse bottleneck optimization problems are given. Cai, Yang and Zhang [1] proved that the inverse center location problem is strongly NP-hard. Zhang et al. [7, 8] presented polynomial algorithms for some inverse max-min (or min-max) optimization problems. Guan and Zhang [3] proposed a general method for the inverse bottleneck problem under weighted l_∞ norm by solving a series of bottleneck cut problems.

In this paper, we consider the inverse bottleneck optimization problems with bound constraints on modification under weighted l_1 norm, weighted sum-Hamming distance and weighted bottleneck-Hamming distance. That is, given a feasible solution F^*, we aim to find a new cost function c^* under some measure such that F^* becomes an optimal solution to the bottleneck optimization problem $\min_{F \in \mathcal{F}} \max_{e \in F} c^*(e)$, i.e., $\max_{e \in F^*} c^*(e) \leq \max_{e \in F} c^*(e)$ for any $F \in \mathcal{F}$. We show that the inverse problem under weighted l_1 norm and weighted sum-Hamming distance can be reduced to $O(m)$ minimum cut problems, while the inverse problem under weighted bottleneck-Hamming distance can be reduced to $O(\log m)$ cut feasibility problems, where $m = |E|$.

In the remainder of the paper, the inverse problems under weighted l_1 norm and Hamming distance are considered in Sections 2 and 3, conclusions and further research are given in Section 4.

2   Inverse Problem Under Weighted l_1 Norm

Now let us consider the inverse bottleneck optimization problem (IBOPs) under weighted l_1 norm. Given a feasible solution F^*, a weight function $w > 0$ and two bound functions $l \geq 0$ and $u \geq 0$ defined on the edge set E, the problem (IBOPs) is to find a new cost function c^* which solves the following problem:

$$\min f_s(E, c^*) := \sum_{e \in E} w(e)|c^*(e) - c(e)|$$

$$(IBOPs) \text{ s.t. } \max_{e \in F^*} c^*(e) \leq \max_{e \in F} c^*(e), \ \forall \ F \in \mathcal{F},$$

$$c - l \leq c^* \leq c + u.$$

Before discussing the inverse bottleneck optimization problem, let us first provide an important property of the bottleneck optimization problem. We define a cut X of a family $\mathcal{F}$ as a subset of E which intersects all the feasible solutions $F \in \mathcal{F}$. For any $F \in \mathcal{F}$, let $g(F, c) := \max_{e \in F} c(e)$ be the bottleneck cost. In fact, we have

Lemma 1. *[3] For any feasible solution F^+, F^+ is an optimal solution to the bottleneck optimization problem $\min_{F \in \mathcal{F}} \max_{e \in F} c(e)$ if and only if $E^+ := \{e \in E | c(e) \geq g(F^+, c)\}$ is a cut of $\mathcal{F}$.*

Next we analyze the properties of the inverse problem (IBOPs), based on which we will present a general method to solve it.

Lemma 2. *If c^* is an optimal solution of problem (IBOPs), $\max_{e \in F^*} c^*(e) = p^*$, and $p^* \leq \min\{c(e) + u(e) | e \in E, c^*(e) \neq c(e)\}$, then for any edge $e \in E$,*

$$c^*(e) = \begin{cases} p^*, & \text{if } c^*(e) \neq c(e), \\ c(e), & \text{otherwise.} \end{cases} \tag{1}$$

Proof. Obviously, for any optimal solution $\bar{c}$, we have $\bar{c}(e) \leq p^*$ for any $e \in F^*$, and $\bar{c}(e) = c(e)$ for any edge $e \notin F^*$ satisfying $c(e) \geq p^*$. Thus it is easy to check that c^* defined by (1) is a feasible solution of problem (IBOPs).

Now we assume there is an optimal solution $\bar{c}$ such that $\bar{c}(e') \neq c(e')$, $\bar{c}(e') \neq p^* = c^*(e')$, and $\bar{c}(e) = c^*(e)$ for any edge $e \neq e'$, we must have $e' \in F^*$ and $\bar{c}(e') < p^*$. If $c(e') \geq p^*$, then $c(e') \geq c^*(e') > \bar{c}(e')$, and thus $f_s(E, c^*) = f_s(E, \bar{c}) - w(e')(p^* - \bar{c}(e')) < f_s(E, \bar{c})$, which contradicts the optimality of $\bar{c}$. On the other hand, if $c(e') < p^*$, let

$$\tilde{c}(e) := \begin{cases} c(e), & \text{if } e = e', \\ c^*(e), & \text{otherwise,} \end{cases}$$

then $\tilde{c}$ is also a feasible solution. Furthermore, we have $\tilde{c}(e') = c(e') < p^* = c^*(e')$. Then it is obvious that $f_s(E, \tilde{c}) = f_s(E, c^*) - w(e')(p^* - c(e')) < f_s(E, c^*)$, which contradicts the optimality of c^*. The proof is completed. $\qquad \square$

Let F_b be an optimal solution to the problem (BOP) $\min_{F \in \mathcal{F}} g(F, c)$. It is clear that $g(F_b, c) \leq g(F^*, c^*) \leq g(F^*, c)$. Let $\underline{c} = \max\{g(F_b, c), g(F^*, c - l)\}$ be the largest lower bound of new bottleneck cost on F^*, then the possible value of $g(F^*, c^*)$ is in the interval $[\underline{c}, g(F^*, c)]$. Consider the collection of distinctive values of the costs c and the upper bounds on costs within the interval $[\underline{c}, g(F^*, c)]$, that is, the set

$$(\{c(e) | e \in E\} \cup \{c(e) + u(e) | e \in E\} \cup \{\underline{c}\}) \cap [\underline{c}, g(F^*, c)].$$

Sort these values in a strictly increasing order, say $\underline{c} = q_1 < q_2 < \cdots < q_t = g(F^*, c)$, where $t = O(m)$ and $m = |E|$. Then we have

Theorem 1. *The problem (IBOPs) has an optimal solution c^* satisfying (1) such that the bottleneck cost of F^* under c^* is one of the components q_h, $1 \leq h \leq t$, that is, there are an optimal solution c^* and an index h $(1 \leq h \leq t)$ such that $p^* = \max_{e \in F^*} c^*(e) = q_h$.*

Proof. Suppose that c^* is an optimal solution of problem (IBOPs), which satisfies $q_{i-1} < p^* = \max_{e \in F^*} c^*(e) < q_i$. Let $H_1 := \{e \in E | c^*(e) < c(e)\}$ and $H_2 := \{e \in E | c^*(e) > c(e)\}$.

Case 1: If $\sum_{e \in H_1} w(e) \geq \sum_{e \in H_2} w(e)$, then define

$$\tilde{c}(e) := \begin{cases} q_i, & \text{if } c^*(e) \neq c(e), \\ c(e), & \text{otherwise.} \end{cases}$$

First note that $c(e) - l(e) \leq p^* < q_i \leq c(e) + u(e)$ for any edge $e \in H_1 \cup H_2$. Next we show that $\max_{e \in F} \tilde{c}(e) \geq \max_{e \in F^*} \tilde{c}(e)$ for any $F \in \mathcal{F}$, then $\tilde{c}$ is a feasible solution. Notice that

$$p^* = \max_{e \in F^*} c^*(e) = \max\{\max_{e \in F^* \cap (H_1 \cup H_2)} c^*(e), \max_{e \in F^* \setminus (H_1 \cup H_2)} c^*(e)\}$$

$$= \max\{p^*, \max_{e \in F^* \setminus (H_1 \cup H_2)} c(e)\}.$$

Then

$$\max_{e \in F^* \setminus (H_1 \cup H_2)} c(e) \leq p^* < q_i \qquad (2)$$

and

$$\max_{e \in F^*} \tilde{c}(e) = \max\{q_i, \max_{e \in F^* \setminus (H_1 \cup H_2)} c(e)\} = q_i.$$

Furthermore, for any $F \in \mathcal{F}$,

$$\max_{e \in F} \tilde{c}(e) = \max\{\max_{e \in F \cap (H_1 \cup H_2)} \tilde{c}(e), \max_{e \in F \setminus (H_1 \cup H_2)} \tilde{c}(e)\}$$

$$= \max\{q_i, \max_{e \in F \setminus (H_1 \cup H_2)} c(e)\} \geq q_i = \max_{e \in F^*} \tilde{c}(e).$$

For the optimality, we have

$$f_s(E, c^*) - f_s(E, \tilde{c})$$

$$= \sum_{e \in H_1} [w(e)(c(e) - p^*) - w(e)(c(e) - q_i)]$$

$$+ \sum_{e \in H_2} [w(e)(p^* - c(e)) - w(e)(q_i - c(e))]$$

$$= \sum_{e \in H_1} w(e)(q_i - p^*) + \sum_{e \in H_2} w(e)(p^* - q_i)$$

$$= [\sum_{e \in H_1} w(e) - \sum_{e \in H_2} w(e)](q_i - p^*)$$

$$\geq 0.$$

Hence, we can construct an optimal solution $\tilde{c}$ which meets the requirement of the theorem. In fact, in this case, we must have $\sum_{e \in H_1} w(e) = \sum_{e \in H_2} w(e)$.

Case 2: If $\sum_{e \in H_1} w(e) < \sum_{e \in H_2} w(e)$, then let

$$\tilde{c}(e) := \begin{cases} q_{i-1}, & if\ c^*(e) \neq c(e), \\ c(e), & otherwise. \end{cases}$$

First note that for any edge $e \in H_2$, $c(e) \leq q_{i-1} < p^* \leq c(e) + u(e)$, and for any edge $e \in H_1$, we have $e \in F^*$. Then $c(e) - l(e) \leq \underline{c} \leq q_{i-1} < p^* \leq c(e)$. Next we show that $\max_{e \in F} \tilde{c}(e) \geq \max_{e \in F^*} \tilde{c}(e)$ for any $F \in \mathcal{F}$. It follows from (2) that

$$\max_{e \in F^* \setminus (H_1 \cup H_2)} c(e) \leq q_{i-1},$$

then for any $F \in \mathcal{F}$,

$$\max_{e \in F} \tilde{c}(e) = \max\{q_{i-1}, \max_{e \in F \setminus (H_1 \cup H_2)} c(e)\}$$

$$\geq q_{i-1} = \max\{q_{i-1}, \max_{e \in F^* \setminus (H_1 \cup H_2)} c(e)\} = \max_{e \in F^*} \tilde{c}(e).$$

Furthermore,

$$f_s(E, c^*) - f_s(E, \tilde{c}) = [\sum_{e \in H_1} w(e) - \sum_{e \in H_2} w(e)](q_{i-1} - p^*) > 0,$$

that is, $f_s(E, c^*) > f_s(E, \tilde{c})$, which contradicts the optimality of c^*. Thus this case can not happen and we complete the proof.   □

For a given value p, we define the restricted inverse bottleneck optimization problem (IBOPs(p)) under weighted l_1 norm as follows:

$$\min \sum_{e \in E} w(e)|c^*(e) - c(e)|$$

$$(IBOPs(p)) \text{ s.t. } \max_{e \in F^*} c^*(e) \leq \max_{e \in F} c^*(e), \ \forall \ F \in \mathcal{F},$$

$$\max_{e \in F^*} c^*(e) = p,$$

$$c - l \leq c^* \leq c + u.$$

According to Theorem 1, we can obtain an optimal solution of problem (IBOPs) by solving all t restricted problems (IBOPs(p)) ($p = q_1, \cdots, q_t$) and then choosing from the t optimal solutions the one that has the minimum objective value.

Now let us handle the restricted inverse problem (IBOPs(p)). Let $F^*(p) := \{e \in F^* | c(e) > p\}$, $E(p) := \{e \in E | c(e) < p\}$ and $\overline{E}(p) := \{e \in E(p) | p \leq c(e) + u(e)\}$. Define a capacity vector $w_p : E(p) \to R$ as follows:

$$w_p(e) := \begin{cases} w(e)(p - c(e)), & \forall e \in \overline{E}(p), \\ +\infty, & \forall e \in E(p) \backslash \overline{E}(p). \end{cases}$$

If $E^+(p) := \{e \in E | c(e) \geq p\}$ is a cut of $\mathcal{F}$, then it follows from Lemma 1 that

$$c^*(e) := \begin{cases} p, & \forall e \in F^*(p), \\ c(e), & otherwise, \end{cases}$$

is an optimal solution of problem (IBOPs(p)), and the optimal value is

$$\sum_{e \in F^*(p)} w(e)(c(e) - p).$$

Otherwise, let $\mathcal{F}(p) := \{F \in \mathcal{F} | F \subseteq E(p)\}$, we aim to increase the cost of at least one edge to p for each set $F \in \mathcal{F}(p)$. That is, we need to find the minimum cut of family $\mathcal{F}(p)$ under the capacity vector w_p defined below.

$$\min \sum_{e \in K} w_p(e)$$

$$(MC(p)) \text{ s.t. } K \subseteq E(p) \text{ is a cut of family } \mathcal{F}(p).$$

Denote by $K(p)$ the optimal solution to problem (MC(p)) and let $w_p(K(p)) := \sum_{e \in K(p)} w_p(e)$ be the corresponding objective value.

If $w_p(K(p)) = +\infty$, then the instance is infeasible; otherwise, define c^* as follows:

$$c^*(e) := \begin{cases} p, & \forall \ e \in K(p) \cup F^*(p), \\ c(e), & otherwise. \end{cases} \tag{3}$$

Then we have

Theorem 2. *If $E^+(p) := \{e \in E | c(e) \geq p\}$ is not a cut of $\mathcal{F}$ and $w_p(K(p)) < +\infty$, then the cost vector c^* defined by (3) is an optimal solution to problem $(IBOPs(p))$, and the corresponding objective value is*

$$\sum_{e \in F^*(p)} w(e)(c(e) - p) + w_p(K(p)).$$

Proof. We first show that c^* is a feasible solution. Obviously, $\max_{e \in F^*} c^*(e) = p$ and $c - l \leq c^* \leq c + u$. Now we claim that $E^+(p) = \{e \in E | c^*(e) \geq p\}$ is a cut of $\mathcal{F}$. Otherwise, there is a feasible solution $F' \in \mathcal{F}$ such that $c^*(e) < p$ for any $e \in F'$. Then $F' \cap (K(p) \cup F^*(p)) = \emptyset$ and $c^*(e) = c(e) < p$, thus $F' \in \mathcal{F}(p)$ and $F' \cap K(p) = \emptyset$, which is impossible since $K(p)$ is a cut of $\mathcal{F}(p)$.

In order to prove the optimality of c^*, we suppose $\tilde{c}$ is an optimal solution to problem (IBOPs(p)), then by Lemma 2, we have

$$\tilde{c}(e) = \begin{cases} p, & if \ \tilde{c}(e) \neq c(e), \\ c(e), & otherwise. \end{cases}$$

Define $\tilde{E}(p) := \{e \in E(p) | \tilde{c}(e) = p\}$. We claim that $\tilde{E}(p)$ is a cut of $\mathcal{F}(p)$. Otherwise, there is a feasible solution $F' \in \mathcal{F}(p)$ such that $F' \cap \tilde{E}(p) = \emptyset$. Then for any edge $e \in F'$, $\tilde{c}(e) = c(e) < p$, and hence $\max_{e \in F'} \tilde{c}(e) < p$, which contradicts the feasibility of $\tilde{c}$. Moreover,

$$\sum_{e \in E(p)} w(e)(\tilde{c}(e) - c(e)) = \sum_{e \in \tilde{E}(p)} w(e)(p - c(e))$$

$$= \sum_{e \in \tilde{E}(p)} w_p(e) \geq \sum_{e \in K(p)} w_p(e) = w_p(K(p)),$$

and

$$\sum_{e \notin E(p)} w(e)(c(e) - \tilde{c}(e)) \geq \sum_{e \in F^*(p)} w(e)(c(e) - p).$$

Thus c^* is an optimal solution to problem $(IBOPs(p))$.   □

Note that when solving the restricted problem (IBOPs(p)), if there exists no cut of $\mathcal{F}(p)$ in $\overline{E}(p)$, then the instance is infeasible. Furthermore, we have

Lemma 3. *[3] If $\overline{E}(p)$ is not a cut of $\mathcal{F}(p)$, then $\overline{E}(q)$ is not a cut of $\mathcal{F}(q)$ for all $q \geq p$.*

It follows from Lemma 3 that if $c_{q_i}(K(q_i)) = +\infty$ for some checking point q_i, then $c_{q_j}(K(q_j)) = +\infty$ for all checking points $q_j \geq q_i$. Hence in real implementation, we may employ the binary search strategy to determine the smallest index r such that $c_{q_r}(K(q_r)) < +\infty$, then check the points in the interval $[q_1, q_r]$. As a result, we can conclude that

Theorem 3. *The inverse problem (ICBPs) can be solved in $O(mT_s)$ operations, where T_s is the complexity to solve the corresponding minimum cut problem. Moreover, if the minimum cut problem can be solved in strongly polynomial time, then the inverse problem can be solved in strongly polynomial time, too.*

3   Inverse Problem Under Weighted Hamming Distance

Now let us consider the inverse bottleneck optimization problem under weighted Hamming distance. Given a feasible solution F^*, a weight function $w > 0$ and two bound functions $l \geq 0$ and $u \geq 0$ defined on the edge set E, the inverse problem (IBOP$_{sH}$) under weighted sum-Hamming distance is to find a new cost function c^* which solves the following problem:

$$\min f_{sH}(E, c^*) := \sum_{e \in E} w(e) H(c^*(e), c(e))$$

$$(IBOP_{sH}) \text{ s.t. } \max_{e \in F^*} c^*(e) \leq \max_{e \in F} c^*(e), \ \forall \ F \in \mathcal{F},$$

$$c - l \leq c^* \leq c + u,$$

where $H(c^*(e), c(e))$ is the Hamming distance between $c^*(e)$ and $c(e)$, that is, $H(c^*(e), c(e)) = 0$ if $c^*(e) = c(e)$ and 1 otherwise.

The inverse problem (IBOP$_{bH}$) under weighted bottleneck-Hamming distance can be defined similarly.

$$\min f_{bH}(E, c^*) := \max_{e \in E} w(e) H(c^*(e), c(e))$$

$$(IBOP_{bH}) \text{ s.t. } \max_{e \in F^*} c^*(e) \leq \max_{e \in F} c^*(e), \ \forall \ F \in \mathcal{F},$$

$$c - l \leq c^* \leq c + u.$$

Note that for the inverse problem under weighted Hamming distance, we only care about whether the cost of an edge is modified or not, but not concern about the magnitude of modification as long as it is restricted in a given interval. Concretely, the inverse problem (IBOP$_{sH}$) under weighted sum-Hamming distance is to minimize the weighted number of modified edges, while the problem (IBOP$_{bH}$) under weighted bottleneck-Hamming distance is to minimize the maximal weight of modified edges.

3.1   Inverse Problem Under Weighted Sum-Hamming Distance

Based on the above analysis, we can easily conclude that

Lemma 4. *If the inverse problem (IBOP$_{sH}$) is feasible, then there is an optimal solution c^* satisfying $\max_{e \in F^*} c^*(e) = p^*$ such that*

$$c^*(e) = \begin{cases} p^*, & if \ c^*(e) \neq c(e), c(e) - l(e) \leq p^* \leq c(e) + u(e), \\ c(e) - l(e), & if \ c^*(e) \neq c(e), p^* < c(e) - l(e), \\ c(e) + u(e), & if \ c^*(e) \neq c(e), p^* > c(e) + u(e), \\ c(e), & otherwise. \end{cases} \tag{4}$$

Consider the collection of distinctive values

$$(\{c(e)|e \in E\} \cup \{c(e) + u(e)|e \in E\} \cup \{\underline{c}\}) \cap [\underline{c}, g(F^*, c)],$$

and sort them in a strictly increasing order, say $\underline{c} = q_1 < q_2 < \cdots < q_t = g(F^*, c)$, where $t = O(m)$, then we can similarly conclude that

Theorem 4. *The problem (IBOP$_{sH}$) has an optimal solution c^* satisfying (4) such that $p^* = \max_{e \in F^*} c^*(e) = q_h$ for some index $h \in \{1, \cdots, t\}$.*

Next we extend the method for the weighted l_1 norm to the weighted sum-Hamming distance. Consequently, we also handle the restricted inverse problem (IBOP$_{sH}(p)$) for a given value p, which is defined as follows:

$$\min \sum_{e \in E} w(e) H(c^*(e), c(e))$$

$$(IBOP_{sH}(p)) \quad \text{s.t.} \ \max_{e \in F^*} c^*(e) \leq \max_{e \in F} c^*(e), \ \forall \ F \in \mathcal{F},$$

$$\max_{e \in F^*} c^*(e) = p,$$

$$c - l \leq c^* \leq c + u.$$

Let $F^*(p) := \{e \in F^*|c(e) > p\}$. If $E^+(p) := \{e \in E|c(e) \geq p\}$ is a cut of $\mathcal{F}$, then it follows from Lemma 1 that

$$c^*(e) := \begin{cases} p, & \forall e \in F^*(p), \\ c(e), & otherwise, \end{cases}$$

is an optimal solution of problem (IBOP$_{sH}(p)$), and $f_{sH}(E, c^*) := \sum_{e \in F^*(p)} w(e)$. Otherwise, let $E(p) := \{e \in E|c(e) < p\}$, $\overline{E}(p) := \{e \in E(p)|p \leq c(e) + u(e)\}$, and $\mathcal{F}(p) := \{F \in \mathcal{F}|F \subseteq E(p)\}$. Define a capacity vector below:

$$w_p(e) := \begin{cases} w(e), & \forall e \in \overline{E}(p), \\ +\infty, & \forall e \in E(p) \backslash \overline{E}(p). \end{cases}$$

Then find the minimum cut $K(p)$ of family $\mathcal{F}(p)$ under the capacity vector w_p. Let $w_p(K(p)) := \sum_{e \in K(p)} w_p(e)$ be the capacity of minimum cut.

If $w_p(K(p)) = +\infty$, then the instance is infeasible; otherwise,

$$c^*(e) := \begin{cases} p, & \forall\, e \in K(p) \cup F^*(p), \\ c(e), & otherwise, \end{cases} \tag{5}$$

is an optimal solution of problem (IBOP$_{sH}(p)$), and the value of objective function is $f_{sH}(E, c^*) := \sum_{e \in K(p) \cup F^*(p)} w(e)$.

Similar to the inverse problem under weighted l_1 norm, the inverse problem (IBOP$_{sH}$) can also be reduced to $O(m)$ minimum cut problems.

3.2   Inverse Problem Under Weighted Bottleneck-Hamming Distance

For a given edge weight $w(e_{i_k})$, let $E_k := \{e \in E | w(e) \le w(e_{i_k})\}$,

$$c^k(e) := \begin{cases} c(e) - l(e), & if\ e \in E_k \cap F^*, \\ c(e) + u(e), & if\ e \in E_k \backslash F^*, \\ c(e), & if\ e \notin E_k, \end{cases} \tag{6}$$

$\max_{e \in F^*} c^k(e) = p_k$, and $E_k^+ := \{e \in E | c^k(e) \ge p_k\}$. Let

$$c^{p_k}(e) := \begin{cases} p_k, & if\ e \in E_k \cap F^*, p_k \le c(e) + u(e), \\ c(e) - l(e), & if\ e \in E_k \cap F^*, p_k > c(e) + u(e), \\ p_k, & if\ e \in E_k \backslash F^*, c(e) - l(e) \le p_k \le c(e) + u(e), \\ c(e) + u(e), & if\ e \in E_k \backslash F^*, p_k < c(e) - l(e)\ or\ p_k > c(e) + u(e), \\ c(e), & if\ e \notin E_k, \end{cases} \tag{7}$$

and $E_{p_k}^+ := \{e \in E | c^{p_k}(e) \ge p_k\}$. Then we have the properties below.

Lemma 5. *If E_k^+ is a cut of $\mathcal{F}$, then $E_{p_k}^+$ is also a cut of $\mathcal{F}$.*

Proof. It is sufficient to show that $E_k^+ \subseteq E_{p_k}^+$, which can be proved by the five cases listed in (7). For any edge $e \in E_k^+$ in Case 1 and 3, $c^k(e) \ge p_k = c^{p_k}(e)$, then $e \in E_{p_k}^+$. For any edge $e \in E_k^+$ in Case 2, we have $c^k(e) \ge p_k > c(e) + u(e)$, which contradicts the feasibility of c^k, and thus the case can not happen. Similarly, we ignore the edges e in Case 4 satisfying $p_k > c(e) + u(e)$. For edges e in Case 4 satisfying $p_k < c(e) - l(e)$ and in Case 5, when $e \in E_k^+$, then $c^{p_k}(e) = c^k(e) \ge p_k$, and thus $e \in E_{p_k}^+$. The proof is completed. $\qquad\square$

Lemma 6. *If $E_{p_k}^+$ is not a cut of $\mathcal{F}$, then for any $\tilde{c}$ with $c - l \le \tilde{c} \le c + u$ and $f_{bH}(E, \tilde{c}) = w(e_{i_k})$, $\widetilde{E}^+ := \{e \in E | \tilde{c}(e) \ge \tilde{p} = \max_{e \in F^*} \tilde{c}(e)\}$ is not a cut of $\mathcal{F}$.*

Proof. First we claim that $\tilde{p} \ge p_k$. It is easy to see that

$$\tilde{p} = \max\{\max_{e \in F^* \cap E_k} \tilde{c}(e), \max_{e \in F^* \backslash E_k} \tilde{c}(e)\} = \max\{\max_{e \in F^* \cap E_k} \tilde{c}(e), \max_{e \in F^* \backslash E_k} c(e)\}$$

$$\ge \max\{\max_{e \in F^* \cap E_k} (c(e) - l(e)), \max_{e \in F^* \backslash E_k} c(e)\} = p_k.$$

Second, it is sufficient to show that $\widetilde{E}^+ \subseteq E_{p_k}^+$, which can be proved in a similar way as in Lemma 5. Here we take Case 4 for example. For any edge $e \in \widetilde{E}^+$ in Case 4 satisfying $p_k < c(e) - l(e)$, then $c^{p_k}(e) = c(e) + u(e) \geq \tilde{c}(e) \geq \tilde{p} \geq p_k$, and thus $e \in E_{p_k}^+$. The lemma holds. $\square$

It follows from Lemma 6 that if $E_{p_k}^+$ is not a cut of $\mathcal{F}$, then there is no feasible solution $\tilde{c}$ of problem (IBOP$_{bH}$) with $f_{bH}(E, \tilde{c}) \leq w(e_{i_k})$. Specially, if $w(e_{i_k}) = \max_{e \in E} w(e)$, then $E_k = E$; moreover, if $E_{p_k}^+$ is not a cut of $\mathcal{F}$ in this case, the instance is infeasible. Next we present a general algorithm to solve the inverse bottleneck optimization problem (IBOP$_{bH}$) under weighted bottleneck-Hamming distance by performing a binary search on the weights $w(e)$ of edges $e \in E$. In each iteration, we are mainly to check whether the set $E_{p_k}^+$ is a cut of $\mathcal{F}$ or not, which is equivalent to checking if $\bar{G}_k = (V, E \backslash E_{p_k}^+)$ contains a feasible solution of $\mathcal{F}$ or not. Let $\lfloor x \rfloor$ be the maximal integer not greater than x.

Algorithm 5. *(A general algorithm to solve problem (IBOP$_{bH}$))*
    Input: *A system* $(V, E; \mathcal{F})$ *with four functions* w, c, l *and* u *defined on* E.
    Step 1. *Sort all the weights* $w(e)$ *of edges* $e \in E$ *in a strictly increasing order,*
i.e., $w(e_{i_1}) < w(e_{i_2}) < \cdots < w(e_{i_\tau})$. *Put* $a := 1$ *and* $b := \tau$.
    Step 2. *If* $E_{p_\tau}^+$ *is not a cut of* $\mathcal{F}$, *then the instance is infeasible, stop.*
    Step 3. *If* $b - a = 1$, *then output the modified cost* c^b *and the objective value*
$f_{bH}(E, c^b) = w(e_{i_b})$, *stop. Otherwise, go to Step 4.*
    Step 4. *Let* $k := \lfloor (a+b)/2 \rfloor$, $E_k := \{e \in E | w(e) \leq w(e_{i_k})\}$, $p_k := \max\limits_{e \in F^*} c^k(e)$
and $E_{p_k}^+ := \{e \in E | c^{p_k}(e) \geq p_k\}$, *where* c^k *and* c^{p_k} *are defined as in (6) and (7),*
respectively. If $E_{p_k}^+$ *is a cut of* $\mathcal{F}$, *then put* $b := k$, *else put* $a := k$. *Return to*
Step 3.

    As a conclusion, we have

Theorem 6. *The inverse problem (IBOP$_{bH}$) under weighted bottleneck-Hamming distance can be solved in* $O(T_{bH} \log m)$ *operations, where* T_{bH} *is the time to check if* $E_{p_k}^+$ *is a cut of* $\mathcal{F}$ *or not. Furthermore, if the cut feasibility problem can be determined in strongly polynomial time, then the inverse problem (IBOP$_{bH}$) can be solved in strongly polynomial time, too.*

For example, the inverse bottleneck assignment problem under weighted bottleneck-Hamming distance can be solved in $O(n^3 \log m)$ operations, where $n = |V|$ [3].

4   Conclusion and Further Research

In this paper, we consider the inverse bottleneck optimization problems with bound constraints on modification under weighted l_1 norm, weighted sum-Hamming distance and weighted bottleneck-Hamming distance. That is, given a feasible solution F^*, we aim to modify the cost function under some measure such that F^* becomes an optimal solution to the bottleneck optimization problem.

We show that the inverse problem under weighted l_1 norm and weighted sum-Hamming distance can be reduced to $O(m)$ minimum cut problems, while the inverse problem under weighted bottleneck-Hamming distance can be reduced to $O(\log m)$ cut feasibility problems.

Guan and Zhang [3] showed that the inverse bottleneck optimization problems with bound constraints on modification under weighted l_∞ norm can be reduced to $O(m^2)$ bottleneck cut problems. Furthermore, Zhang, Yang and Cai [8] clarified that whether there is a polynomial algorithm for the inverse min-max spanning tree problem under weighted l_2 norm is a promising problem, although we can deduce from the results in this paper and in [8] that the inverse min-max spanning tree problems under weighted l_∞ norm, weighted l_1 norm, weighted sum-Hamming distance and weighted bottleneck-Hamming distance can all be solved in strongly polynomial time.

Therefore, as a further research topic, we can try to find a general method to solve the inverse bottleneck optimization problem under weighted l_2 norm, and present polynomial algorithms for some specific inverse bottleneck optimization problems according to their combinatorial properties.

References

1. Cai, M., Yang, X., Zhang, J.: The complexity analysis of the inverse center location problem. J. Global Optim. **5** (1999) 213–218
2. Duin, C.W., Volgenant, A.: Some inverse optimization problems under the Hamming distance. European J. Oper. Res. **170** (2006) 887–899
3. Guan, X., Zhang, J.: Inverse Constrained Bottleneck Problems Under Weighted l_∞ Norm. Comput. and Oper. Res. Available online 2 February 2006
4. Heuberger, C.: Inverse optimization, a survey on problems, methods, and results. J. Comb. Optim. **329** (2004) 329–361
5. He, Y., Zhang, B., Yao, E.: Weighted Inverse Minimum Spanning Tree Problems under Hamming Distance. J. Comb. Optim. **9** (2005) 91–100
6. He, Y., Zhang, B., Zhang, J.: Constrained Inverse Minimum Spanning Tree Problems under the bottleneck-type Hamming Distance. J. Global Optim. **34** (2006) 467–474
7. Yang, C., Zhang, J.: Inverse maximum capacity problems. OR Spektrum. **20** (1998) 97–100
8. Zhang, J., Yang, X., Cai, M.: Some inverse min-max network problems under weighted l_1 and l_∞ norms. Working Paper. Department of Mathematics, City University of Hong Kong (2004)
9. Zhang, J., Yang, X.: Inverse Maximum Flow Problems Under l_∞ Norm, l_2 Norm and Hamming Distance. Working Paper. Department of Mathematics, City University of Hong Kong (2005)

An Efficient Algorithm for Evacuation Problems in Dynamic Network Flows with Uniform Arc Capacity

Naoyuki Kamiyama, Naoki Katoh, and Atsushi Takizawa

Department of Architecture and Architectural Engineering, Kyoto University
Kyotodaigaku-Katsura, Nishikyo-ku, Kyoto, 615-8540, Japan
{is.kamiyama, naoki, kukure}@archi.kyoto-u.ac.jp

Abstract. In this paper, we consider the quickest flow problem in a network which consists of a directed graph with capacities and transit times on its arcs. We present an $O(n \log n)$ time algorithm for the quickest flow problem in a network of grid structure with uniform arc capacity which has a single sink where n is the number of vertices in the network.

1 Introduction

It is very important to establish crisis management systems against large-scale disasters such as big earthquakes, conflagrations and tsunamis. We need to consider the crisis management against disasters to secure evacuation pathways and to effectively guide residents to a safe place. In our work, we adopt dynamic network flows as a model for evacuation. A dynamic network flow is defined on a network which consists of a directed graph $D = (V, A)$ with capacity $c(e)$ and transit time $\tau(e)$ on every arc $e \in A$. For example, if we consider urban evacuation, vertices model buildings, rooms, exits and so on, and an arc models a pathway or a road connecting vertices. For an arc e, capacity $c(e)$ represents the number of people which can traverse the arc e per unit time, and $\tau(e)$ denotes the time it takes to traverse e. Given a network with initial supplies at vertices, the problem is to find an optimal dynamic network flow such that we can send all the initial supplies to sinks as quickly as possible. In the case where a network has several sources and sinks which have specified supply or demand respectively, this problem can be solved by the algorithm of Hoppes and Tardos [1] in polynomial time. However their running time is high-order polynomial, and hence is not practical in general. So it is necessary to devise a faster algorithm for a tractable and practically useful subclass of this problem.

In this paper, we restrict our attention to grid networks with uniform arc capacity. The condition that arc capacity is uniform is practically acceptable because the width of road or corridor is generally standardized. Restriction of network structure to grid networks is useful since such structure often appears in modelling building corridors and city streets. We present an $O(n \log n)$ time algorithm for the quickest flow problem in a network of grid structure with

S.-W. Cheng and C.K. Poon (Eds.): AAIM 2006, LNCS 4041, pp. 231–242, 2006.

uniform arc capacity which has a single sink where n is the number of vertices in the network.

Previous Works. As mentioned above, Hoppes and Tardos proposed a polynomial time algorithm for the problem [1]. As a special class of networks, Mamada et al. [2] considered tree networks with a single sink and presented an $O(n \log^2 n)$ time algorithm. For the case of tree networks with multiple sinks, Mamada et al. [3] presented an $O(n \log^3 n)$ time algorithm for two-sink case and Mamada et al. [4] presented an $O(n^2 k \log^2 n)$ time algorithm for k-sink case under the restriction that all the supplies going through a common vertex are sent to a single sink. However, to the authors' knowledge, no one has ever studied special class of networks other than tree networks for the evacuation problem.

Organization. Section 2 gives necessary definitions and preliminaries. Section 3 considers the quickest flow problem for grid networks with uniform arc capacity and proposes an $O(n \log n)$ time algorithm. Section 4 concludes the paper.

2   Problem Formulation and Notations

We consider a network $\mathcal{N} = (D = (V, A), c, \tau, b_v, V^*)$, where D is a directed graph, V is a set of vertices, A is a set of arcs, $c \colon A \to \mathbf{R}_+$ is the upper bound for the rate of flow that enters each arc per unit time, $\tau \colon A \to \mathbf{Z}_+$ is a transit time function, $b_v \in \mathbf{R}_+$ gives an initial supply of $v \in V$, and $V^* \subset V$ is a set of sinks. Here $\mathbf{R}_+$ denotes the set of nonnegative reals and $\mathbf{Z}_+$ denotes the set of nonnegative integers. For simplicity, we write $c(v, w)$ and $\tau(v, w)$ instead of $c((v, w))$ and $\tau((v, w))$ respectively for any $(v, w) \in A$. Given a network, our problem is to compute the minimum time required to send all supplies to sinks.

Here we define a discrete-time dynamic network flow $f \colon A \times \mathbf{Z}_+ \to \mathbf{R}_+$. For any arc $e \in A$ and $\theta \in \mathbf{Z}_+$, we denote by $f(e, \theta)$ the flow rate entering the arc e at time θ which arrives at the head of e at time $\theta + \tau(e)$. We call $f \colon A \times \mathbf{Z}_+ \to \mathbf{R}_+$ a *feasible dynamic flow* in $\mathcal{N}$ if it satisfies the following three conditions, i.e., capacity constraint, flow conservation, and demand constraint [2].

Capacity constraint: For any arc $e \in A$ and $\theta \in \mathbf{Z}_+$,

$$0 \le f(e, \theta) \le c(e). \tag{1}$$

Flow conservation: For any $v \in V$ and $\Theta \in \mathbf{Z}_+$,

$$\sum_{e \in \Delta^+(v)} \sum_{\theta=0}^{\Theta} f(e, \theta) - \sum_{e \in \Delta^-(v)} \sum_{\theta=\tau(e)}^{\Theta} f(e, \theta - \tau(e)) \le b_v. \tag{2}$$

Demand constraint: There exists a time $\Theta \in \mathbf{Z}_+$ such that

$$\sum_{e \in \Delta^-(V^*)} \sum_{\theta=\tau(e)}^{\Theta} f(e, \theta - \tau(e)) - \sum_{e \in \Delta^+(V^*)} \sum_{\theta=0}^{\Theta} f(e, \theta) = \sum_{v \in V \setminus V^*} b_v. \tag{3}$$

Here $\Delta^+(V') \equiv \{(v,w) \in A, \,|\, v \in V', w \notin V'\}$, and $\Delta^-(V') \equiv \{(v,w) \in A \,|\, v \notin V', w \in V'\}$ for any $V' \subseteq V$. For simplicity, we write $\Delta^+(v)$ and $\Delta^-(v)$ instead of $\Delta^+(\{v\})$ and $\Delta^-(\{v\})$, respectively. For a feasible dynamic flow f, let $\Theta(f)$ denote the completion time for f, i.e., the minimum time Θ satisfying (3), and let $\mathcal{F}_{\mathcal{N}}$ denote the set of all feasible dynamic flows in $\mathcal{N}$. The quickest flow problem asks to find a feasible dynamic flow f that minimizes $\Theta(f)$.

Here we define *flow-table* [2] which is a function from $\mathbf{Z}_+$ to $\mathbf{R}_+$. There are two kinds of flow-tables, *arriving-table* AT_v for each vertex $v \in V$, and *sending-table* ST_e for each arc $e \in A$. Arriving-table AT_v represents the sum of the flow rates arriving at the vertex v as a function of time θ, i.e.,

$$AT_v(\theta) = B_v(\theta) + \sum_{e \in \Delta^-(v)} f(e, \theta - \tau(e)) \qquad (4)$$

where we regard the initial supply b_v as a flow-table B_v as follows: $B_v(0) = b_v$ and $B_v(\theta) = 0$ if $\theta \neq 0$. Sending table ST_e represents the flow rate entering the arc e as a function of time θ, i.e.,

$$ST_e(\theta) = f(e, \theta). \qquad (5)$$

We define T as a function of flow-table FT as follows: $T(FT) = \max\{\theta \in \mathbf{Z}_+ \,|\, FT(\theta) > 0\}$. $T(ST_e) + \tau(e)$ represents the time to complete the evacuation from e.

In this paper, we will focus our attention on *grid graph* as an underlying graph of a network. For simplicity, we assume a grid graph is on N^2 grid points $\{1, \ldots, N\} \times \{1, \ldots, N\}$ in the plane, and let $n = N^2$. Here a vertex is identified with (i,j) with $1 \leq i \leq N$ and $1 \leq j \leq N$. A sink r is specified as one of vertices. The distance between two vertices (i,j) and (i',j') is defined as $|i - i'| + |j - j'|$. Two vertices (i,j) and (i',j') are connected by an edge if and only if $|i - i'| + |j - j'| = 1$ (Fig. 1(a)). The edge which connects v and v' is directed from v to v' if and only if the distance from v' to r is smaller than that from v to r (Fig. 1(b)). A network defined on a grid graph is called *grid network*. We assume throughout this paper that, in networks we are concerned with, the capacities of all arcs have the same value $c \in \mathbf{R}_+$ and the transit times of all arcs take the same value $\tau \in \mathbf{Z}_+$. Notice that we define c and τ as not a function but an integer here. From this assumption, we use the notation $\mathcal{N} = (D = (V, A), b_v, V^*)$ for simplicity by omitting the capacity function and the transit time function. In addition to the above assumption for the capacities and the transit times of arcs, we assume a sink is an inner vertex, i.e. the in-degree of a sink is four (the other case can be similarly treated).

Given a grid network $\mathcal{N} = (D = (V, A), b_v, V^* = \{r\})$, we consider the quickest flow problem QF formally defined as follows:

$$\text{QF: minimize } T(AT_r^f) \text{ subject to } f \in \mathcal{F}_{\mathcal{N}}$$

where AT_r^f is the arriving-table at the sink r with respect to f.

For any vertices v and w such that there exists a directed path from v to w, we define $l(v, w)$ as the sum of transit times of arcs on the path. Vertex set V is partitioned into layers according to the distance from r. Thus, a directed graph D can be viewed as *a layered graph*. A layered graph $D = (V, A)$ is a directed graph consisting of several layers which partition V into subsets $V^0(= \{r\}), V^1, V^2, \ldots$ such that vertices $v \in V^i$ and $w \in V^j$ are connected by a directed arc (v, w) only if $i - j = 1$, and V^p (p-th layer) denotes the set of all of vertices satisfying $l(v, r) = p\tau$ (Fig. 1(c)). A network defined on a layered graph is called *a layered network*.

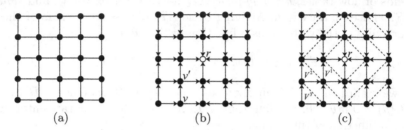

(a)   (b)   (c)

Fig. 1. (a)Grid network (b)Sink r and direction of arcs (c)Layers of grid network

Now we define $m\tau = \max\{l(v, r) \mid b_v > 0, v \in V\}$ for a grid network. A vertex $v \in V^p$ is said to be at level p, and an arc connecting between V^p and V^{p-1} is said to have a level p. For any $v \in V$, let CH_v denote the set of children of v (i.e., $w \in CH_v$ has a level higher than v by one), and let PA_v denote the set of parents (i.e., $w \in PA_v$ has a level smaller than v by one).

3   Quickest Flow Problem for Grid Networks

In this section, we consider the quickest flow problem for a grid network $\mathcal{N} = (D = (V, A), b_v, V^* = \{r\})$. First we explain the overall idea of our algorithm.

Our algorithm benefits from the structure of a grid graph. Let $CH_r = \{u_1, u_2, u_3, u_4\}$. By the way of directing arcs of a grid graph, we can decompose V into eight subsets, U_1, U_2, U_3, U_4 and W_1, W_2, W_3, W_4 as in Fig. 2 where U_i denotes the set of vertices on horizontal or vertical axis whose supplies are all sent to sink r through arc (u_i, r) and W_i denotes the set of vertices whose supplies are sent to sink r through either (u_i, r) or (u_{i+1}, r) (we assume throughout the paper that the index i is given as $(i \bmod 4) + 1$). Here let $H_i, i = 1, 2, 3, 4$ be a subgraph induced by $W_{i-1} \cup U_i \cup W_i \cup \{r\}$. For an optimal dynamic flow f for problem QF, it can be decomposed into four flows $f_i, i = 1, 2, 3, 4$ such that each f_i represents the flow of supplies which reaches r through arc (u_i, r). The sub-flow $f_i, i = 1, 2, 3, 4$ induces a rooted graph D_i such that its vertex set and arc set are defined as those which a positive amount of f_i passes through. Notice that D_i contains an arc (u_i, r) and its vertex set is a subset of $W_{i-1} \cup U_i \cup W_i \cup \{r\}$ (D_i is clearly a subgraph of H_i).

The proposed algorithm is based on the following four ingredients.

Theorem 1. *There exists a subgraph H_i' of H_i which spans $W_{i-1} \cup U_i \cup W_i \cup \{r\}$ for $i = 1, 2, 3, 4$ such that H_i' are arc disjoint for $i \neq j$.*

It is easy to see that the above theorem holds from Fig 3. Notice that that arc-disjoint subgraph H'_i are not uniquely determined.

Now suppose that for every $v \in W_i$ with $i = 1, 2, 3, 4$, the amounts of supply (denoted by $b_{v,i}$ and $b_{v,i+1}$ respectively) which reach r via arcs (u_i, r) and (u_{i+1}, r) respectively are fixed.

Theorem 2. *Let us consider dynamic flow problems QF_i defined on H'_i such that the supply of $v \in W_{i-1} \cup U_i \cup W_i$ is $b_{v,i}$. The optimal objective value for QF_i for every i does not depend on the choice of arc-disjoint subgraphs H'_i, but remains the same.*

Theorem 3. *There exists an optimal dynamic flow f such that f_i and f_j does not share any arc for every $i \neq j$.*

From these facts, when b_i and b_{i+1} are fixed for every $v \in W_i$ and every i with $i = 1, 2, 3, 4$, an optimal flow of QF can be found by independently obtaining an optimal flow f_i^* for QF_i for each i. Since the subgraph H'_i is a rooted tree, the solution of QF_i can be given by simply specifying the supply at each $v \in W_{i-1} \cup U_i \cup W_i$. Therefore, the problem QF reduces to finding an optimal allocation of b_v to $b_{v,i}$ and $b_{v,i+1}$ for each $v \in W_i$ with $i = 1, 2, 3, 4$, and we call this problem *the optimal allocation problem for supplies*. Moreover, we prove the following theorem. Consequently, we can solve the quickest flow problem for grid networks with uniform arc capacity efficiently as will be shown in Section 3.2.

Theorem 4. *The optimal allocation problem for supplies can be transformed into the min-max resource allocation problem under network constraints [5, 6, 7].*

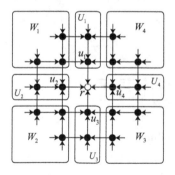

Fig. 2. Decomposition of $\mathcal{N}$

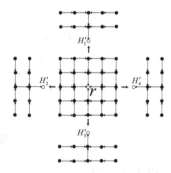

Fig. 3. H'_1, H'_2, H'_3, H'_4

From the above discussion, our algorithm consists of two phases: (1) The first phase is to reduce the quickest flow problem QF to the optimal allocation problem for supplies, and (2) the second phase is to reduce the optimal allocation problem for supplies to the min-max resource allocation problem under network constraints [5, 6, 7].

3.1 Reduction the Quickest Flow to the Optimal Allocation Problem for Supplies

In this subsection, we prove that the quickest flow problem QF can be reduced to the optimal allocation problem for supplies. From the above discussion, the reduction is done by proving Theorem 2 and Theorem 3.

3.1.1 Proof of Theorem 2

Theorem 2 is proved after showing Lemma 1 and Lemma 2. We prove these lemmas by using properties of flow-tables. Thus, before proving these lemmas we introduce operators concerning flow-tables: *shifting*, and *ceiling* [2], and we will then show some basic properties of those operations.

Definition 1 (table shifting). *For any flow-table FT, $\tau \in \mathbf{Z}_+$ and $\theta \in \mathbf{Z}_+$, we define $S_\tau(FT)$ as follows : $S_\tau(FT)(\theta) = 0$ if $\theta < \tau$ and $S_\tau(FT)(\theta) = FT(\theta - \tau)$ if $\theta \geq \tau$.*

It is easy to see that for any flow-tables FT_1, FT_2 and $\tau_1, \tau_2 \in \mathbf{Z}_+$, $S_{\tau_1}(FT_1 + FT_2) = S_{\tau_1}(FT_1) + S_{\tau_2}(FT_2)$ and $S_{\tau_1+\tau_2}(FT_1) = S_{\tau_1}(S_{\tau_2}(FT_1))$ hold. From the above definitions and (4), (5) can be rewritten as

$$AT_v = B_v + \sum_{e \in \Delta^-(v)} S_{\tau(e)}(ST_e). \tag{6}$$

Definition 2 (table ceiling). *For any flow-table FT and $c \in \mathbf{R}_+$, $[FT]_c$ is a flow-table obtained by carrying over the excess of $FT(\theta)$ (i.e. $FT(\theta) - c$) to $FT(\theta + 1)$ in the order of $\theta = 0, 1, \ldots$.*

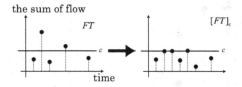

Fig. 4. Table ceiling

Here we show two facts concerning flow-tables.

Fact 1. *For any flow-tables FT_1, FT_2 and for any $c \in \mathbf{R}_+$, $[[FT_1]_c + FT_2]_c = [FT_1 + FT_2]_c$.*

Fact 2. *For any flow-table FT, $c \in \mathbf{R}_+$ and $\tau \in \mathbf{Z}_+$, $S_\tau([FT]_c) = [S_\tau(FT)]_c$ holds.*

Here it should be noted that given a network $\mathcal{N} = (D = (V, A), b_v, V^*)$ and AT_v for a $v \in V$ and the distribution of AT_v to each $e \in \Delta^+(v)$ as AT_v^e, we can assume $ST_e = [AT_v^e]_c$ holds for any $e \in \Delta^+(v)$ in order to attain an optimal

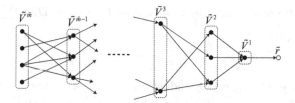

Fig. 5. Layered network $\mathcal{L}$

solution [8]. Thus, throughout this paper, we restrict the set of feasible dynamic flows to those which satisfy this condition.

Let us now return to Theorem 2 again. Here we consider a layered network $\mathcal{L} = (\tilde{D} = (\tilde{V}, \tilde{A}), \tilde{b}_v, \tilde{V}^* = \{\tilde{r}\})$ with $|\Delta^-(\tilde{r})| = 1$, where $\tilde{V}^p = \{v \in \tilde{V} \mid l(v, \tilde{r}) = p\tau\}$, $\tilde{m}$ is the number of layers in $\mathcal{L}$, $\tilde{B}_v$ denotes the extension of $\tilde{b}_v$ to flow-table, $\tilde{\boldsymbol{B}}^p = \sum_{v \in \tilde{V}^p} \tilde{B}_v$ for any $p \in \{1, 2, \ldots, \tilde{m}\}$ (Fig. 5). We show the following lemmas concerning $\mathcal{L}$. Lemma 1 shows the relationship between the arriving-tables of vertices whose level is p and those of vertices whose level is $p + 1$.

Lemma 1. *For any feasible flow of $\mathcal{L}$ we have*

$$[\sum_{v \in \tilde{V}^p} AT_v]_c = [\tilde{\boldsymbol{B}}^p + S_\tau(\sum_{u \in \tilde{V}^{p+1}} AT_u)]_c.$$

Proof. Consider AT_v for $v \in \tilde{V}^p$. Then

$$AT_v = \tilde{B}_v + \sum_{u \in CH_v} S_\tau(ST_{(u,v)}) \tag{7}$$

holds by (6). Here we define $\{AT_u^{(u,v)} \mid v \in PA_u\}$ as a distribution of AT_u for $u \in \tilde{V}^{p+1}$ such that $\sum_{v \in PA_u} AT_u^{(u,v)} = AT_u$ and $ST_{(u,v)} = [AT_u^{(u,v)}]_c$ hold. It is clear that such distribution always exists. Thus, we have

$$[\sum_{v \in \tilde{V}^p} AT_v]_c$$
$$= [\sum_{v \in \tilde{V}^p}(\tilde{B}_v + \sum_{u \in CH_v} S_\tau(ST_{(u,v)}))]_c \text{ (by (7))}$$
$$= [\sum_{v \in \tilde{V}^p} \tilde{B}_v + \sum_{v \in \tilde{V}^p} \sum_{u \in CH_v} S_\tau(ST_{(u,v)})]_c$$
$$= [\tilde{\boldsymbol{B}}^p + \sum_{u \in \tilde{V}^{p+1}} \sum_{v \in PA_u} S_\tau(ST_{(u,v)})]_c$$
$$= [\tilde{\boldsymbol{B}}^p + \sum_{u \in \tilde{V}^{p+1}} \sum_{v \in PA_u} S_\tau([AT_u^{(u,v)}]_c)]_c \text{ (by } ST_{(u,v)} = [AT_u^{(u,v)}]_c)$$
$$= [\tilde{\boldsymbol{B}}^p + \sum_{u \in \tilde{V}^{p+1}} \sum_{v \in PA_u} [S_\tau(AT_u^{(u,v)})]_c]_c \text{ (by Fact 2)}$$
$$= [\tilde{\boldsymbol{B}}^p + \sum_{u \in \tilde{V}^{p+1}} \sum_{v \in PA_u} S_\tau(AT_u^{(u,v)})]_c \text{ (by Fact 1)}$$
$$= [\tilde{\boldsymbol{B}}^p + S_\tau(\sum_{u \in \tilde{V}^{p+1}} \sum_{v \in PA_u} AT_u^{(u,v)})]_c$$
$$= [\tilde{\boldsymbol{B}}^p + S_\tau(\sum_{u \in \tilde{V}^{p+1}} AT_u)]_c \text{ (by } \sum_{v \in PA_u} AT_u^{(u,v)} = AT_u). \qquad \square$$

The following lemma is immediate from Lemma 1. This lemma says that the minimum completion time remains the same for any layered network with a single sink whose in-degree is one, and thus it does not change as long as the initial supply and the level of every vertex remain the same. This proves Theorem 2.

Lemma 2. *In the layered network $\mathcal{L}$ of Lemma 1, we have $AT_{\tilde{r}} = [\sum_{i=1}^{\tilde{m}} S_{i\tau}(\tilde{\boldsymbol{B}}^i)]_c$.*

Proof. We first prove

$$[\sum_{v \in \tilde{V}^p} AT_v]_c = [\sum_{i=p}^{\tilde{m}} S_{(i-p)\tau}(\tilde{\boldsymbol{B}}^i)]_c \tag{8}$$

holds for any $p \in \{1, \ldots, \tilde{m}\}$ by induction on p. Let us first consider the case of $p = \tilde{m}$. Since $AT_v = \tilde{B}_v$ holds for any $v \in \tilde{V}^{\tilde{m}}$, we have $[\sum_{v \in \tilde{V}^{\tilde{m}}} AT_v]_c = [\sum_{v \in \tilde{V}^{\tilde{m}}} \tilde{B}_v]_c$. Next assume that the lemma is true for $p = t + 1$. Thus, by Lemma 1 and the induction hypothesis, we have

$$[\sum_{v \in \tilde{V}^t} AT_v]_c$$
$$= [\tilde{\boldsymbol{B}}^t + S_\tau(\sum_{v \in \tilde{V}^{t+1}} AT_v)]_c \text{ (by Lemma 1)}$$
$$= [\tilde{\boldsymbol{B}}^t + S_\tau([\sum_{v \in \tilde{V}^{t+1}} AT_v]_c)]_c \text{ (by Fact 1 and Fact 2)}$$
$$= [\tilde{\boldsymbol{B}}^t + S_\tau([\sum_{i=t+1}^{\tilde{m}} S_{(i-(t+1))\tau}(\tilde{\boldsymbol{B}}^i)]_c)]_c \text{ (by the induction hypothesis)}$$
$$= [\sum_{i=t}^{\tilde{m}} S_{(i-t)\tau}(\tilde{\boldsymbol{B}}^i)]_c. \text{ (by Fact 1 and Fact 2).}$$

This completes the proof of (8). Let $CH_{\tilde{r}} = \{v_1\}$. $[AT_{v_1}]_c = [\sum_{i=1}^{\tilde{m}} S_{(i-1)\tau}(\tilde{\boldsymbol{B}}^i)]_c$ holds from (8). Thus, from $AT_{\tilde{r}} = S_\tau([AT_{v_1}]_c)$, the lemma follows. ☐

Notice that the lemma does not always hold if the in-degree of $\tilde{r}$ is more than one.

3.1.2 Proof of Theorem 3

Next let us consider Theorem 3. There may be an arc e such that both f_i and $f_j (i \neq j)$ share. If there is such an arc e, it is called a *mixed arc* with respect to f, and such flow f is called a *mixed flow*.

Proof. **(Theorem 3)** Let us consider an optimal dynamic flow $\hat{f}$, and assume that it is a mixed flow. Let us decompose $\hat{f}$ into $\hat{f}_i, i = 1, 2, 3, 4$, and D_i for $\hat{f}_i$ be $\hat{D}_i \equiv (\hat{V}_i, \hat{A}_i)$.

From the proof assumption, $\hat{A}_i \cap \hat{A}_j \neq \emptyset$ for some $i \neq j$. Let us define a network $\hat{\mathcal{N}}_i$ for $\hat{D}_i$ such that the arc capacity and the transit time of all $e \in \hat{A}_i$ remain the same as the original problem, and the initial supply of $v \in V$ is equal to $b_{v,i}$. Now, it holds for $i = 1, 2, 3, 4$ that $\hat{f}_i$ is a feasible dynamic flow of $\hat{\mathcal{N}}_i$. Here let us define a network $\mathcal{N}_i$ such that initial supply of vertices, the arc capacity and transit time are the same as $\hat{\mathcal{N}}_i$ and the underlying graph is $H_i' \equiv (V_i, A_i)$ as the one shown in Fig. 3. Notice that $\hat{V}_i \subseteq V_i$ holds.

If we independently consider the dynamic flow problems for $\hat{\mathcal{N}}_i, i = 1, 2, 3, 4$, the optimal objective values for $\mathcal{N}_i$ and $\hat{\mathcal{N}}_i$ are the same for each $i = 1, 2, 3, 4$ from Lemma 2. Let f_i^* for $i = 1, 2, 3, 4$ denote an optimal dynamic flow for $\mathcal{N}_i, i = 1, 2, 3, 4$ respectively, and let $\hat{f}_i^*$ for $i = 1, 2, 3, 4$ denote an optimal dynamic flow for $\hat{\mathcal{N}}_i, i = 1, 2, 3, 4$, respectively. Then, we have

$$\Theta(f_i^*) = \Theta(\hat{f}_i^*) \leq \Theta(\hat{f}_i).$$

This proves the theorem because $\mathcal{N}_i$ with $i = 1, 2, 3, 4$ are arc-disjoint. ☐

3.1.3 Problem Reformulation

Let us consider the four arc-disjoint networks $\mathcal{N}_i$ defined in the proof of Theorem 3. Recall that for each $v \in W_i, i = 1, 2, 3, 4$, $b_{v,i}$ and $b_{v,i+1}$ denotes the amounts of supply which go to r through $\mathcal{N}_i$ and $\mathcal{N}_{i+1}$ respectively. We say that $b_{v,i}$ and $b_{v,i+1}$ are allocated to $\mathcal{N}_i$ and $\mathcal{N}_{i+1}$, respectively. For a vertex v on U_i with $i = 1, 2, 3, 4$, all the supply goes to v using $\mathcal{N}_i$. In this case, we say that $b_{v,i}(= b_v)$ is assigned to $\mathcal{N}_i$. In general, let B_v and $B_{v,i}$ denote the flow-table corresponding to b_v and $b_{v,i}$ respectively. Then the quickest flow problem QF can be written as follows:

$$\begin{aligned}
&\text{minimize} \quad \max_{i=1,2,3,4} T([\textstyle\sum_{v \in V} S_{l(v,r)}(B_{v,i})]_c) \\
&\text{subject to} \quad B_{v,i} = B_v \text{ and } B_{v,j} = \mathbf{0}, \ j \neq i \text{ for } v \in U_i, \\
&\qquad\qquad\quad B_{v,i} + B_{v,i+1} = B_v \text{ and } B_{v,j} = \mathbf{0}, \ j \neq i, i+1 \text{ for } v \in W_i,
\end{aligned}$$

For every p and i, let $b_i^p = \sum_{v \in V^p \cap (W_{i-1} \cup U_i \cup W_i)} b_{v,i}$ which represents the amounts of supply of vertices at level p allocated to $\mathcal{N}_i$, and let B_i^p denote its flow-table extension. Then from Lemma 2, the completion time for $\mathcal{N}_i$ is expressed as $T([\sum_{p=1}^{m} S_{p\tau}(B_i^p)]_c)$. Therefore, the allocation of b_v of a particular vertex $v \in V^p \cap (W_{i-1} \cup W_i)$ does not affect the minimum completion time for $\mathcal{N}_i$ but only the total amounts b_i^p allocated to $\mathcal{N}_i$ does affect it. From this observation, we contract the set $V^p \cap W_i$ into a single vertex w_i^p for each p with $1 \leq p \leq m$ and i with $1 \leq i \leq 4$ such that the initial supply of w_i^p is equal to $\sum_{v \in V^p \cap W_i} b_v$ which is simply denoted by a_i^p. Let u_i^p denote a single vertex corresponding to $V^p \cap U_i$. The initial supply of u_i^p is denoted by g_i^p. Let the allocation of a_i^p to $\mathcal{N}_i$ and $\mathcal{N}_{i+1}$ be $a_{i,i}^p$ and $a_{i,i+1}^p$, respectively. Therefore, defining the flow-table FT_i such as

$$FT_i(p\tau) = g_i^p + a_{i,i}^p + a_{i-1,i}^p, \quad p = 1, 2, \ldots, m, \tag{9}$$

the minimum completion time of $\mathcal{N}_i$ is equal to $T([FT_i]_c)$. Here we introduce the following theorem to calculate $T([FT_i]_c)$ efficiently.

Lemma 3. *For any flow-table FT and $c \in \mathbf{R}_+$,*

$$T([FT]_c) = \max_{0 \leq \theta \leq T(FT)} \left\lceil \{c\theta + \textstyle\sum_{t=\theta}^{T(FT)} FT(t)\}/c \right\rceil - 1$$

Proof. We give a sketch of the proof. The idea is to prove the time that satisfies $\max_{0 \leq \theta \leq T(FT)} \{c\theta + \sum_{t=\theta}^{T(FT)} FT(t)\}$ is equal to

$$\max \left\{ \theta \in \mathbf{Z}_+ \ \middle|\ \textstyle\sum_{t=0}^{\theta} FT(t) = \sum_{t=0}^{\theta} [FT]_c, \theta < T([FT]_c) \right\} + 1.$$

This claim can be proved by the properties of any flow-table FT as follows:

$\sum_{t=0}^{\theta} FT(t) \geq \sum_{t=0}^{\theta} [FT]_c(t)$ for any $\theta \in \mathbf{Z}_+$, and
$\sum_{t=0}^{\theta} FT(t) = \sum_{t=0}^{\theta} [FT]_c(t)$ for any $\theta \in \mathbf{Z}_+$ with $[FT]_c(\theta) < c$. $\square$

Thus, from Lemma 3 and (9), QF can be reduced to the following problem QF′.

$$\text{QF}' : \text{minimize} \max_{1 \leq i \leq 4} \max_{1 \leq p \leq m} \{cp\tau + \textstyle\sum_{k=p}^{m} FT_i(k\tau)\}$$

3.2   Reduction to Min-Max Resource Allocation Problem Under Network Constraints

The problem QF$'$ can be reduced to the min-max resource allocation problem under network constraints as will be shown below. This problem is a kind of min-max flow problem with multiple sources and sinks in a static network [5, 6, 7] which is defined as follows. Suppose we are given a network with multiple sources and sinks such that a fixed amount of supply is associated with each source, and the cost function $\gamma_t(x_t)$ which is nondecreasing in x_t is associated with each sink t where x_t denotes the amount of flow entering t. Then the problem asks to find a (static) flow that minimizes the maximum of the cost functions of sinks.

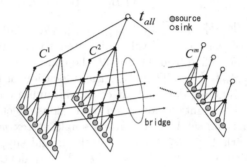

Fig. 6. Illustration of the entire network constructed in Section 3.2

We will explain how we construct a (static) network (see Fig. 6) for which finding an optimal solution for the min-max resource allocation problem produces an optimal solution of the problem QF$'$. The network to be constructed consists of m components $C^1, C^2, \ldots, C^m$. Each component C^p except C^m has four layers while C^m has three layers. The first layer of each component C^p has eight sources which correspond to vertices $w_i^p, u_i^p, i = 1, 2, 3, 4$ defined in the previous subsection. The second and third layers consists of four vertices denoted by $v_{1,i}^p, v_{2,i}^p, i = 1, 2, 3, 4$. The fourth layer consists of a single vertex v_3^p.

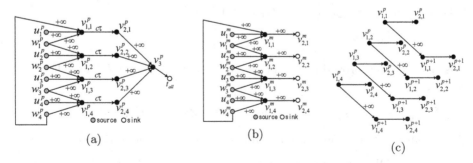

Fig. 7. (a)p-th component C^p (b)m-th component C^m (c)p-th bridges

The connection between the layers are as shown in Fig. 7(a). Vertex w_i^p is connected to $v_{1,i}^p$ and $v_{1,i+1}^p$ such that the flows on $(w_i^p, v_{1,i}^p)$ and $(w_i^p, v_{1,i+1}^p)$ represent the allocation of a_i^p to $\mathcal{N}_i$ and $\mathcal{N}_{i+1}$, respectively. The vertex u_i^p is connected to $v_{1,i}^p$ and the flow on $(u_i^p, v_{1,i}^p)$ represents the supply g_i^p allocated to $\mathcal{N}_i$. In general, if p is large, $V^p \cap U_i$ may become empty. This case can be treated by letting $g_i^p = 0$. Only the arcs from the second to third layer have finite capacity $c\tau$ in C^p with $1 \le p \le m-1$ while the arcs in C^m have infinite capacity. The capacity of the other arcs is ∞. All vertices v_3^p with $1 \le p \le m-1$ are connected to t_{all}. The vertices $v_{2,i}^m, i = 1,2,3,4$ of C^m as well as t_{all} are sinks of this network which are associated with a cost function. The actual cost function for each $v_{2,i}^m, i = 1,2,3,4$ is equal to the amount of the flow entering it. The cost function associated with t_{all} takes zero irrespective of the flow value entering it. In addition to this, we prepare arcs between consecutive components. More precisely, as shown in Fig. 7(c), there is an arc from $v_{1,i}^p$ to $v_{1,i}^{p+1}$ for each p with $1 \le p \le m-1$ and i with $1 \le i \le 4$. The capacity of this arc is defined to be ∞. This arc is called a bridge.

The meaning of the capacity $c\tau$ on the arcs from the second to the third layer in C^p with $1 \le p \le m-1$ is as follows. Let us consider FT_i of (9) and perform ceiling operation to obtain the completion time of $\mathcal{N}_i$. If the amounts of supply carried over to time $p'\tau$ plus the amount $FT_i(p'\tau)$ is less than or equal to $c\tau$, $\max_{1 \le p \le m}\{cp\tau + \sum_{k=p}^m FT_i(k\tau)\}$ is attained for $p > p'$. Thus, a positive flow going through a bridge stands for the situation that the amounts of supply carried over to time $p'\tau$ plus the amount $FT_i(p'\tau)$ is larger than $c\tau$ and thus a positive amount of flow will be carried over to $FT_i((p'+1)\tau)$. The cost function associated with the min-max resource allocation problem here associated with each sink $v_{2,i}^m, i = 1,2,3,4$ is the amount of the excess carried over to time $m\tau$ when performing ceiling operation to FT_i which is equivalent to $\max_{1 \le p \le m}\{cp\tau + \sum_{k=p}^m FT_i(k\tau)\} - cm\tau$. Since $cm\tau$ is constant, the min-max resource allocation problem defined in this subsection solves the problem QF'.

It is known that the min-max resource allocation problem for the network with $|V|$ vertices, $|A|$ arcs and $|T|$ sinks can be solved in $O(|T|(|V||A|\log|V| + |T|\log\frac{M}{|T|}))$ time where M denotes the sum of supplies [5, 6, 7]. The second term in the parenthesis, i.e., $O(|T|\log\frac{M}{|T|})$, is the time required to solve the resource allocation problem without the network constraints. Since our cost function associated with $v_{2,i}^m, i = 1,2,3,4$ is linear, we can reduce the time to $O(|T|)$ (the details are omitted). In our case, $|T|$ is constant and $|V| = O(\sqrt{n}), |A| = O(\sqrt{n})$, thus the running time becomes $O(n\log n)$.

4   Conclusion

We have presented an $O(n\log n)$ time algorithm for the quickest flow problem in a grid network with uniform arc capacity. The algorithm proposed in this paper can be extended to a general layered network $\mathcal{N}$ such that (1) the transit time from a vertex v to a sink r dose not depend on the choice of a path, and (2) the underlying layered graph D can be decomposed into arc-disjoint layered graphs $D_1, D_2, \ldots, D_k$ which spans $V_1, V_2, \ldots, V_k$ respectively, where $CH_r = \{v_1, v_2, \ldots, v_k\}$ and V_i is the set of vertices from which v_i is reachable in D. Thus, the result can also be generalized to the case where the arc capacity is a multiple of c by regarding the arc as multiple ones as long as the resulting layered graph satisfies the requirement just mentioned above.

Acknowledgements

This research is supported by JSPS Grant-in-Aid for Scientific Research on priority areas of New Horizons in Computing.

References

1. Hoppes, B., Tardos, É.: The quickest transshipment problem. Mathematics of Operations Research **25** (2000) 36–62
2. Mamada, S., Uno, T., Makino, K., Fujishige, S.: An $O(n \log^2 n)$ algorithm for a sink location problem in dynamic tree networks. Discrete Applied Mathematics (to appear)
3. Mamada, S., Makino, K., Fujishige, S.: Evacuation problems and dynamic network flows. In: Proc. SICE Annual Conference 2004. (2004) 530–535
4. Mamada, S., Uno, T., Makino, K., Fujishige, S.: A tree partitioning problem arising from an evacuation problem in tree dynamic networks. Journal of the Operations Research Society of Japan **48** (2005) 196–206
5. Ibaraki, T., Katoh, N.: Resource allocation problems under submodular constraints. In: Resource Allocation Problems : Algorithmic Approaches. MIT Press, Cambridge, MA (1988) 144–176
6. Fujishige, S.: Nonlinear optimization with submodular constraints. In: Submodular Functions and Optimization. 2nd edn. Elsevier Science Ltd, North-Holland (2005) 223–250
7. Fujishige, S.: Lexicographically optimal base of a polymatroid with respect to a weight vector. Mathematics of Operations Research **5** (1980) 186–196
8. Mamada, S., Makino, K., Fujishige, S.: Optimal sink location problem for dynamic flows in a tree network. IEICE Transactions on Fundamentals **E85-A** (2002) 1020–1025

Connected Set Cover Problem and Its Applications*

Tian-Ping Shuai[1] and Xiao-Dong Hu[2]

[1] Department of Mathematics,
Beijing University of Post and Telecom., Beijing 100876, China
stpmath@sohu.com
[2] Institute of Applied Mathematics
Chinese Academy of Sciences
P. O. Box 2734, Beijing 100080, China
xdhu@amss.ac.cn

Abstract. We study an extension of the set cover problem, the connected set cover problem, the problem is to find a set cover of minimal size that satisfies some connectivity constraint. We first propose two algorithms that find optimal solutions for two cases, respectively, and then we propose one approximation algorithm for a special case that has the best possible performance ratio. At last we consider how to apply the obtained result to solve a wavelength assignment problem in all optical networks.

Keywords: Set cover, Approximation algorithm, Performance ratio, Wavelength assignment.

1 Introduction

Given a set system $(U, \mathcal{F})$, where U contains n elements and $\mathcal{F}$ is a family of m subsets of U such that every element of U belongs to at least one subset in $\mathcal{F}$, here each subset in $\mathcal{F}$ has a positive weight, a *set cover* C of U is a subfamily of $\mathcal{F}$ such that every element in U is in at least one of the subsets in C. The set cover problem is to find a set cover with the minimal total weight of subsets in the set cover. For this famous NP-hard problem, Johanson [4] proposed a simple greedy algorithm for the unweighted case (or equivalently all weights are the same) with approximation ratio upper bounded by $1 + \ln n$, and later Chvátal [2] generalized their algorithms to the weighted case and proved the same result.

The set cover problem has many applications in practice. For example, Ruan et al [7] studied how to route and allocate wavelengths to a broadcast connection so that the total wavelength conversions required is minimized. Under some conditions they formulate this problem as two closely related set cover problems, the minimum wavelength-covering problem and the minimum vertex-wavelength-covering problem. But some practical problems may have special

---

* This work was supported in part by the National Natural Science Foundation of China under Grant No. 70221001, 60373012 and 10531070.

configurations and the set cover model may not appropriate for them. In particular, the set cover model proposed by Ruan el al [7] is not applicable to the case of limited wavelength conversions. In this case, connectedness of a set cover appears to be an important requirement.

We therefore in this paper consider a natural extension of the set cover problem, the *connected set cover problem*. Besides the universal set U and a family $\mathcal{F}$ of U, we are also given a graph G with vertex-set $V(G) = \mathcal{F}$ and edge-set $E(G)$ consisting some edges between some pairs of subsets in $\mathcal{F}$. A set cover $C \subseteq \mathcal{F}$ of U is called connected if the induced subgraph $G(C)$ is connected, where $G(C)$ is a subgraph of G that consists of all edges whose two endpoints are both in C. The problem is to find a connected set cover with the minimal number of subsets. It is easy to see that the classic set cover problem is a special case of the connected set cover problem with a completed graph.

In this paper we will first show that the connected set cover problem is NP-hard even if at most one vertex of the given graph has degree greater than two, and it cannot be solved in polynomial-time. We then propose two polynomial-time algorithms for the case where every vertex in the graph has degree at most two. For the case where at most one vertex has degree greater than two, we propose an approximation algorithm with performance ratio at most $1 + \ln n$ that is the best possible. In the end we discuss an application of the connected set cover problem to the wavelength assignment of broadcast connections in the optical networks.

2   Complexity Study

In this section, we study the complexity of the connected set cover problem for some special graphs. We shall see that the difficulty in solving the connected set cover problem not only lies in the structure of $(U, \mathcal{F})$ system but also related to the property of give graph G. Graph G is called a *line graph* if two vertices in $V(G)$ have degree one and all others have degree two. Graph G is called a *ring graph* if it is connected and every vertex in $V(G)$ has degree two. Graph G is called a *spider graph* if G is a tree and only one vertex has degree greater than two, a spider graph is particularly called a *star graph* if one vertex has degree greater than one while all others have degree one.

Theorem 1. *The connected set cover problem on star graphs is NP-hard.*

Proof. Given an instance of the set cover problem of uniform weight, $(U, \mathcal{F})$, we construct an instance of the connected set cover problem, $(U', \mathcal{F}')$ and a graph G on $\mathcal{F}'$ as follows: the universal set $U' = U \cup \{u_0\}$, where $u_0 \notin U$, the family $\mathcal{F}' = \mathcal{F} \cup \{u_0\}$, and graph G has edge-set $E(G) = \{(u_0, f) \mid f \in \mathcal{F}\}$. Clearly, G is a star graph and every set cover of U' must include subset $\{u_0\}$ of U'. Thus the set cover problem has a set cover C if and only if the connected set cover problem has a set cover $C \cup \{u_0\}$.

Theorem 2. *The connected set cover problem on line or ring graphs can be solved in polynomial time.*

Proof. Every line graph G can be represented by a path $(f_1 f_2 \cdots f_{m-1} f_m)$, where each vertex f_i corresponds to a subset in $\mathcal{F}$, and f_1 and f_m have degree one (they are two ends of the path). Notice in this case that any connected set cover consists of subsets whose corresponding vertices make a subpath $(f_i f_{i+1} \cdots f_{j-1} f_j)$ for some i, j with $i \leq j$. Thus to find the minimum connected set cover we just need to check all $\binom{m}{2}$ possible solutions (some of them may not be set covers) and then choose the minimum one. This method requires time $O(m^3 n)$.

Similarly, for ring graphs we just need to check all $m(m-1)$ possible solutions and then choose the minimum one. This method also requires time $O(m^3 n)$.

3   Efficient Algorithms for Line and Ring Graphs

In the proof of Theorem 2 we have described a simple algorithm for the connected set cover problem for line and ring graphs, respectively, both have time-complexity of $O(m^3 n)$. In this section, we will propose more efficient algorithms for these two special cases.

We first study how to find the minimum connected set cover in line graphs in an efficient way. The basic idea is to delete as many vertices as possible until the remaining vertices cannot constitute a set cover. This can be carried out as follows: (1) Delete the vertices from the leftmost to the right one by one until the remaining vertices can not constitute a set cover, and then delete the vertices from the rightmost to left one by one until the remaining vertices can not constitute a set cover. (2) Do the same operations as in (1) but in the reverse order, that is, deleting first from the rightmost to left and then from the leftmost to right. (3) Delete the rightmost and then the leftmost vertices alternatively until the remaining vertices can not constitute a set cover. When the process is stopped, if the last vertex is deleted from the left (right) side then repeat delete the vertices from left (right) until the remaining vertices can not constitute a set cover. (4) Choose the best of these three solutions obtained.

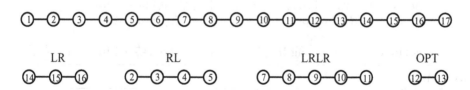

Fig. 1. An counterexample

Denote the above three operations by LR, RL, and LRLR. Unfortunately, the above described method could not find the optimal solution. Fig. 1 gives such an example with $U = \{i \mid i = 1, 2, \cdots, 9\}$ and $\mathcal{F} = \{f_j \mid j = 1, 2, \cdots, 17\}$, where f_i is defined as follows.

$$f_1 = \{4,5,9\}, \qquad f_2 = \{3,9\}, \qquad f_3 = \{6,8\}, \qquad f_4 = \{3,4,7\},$$
$$f_5 = \{1,2,5\}, \qquad f_6 = \{7,9\}, \qquad f_7 = \{6,7\}, \qquad f_8 = \{8\},$$
$$f_9 = \{2,3\}, \qquad f_{10} = \{4,5\}, \quad f_{11} = \{5,9\}, \quad f_{12} = \{1,2,3,4,5\},$$
$$f_{13} = \{5,6,7,8,9\}, f_{14} = \{1,2,3\}, f_{15} = \{4,5,6\}, f_{16} = \{7,8,9\},$$
$$f_{17} = \{1,4,7\}.$$

It can be verified that LR, RL,and LRLR produce three different solutions, but neither of them is optimal. Observe that the optimal solution $\{f_{12}, f_{13}\}$ is on the right half side of the line graph and does not contain the central vertex f_9. However, we shall see that if the optimal solution contain the central vertex, then the above method can be modified to find the optimal solution.

The above example and analysis suggest that we should first find such an optimal solution that contains the central vertex, and then find those two optimal solutions that belong to the left and right half sides of line graph, respectively. In the end we just choose the best solution among these three solutions.

To implement this method, we can modify operations LR and RL as follows: deleting vertices from left to right, or from right to left is stopped until the remaining vertices can not constitute a set cover or the central vertex is reached, and modify LRLR as follows:adding neighbor vertex (and its adjacent edge) of left endnodes of current path and deleting vertices from rightmost to left one by one until the remaining vertices can not constitute a set cover. We use RR represent the operation deleting vertices from rightmost to left is stopped until the remaining vertices can not constitute a set cover. The algorithm is described below as a recursive procedure, where initially $i = 1$ and $j = m$.

Algorithm A. Finding an Optimal Set Cover in Line Graphs

---

procedure LineCover(i, j):
    if $j - i \leq 2$ **then** **return** f_i or f_j if one of them covers U
    else find a cover $F_1 = \{f_{l_1}, f_{l_1+1}, \cdots, f_{r_1}\}$ applying modified LR on path
        between f_i and f_j;
        find a cover $F_2 = \{f_{l_2}, f_{l_2+1}, \cdots, f_{r_2}\}$ applying modified RL on path
        between f_i and f_j;
        find a cover $F_3 = \{f_{l_3}, f_{l_3+1}, \cdots, f_{r_3}\}$ applying procedure
        Modified LRLR on path between f_{l_1} and f_{r_1} and path
        between f_{l_2} and f_{r_2};
        return the best among $\{F_1, F_2, F_3, \text{LineCover}(i, \frac{i+i}{2}), \text{LineCover}(\frac{i+i}{2}, j)\}$.

---

procedure Modified LRLR

Input: $F_1 = \{f_{l_1}, f_{l_1+1}, \cdots, f_{r_1}\}$ and $F_2 = \{f_{l_2}, f_{l_2+1}, \cdots, f_{r_2}\}$
    if $l_1 - l_2 \leq 1$ *or* $r_1 - r_2 \leq 1$ **then return** the best among $\{F_1, F_2\}$
    else for $j = 1, 2, \cdots, l_1 - l_2 - 1$ **do**
        find a cover $F_j = \{f_{l_1-j}, \cdots, f_{r'_j}\}$ by applying RR on path
        between f_{l_1-j} and $f_{r'_j-1}$.
        return the best among $\{F_j | j = 1, 2, \cdots, l_1 - l_2 - 1\}$.

Theorem 3. *Given an instance of the connected set cover problem, $(U, \mathcal{F})$ and a line graph G on $\mathcal{F}$,* **Algorithm A** *finds an optimal solution to the problem in $O(nm^2)$ time.*

Proof. Notice that if the optimal solution does not include the central vertex in the line graph, then it must be in either the left half or the right half of the line graph. Thus to prove that the algorithm returns an optimal solution, it suffices to show that if there exists an optimal solution $F^* = \{f_l, f_{l+1}, \cdots, f_{r-1}, f_r\}$ that contains the central vertex, then it can be found by one of the three operations.

Let us denote the solutions obtained by applying modified operations LR, RL, and LRLR with $i = 1$ and $j = m$, by F_i for $i = 1, 2, 3$, respectively. By the rules of operations LR, RL, and LRLR, we have $l_2 \leq l_3 \leq l_1$ and $r_1 \geq r_3 \geq r_2$. See Fig. 2.

Fig. 2. For the proof of Theorem 3

It is easy to verify that when $l = l_1$, $F^* = F_1$, when $r = r_2$, $F^* = F_2$, and when $l < l_1$ and $r > r_2$, $F^* = F_3$. Thus the solution returned by the algorithm is optimal.

For the time-complexity of the algorithm, notice that in invoking the **procedure** LineCover(i, j) each of the two operations LR, and RL deletes at most $O(j - i)$ vertices and operation LRLR add and deletes total at most $O(j - i)$ vertices, and to check if the remaining vertices make a set cover needs time $O(mn)$. In addition, the **procedure** LineCover(i, j) is invoked at most $\cdot 2^k$ times for subpaths of length $m/2^k$, each time produces 3 solutions; In the total at most $3 \sum_{k=0}^{\log_2 m} 2^k = 3m$ solutions are produced, this requires time bounded by

$$3 \sum_{k=0}^{\log_2 m} 2^k (\frac{m}{2^k})^2 n \leq 6m^2 n.$$

To find the best solution among $3m$ ones requires times $O(m)$. Thus the algorithm has the running time at most $O(m^2 n)$.

We now study how to find the minimum connected set cover in ring graphs in an efficient way. For the simplicity of the presentation, we just consider the case of even m. The basic idea is to make the given ring graph into a line graph by removing a vertex, say f_1, and then apply **Algorithm A** to the resulting line graph. Notice however that the minimum connected set cover may include vertex f_1 excluded from the line graph. Thus we need to apply **Algorithm A** to the line graph obtained by removing the vertex $f_{\frac{m}{2}+1}$, which is on the opposite side of vertex f_1. As a result, we find two connected set covers. See Fig. 3.

If the better of these two solutions has size less than $m/2$, then this must be the optimal solution. Otherwise it has size greater than $m/2$. In this case, the optimal solution must contain both vertices f_1 and $f_{\frac{m}{2}+1}$ and includes either all vertices in $\{f_1, f_2, \cdots, f_{\frac{m}{2}+1}\}$ or all vertices in $\{f_{\frac{m}{2}+1}, f_{\frac{m}{2}+2}, \cdots, f_m, f_1\}$. Therefore we

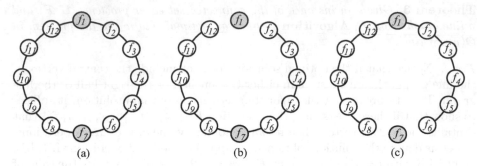

Fig. 3. (a) The original ring graph, (b) and (c) ling graphs obtained by removing two vertices oppositive to each other on the ring

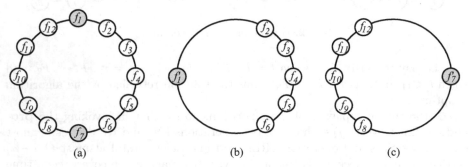

Fig. 4. (a) The original ring graph, (b) and (c) ring graphs obtained by merging half number of vertices

can shrink the ring graph of size m to two ring graphs of size $m/2$ by merging all vertices in $\{f_1, f_2, \cdots, f_{\frac{m}{2}+1}\}$ and $\{f_{\frac{m}{2}+1}, f_{\frac{m}{2}+2}, \cdots, f_m, f_1\}$ into one vertex f_1' and $f_{\frac{m}{2}+1}'$, respectively. See Fig. 4. In the end, we return the best of the four obtained solutions.

The problem is now reduced to two subproblems such that optimal solutions must contain the new vertex f_1' ($f_{\frac{m}{2}+1}'$) in two ring graphs of half size, where the universal set also becomes smaller since those vertices covered by $\{f_1, f_2, \cdots, f_{\frac{m}{2}+1}\}$ and $\{f_{\frac{m}{2}+1}, f_{\frac{m}{2}+2}, \cdots, f_m, f_1\}$ should be removed. This process is repeated until the optimal solution is found. The algorithm is again more formally described as a recursive procedure, where initially $i = 1$, $j = m$ and $U' = U$.

Theorem 4. *Given an instance of the connected set cover problem, $(U, \mathcal{F})$ and a ring graph G on $\mathcal{F}$, **Algorithm B** finds an optimal solution to the problem in $O(m^2 n)$ time.*

Proof. The correctness of the proof follows from two facts: (1) If the optimal solution F_{opt} has size less than $m/2$, then it must be either F_1 or F_2 since it must be included in either $\mathcal{F} \setminus \{f_1\}$ or $\mathcal{F} \setminus \{f_{\frac{m}{2}+1}\}$ (maybe in both of them). (2) If F_{opt}

has size greater than $m/2$ and it is not equal to F_1 or F_2, then it must be either
$\{f_{\frac{m}{2}}, \cdots, f_1\} \cup \text{RingCover}(1, m/2, U_L)$ or $\{f_1, \cdots, f_{\frac{m}{2}}\} \cup \text{RingCover}(m/2, 1, U_R)$.

Algorithm B. Finding an Optimal Set Cover in Ring Graphs

---

procedure RingCover($1, m, U$):
    if $m - 1 \leq 2$ **then** **return** f_1 or f_2 or $\{f_1, f_2\}$ if one of them covers U.
    else find the optimal cover F_1 with f_1 removed using **Algorithm A**,
        find the optimal cover F_2 with f_m removed using **Algorithm A**,
        find the optimal cover F_3 with f_1 included using
        Procedure RingCover(i, j, U', f_i) on ring $(1, m)$.
        return the best of three covers, F_1, F_2, F_3.

---

procedure RingCover(i, j, U', f_i):
    if $j - i \leq 2$ **then** **return** f_i or f_j if one of them covers U.
    else find the optimal cover F_1 with $f_{\frac{i+j}{2}}$ removed using **Algorithm A**,
        if the cover has size less than $\frac{j-i}{2}$ **then** **return** it
        else set U_L be the set consisting of vertices in U not covered by
            $\{f_{\frac{i+j}{2}}, \cdots, f_i\}$, $f_i = \emptyset$
        set U_R be the set consisting of vertices in U not covered by
            $\{f_i, \cdots, f_{\frac{i+j}{2}}\}$, $f_i = \emptyset$.
        return the best of three covers, F_1,
        $\{f_{\frac{i+j}{2}}, \cdots, f_i\} \cup \text{RingCover}(i, \frac{j+i}{2} - 1, U_L, f_i)$, and
        $\{f_i, \cdots, f_{\frac{i+j}{2}}\} \cup \text{RingCover}(\frac{i+j}{2} + 1, i, U_R, f_i)$.

For the time-complexity of the algorithm, notice that each time when we invoke the **procedure** RingCover(i, j, U'), we produce F_1 using **Algorithm A**, this requires time $O(n(j - i)^2)$ by Theorem 3. In addition, the **procedure** LineCover(i, j) is invoked at most $\cdot 2^k$ times for subrings of length $m/2^k$, each time produces 2 solutions; In the total at most $2 \sum_{k=0}^{\log_2 m} 2^k = 2m$ solutions are produced, this requires time bounded by

$$2 \sum_{k=0}^{\log_2 m} 2^k 6(\frac{m}{2^k})^2 n \leq 24 m^2 n.$$

To find the best solution among $2m$ ones requires times $O(m)$. Thus the algorithm has the running time at most $O(m^2 n)$.

4 Approximation Algorithm for Spider Graphs

As in the previous section we have proved that the connected set cover problem is NP-hard even for star graphs, thus in this section we will propose an approximation algorithm for the problem in spider graphs. We will show that this algorithm has almost the best possible approximation ratio.

Suppose that G is a spider graph with central vertex $f_0 \in V(G)$ having degree $k > 2$. We now decompose the graph into line subgraphs $L_1, L_2, \cdots, L_k$, which have a common end f_0. Then there are two possible cases for an optimal set cover F^*, (1) it consists of only vertices in one of the line subgraphs, and (2) it includes at least two vertices belonging to different line graphs. For case (1) we can find the optimal solution by running **Algorithm A** k times (just choose the best among k solutions). For case (2) we can first transform the problem into the set cover problem, and then solve the problem approximately by the generalized greedy algorithm [2].

The transformation can be done as follows: For each subset $f \in \mathcal{F}$ with $f \neq f_0$, which corresponds to a vertex in graph G, we define a new subset f' that is the union of the subsets whose corresponding vertices in G are on the path $p(f, f_0)$ between f and f_0 and delete the elements contained in f_0, and define $f_0' = f_0$. We then construct a new family of subsets $\mathcal{F}' = \{f' \mid f \in \mathcal{F}\}$. We also assign a weight $w(f')$ to f' which is equal to the number of edges in $p(f, f_0)$, $f \neq f_0$, and $w(f_0') = 1$. The following lemma shows that the new set system $(U, \mathcal{F}')$ has the property that we need for our algorithm.

Lemma 1. *The set system $(U, \mathcal{F})$ with a graph G on $\mathcal{F}$ has a minimum connected set cover C of size $|C|$ that includes subset f_0 if and only if the set system $(U, \mathcal{F}')$ has a minimum weighted set cover C' with weight $w(C') = |C|$.*

Proof. "Only-if": For $i = 1, 2, \cdots, k$, let $\overline{f}_i \in C$ be the subset whose corresponding vertex in L_i has the longest path to f_0. Then C contains every subset whose corresponding vertex is on path $p(\overline{f}_i, f_0)$. Clearly, the size of $C \setminus \{f_0\}$ is equal to the sum of number of edges in path $p(\overline{f}_i, f_0)$ for $i = 1, 2, \cdots, k$. Hence $\overline{f}_0'$ and $\overline{f}_i'$ for $i = 1, 2, \cdots, k$ constitute a set cover of U that has weight $|C|$.

"If": Notice that C' does not contain two subsets f' and g' such that the corresponding vertices f and g are in the same line graph L_i for some i, otherwise either $f' \subset g'$ or $g' \subset f'$, thus one of them is redundant contradicting that C' is a minimum set cover. Let $f_i' \in C$ be the subset which includes $w(f_i')$ vertices in $L_i \setminus \{f_0\}$, $i = 1, 2, \cdots, k$. Hence the union of the subsets whose corresponding vertices are within $w(f_i')$ distance from f_0 on line graph L_i, for $i = 1, 2, \cdots, k$, makes a connected set cover C of U with size $|C| = \sum_i w(f_i')$.

Algorithm C. Finding Connected Set Covers in Spider Graphs

Decompose the spider graph into line graphs L_i's.
Find the optimal set cover F_i for each i using **Algorithm A**.
Construct a new set system $(U, \mathcal{F}')$.
Find a set cover F_0' of U using the generalized greedy algorithm.
Produce the corresponding set cover F_0 of $(U, \mathcal{F}')$ from F_0'.
Return the best set cover among $\{F_i \mid i = 0, 1, 2 \cdots\}$.

Theorem 5. *Given an instance of the connected set cover problem, $(U, \mathcal{F})$ and a spider graph G on $\mathcal{F}$,* **Algorithm C** *returns a solution in time of $O(m^2 n)$ whose size is at most $\log n$ times that of the optimal solution.*

Proof. Let F_{opt} be an optimal solution, and F_C be a solution returned by **Algorithm C**. If a line graph L_i contains F_{opt} for some i, then $F_C = F_{opt}$. If not (there are two sets in F_{opt} such that one is in line graph L_i while the other in line graph L_j), $w(F_C) \leq w(F_0') \leq \log n w(F_{opt})$, the last inequality comes from Chvátal's result [2].

In fact, we will show that **Algorithm C** is the best possible approximation algorithm for the connected set cover problem in spider graphs. To prove this we need the following lemma due to Feige [3].

Lemma 2. *For any $0 < \rho < 1$, there is no approximation algorithms with performance ratio $\rho \ln n$ for the set cover problem unless $NP \subset D_{TIME}(n^{poly \log n})$.*

Theorem 6. *For any $0 < \rho < 1$, there is no approximation algorithms with performance ratio $\rho \ln n$ for the connected set cover problem in spider graphs unless $NP \subset D_{TIME}(n^{poly \log n})$.*

Proof. Suppose, by contradiction argument, that there exists an algorithm $A_{\rho'}$ with approximation performance ratio $\rho' < 1$ for the connected set cover problem in spider graphs. We now design an algorithm A_ρ for the set cover problem using algorithm $A_{\rho'}$ as a subroutine.

Given an instance I of the set cover problem, a set system $(U, \mathcal{F})$, construct an instance I' of the connected set cover problem, a set system $(U', \mathcal{F}')$ and a graph G on $\mathcal{F}'$ as follows: Let $U = \{u_i \mid i = 1, 2, \cdots, n\}$, $\mathcal{F} = \{f_i \mid i = 1, 2, \cdots, m\}$, and take $k = \lceil \rho' / (1 - \rho') \rceil$. w.l.o.g suppose that $k < m$. Set $U' = U \cup \{u_0\} \cup W$ and $\mathcal{F}' = \mathcal{F} \cup \{f_{ij} \mid i = 1, 2, \cdots, m, \ j = 1, 2, \cdots, k\} \cup \{u_0\}$, and spider graph G has central vertex u_0 and m paths $< u_0 f_{i1} f_{i2} \cdots f_{ik} f_i >$ are attached to u_0 for $i = 1, 2, \cdots, m$, where $W = \{i_1, i_2, \cdots, i_m\}$, $\bigcup_{j=1}^{k} f_{ij} = W$, $i = 1, \cdots, m$, and for every pair $(i, j) \neq (i', j')$, $f_{ij} \neq f_{i'j'}$. It is easy to see that instance I has a set cover C with $|C| > 1$ if and only if instance I' has a set cover C' with $|C| > 2$ and $C' = C \cup \{f_{ij} \mid f_i \in C\} \cup \{u_0\}$. Let C_{opt} and C'_{opt} be optimal solutions to instances I and I', respectively, then $|C'_{opt}| = k|C_{opt}| + 1$ if C_{opt} is not a singleton. We now apply algorithm $A_{\rho'}$ to instance I' and obtain a set cover C' satisfying $|C'| \leq \rho'(\ln n)|C'_{opt}|$. Thus we have

$$k|C| + 1 \leq \rho'(\ln n)(k|C_{opt}| + 1) = k\rho'(\ln n)|C_{opt}| + \rho' \ln n,$$

from which we deduce

$$|C| < \rho'(\ln n)|C_{opt}| + \frac{\rho'}{k} \ln n \leq \rho'(\ln n)(1 + \frac{1}{k})|C_{opt}|.$$

This contradicts Lemma 2 since $\rho'(1 + 1/k) < 1$.

5   Application

In this section we shall study the wavelength assignment problem of broadcast connections in optical networks, which is a special case of the connected set cover problem. An optical network can be modelled as a connected graph $G(V, E, w)$, where V is the vertex-set of graph G representing the set of routing nodes in the network, E is the edge-set of graph G corresponding to optical fiber links between nodes in the network, and $w(e)$ represents the wavelengths available on edge $e \in E$. Fig. 5(a) shows an example of such a network of 10 vertices. Observe that 5 wavelengths $\{w_1, w_2, \cdots, w_5\}$ are used in the network, but only two wavelengths w_2 and w_3 are available on edge between v_4 and v_8.

In multi-hop optical networks where wavelength converters are equipped at routing nodes, a broadcast connection between communication nodes consists of one or more light-trees. A wavelength conversion is required at the joint of two light-trees if they use different wavelengths. In an all optical network, the optical signal is allowed in the optical domain throughout the conversion process, however shifting wavelength channels from one to another makes routing/switching complicated. Thus an incoming wavelength at a routing node is allowed to convert to a subset of available wavelengths [6]; In particularly, it is allowed to be shifted only to neighboring wavelengths. For example, w_3 can only be shifted to w_2 or w_4. Thus it is desirable to minimize the number of wavelength used to reduce the conversion delay and workload of routing nodes.

Our problem here is how to construct a spanning tree T of given graph $G(V, E, w)$ such that the number of wavelengths used is minimized. Fig. 5(b) shows an optimal wavelength assignment for the example given in Fig. 5(a), where four wavelengths are needed and wavelength conversions are required at vertices v_4 and v_8.

Let us see how to formulate this problem as a connected set cover problem. Let U be the vertex-set V, and f_i be the set of vertices in V that are incident to some edges e with $w_i \in w(e)$, that is the set of vertices covered by wavelength w_i. For the example of Fig. 5(a), $f_1 = \{v_2, v_3, v_4, v_5\}$. First we assume that the vertices in f_i induce a connected subgraph of $G(V, E, w)$. The network of Fig. 5(a) satisfies this assumption. Now define a graph G_w on the set system

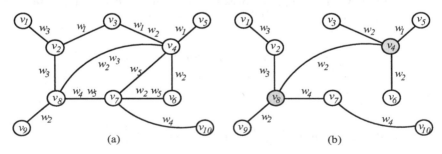

Fig. 5. (a) An optical networks, and (b) an optimal wavelength assignment

$(U, \mathcal{F} = \{f_i\})$ such that there is an edge between f_i and f_{i+1} if and only if $f_i \cap f_{i+1} \neq \emptyset$. Clearly, G_w is a line graph of union of two or more line graphs.

Thus we can use **Algorithm A** to find the minimal number of wavelengths to cover all vertices in V. As a result, we obtain a subgraph of $G(V, E, w)$ with only selected wavelengths on its edges. Notice that the subgraph may not be a tree, so we need to remove some edges. Moreover, there may exist some edge e with $w(e)$ including more than one wavelengths selected, so we must determine which one to use. These two tasks can be carried out as follows: remove edges and the wavelength w_i such that the vertices in the resulting f_i still constitute a connected subgraph of $G(V, E, w)$. Fig. 6 shows the obtained subgraph of example Fig. 5(a). After removing two edges (v_2, v_3), (v_6, v_7), and wavelengths w_2 and w_3 on (v_3, v_4) and (v_4, v_8), respectively, we get the optimal solution as shown in Fig. 5(b).

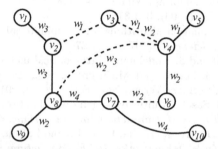

Fig. 6. Establishing a broadcast connection from the obtained subgraph

In the end let us consider the case where the vertices in f_i do not induce a connected subgraph of $G(V, E, w)$ for some i. In this case we can modify the original graph $G(V, E, w)$ as follows: Suppose that the vertices in f_i form k disjoint connected components $C_1, C_2, \cdots, C_k$ for some $k > 1$. We can replace one wavelength w_i by k different dummy wavelengths $w_{i1}, w_{i2}, \cdots, w_{ik}$ in such a way that wavelength w_{ij} is available on all edges in C_j. These k new wavelengths are not introduced physically in the network, they are just wavelength w_i and used only for the simplicity of discussion. After such a modification, we are able to use the above described method.

6   Conclusions

We have studied the connected set cover problem and also discussed its application to the wavelength assignment problem of broadcast connections in all optical networks.

Another possible application comes from the biological conservation [1,5]. The problem concerned is how to establish a series of protected areas or reserves in order to conserve species or habitat types. The objective is to select the minimal number of sites (from some candidate sites) to represent all species in the area. Clearly, this can be modeled as a set cover problem. The obtained solution, however, often produces a highly fragmented network since the solution generally neglect the spatial location of sites. This restricts the possibility of dispersal between sites, which for many species may be essential for long-term persistence. When incorporating considerations of reserve connectivity and the cost, we will get a weighted version of the connected set cover problem, which is more difficulty to solve.

References

1. A. Briers, Incorporating connectivity into reserve selection procedures, *Biological Conservation*, 14(2000) 342-355.
2. V. Chvátal, A greedy heuristic for the set-covering problem, *Mathematics of Operations Research*, 4(1979), 233-235.
3. U. Feige, A threshold of ln n for aproximating set cover, *Proc. 28th ACM Symposium on Theory of Computing*, (1996), 314-318.
4. D. S. Johnson, Approximation algorithms for combinatorial problems, *Journal of Computer and System Sciences*, 9(1974), 256-278.
5. H. Possingham, I. Ball and S. Andelman, Mathematical methods for representative reserve networks, In: S. Ferson and M.Burgman, eds., *Quantitative Methods for Conservation Biology*, Springer-Verlag, New York, (2000), 291-306.
6. R. Ramaswami and G. Sasaki, Multiwavelength optical networks with limited wavelength conversion, *IEEE/ACM Transactions on Networking*, 6(6)(1998), 744-754.
7. L. Ruan, D.-Z. Du, X.-D. Hu, X.-H. Jia, D.-Y. Li, and Z. Sun, Converter placement supporting broadcast in WDM networks, *IEEE Transactions on Computers*, 50 (7) (2001), 750-758.

A Branch and Bound Algorithm for Exact, Upper, and Lower Bounds on Treewidth*

Emgad H. Bachoore and Hans L. Bodlaender

Institute of Information and Computing Sciences, Utrecht University,
P.O. Box 80.089, 3508 TB Utrecht, The Netherlands

Abstract. In this paper, a branch and bound algorithm for computing the treewidth of a graph is presented. The method incorporates extensions of existing results, and uses new pruning and reduction rules, based upon properties of the adopted branching strategy. We discuss how the algorithm can not only be used to obtain exact bounds for the treewidth, but also to obtain upper and/or lower bounds. Computational results of the algorithm are presented.

1 Introduction

The notions treewidth, pathwidth and branchwidth have received a growing interest in recent years not only because of their theoretical significance in (algorithmic) graph theory, but also because many problems that are intractable on graphs, including a large number of well-known NP hard problems, have been shown to be polynomial-time and even linear-time solvable on graphs that are given together with a tree decomposition of width at most some constant k. (See e.g., [6, 8].)

Arnborg et al. [2] proved that computing the treewidth of a graph is an NP-hard problem. So, in recent years, many different algorithms have been designed to compute the treewidth exactly, approximately, or do preprocessing. Many of these algorithms have been designed for special classes of graphs, but in this paper, we will focus on algorithms that work on general undirected graphs.

A number of these algorithms were approximation algorithms. These are polynomial time approximation algorithms for treewidth, that have approximation ratio $O(\log n)$ [6] or $O(\log k)$ [1], where k is the actual treewidth of the input graph. Other algorithms [1] have a constant approximation ratio, but their running time is exponential in the treewidth. On the other hand there were several proposals for heuristic algorithms for upper and lower bound on the treewidth. Some of these algorithms are based on the concept of graph triangulation. These are Maximum Cardinality Search, Lexicographic Breadth First search algorithms, Minimum Degree, Minimum Fill-in, MFEO1, MFEO2, RATIO1 and RATIO2. Other heuristics are based on other ideas, e.g., the Minimum Separating Vertex Set algorithm. Some examples of lower bound methods are Maximum Minimum Degree, MMD+, D-LB, and contraction and treewidth lower bounds algorithms. See e.g. [3, 6, 8, 10, 12, 14].

Using the above techniques may help to find a close value for the treewidth of a graph, but in many cases, not the exact one. Therefore, there is a need for algorithms

---

* This work has been supported by the Netherlands Organization for Scientific Research NWO (project TACO: 'Treewidth And Combinatorial Optimization').

S.-W. Cheng and C.K. Poon (Eds.): AAIM 2006, LNCS 4041, pp. 255–266, 2006.
© Springer-Verlag Berlin Heidelberg 2006

that produce the exact treewidth, at least for some graphs. Bodlaender [7] invented a linear-time algorithm to decide whether the treewidth of a graph is at most a constant k. Unfortunately, this algorithm is exponential in a polynomial in k and hence appears to be impractical even for $k = 4$, see [16].

Well known techniques that we can use to design an algorithm for finding the exact treewidth are for example Branch-and-Bound and Integer Linear Programming. In this paper we introduce an algorithm for finding the treewidth of a graph using a Branch-and-Bound technique. The results of this study may suggest further research into the effects of using branch and bound technique to find the treewidth of graphs.

In 2003, Gogate and Dechter [13] reported on work on a branch and bound algorithm on treewidth. Independently, part of the work reported in the current paper had been done before publication of [13]. There are some significant differences in details and results between [13] and this paper. We report here also on these differences.

A generalization of the algorithm can be used to compute the weighted treewidth of weighted graphs. In this paper, we focus on the unweighted case. Furthermore, more than one variant of the algorithm has been developed and implemented to test the effect of using different pruning rules on the efficiency of the algorithm. Finally, it is interesting to note that, for some large graphs, it is infeasible to find the exact treewidth of the graph in a reasonable time by using branch and bound. Therefore, we developed our algorithm such that it yields better lower and/or upper bounds on the treewidth in such cases.

Several proofs are skipped in this extended abstract due to space constraints.

2 Definitions and Preliminary Results

Throughout this paper $G = (V, E)$ denotes a finite, simple and undirected graph, where V is the set of vertices of the graph and E the set of edges of the graph. A subgraph of a graph $G(V, E)$, induced by a set of vertices $W \subseteq V$, is denoted by $G[W] = (W, \{\{v, w\} \in E | v, w \in W\})$. A graph H is a **minor** of graph G, if H can be obtained from G by zero or more vertex deletions, edge deletions, and edge contractions. **Edge contraction** is the operation that replaces two adjacent vertices v and w by a single vertex that is connected to all neighbors of v and w. The set of neighbors of a vertex v is denoted $N(v) = \{w \in V | \{v, w\} \in E\}$. The set of neighbors of v plus v itself is denoted $N[v] = N(v) \cup \{v\}$. In the same manner we define $N^0[v] = \{v\}$, $N^{i+1}[v] = N[N^i[v]]$, $N^{i+1}(v) = N^{i+1}[v] \setminus N^i[v]$. We can extend the above definition to a set of vertices instead of one vertex. Suppose that S is a set of vertices, then $N^0[S] = S$, $N^{i+1}[S] = N[N^i[S]]$, $N^{i+1}(S) = N^{i+1}[S] \setminus N^i[S]$, $N[S] = \bigcup_{v \in S} N[v]$, $N(S) = N[S] \setminus S$, $i \in \mathcal{N}$. A vertex v in G is called **simplicial**, if its neighbors $N(v)$ form a clique in G. A vertex v in G is called **almost simplicial**, if its neighbors except one form a clique in G, i.e., if v has a neighbor w such that $N(v) - \{w\}$ is a clique. A graph G is called **triangulated** (or: chordal) if every cycle of length four of more possesses a chord. A **chord** is an edge between two non consecutive vertices of the cycle. A graph $H = (V, F)$ is a **triangulation of graph** $G = (V, E)$, if G is a subgraph of H and H is a triangulated graph.

Definition 1. *Let x be a vertex in a graph $G = (V, E)$. The fill-in of x in a graph G, is the number of edges that must be added between the neighbors x, $N(x)$, to make x simplicial, i.e.,*

$$\text{fill-in}(x) = |\{\{v, w\}|v, w \in N(x), \{v, w\} \notin E\}|$$

The fill-in-excluding-one neighbor of x in a graph G is the minimum number of edges that must be added between vertices in $N(x)$ (minus one vertex), such that x is almost simplicial, i.e.,

$$\text{fill-in-excl-one}(x) = min_{z \in N(x)}|\{\{v, w\}|v, w \in N(x) - \{z\}, \{v, w\} \notin E\}|$$

Definition 2. *A **tree decomposition** of the graph $G = (V, E)$ is a pair (X, T) in which $T = (I, F)$ a tree, and $X = \{X_i | i \in I\}$ a collection of subsets of V, one for each node of T, such that $\bigcup_{i \in I} X_i = V$, for all $\{u, v\} \in E$, there exists an $i \in I$ with $u, v \in X_i$, and for all $i, j, k \in I$: if j is on path from i to k in T, then $X_i \cap X_k \subseteq X_j$. The **width** of the tree decomposition $((I, F), \{X_i | i \in I\})$ is $max_{i \in I}|X_i - 1|$. The **treewidth** of a graph G is the minimum width over all tree decompositions of G.*

3   The Branch and Bound Algorithm for Treewidth BB-tw

The main elements of the branch and bound algorithm for finding the treewidth of a graph, BB-tw, are: The space of all feasible solutions, the upper and lower bounds on the treewidth, and the rules for pruning the feasible solutions that do not contain optimal solution. These are described in Sections 3.1 and 3.2. More details are discussed in Sections 3.3 – 3.6.

3.1   Problem Description

Several algorithms for determining or approximating the treewidth of a graph are based on triangulations formed from vertex orderings.

Definition 3. *A **linear ordering** of a graph $G = (V, E)$ is a bijection $f : V \rightarrow \{1, 2, \cdots, |V|\}$, also denoted as $[f(1), \cdots, f(|V|)]$. A linear ordering of the vertices of a graph G, $\sigma = [v_1, \cdots, v_n]$ is called a **perfect elimination order (p.e.o.)** of G, if for every $1 \leq i \leq n$, v_i is a simplicial vertex in $G[v_i, \cdots, v_n]$, i.e., the higher numbered neighbors of v_i form a clique.*

Lemma 1. *(See [8].) A graph G is triangulated, if and only if G has a perfect elimination order (p.e.o).*

Definition 4. *Let v be a vertex in a graph G. **Eliminating** a vertex v from a graph G, $eliminate(v_G)$, is the procedure of adding an edge between every pair of non-adjacent neighbors of v in G, and then removing v and its incident edges from G. We call the graph G' obtained from eliminating a vertex v from a graph G, a **temporary graph** of G. I.e., $G'(W, F) = G(V - \{v\}, E - \{\{v, w\}|\{v, w\} \in E, \{v, w\} \notin F\})$.*

Lemma 2. *Let* $G' = (V', E')$ *be a minor of graph* $G = (V, E)$. *Then* $treewidth(G') \leq treewidth(G)$.

Let $\sigma = [v_1, \cdots, v_n]$ be a linear ordering of $G = (V, E), n = |V|$. Construct a graph H as follows: Set $H = G$; for $i = 1$ to n, add to H an edge between each pair of non-adjacent higher numbered neighbors (in H) of v_i. H is the graph obtained from G by the fill-in procedure with respect to σ; σ is a perfect elimination scheme for H, hence H is chordal. The following theorem is well known.

Theorem 1. *Let* $G = (V, E)$ *be a graph,* $k = |V|$. *The following statements are equivalent.*

- *The treewidth of G is at most k.*
- *G has a triangulation with maximum clique size at most* $k + 1$.
- *There is a linear ordering* σ *of width at most k.*

Lemma 3. *Let H be a triangulation obtained from applying the* fill-in *procedure on a graph G due to an ordering* σ *of the vertices of G. Then, the treewidth of G is at least the size of the maximum clique in H minus* 1.

Thus, we take as the space of all feasible solutions for the computation of the treewidth of G the set of all possible linear orderings of the vertices of G. We represent these by a *search tree of all possible solutions* T_r by taking a root r, and having for each node a child for each possible choice of the next vertex in the elimination ordering. Thus, a node in the search tree represents a fixed initial part of the linear ordering. From this point, we look for the best ordering of the vertices not in the initial part. This is equivalent to looking for a linear ordering for the graph, obtained by eliminating all vertices in this initial part.

Initializing the space of all feasible solutions. Suppose that the given graph is as in Figure 1 and the initial value for the upper bound on the treewidth of this graph equals 9. Furthermore, suppose we use the above method for building the search tree, and we use only one pruning rule in the branch and bound algorithm, that is, if the degree of the current elimination vertex in the temporary graph is greater than the value of the reported upper bound, then we skip the search operation from the current node in the search tree to its next right sibling. Hence, the first elimination ordering that the algorithm will check is [1, …,27]. This means that we have to visit more nodes in the space of all feasible solutions than that we have to visit if we have build this space in the following manner. Instead of arranging the available nodes in the tree in ascending order, due to their labels, from left to right in each level of each subtree of the search tree, we arrange these nodes in each level due to their sequence in the perfect elimination ordering for finding the best upper bound on the treewidth. Therefore, the initial values of the first elimination ordering for implementing BB-tw algorithm on the graph in Figure 1 will be [10,11,12,13,14,15,16,17,18,19,20,21,22,23,24,25,26,27,2,3,4,5,6,7,8,9,1] when we use for example the elimination ordering obtained from finding the upper bound on the treewidth by using the Minimum Fill-in Heuristic. Thus, the number of visited nodes in the tree becomes smaller by using this method for building the search tree. As a result, the running time of the branch and bound algorithm for finding the treewidth of a graph becomes also smaller.

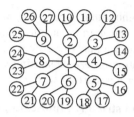

Fig. 1.

3.2 Pruning Rules

In a branch and bound algorithm, we do not search the entire search tree, but speed up the computation by omitting several parts of the search tree, of which we have established that we do not need these parts for finding the optimal solution. We consider a number of pruning rules.

Pruning Rule 1: The upper bound equals to the lower bound. Before starting the branch-and-bound search, we compute the upper bounds on the treewidth of the given graph by using the heuristics introduced in [3], and the lower bounds on the treewidth by using two heuristics, namely, the Degeneracy heuristic and Ramachandramurthi γ parameter of the graph [15]. We report the best upper and lower bounds obtained from these heuristics. Then, we check whether the best lower bound equals the best upper bound, and if so, then this value is returned as the treewidth of the given graph; otherwise we start the branch and bound. This comparison is also done when a new (better) upper or lower bound for the treewidth is found during the branch-and-bound search.

Moreover, in nodes in the search tree, we compute the degeneracy lower bound, and prune when this value is not smaller than the best known upper bound.

Pruning Rule 2: Number of vertices in the temporary graph. Let $G' = (V', E')$ be the temporary graph obtained by eliminating a set of vertices $X \subseteq V$ from a graph $G = (V, E)$. Let max be the maximum degree of all vertices in the initial part σ^0 of the elimination ordering, as represented by the node the search currently is on. Let α be the maximum degree of the vertices in σ^0 at the moment of their elimination. As any linear ordering that starts with σ^0 has width at most $\beta = max(\alpha, |V'| - 1)$, we prune when $\beta < ub$ with ub the currently best known upper bound, and set ub to β.

Pruning Rule 3: The degree of the eliminated vertex. If the degree of a vertex v in the temporary graph is larger than or equal to the best upper bound known for the treewidth of the input graph, then we know that eliminating v will give an elimination ordering whose width is at least the degree of v, hence will not yield an improvement to the upper bound. Thus, the branch which selects v as next vertex to be eliminated can be pruned at this point.

Pruning Rule 4: Equivalent elimination orderings. Gogate and Dechter [13] observed that in some cases, swapping two successive vertices does not affect the width of a linear ordering. We use a simpler but equivalent test. Suppose v and w are succesive vertices in a linear ordering σ, and v and w are not adjacent or v and w are adjacent and each has a higher numbered neighbor that is not a neighbor of the other, then the ordering σ', obtained by swapping v and w in σ, has the same width as σ. Thus, we prune the search tree as follows: for such a pair of vertices v, w, when we have looked at a branch representing the elimination orderings starting with $x_1, \ldots, x_i, v, w$, we prune the branch representing the orderings starting with $x_1, \ldots, x_i, w, v$.

Pruning Rule 5: Simplicial and strongly almost simplicial vertices. Given a lower bound β for the treewidth (of the original graph), a vertex is strongly almost simplicial if it is almost simplicial and its degree is at most β. It is known (see [5]) that for each simplicial or strongly almost simplicial vertex v, there is always a linear ordering of minimum width that starts with v. Thus, at each point in the search tree, we check if there is a simplicial or strongly almost simplicial vertex v. If so, we have only one branch, selecting v as the vertex to be eliminated, and we prune all sibling branches selecting a vertex $\neq v$.

3.3   The Edge Addition Rule

Gogate and Dechter use another rule, based on the notion of improved graph. If two vertices v and w have at least $ub + 1$ common neighbors, where ub is an upper bound for the treewidth, then adding the edge $\{v, w\}$ does not increase the treewidth (see [9]). Thus, the algorithm of [13] has at nodes a step that looks for such pairs of nonadjacent vertices with many common neighbors, and if found, adds the edge. The hope is that this leads to larger degree vertices that may be pruned by Rule 3.

However, a close analysis of the step shows that it will not help to prune the search tree, and hence only unnecessary spends time: Suppose v and w have at least $ub + 1$ common neighbors, but are not adjacent. As long as no common neighbor of v and w has been eliminated, the degrees of v and w are too large to have these vertices selected; after a common neighbor is eliminated, v and w are anyhow adjacent. Thus, we save time, and do not use this edge addition rule.

3.4   Balancing the Use of Pruning Rules

Using a pruning rule can have a positive or a negative effect on the running time: the time saved by the reduction of the number of considered nodes in the search tree should be less than the time used for testing the validity of the rules. Also, the gain obtained by using some pruning rule may depend on what other rules are also used. Pruning rules 2 and 3 should always be used. For the other rules, we have tested their effect on the running time when used separately and when used in combination with other pruning rules on a large number of graphs.

3.5   The Algorithm

In Figure 2, we give one version of the BB-tw algorithm. In the first step of the algorithm, we check if the best upper and lower bounds for the treewidth of the given graph, obtained form the heuristics, are equal. If so, we return this value as the treewidth of the graph. Otherwise, we initialize the best upper bound found so far to the treewidth and the perfect elimination ordering for the best upper bound to the elimination ordering

Algorithm BB-tw(S, σ, ub, i)

  1   **if** $((n - i) < max)$ /* PR2 */

  2   **set** $ub \leftarrow max; max \leftarrow max'; \sigma \leftarrow \sigma'; G' \leftarrow G;$

  3   **if** $(ub = lb)$ { return(ub); exit(); } /* PR1 */

  4 **else**

  5   **foreach** v in G' **do**

  6   **if** $(degree(v) < ub)$ /* PR3 */

  7   **if** $(not((\exists$ ordering O have been tested before the current

  8   elimination ordering σ such that, $O[k] = \sigma[k], k = 1, \cdots i - 2\&$

  9         $\sigma[i - 1] = w \ \& \ O[i - 1] = v \ \& \ O[i] = w) \ \& \ ((\{v, w\} \notin E(G'')))$

 10   or $(\{v, w\} \in E(G'') \ \& \ |N(v)| = 0$ or $|N(w)| = 0))$ /* PR4 */

 11   **if** $(degeneracy(G') < ub)$

 12   **set** $G'' \leftarrow G'$; **eliminate** v from G';

 13   **set** $S' \leftarrow S$; **remove** v from S;

 14   **set** $\sigma' \leftarrow \sigma$; **add** v to position i in the ordering σ;

 15   **if** $(degree(v) > max)$

 16   **set** $max' \leftarrow max; max \leftarrow degree(v)$;

 17   **BB-tw** $(S, \sigma, ub, i + 1)$;

 18            $G' \leftarrow G''; max \leftarrow max'; S \leftarrow S'; \sigma \leftarrow \sigma'$;

 19   if($max > ub$) return;

 20 **endif**

end **BB-tw**;

Fig. 2. A general scheme for the BB-tw Algorithm

for the treewidth. Next, we look for the linear ordering with a better upper bound in the search tree. We prune any ordering from the search tree, which does not yield a better upper bound than that we have reported yet.

We have developed several different versions of the algorithm depending on which pruning rules we incorporate and how we incorporate them in the algorithm. This allows us to see which of the different setups is most effective.

Note: We declare G' and lb in the algorithm as global parameters. The initial value of G' equals the given graph G and the initial value of the lb is lb_h. The values of the parameters of the first call for the algorithm are BB-tw $(peo_{ub_h}, [\,], ub_h, 0)$, where ub_h and lb_h are the best upper and lower bounds obtained from the upper bound and lower bound heuristics for the the treewidth of G, and peo_{ub} is the perfect elimination ordering corresponds to the best upper.

Theorem 2. *If the BB-tw algorithm terminates normally, then the upper bound obtained from this algorithm equals the exact treewidth of the graph.*

Incorporating the Simplicial and Strongly Almost Simplicial Pruning Rule in the BB-tw Algorithm

We have tested two methods for incorporating the Simplicial and Strongly Almost Simplicial pruning rule in the branch and bound algorithm. In the first method, we check at each visited node v in the search tree, whether or not v is simplicial or strongly almost simplicial in the temporary graph G'. In the case that v is simplicial or strongly almost simplicial, we test if the degree of v is less than the best upper bound reported up to now, ub. If it satisfies this condition too, then we prune the subtrees rooted at the siblings of this node from the search tree, eliminate the vertex from the graph G' and set the value of the parameter max to the maximum value of its current value and the degree of the eliminated vertex. However, if v is simplicial or strongly almost simplicial and its degree in G' is greater than or equal to the ub, then we prune the subtree rooted at the parent of v from the search tree.

In the second method, at each visited node v in the search tree, we find all simplicial and strongly almost simplicial vertices in G'. Then, if there is no vertex amongst these whose degree is greater than or equal to the reported upper bound, ub, then we eliminate all these vertices from the graph, set the value of max to the maximum value of its current value and the maximum degree of these vertices, and prune all the subtrees rooted at the sibling of each of these vertices, such that if $v_1, \cdots, v_m$ are simplicial vertices in G' and they are siblings of the current visited node v in the search tree, then we eliminate the subtrees rooted at the siblings of v_1, then we eliminate the subtrees rooted at the children of v_1 except the subtree rooted at the node v_2, and so on until the subtree rooted at v_m. However, if there is a vertex amongst these simplicial or strongly almost simplicial vertices its degree is greater than or equal to ub, then we eliminate the subtree rooted at the parent of the node v from the search tree.

3.6 A Memory Friendly Data Structure

In order to save memory and time, we use an elegant data structure, which is a variant of the adjacency matrix representation of graphs. Assume the vertices are numbered $1, 2, \ldots, n$. We have an n by n integer matrix A, with for $i < j$, $A_{ij} = 1$ if $\{i, j\} \in E$, and $A_{ij} = 0$ otherwise. The diagonal entries A_{ii} are used to denote if i is already eliminated; $A_{ii} = -1$ if i is not eliminated, and $A_{ii} = k$ if i is the kth eliminated vertex. The lower half of the matrix is used to give fill-in edges: for $j < i$, A_{ij} is initially 0, and becomes k if the edge $\{i, j\}$ is created when eliminating vertex k. We use this matrix as a global variable, and thus have very little parameters to pass on when doing recursive calls in the search, thus giving a considerable gain in speed.

4 Using the Branch and Bound Algorithm for Finding Better Upper and Lower Bounds for the Treewidth

If the branch and bound algorithm does not terminate within reasonable time, then it still often can be used as an upper bound or lower bound heuristic. For this, we initialize the upper bound for the treewidth in the BB-tw algorithm with the best upper bound value obtained from heuristic algorithms. When we terminate the algorithm (e.g., after a pre-determined amount of time), we report the best upper bound found by the algorithm, with its corresponding linear ordering. Thus, the algorithm either terminates normally, and yields the exact treewidth, or gives an upper bound.

To use the algorithm as a lower bound method, we give the algorithm as upper bound value some integer α. α can be any value, and does not need to be a real upper bound on the treewidth. Now, the algorithm may possibly find a solution of width less than α. If it terminates without having found such a solution, we know that α is a lower bound to the treewidth of G.

We have employed a scheme where we restart the branch and bound algorithm with several different values for the upper bound; each restart is made after the algorithm has run for some fixed amount of time (e.g., 10 minutes or an hour).

Branch and bound flexibility. The branch and bound technique has a clear advantage over other treewidth lower and upper bounds techniques: it is often known as the 'any time' property. When given enough time, the algorithm can find the exact bound, but using more time allows to obtain better bounds. Lower or upper bound heuristic give one lower or upper bound value, and when this is not the exact value, giving the heuristic more time does not help to get closer or equal to the exact value.

5 Experimental Results

In this section, we report on computational experiments for the branch and bound algorithm BB-tw. Our experiments are conducted on a large number of graphs and networks. Because of space constraints, we will show only some of these results in two tables.

Table 1 shows 15 instances of sizes between 21 and 50 vertices. Table 2 shows 12 instances of sizes between 67 and 450 vertices. The Alarm, Oesoca, Vsd and Wilson are probabilistic networks taken from medical applications; several versions exist of the Myciel networks. The Barley and Mildew networks are used for agricultural purposes, the Water network models a water purification process and Oow-trad, Oow-bas, Oow-solo, and Ship-ship networks are developed for maritime use. The other graphs are obtained from the well-known DIMACS benchmarks for vertex coloring. [1]

The algorithm was implemented using C++ on Windows 2000 PC with Pentium 4, 2.8 GHz processor. The tables shown in this section include besides the basic information for the graph, also columns for the treewidth of the graph (tw), the initial upper bound (ub), the method used for finding the initial upper bound ($ub - heuristic$), and the running time of the algorithm ($time$). We use the character "*" in the column $time$ to indicate that the algorithm did not terminate normally, namely, the algorithm ran out of time. We defined one hour as the maximum limit time for running the algorithm on the input graph, i.e., if the algorithm did not find the exact treewidth within one hour, then it was ended and returned an upper bound value for the treewidth.

All the instances we have chosen to show in Tables 1 and 2 have the following property. The best known upper bound on treewidth of each instance does not equal the best known lower bound on the treewidth of the same instance, obtained from upper and lower bound heuristics. In other words, we have excluded each instance from Tables 1 and 2 whose known upper bound equals its known lower bound. The BB-tw

---

[1] http://www.cs.uu.nl/people/hansb/treewidthlib, 2004 - 2005.

algorithm needs less than one second to determine the exact treewidth of any graph when the upper bound on the treewidth equals the lower bound on the treewidth of a graph.

We observe that BB-tw algorithm was able to determine the exact treewidth for all instances of sizes less than 50 vertices except one, namely, Ship-ship, within one hour time. Whereas, the algorithm was able to determine the exact treewidth of only 4 graphs of the large size given in Table 2 within one hour time.

We confirm the observation of Gogate and Dechter [13], that graphs of small treewidth (close to 1) and large (close to n) are easy for BB-tw algorithm also.

Table 1.

Graphname	$\|V\|$	$\|E\|$	lb	ub	ub-heuristic	tw	time
Alarm	37	65	2	4	MFEO1	4	0
Barley	48	126	3	7	MFEO1	7	30.443
Mildew	35	80	2	4	MFEO1	4	0
myciel4	24	71	8	11	MF	10	0.2
myciel5	47	236	14	20	MFEO1	19	178.7
Oesoca+	67	208	9	11	MFEO1	11	0.02
Oow-bas	27	54	2	4	MFEO1	4	0
Oow-solo	40	87	4	6	MFEO1	6	275.608
Oow-trad	33	72	4	6	MFEO1	6	86.772
Queen5_5	25	320	12	18	MFEO1	18	0.62
Queen6_6	36	580	15	26	MF	25	10.62
Ship-ship	50	114	4	8	MFEO1	8	*
VSD	38	62	2	4	MFEO1	4	0
Water	32	123	8	10	MF	9	0.07
Wilson	21	27	2	3	MFEO1	3	0

Table 2.

Graphname	$\|V\|$	$\|E\|$	lb	ub	ub-heuristic	tw	t
anna	138	986	11	12	MFEO1	12	0.911
david	87	812	11	13	MFEO1	13	56.72
dsjc125_1	125	736	15	64	MFEO1	64	*
dsjc125_5	125	3891	55	109	RATIO2	109	*
dsjc250_1	250	3218	43	177	MFEO1	177	*
inithx.i.2	645	13980	31	35	MF	31	0.01
inithx.i.3	621	13969	31	35	MF	31	0.02
games120	120	1276	10	38	RATIO2	38	*
LE450_5A	450	5714	53	304	RATIO2	304	*
myciel6	95	755	12	35	MFEO1	35	*
myciel7	191	2360	31	66	MFEO1	66	*
school1	385	19095	80	209	RATIO2	209	*

6   Conclusions

We can summarize the main results from implementing the BB-tw algorithm as follows:

1. The branch and bound algorithm is efficient for finding the treewidth, namely, it gives the treewidth of a graph in reasonable time, if the given graph has one or more of the following properties:
   - The graph consists of at most 20 vertices.
   - The graph is triangulated or almost triangulated, namely, includes many cliques.
   - The differences between the degrees of the vertices of the graph are relatively large. In other words, the distribution of the degrees of the vertices is irregular.
   - The treewidth of the graph is small (less than 7).
   - The graph has a treewidth which is close to the graph cardinality.

2. The branch and bound algorithm is not efficient for finding the treewidth, if the given graph has one or more of the following properties:
   - The graph consists of a large number of vertices and a large number of edges.
   - If the distribution of the degrees of the vertices of the graph is regular, namely, the degrees of the vertices of the graph are close to each other.

3. It is difficult to predict the running time of the algorithm when the given graph has a large number of vertices and edges.

4. The ordering of the pruning rules in the BB-tw algorithm has a significant effect on the running time of the algorithm. Two different ordering of the pruning rules in the algorithm may guide, in many instances, to two different running times of the algorithm for the same instance.

5. The efficiency of the algorithm depends critically on the effectiveness of the branching and bounding rules used; bad choices could lead to repeated branching, without any pruning, until the temporary graph becomes very small. In that case, the method would be reduced to an exhaustive enumeration of the domain, which is often impractically large.

6. The branch and bound algorithm BB-tw can be improved in two directions. The first one has to do with the data type we use for representing the graph in the memory. In the current version of the algorithm, we have defined all the values of the adjacency matrix as integers, whereas we believe that the values of the upper part can be defined as bits. The second direction has to do with incorporating more effective pruning and reduction rules in the algorithm. A third approach is to look for a large clique in the input graph, and using the fact that there is an elimination ordering with optimal width that ends with the vertices on the clique.

7. A generalization of the algorithm can be used to compute the weighted treewidth of weighted graphs. This will be reported elsewhere, with other results on weighted treewidth.

Acknowledgments. We would like to thank Gerard Tel for useful comments on earlier versions of this paper, and Arie Koster for joint work on the data structure of Section 3.6.

References

1. E. Amir. Efficient approximation for triangulation of minimum treewidth. *Proceedings of the 17th Conference on Uncertainty in Artificial Intelligence, UAI 2001*, pages 7-15, Seattle, Washington, USA, 2001.
2. S. Arnborg, D. G. Corneil, and A. Proskurowski. Complexity of finding embeddings in a k-tree. *SIAM Journal on Algebraic and Discrete Methods*, 8:277-284, 1987.
3. E. Bachoore and H. L. Bodlaender. New heuristics for upper bound of treewidth. *Proceedings of the 4th International Workshop, WEA 2005*, pages 216-227, Santorini Island, Greece 2005, Springer Verlag, Lecture Notes in Computer Science, 3503.
4. A. Becker and D. Geiger. A sufficiently fast algorithm for finding close to optimal clique trees. *Artificial Intelligence Journal*, 125:3-17, 2001.
5. H. L. Bodlaender, A. M. C. A. Koster, and F. van den Eijkhof. Pre-processing rules for triangulation of probabilistic networks. *Computational Intelligence*, 21(3):286–305, 2005.
6. H. L. Bodlaender. Discovering treewidth. *Proceedings SOFSEM 2005: Theory and Practice of Computer Science*, pages 1-16, Liptovsky Jan, Slovak Republic, 2005. Springer Verlag, Lecture Notes in Computer Science, vol. 3381.
7. H. L. Bodlaender. A linear time algorithm for finding tree-decompositions of small treewidth. *SIAM J. Comput.*, 25:1305–1317, 1996.
8. H. L. Bodlaender. A tourist guide through treewidth. *Acta Cybernetica*, 11(1-2): 1-21,1993.
9. H. L. Bodlaender. Necessary edges in k-chordalizations of graphs. *Journal of Combinatorial Optimization*, 7:283–290, 2003.
10. H.L. Bodlaender, A.M.C.A. Koster and T. Wolle. Contraction and treewidth lower bounds. *Proceedings of 12th Annual European Symposium, Algorithms-ESA*, pages 628-639, Bergen, Norway 2004, Springer Verlag, Lecture Notes in Computer Science, 3221.
11. F. van den Eijkhof, H.L. Bodlaender and A.M.C.A. Koster. Safe reduction rules for weighted treewidth. *Proceedings of the 28th International Workshop in Graph-Theoretic Concepts in Computer Science (WG 2002)*, pages 176-185, Berlin, Germany, 2002. Springer Verlag, Lecture Notes in Computer Science, vol. 2573.
12. F. Clautiaux, A. Moukrim, S. Négre, and J. Carlier. Heuristic and meta-heuristic methods for computing graph treewidth. *RAIRO Operations Research*, 38:13–26, 2004.
13. V. Gogate and R. Dechter. A complete any time algorithm for treewidth. *Proceedings of the 20th Conference on Uncertainty in Artificial Intelligence*, pages 201–208, Banff, Canada, 2004, AUAI Press, Arlington, Virginia.
14. A. M. C. A. Koster, H. L. Bodlaender, and S. van Hoesel. Treewidth: Computational experiments. In *Electronic Notes in Discrete Mathematics*, volume 8. Elsevier Science Publishers, 2001.
15. S. Ramachandramurthi. The structure and number of obstructions to treewidth. *SIAM J. Disc. Math.*, 10:146–157, 1997.
16. H. Röhrig. Tree decomposition: A feasibility study. Master's thesis, Max-Planck-Institut für Informatik, Saarbrücken, Germany, 1998.

Recognition of Probe Cographs and Partitioned Probe Distance Hereditary Graphs

David B. Chandler[1], Maw-Shang Chang[2,*], Ton Kloks[**],
Jiping Liu[3], and Sheng-Lung Peng[4]

[1] Institute of Mathematics, Academia Sinica
Nangang, Taipei 11529, Taiwan
chandler@math.sinica.edu.tw
[2] Department of Computer Science and Information Engineering
National Chung Cheng University
Chiayi 62107, Taiwan
mschang@cs.ccu.edu.tw
[3] Department of Mathematics and Computer Science
The University of Lethbridge
Alberta, T1K 3M4, Canada
[4] Department of Computer Science and Information Engineering
National Dong Hwa University
Hualien 974, Taiwan
lung@csie.ndhu.edu.tw

Abstract. Given a class of graphs $\mathcal{G}$, a graph G is a *probe graph of* $\mathcal{G}$ if its vertices can be partitioned into two sets $\mathbb{P}$ (the probes) and $\mathbb{N}$ (nonprobes), where $\mathbb{N}$ is an independent set, such that G can be embedded into a graph of $\mathcal{G}$ by adding edges between certain vertices of $\mathbb{N}$. If the partition of the vertices into probes and nonprobes is part of the input, then we call the graph a *partitioned probe graph of* $\mathcal{G}$. We give the first polynomial-time algorithm for recognizing partitioned probe distance-hereditary graphs. By using a novel data structure for storing a multiset of sets of numbers, the running time of this algorithm is $O(n^2)$, where n is the number of vertices of the input graph. We also show that the recognition of both partitioned and unpartitioned probe cographs can be done in $O(n^2)$ time.

1 Introduction

To analyze long DNA sequences, restricted enzymes are used to cut the DNA into smaller fragments called *clones*. The clones are then reproduced many times for further research. To resequence the DNA strand, tests are performed to determine whether a pair of clones overlap in the longer DNA sequence. In DNA

---

* The author thanks the Institute of Information Science of Academia Sinica of Taiwan for their hospitality and support where part of this research took place, and is partially supported by the National Science Council of Taiwan, NSC94-2213-E-194-009.
** Supported by the National Science Council of Taiwan, grant NSC94-2627-B-007-001.

S.-W. Cheng and C.K. Poon (Eds.): AAIM 2006, LNCS 4041, pp. 267–278, 2006.

physical mapping, one wishes to find the linear order of the clones based upon experimental information. To save some experimental cost to test the overlap between clones, the following algorithm is proposed by Zhang *et al.* [19, 22]. The clones are distinguished as being either probes or nonprobes. No experiments are performed to test whether pairs of nonprobes overlap, but a probe is tested against each other probe or nonprobe.

In graph terminology, we are given a graph G whose vertices are distinguished as being either probes or nonprobes. The set of all nonprobes is an independent set of G. A certain property π defines a class of graphs $\mathcal{G}$. We want to construct a graph H having property π by adding certain edges to G between vertices identified as nonprobes. We call it the *partitioned probe-$\mathcal{G}$-graph recognition problem*. In the original DNA mapping problem, the graph H should be an *interval graph*. A graph $G = (V, E)$ is called an interval graph if each vertex $v \in V$ can be assigned an interval I_v on the real line such that $(x, y) \in E$ if and only if $I_x \cap I_y \neq \emptyset$. Thus we formulate the above DNA physical mapping based on incomplete overlapping information as the *partitioned probe-interval-graph recognition problem*.

Definition 1 ([5]). *Let $\mathcal{G}$ be a class of graphs. A graph $G = (V, E)$ is a* probe *graph of $\mathcal{G}$ if its vertex set can be partitioned into a set of* probes $\mathbb{P}$ *and an independent set of* nonprobes $\mathbb{N}$, *such that G can be made into a graph $H \in \mathcal{G}$ by adding edges between certain nonprobes. We call H an* embedding *of G.*

If the partition of the vertices of a graph G into a set of probes $\mathbb{P}$ and a set of nonprobes $\mathbb{N}$ is part of the input, then we call G a *partitioned probe graph of $\mathcal{G}$*, if G can be embedded into a graph of $\mathcal{G}$ by adding edges between certain vertices of $\mathbb{N}$. In this paper we denote a partitioned graph as $G = (\mathbb{P} + \mathbb{N}, E)$, and when this notation is used it is to be understood that $\mathbb{N}$ is an independent set. We will refer to the class of (partitioned) probe graphs of the class of (XXX) graphs as (partitioned) probe (XXX) graphs where (XXX) is the name of a graph class.

Efficient algorithms for the recognition of partitioned probe interval graphs appeared in [16, 18]. Partitioned and unpartitioned probe chordal graphs were handled along the way in [2, 12], starting the research into other graph classes. A recognition algorithm for unpartitioned probe interval graphs appeared in [5]. Probe interval bigraphs were studied in [4]. Cycle-free probe interval graphs were addressed in [21]. If the complement of any graph G in graph class $\mathcal{G}$ is also a graph in $\mathcal{G}$, we call graph class $\mathcal{G}$ a *self-complementary* class of graphs. The recognition of partitioned and unpartitioned probe graphs of some self-complementary classes, such as cographs and split graphs, were studied in [6].

The partitioned probe-$\mathcal{G}$ recognition problem is a special case of the *graph sandwich problem for graph class $\mathcal{G}$*, defined as follows. Given two graphs $G_1 = (V, E_1)$ and $G_2 = (V, E_2)$ with $E_1 \subseteq E_2$, determine whether there exists a graph $G = (V, E)$ with $E_1 \subseteq E \subseteq E_2$ such that $G \in \mathcal{G}$ [11]. When $E_2 - E_1$ is the set of edges of a clique, we obtain the partitioned probe-$\mathcal{G}$ recognition problem. In [11] it is shown that the sandwich problem can be solved in polynomial time for threshold graphs, splitgraphs, and cographs. The problem is NP-complete for comparability graphs, permutation graphs, and for several other graph classes.

In this paper, we solve the partitioned probe recognition problem for two graph classes, namely, for the class of distance hereditary graphs and for the class of cographs. We give the first polynomial-time recognition algorithm for probe distance-hereditary graphs. By using a novel data structure for storing a multiset, the running time of this algorithm is $O(n^2)$, where n is the number of vertices of the graph. We also show that the recognition of both partitioned and unpartitioned probe cographs can be done in $O(n^2)$ time.

2   Recognizing Partitioned Probe Distance Hereditary Graphs

A graph G is a pair $G = (V, E)$, where the elements of V are called the vertices of G, and where E is a family of two-element subsets of V called the edges. We denote edges of a graph G as (x, y) and we call x and y the endvertices of the edge. Unless stated otherwise, a graph is regarded as undirected. For a vertex x we write $N(x)$ for its set of neighbors, and $N[x] = N(x) \cup \{x\}$. We write $n = |V|$ for the number of vertices and $m = |E|$ for the number of edges. For a graph $G = (V, E)$ and a subset $S \subseteq V$ of vertices, we write $G[S]$ for the subgraph of G *induced* by S. For a subset $W \subseteq V$ of vertices of a graph $G = (V, E)$ we write $G - W$ for the graph $G[V - W]$, *i.e.*, the subgraph induced by $V - W$. For a vertex x we write $G - x$ rather than $G - \{x\}$.

Definition 2 ([14]). *A graph is distance hereditary if the distance between any two vertices in any connected induced subgraph equals the distance in the original graph.*

Definition 3. *A pendant vertex in a graph is a vertex of degree 1.*

Definition 4. *A twin in a graph is a module[1] with two vertices. A twin is true if the vertices are adjacent. Otherwise the twin is false.*

Theorem 1 ([1]). *Let G be a graph. The following conditions are equivalent:*

1. G *is distance hereditary.*
2. G *does not contain a house, hole, domino, or gem as an induced subgraph.*
3. *Every connected induced subgraph of G with at least two vertices has a pendant vertex or a twin.*
4. *For every pair of vertices x and y with $d(x, y) = 2$, there is no induced x, y-path of length greater than 2.*

Affirmation of membership in the class of distance-hereditary graphs can be obtained using a linear-time algorithm [7, 10, 15].

Definition 5. *Let $G = (\mathbb{P}+\mathbb{N}, E)$ be a partitioned graph. A pair of vertices $\{x, y\}$ is a* probe twin *if one of the following holds:*

---

[1] A *module* is a subset M of vertices such that for all vertices $x, y \in M$ and $z \in V - M$ $(x, z) \in E$ if and only if $(y, z) \in E$.

1. $x, y \in \mathbb{P}$ and $\{x, y\}$ is a module in G, or
2. $x, y \in \mathbb{N}$ and $\{x, y\}$ is a module in G, or
3. $x \in \mathbb{P}$, $y \in \mathbb{N}$ and $N(y) - x = (N(x) - y) \cap \mathbb{P}$.

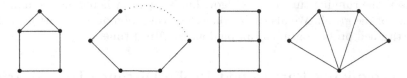

Fig. 1. A house, a hole, a domino, and a gem

Lemma 1. *Assume x is a pendant vertex in G. Then G is probe distance hereditary if and only if $G - x$ is probe distance hereditary.*

Proof. Consider an embedding of $G - x$ into a distance-hereditary graph. Obviously, adding x as a pendant vertex to the embedding does not introduce a house, hole, domino, or gem.   □

Theorem 2. *A partitioned graph $G = (\mathbb{P} + \mathbb{N}, E)$ is partitioned probe distance hereditary if and only if every connected induced subgraph with at least two vertices has a pendant vertex or a probe twin.*

Proof. Assume that G is partitioned probe distance hereditary. We may assume that G is connected. We show that G has a pendant vertex or a probe twin. Let H be an embedding of G. Assume H has a pendant vertex x. Since G is connected, x is also pendant in G. If H has a twin, say $\{x, y\}$, then $\{x, y\}$ is obviously a probe twin in G.

Assume that every connected induced subgraph of G has a pendant vertex or a probe twin. Let x be a pendant vertex in G. By induction, $G - x$ is partitioned probe distance hereditary. By Lemma 1, G is also partitioned probe distance hereditary.

Assume G has a probe twin $\{x, y\}$. If at least one of them is a nonprobe, then let x be a nonprobe. By induction, $G - x$ is partitioned probe distance hereditary. We may add x to an embedding of $G - x$ as a true twin or a false twin of y. We do not create a house, hole, domino, or gem, since these forbidden graphs are twin-free. Therefore, the graph is an embedding of G.   □

By Theorem 2 we can test whether a partitioned graph G is a partitioned probe distance-hereditary graph by checking whether we can repeatedly remove a pendant vertex or a vertex of a probe twin from G until G becomes empty. It is fairly easy to see that finding a pendant vertex or a probe twin can be done in $O(n + m)$ time. We get an $O(n^2 + nm)$-time algorithm for the recognition of partitioned probe distance-hereditary graphs. In the following we show that this algorithm can be implemented to run in $O(n^2)$ time using a data structure to speed up the procedure of finding probe twins. The data structure is a binary tree that stores a multiset. It is similar to a *trie* (see [20, p. 104]) and is used to test whether there are two sets that are equal in the multiset. In the following we describe the data structure first, and then the implementation of the recognition algorithm using the data structure.

A *multiset* differs from a set in that each element has a multiplicity, which is a natural number indicating (loosely speaking) how many times it is in the multiset. In this paper we consider multisets whose elements are subsets of $V = \{1, \ldots, n\}$ and denote them by calligraphic capitals, like $\mathcal{S}$. For clarity of notation, we use $\{1, 2, 3\}$ for a normal set with elements 1, 2, and 3 and use $\{* \; 2, 2, 3 \; *\}$ for a multiset with elements 2, 2, and 3. It is possible that the empty set is an element of a multiset $\mathcal{S}$.

Definition 6. *The* representative *of a nonempty multiset $\mathcal{S}$, denoted by $r(\mathcal{S})$, is defined as follows. If $|\mathcal{S}| = 1$ or when all sets in $\mathcal{S}$ are equal then let $r(\mathcal{S}) = n+1$. Otherwise, $r(\mathcal{S})$ is the smallest number r such that there exist $U, U' \in \mathcal{S}$, with $r \in U$ and $r \notin U'$.*

We recursively define as follows a binary tree, $\mathsf{RT}(\mathcal{S})$, for a multiset $\mathcal{S}$. The binary tree $\mathsf{RT}(\mathcal{S})$ is a rooted tree. Each node x in RT corresponds with a multiset $\mathcal{X}(x)$, and is associated with a list $\Upsilon(x)$ (possibly NIL), which is the set of "universal" elements less than $r(\mathcal{X}(x))$ contained in all elements of $\mathcal{X}(x)$, in increasing order. We also use $\Upsilon(x)$ to denote the corresponding set. The label $\ell(x)$ is the representative $r(\mathcal{X}(x))$.

1. If $r(\mathcal{S}) = n+1$, then RT is a tree consisting of one node x which is the root. In this case, $\mathcal{X}(x) = \mathcal{S}$, $\Upsilon(x)$ is the ordered list of elements of any set in $\mathcal{S}$, and $\ell(x) = r(\mathcal{S})$.
2. Suppose $r(\mathcal{S}) \leq n$. For the root x of RT, $\mathcal{X}(x) = \mathcal{S}$, $\Upsilon(x)$ is the ordered list of $\{s \mid s < r(\mathcal{S}) \text{ and } s \in U \text{ for all } U \in \mathcal{S}\}$, and $\ell(x) = r(\mathcal{S})$. Let $\mathcal{S}_L$ and $\mathcal{S}_R$ be two multisets where

$$\mathcal{S}_L = \{* \; U - \Upsilon(x) \mid U \in \mathcal{S} \text{ and } r(\mathcal{S}) \notin U \; *\}, \text{ and}$$
$$\mathcal{S}_R = \{* \; U - (\Upsilon(x) + \ell(x)) \mid U \in \mathcal{S} \text{ and } r(\mathcal{S}) \in U \; *\}.$$

The left and right children of x are the root of $\mathsf{RT}(\mathcal{S}_L)$ and the root of $\mathsf{RT}(\mathcal{S}_R)$, respectively.

Denote the root of $\mathsf{RT}(\mathcal{S})$ by $\rho(\mathcal{S})$. For each leaf node x of RT, let $P(x)$ be the path in RT from x to ρ. For each edge (y, z) of $P(x)$, let $S(y, z) = \Upsilon(z)$ if z is the left child of y and $S(y, z) = \ell(y) + \Upsilon(z)$ if z is the right child of y. Let $S(x)$ denote the set that is the union of $S(y, z)$'s of all edges (y, z) visited by $P(x)$, where z is a child of y, and $\Upsilon(\rho(\mathcal{S}))$.

For example, let

$$n = 6, \quad \mathcal{S} = \{* \; \{1, 2, 3\}, \{1, 2, 3\}, \{1, 2, 4, 5\}, \{1, 4, 5, 6\}, \{1, 4, 5, 6\} \; *\}.$$

Then $\mathsf{RT}(\mathcal{S})$ is a tree with five vertices, x_1, x_2, x_3, x_4, and x_5, where x_1 is the root, x_2 and x_3 are the left and right children of x_1, respectively, and x_4 and x_5 are the left and right children of x_3, respectively. Then $\ell(x_1) = 2$,

$\Upsilon(x_1) = \{1\}$, $\mathcal{X}(x_2) = \{* \{4,5,6\}, \{4,5,6\} *\}$, $\ell(x_2) = 7$, $\Upsilon(x_2) = \{4,5,6\}$, $\mathcal{X}(x_3) = \{* \{3\}, \{3\}, \{4,5\} *\}$, $\ell(x_3) = 3$, $\Upsilon(x_3) = \varnothing$, $\mathcal{X}(x_4) = \{* \{4,5\} *\}$, $\ell(x_4) = 7$, $\Upsilon(x_4) = \{4,5\}$, $\mathcal{X}(x_5) = \{* \varnothing, \varnothing *\}$, $\ell(x_5) = 7$, $\Upsilon(x_5) = \varnothing$. Also, $S(x_2) = \{1,4,5,6\}$, $S(x_4) = \{1,2,4,5\}$, and $S(x_5) = \{1,2,3\}$.

There is a surjective mapping from the elements in S to the leaf nodes of $\mathsf{RT}(S)$. Let x be a leaf node of $\mathsf{RT}(S)$. It is easy to see that $|\mathcal{X}(x)| > 0$, and there are $|\mathcal{X}(x)|$ copies of the set $S(x)$ in S. Conversely, if $U \in S$, then there exists a leaf node x of RT such that $U = S(x)$. In our application, in each leaf node x we store $|\mathcal{X}(x)|$ pointers with each of them pointing to a copy of the set $S(x)$ in S. These pointers are bidirectional. The tree $\mathsf{RT}(S)$ is a binary tree and every internal node has two children. If x is a leaf node, then $|\mathcal{X}(x)| \geq 1$. Therefore the number of nodes of $\mathsf{RT}(S)$ is $O(|S|)$.

Definition 7. *1. Denote the size of S as $\|S\|$, where $\|S\| = \sum_{U \in S} |U|$.*
2. Denote the number of nodes in $\mathsf{RT}(S)$ by $n(\mathsf{RT}(S))$.
3. Let W denote the set of leaves of $\mathsf{RT}(S)$. Define $\|\mathsf{RT}\|$, the size of $\mathsf{RT}(S)$, as

$$\|\mathsf{RT}\| = n(\mathsf{RT}(S)) + \sum_{x \in V(\mathsf{RT}(S))} |\Upsilon(x)| + \sum_{x \in W} |\mathcal{X}(x)|.$$

4. Let x be a leaf node of $\mathsf{RT}(S)$. We say that a set $U \in S$ is in $\mathcal{X}(x)$, denoted by $U \in \mathcal{X}(x)$, if $U = S(x)$.

Notice that we don't store $\mathcal{X}(x)$ of any node x. We only store $|\mathcal{X}(x)|$ pointers, each of them pointing to a copy of the set $U = S(x)$ in each leaf node x of S.

Lemma 2. $\|\mathsf{RT}(S)\| = O(\|S\| + |S|)$.

Lemma 3. *Two sets S and S' in S are equal if and only if they are in $\mathcal{X}(x)$ for some leaf node x of $\mathsf{RT}(S)$.*

Hence, by Lemma 3, there exist two equal sets in S if and only if there exists a leaf node x of $\mathsf{RT}(S)$ such that both sets are in $\mathcal{X}(x)$.

In our applications, we need pointers to speed up the operations on $\mathsf{RT}(S)$. For each $s \in \{1, \ldots, n\}$ we maintain a list of pointers that point to the position in lists $\Upsilon(x)$ of all $\mathcal{X}$ that contain s. For each $s \in \{1, \ldots, n\}$ we also maintain a list of pointers that point to nodes x with $\ell(x) = s$. These pointers are bidirectional, *i.e.*, for each $s \in \Upsilon(x)$ or $s = \ell(x)$ of a node x there is a pointer that points back to the pointer of s pointing to it. For each set $U \in S$ we maintain a pointer that points to the leaf node x where U is in $\mathcal{X}(x)$. In the following we assume all binary trees $\mathsf{RT}(S)$ contain these pointers, and all operations performed on the binary trees of multsets update pointers accordingly.

In the following, we describe a basic procedure that merges two binary trees $\mathsf{RT}(S)$ and $\mathsf{RT}(S')$ of multisets S and S', respectively. Let $x = \rho(S)$ and $x' = \rho(S')$. Given $\mathsf{RT}(S)$ and $\mathsf{RT}(S')$, procedure MERGE(x, x') constructs $\mathsf{RT}(S+S')$ in time proportional to $\|\mathsf{RT}(S)\| + \|\mathsf{RT}(S')\| - \|\mathsf{RT}(S+S')\| + |S| + |S'|$. By definition, $\ell(x) = r(S)$ and $\ell(x') = r(S')$. For notational brevity, use r and r' for $r(S)$ and

$r(\mathcal{S}')$, respectively. We describe procedure MERGE(x, x') recursively as follows. Assume $r \leq r'$. The case where $r > r'$ can be done in a similar way.

Compute $r^* = r(\{\Upsilon(x), \Upsilon(x')\})$. Since $\Upsilon(x)$ and $\Upsilon(x')$ are in increasing order, r^* can be computed in $O(|\{s \mid s \in \Upsilon(x), s \leq r^*\}| + 1)$ time. There are several cases. In the following, for an internal node z let z_L and z_R denote the left and right children of z. If z is not the root, denote the parent of z by z_P.

1. $r^* = n + 1$. By definition, $\Upsilon(x) = \Upsilon(x')$. All sets in $\mathcal{S} + \mathcal{S}'$ contain $\Upsilon(x)$.
    (a) $r = n + 1$. By the assumption, $r = r' = r^* = n + 1$. Hence all sets in $\mathcal{S} + \mathcal{S}'$ are equal. Both RT$(\mathcal{S})$ and RT$(\mathcal{S}')$ are trees of a single node. Output the tree obtained from RT$(\mathcal{S})$ by replacing $\mathcal{X}(x)$ with $\mathcal{X}(x) + \mathcal{X}'(x')$ as RT$(\mathcal{S} + \mathcal{S}')$. The list of $\mathcal{X}(x) + \mathcal{X}'(x)$ can be obtained in $O(1)$ time by concatenating the two lists.
    (b) $r < n + 1$. There are two cases.
        i. $r = r'$. Some sets in $\mathcal{S}$ and in $\mathcal{S}'$, but not all of them, contain r. Call MERGE(x_L, x'_L) and MERGE(x_R, x'_R), recursively. Then replace the subtrees of RT$(\mathcal{S})$ rooted at x_L and x_R with the trees returned by MERGE(x_L, x'_L) and MERGE(x_R, x'_R), respectively, and output the tree rooted at x.
        ii. $r < r'$. All sets in $\mathcal{S}'$ do contain $\Upsilon(x) = \Upsilon(x')$ but do not contain any number k where $r \leq k < r'$. Replace $\Upsilon(x')$ with the empty set. Call MERGE(x_L, x'), recursively. Replace the subtree of RT$(\mathcal{S})$ rooted at x_L with the tree MERGE(x_L, x'), and output the tree rooted at x.
2. $r < r^* < n + 1$. Clearly, $r^* \notin \Upsilon(x)$ but $r^* \in \Upsilon(x')$. All sets in $\mathcal{S}'$ contain $\Upsilon(x)$ but do not contain any number k where $r \leq k < r^*$. Replace $\Upsilon(x')$ with $\{s \in \Upsilon(x') \mid s \geq r^*\}$. Call MERGE$(x_L, x')$, recursively. Replace the subtree of RT$(\mathcal{S})$ rooted at x_L by MERGE(x_L, x'). Output the tree rooted at x.
3. $r = r^* < n + 1$. All sets in $\mathcal{S}'$ contain both r and $\Upsilon(x)$. Replace $\Upsilon(x')$ by $\{s \in \Upsilon(x') \mid s > r\}$. Call MERGE$(x_R, x')$, recursively. Replace the subtree of RT$(\mathcal{S})$ rooted at x_R with the tree MERGE(x_R, x'). Output the tree rooted at x.
4. $r^* < r$. Assume $r^* \in \Upsilon(x)$. Then $r^* \in S$ for all $S \in \mathcal{S}$ and $r^* \notin S'$ for any $S' \in \mathcal{S}'$. Create a new node y as the root of RT$(\mathcal{S} + \mathcal{S}')$. Let $\Upsilon(y) = \{s \in \Upsilon(x) \mid s < r^*\}$ and $\ell(y) = r^*$. Let x be the right child of y and $\Upsilon(x) = \{s \in \Upsilon(x) \mid s > r^*\}$. Let x' be the left child of y, and $\Upsilon(x') = \{s' \in \Upsilon(x') \mid s' > r^*\}$. Output the tree rooted at y. The case where $r^* \in \Upsilon(x')$ is similar.

Lemma 4. *One can merge two binary trees* RT$(\mathcal{S})$ *and* RT$(\mathcal{S}')$ *of two multisets $\mathcal{S}$ and $\mathcal{S}'$, respectively, into the binary tree* RT$(\mathcal{S} + \mathcal{S}')$ *of multiset $\mathcal{S} + \mathcal{S}'$ in* $O(\|$RT$(\mathcal{S})\| + \|$RT$(\mathcal{S}')\| - \|$RT$(\mathcal{S} + \mathcal{S}')\| + |\mathcal{S}| + |\mathcal{S}'|)$ *time.*

Proof. The proof is by induction. There are several cases in the merge procedure. We present a tithe in this extended abstract. The whole hog can be found in the full version.

Since numbers in $\Upsilon(x)$ and $\Upsilon(x')$ are in increasing order, $r^*, \{s \in \Upsilon(x) \mid s < r^*\}$, $\{s \in \Upsilon(x) \mid s > r^*\}$, and $\{s' \in \Upsilon(x') \mid s' > r^*\}$ can be obtained in $O(|\{s \in \Upsilon(x) + \Upsilon(x') \mid s \leq r^*\}|)$ time if $r^* < n + 1$ and in $O(|\Upsilon(x)| + 1)$ time otherwise.

Consider Case 1(a). It takes $O(|\Upsilon(x)|+1)$ time to compute r^*. The sizes of $RT(\mathcal{S})$ and $RT(\mathcal{S}')$ are $O(1+|\mathcal{S}|+|\Upsilon(x)|)$ and $O(1+|\mathcal{S}'|+|\Upsilon(x')|)$, respectively. The size of $RT(\mathcal{S}+\mathcal{S}')$ is $O(1+|\mathcal{S}|+|\mathcal{S}'|+|\Upsilon(x)|)$. Since $\Upsilon(x)=\Upsilon(x')$, $RT(\mathcal{S})+RT(\mathcal{S}')-RT(\mathcal{S}+\mathcal{S}')=|\Upsilon(x)|+1$, and it takes $O(|\Upsilon(x)|+1+n(RT(\mathcal{S}))+n(RT(\mathcal{S}')))$ time to obtain $RT(\mathcal{S}+\mathcal{S}')$, the lemma is clearly true in this case. □

Lemma 5. *Given $\mathcal{S}$, $RT(\mathcal{S})$ can be constructed in $O(\|\mathcal{S}\|+|\mathcal{S}|^2+n|\mathcal{S}|)$ time.*

Proof. First we sort the integers in each set in $\mathcal{S}$ in increasing order, which can be done in $O(n)$ time for each set, since integers are not greater than n. Hence, the sorting step takes $O(n|\mathcal{S}|)$ time. Then, we build the binary tree $RT(\{S\})$ for each set $S \in \mathcal{S}$. The binary tree of each multiset $\{S\}$ can be built in $O(1)$ time, since the integers in S are sorted. The total size of these trees is $O(\|\mathcal{S}\|+|\mathcal{S}|)$. Let $\mathcal{T}$ be the set of these trees. While $\mathcal{T}$ consists of more than one tree, pick any two trees in $\mathcal{T}$ and merge them. Eventually, the only tree left in $\mathcal{T}$ is $RT(\mathcal{S})$. By Lemma 4, it takes $O(\|\mathcal{S}\|+|\mathcal{S}|^2)$ to merge the initial $|\mathcal{S}|$ binary trees into one. Therefore $RT(\mathcal{S})$ can be constructed in $O(\|\mathcal{S}\|+|\mathcal{S}|^2+n|\mathcal{S}|)$ time. □

In our applications of the binary tree of a multiset, we never insert any new set into the tree. The operations needed in our applications are deleting a set from the tree and deleting an element s from all the sets that contain s. These two deletion operations reduce the size of the tree. For $U \in \mathcal{S}$ we write $\mathcal{S}-U$ for the multiset obtained from $\mathcal{S}$ by deleting one copy of U from it. If U is not in $\mathcal{S}$, $\mathcal{S}-U=\mathcal{S}$. For $s \in \{1,\ldots,n\}$, let $\mathcal{S}-s$ be the multiset $\{* U-s \mid U \in \mathcal{S} *\}$. The binary tree supports the operation of deleting a set U from $\mathcal{S}$, *i.e.*, obtaining $RT(\mathcal{S}-U)$ from $RT(\mathcal{S})$ as follows. Follow the pointer of U that points to the position of U in the list storing $\mathcal{X}(x)$ of leaf node x where $U \in \mathcal{X}(x)$. If $|\mathcal{X}(x)|>1$, we simply remove U from $\mathcal{X}(x)$.[2] In the following we assume $|\mathcal{X}(x)|=1$. If x is a root, *i.e.*, $RT(\mathcal{S})$ is a single node, then $RT(\mathcal{S})$ becomes empty. In the following assume x is a leaf node, x is not the root, and y is the parent of x.

   i. Assume $x=y_L$. If y is the root, then remove both x and y from $RT(\mathcal{S})$; let y_R be the new root; and replace $\Upsilon(y_R)$ by the concatenation of $\Upsilon(y)+\ell(y)+\Upsilon(y_R)$. Otherwise, remove x and y from $RT(\mathcal{S})$, let y_R be the child of y_P instead of y, and replace $\Upsilon(y_R)$ by $\Upsilon(y)+\ell(y)+\Upsilon(y_R)$.

   ii. Assume $x=y_R$. If y is the root, then remove both x and y from $RT(\mathcal{S})$; let y_L be the new root; and replace $\Upsilon(y_L)$ by $\Upsilon(y)+\Upsilon(y_L)$. Otherwise, remove x from $RT(\mathcal{S})$, let y_L be the child of y_P instead of y, and replace $\Upsilon(y_L)$ by $\Upsilon(y)+\Upsilon(y_L)$.

Lemma 6. *For any set $U \in \mathcal{S}$, one can obtain $RT(\mathcal{S}-U)$ from $RT(\mathcal{S})$ in $O(1)$ time.*

---

[2] The removal of leaf x from $RT(\mathcal{S})$ implies also the deletion all those pointers that point to leaf x and the integers in $\Upsilon(x)$. Likewise, the removal of a set U from $\mathcal{X}(x)$ not only means the deletion of U from $\mathcal{X}(x)$ but also the deletion of the pointer of $U \in \mathcal{S}$ that points to the position of U in $\mathcal{X}(x)$.

Lemma 7. *Given* $\mathrm{RT}(\mathcal{S})$, *one can determine in* $O(|\mathcal{S}|)$ *time whether there exist two sets in* $\mathcal{S}$ *that are equal, and find them if they exist.*

Proof. We can check in $O(|\mathcal{S}|)$ time whether there exists a leaf node x of $\mathrm{RT}(\mathcal{S})$ with $|\mathcal{X}(x)| > 1$. □

Lemma 8. *For any integer* $s \in \{1, \ldots, n\}$, *one can obtain* $\mathrm{RT}(\mathcal{S}-s)$ *from* $\mathrm{RT}(\mathcal{S})$ *in* $O(\|\mathrm{RT}(\mathcal{S})\| - \|\mathrm{RT}(\mathcal{S}-s)\| + |\mathcal{S}|)$ *time.*

Proof. Follow the pointers of s to finds all x's with $\ell(x) = s$ or $s \in \Upsilon(x)$. For a node x with $s \in \Upsilon(x)$, remove s from $\Upsilon(x)$. For a node x with $\ell(x) = s$, clearly x is not a leaf node. Replace the subtree rooted at x with $\mathrm{MERGE}(x_L, x_R)$. Let y be the root of $\mathrm{MERGE}(x_L, x_R)$. Replace $\Upsilon(y)$ with $\Upsilon(y) + \Upsilon(x)$. Suppose both nodes u and v have the property that either the label of the node is s or the list associated with the node contains s. Then it is not hard to verify that neither u is an ancestor of v nor v is an ancestor of u. Thus all subtrees of $\mathrm{RT}(\mathcal{S})$ rooted at x_L, x_R, y_L, and y_R are distinct. Therefore the procedure MERGE always merges distinct subtrees, and the number of nodes visited by all MERGE procedures is $O(|\mathcal{S}|)$. The total size reduction for all MERGE procedures is $O(\|\mathrm{RT}(\mathcal{S})\| - \|\mathrm{RT}(\mathcal{S}-s)\|)$. By Lemma 4, the deletion of s from $\mathcal{S}$ takes $O(\|\mathrm{RT}(\mathcal{S})\| - \|\mathrm{RT}(\mathcal{S}-s)\| + |\mathcal{S}|)$ time. □

Our algorithm maintains four multisets of sets of vertices: $\mathcal{S}^1 = \{* N[x] \mid x \in \mathbb{P} *\}$, $\mathcal{S}^2 = \{* N(x) \mid x \in \mathbb{P} *\}$, $\mathcal{S}^3 = \{* N[x] \cap \mathbb{P} \mid x \in \mathbb{P}+\mathbb{N} *\}$, and $\mathcal{S}^4 = \{* N(x) \cap \mathbb{P} \mid x \in \mathbb{P} + \mathbb{N} *\}$. If one of these has repeated elements, by Lemma 7 a probe twin can be found in $O(n)$ time, once the four binary trees for these multisets have been built. The total size of the four binary trees is $O(n + m)$. In addition to the trees, we maintain the degree of all vertices and a list P of pendant vertices, vertices of degree one. A pendant vertex can be found in $O(1)$ time from list P. Whenever we remove a pendant vertex or a probe twin v, we update the degrees of all vertices and list P in $O(|N(v)|)$ time. The four binary trees are also updated by deleting the neighborhood sets of the removed pendant or probe twin vertex and obtaining $\mathrm{RT}(\mathcal{S}^i - v)$ from $\mathrm{RT}(\mathcal{S}^i)$, $1 \leq i \leq 4$. By Lemma 8, the total time in deleting vertices from the four binary trees is $O(n^2)$, since the initial size of the four binary trees is $O(n + m)$ and they become empty at the end of the algorithm. Thus the algorithm can be implemented to run in $O(n^2)$ time and we have the following theorem.

Theorem 3. *There exists an* $O(n^2)$ *algorithm to test whether a partitioned graph* G *is a partitioned probe distance-hereditary graphs and to find an embedding for* G *in the affirmative case.*

3   Recognizing Probe Cographs

A cograph is a graph without an induced P_4, *i.e.*, an induced path with four vertices. Since the complement of a P_4 is again a P_4, it follows that cographs form a self-complementary class of graphs. By now there are many characterizations known, and various characterizations are used in the literature to define the class.

Theorem 4 ([8]). *Cographs can be characterized as follows:*

1. *A graph consisting of a single vertex is a cograph.*
2. *Let G_1 and G_2 be cographs. Then the join of G_1 and G_2, obtained by making every vertex of G_1 adjacent to every vertex of G_2 is again a cograph.*
3. *Let G_1 and G_2 be cographs. Then the (disjoint) union of G_1 and G_2 is again a cograph.*
4. *There are no other cographs.*

Notice that this decomposition recursively defines a cotree in which leaves correspond with the vertices of the graph and internal vertices are labeled as joins or unions. There is a wide variety of linear time cograph recognition algorithms. To mention just a few, see, *e.g.*, [9, 13].

In the following we show that the problem of recognizing probe cographs can be reduced to the problem of recognizing partitioned probe cographs in $O(n+m)$ time.

Theorem 5. *The problem of recognizing probe cographs can be reduced to the problem of recognizing partitioned probe cographs in $O(n + m)$ time.*

Proof. If G is disconnected, then the problem reduces to the problem of testing whether each connected component of G induces a probe cograph in G. If $\overline{G}$ is disconnected, then all the nonprobes must lie in one component of $\overline{G}$. The problem reduces to the problem of testing whether each connected component of $\overline{G}$ is a cograph, except possibly one which is a probe cograph. Using the modular decomposition tree [17] of G, we can locate in linear time a set of modules which partition $V(G)$, such that each module is connected and the complement is connected. We call a graph *coconnected* if its complement is connected. The graph G is a probe cograph if and only if the graph $G[C]$ is a probe cograph for each such module C, with one additional restriction. For each module in the decomposition tree which induces a graph which is not coconnected, there can be only one coconnected component which is not a cograph. (This information can be read from the tree.)

In the following assume G and $\overline{G}$ are connected. First we run the cograph recognition algorithm of, *e.g.*, [9, 13] which tests whether G is a cograph (we assumed it was not) and produces an induced P_4 in G if it is not. Let the induced P_4 found by the algorithm be $P = [a, b, c, d]$. We distinguish two possibilities.

Case 1. Assume $a, c \in \mathbb{N}$ and $b, d \in \mathbb{P}$. First, test whether $G = (\mathbb{P} + \mathbb{N}, E)$, where $\mathbb{P} = N(a) + N(c)$ and $\mathbb{N} = V(G) - (N(a) + N(c))$, is a partitioned probe cograph. If it is then we are done. Otherwise consider an embedding H of G, which must be the join of two cographs H_1 and H_2. It is not hard to see that $\{a, b, c, d\} \subseteq V(H_1)$ or $\{a, b, c, d\} \subseteq V(H_2)$. Assume the former is the case. There must exist a nonprobe $\alpha \in V(H_2)$ since otherwise $\overline{G}$ would be disconnected. Then α is adjacent to b and to d and not adjacent to a nor c. Consider

$$\Omega = \{\alpha \mid [a, b, \alpha, d] \text{ is an induced } P_4 \text{ in } G\}$$

The vertices of Ω must all be nonprobes (or we have a P_4 we cannot destroy) and we have $\mathbb{P} = N(a) + N(\Omega)$. Note that this case includes the possibility that $\mathbb{P} = N(a) + N(c)$. Since Ω can be found in $O(n+m)$ time, so can $\mathbb{P}$. The feasible partition can be tested by an algorithm recognizing partitioned probe cographs. The case where $a, c \in \mathbb{P}$ and $b, d \in N$ is similar.

Case 2. Assume $a, d \in N$ and $b, c \in \mathbb{P}$. Similarly to Case 1, if $N(a) + N(d)$ is not the complete set of probes, then the vertices $\{a, b, c, d\}$ cannot lie in both H_1 and H_2. Assume $\{a, b, c, d\} \subseteq V(H_1)$. There exists a nonprobe $\alpha \in V(H_2) \cap N$. In G, α is adjacent to b and c and not adjacent to a and d. Now, it is easy to see that $\mathbb{P} = N(a) + N(\alpha)$. Find the set

$$\Omega = \{\alpha \mid b, c \in N(\alpha) \text{ and } a \notin N(\alpha) \text{ and } d \notin N(\alpha)\}$$

All vertices of Ω must be nonprobes; otherwise, there exists a house in any embedding. Thus in this case $\mathbb{P} = N(a) + N(d) + N(\Omega)$. Since Ω can be found in $O(n+m)$ time, so can $\mathbb{P}$. $\qquad\square$

Obviously, by Theorem 1, cographs are distance hereditary. Cographs can be characterized as those graphs for which every nontrivial induced subgraph has a twin [3, Theorem 11.3.3]. Thus, partitioned probe cographs are those partitioned graphs for which every connected induced subgraph with at least two vertices has a probe twin. Therefore the recognition algorithm for partitioned probe distance hereditary graphs can be easily modified to recognize partitioned probe cographs. By Theorem 5 we obtain the following theorem.

Theorem 6. *There exists an* $O(n^2)$ *algorithm to test whether a graph G is a (partitioned or unpartitioned) probe cograph and to find an embedding for G in the affirmative case.*

In Memoriam

It is with deep sadness that we report the death of our friend Jiping (Jim) Liu on 14 January 2006, as the result of an automobile accident.

References

1. Bandelt, H. J. and H. M. Mulder, Distance-hereditary graphs, *Journal of Combinatorial Theory, Series B* **41** (1989), pp. 182–208.
2. Berry, A., M. C. Golumbic, and M. Lipshteyn, Two tricks to triangulate chordal probe graphs in polynomial time, *Proceedings* 15th *ACM–SIAM Symposium on Discrete Algorithms* (2004), pp. 962–969.
3. Brändstadt, A., Van Bang Le, and J. P. Spinrad, *Graph Classes: A Survey*, SIAM Monographs on Discrete Mathematics and Applications, Philadelphia, 1999.
4. Chang, Gérard Jennwha, Ton Kloks, and Sheng-Lung Peng, Probe interval bigraphs (extended abstract), *Electronic Notes in Discrete Mathematics* **19** (2005), pp. 195–201.
5. Chang, Gérard Jennwha, Antonius J. J. Kloks, Jiping Liu, and Sheng-Lung Peng, The PIGs full monty - a floor show of minimal separators, *Proceedings STACS'05*, LNCS 3404 (2005), pp. 521–532.

6. Chang, Maw-Shang, Ton Kloks, Dieter Kratsch, Jiping Liu, and Sheng-Lung Peng, On the recognition of probe graphs of some self-complementary graph classes, *Computing and Combinatorics, Proceedings of the 11th Annual International Conference COCOON 2005*, LNCS 3595, pp. 808–817.
7. Chang, Maw-Shang, S. Y. Hsieh, and G. H. Chen, Dynamic programming on distance-hereditary graphs, *Proceedings of ISAAC'97*, LNCS 1350 (1997), pp. 344–353.
8. Corneil, D. G., H. Lerchs, and L. Stewart-Burlingham, Complement reducible graphs, *Discrete Applied Mathematics* **3** (1981), pp. 163–174.
9. Corneil, D. G., Y. Perl, and L. K. Stewart, A linear recognition algorithm for cographs, *SIAM Journal on Computing* **14** (1985), pp. 926–934.
10. Damiand, G., M. Habib, and C. Paul, A simple paradigm for graph recognition: application to cographs and distance-hereditary graphs, *Theoretical Computer Science* **263** (2001), pp. 99–111.
11. Golumbic, M. C., H. Kaplan, and R. Shamir, Graph sandwich problems, *J. of Algorithms* **19** (1995), pp. 449–473.
12. Golumbic, M. C. and M. Lipshteyn, Chordal probe graphs, *Discrete Applied Mathematics*, **143** (2004), pp. 221–237.
13. Habib, M. and C. Paul, A simple linear time algorithm for cograph recognition, *Discrete Applied Mathematics* **145** (2005), pp. 183–197.
14. Howorka, E., A characterization of distance-hereditary graphs, *The Quarterly Journal of Mathematics* **28** (1977), pp. 417–420.
15. Hsieh, S.-Y., C.-W. Ho, T.-S. Hsu, M.-T. Ko, and G.-H. Chen, A faster implementation of a parallel tree contraction scheme and its application on distance-hereditary graphs, *Journal of Algorithms* **35** (2000), pp. 50-81.
16. Johnson, J. L. and J. Spinrad, A polynomial-time recognition algorithm for probe interval graphs, *Proceedings 12th ACM–SIAM Symposium on Discrete Algorithms* (2001), pp. 477–486.
17. McConnell, R. M. and J. P. Spinrad, Modular decomposition and transitive orientation, *Discrete Mathematics* **201** (1999), pp. 189–241.
18. McConnell, R. M. and J. Spinrad, Construction of probe interval graphs, *Proceedings 13th ACM–SIAM Symposium on Discrete Algorithms* (2002), pp. 866–875.
19. McMorris, F.R., C. Wang, and P. Zhang, On probe interval graphs, *Discrete Applied Mathematics* **88** (1998), pp. 315–324.
20. Mehlhorn, Kurt, *Data Structures and Algorithms 1: Sorting and Searching*, Monographs in Theoretical Computer Science, an EATCS series, Vol. 1, Springer, 1984.
21. Sheng, L., Cycle-free probe interval graphs, *Congressus Numerantium* **140** (1999), pp. 33–42.
22. Zhang, P., E. A. Schon, S. G. Fisher, E. Cayanis, J. Weiss, S. Kistler, P. E. Bourne, An algorithm based on graph theory for the assembly of contigs in physical mapping of DNA, *CABIOS* **10** (1994), pp. 309–317.

A New Approach for Solving the Maximum Clique Problem

P.J. Taillon

Carleton University,
School of Computer Science
Ottawa, Ontario, Canada
ptaillon@scs.carleton.ca

Abstract. We describe an improved algorithm for solving the MAX-
IMUM CLIQUE problem in a graph using a novel sampling technique
combined with a parameterized k-vertex cover algorithm. Experimental research shows that this approach greatly improves the execution
time of the search, and in addition, provides intermediate results during computation. We also examine a very effective heuristic for finding
a large clique that combines our sampling approach with fast independent set approximation. In experiments using the DIMACS benchmark,
the heuristical approach established new lower bounds for four instances
and provides the first optimal solution for an instance unsolved until
now. The heuristic competitively matched the accuracy of the current
best exact algorithm in terms of correct solutions, while requiring a fraction of the run time. Ideally such an approach could be beneficial as a
preprocessing step to any exact algorithm, providing an accurate lower
bound on the maximum clique, in very short time.

1   Introduction and Background

The problem of finding a *vertex cover* in a graph, $G = (V, E)$, that is to say a
subset of vertices $VC \subseteq V$, $|VC| \leq K$, such that every edge is adjacent to one of
the vertices in VC, is one of the original six problems shown to be *NP*-complete
by Karp [17]. This problem, among its many applications, figures prominently
in VLSI design, computational biology, to name a few.

Downey and Fellows [11, 12] developed parameterized tractability as a framework for studying classes of problems that are *NP*-complete and yet for which
there exist efficient algorithms that decide the problem in time bounded by an
exponential function of some fixed parameter. Such problems are called *fixed-
parameter tractable*, or FPT. Problems that are in the class *FPT* have the property that their instances can be reduced in polynomial time to instances of size
bounded by a function of a fixed parameter, k, and thus FPT algorithms have
a complexity described by $O(n^{O(1)} + f(k))$, where n is the size of the problem
instance and f is an arbitrary function. For example, the best FPT algorithm
that solves the k-VERTEX COVER problem, i.e., we wish to determine if a graph,
$G = (V, E)$, has a vertex cover of size bounded by a constant, k, has complex-
ity $O(k|V| + 1.2852^k)$[9]. The W-hierarchy consists of classes of problems that

S.-W. Cheng and C.K. Poon (Eds.): AAIM 2006, LNCS 4041, pp. 279–290, 2006.

are not likely to be fixed-parameter tractable, with the class $W[1]$ representing the lowest level of intractability [13, 12]. For example, k-INDEPENDENT SET is complete for $W[1]$.

A *clique* is a subset, $C \subseteq V$, such the vertices of C form a complete subgraph in G. The CLIQUE problem, like VERTEX COVER, is one of the original *NP*-complete problems. A wealth of different approaches have been used to tackle the problem, including graph coloring [28] and integer programming formulations [26], not to mention heuristics (for a survey, see Pelillo [23]). The *complement graph* of G is the graph $\overline{G} = (V, \overline{E})$, where $\overline{E} = \{(i,j)|i, j \in V, (i,j) \notin E\}$. C is a clique in G if and only if C is an independent set in $\overline{G}$. Finally, a graph G has a vertex cover of size k if and only if the complement graph $\overline{G}$ has a clique of size $n - k$.

In parameterized complexity theory, the k-CLIQUE problem is $W[1]$-complete [12]. Abhu-Khzam, Langston, Shanbhag [1], Abhu-Khzam, et al. [2], and later Baldwin, et al. [4], describe experiments that use a parallel FPT k-vertex cover implementation for solving the k-CLIQUE problem by determining a minimum vertex cover in the complemented input graph.

In the conventional realm, the algorithm of Wood [27] is based on a branch-and-bound approach that uses a fractional coloring procedure [3] as an upper bound heuristic. Östegard [19] uses a strategy that is similar to dynamic programming, where the problem is solved for subsequently increasing numbers of nodes. An optimal value from previous computations is used as the minimal value for the current problem. The algorithm of Fahle [14] introduces simple cost-based domain filtering [15]. The approach restricts the candidate set by eliminating vertices that can not be used to extend the clique under construction, or by fixing certain vertices that will be in the current clique. The algorithm of Regin [24], also based on constraint programming, proposed two new upper bounds on the largest clique and a new search strategy used to control the search for an optimal solution. We direct the reader to Bomze, et al. [6] for a thorough survey on exact algorithms.

2   Maximum Clique Algorithms

In this section, we introduce a graph sampling technique that we combine with a k-vertex cover algorithm to develop an improved algorithm for a finding maximum clique in a graph. We will later describe a heuristic for the maximum clique problem that is not based on determining a vertex cover, but instead on calculating an independent set.

Let $G = (V, E)$ be a graph and assume $v \in V$ is a member of a clique, $C \subseteq V$, of G. In what follows, we use $N(v)$ to denote the set of vertices adjacent to some vertex $v \in V$, $N[v]$ to denote the set $N(v) \cup \{v\}$, and $G(N[v])$ is the subgraph induced by $N[v]$. The algorithms exposited in this section are based on the following theorem.

Theorem 1. *Let $G = (V, E)$ be a graph and assume $v \in V$ is a member of a clique, $C \subseteq V$, of G. If C is the largest clique of which v is a member, then* *the maximum independent set of the complement of the subgraph $G(N[v])$ is exactly C.*

Proof. The proof follows from the definition of a clique and the relationship between the MINIMUM VERTEX COVER, MAXIMUM INDEPENDENT SET and MAXIMUM CLIQUE problems.

What makes Theorem 1 effective is that the parameterized approaches (for example, Abhu-Khzam, Langston, Shanbhag [1]) solve the MAXIMUM CLIQUE problem by searching the entire complemented graph. Depending on the maximum degree of the graph, $\Delta(G)$, it can be asymptotically more efficient to process some subset of complemented subgraphs.

2.1   Vertex Cover-Based Maximum Clique Algorithm

We can in fact derive an algorithm whereby we extract the subgraph $G(N[v])$, form its complement, $\overline{G}(N[v])$, and determine a minimum vertex cover for this subgraph. A maximum independent set for $\overline{G}(N[v])$ is the largest of all cliques in G of which v is a member. Although determining such a local maximum clique remains of exponential time complexity, the benefit arises from limiting the size of the complemented subgraph that is searched. If our candidate vertex v is a member of the maximum clique then we can determine the maximum clique without having to search the entire complemented graph $\overline{G} = (V, \overline{E})$.

Algorithm description. The algorithm consists of the following steps:

1. Given an input graph, $G = (V, E)$, randomly choose some vertex, v, $\deg(v) > 2$, and extract the subgraph $G(N[v])$;
2. generate the complement of subgraph $G(N[v])$, denoted $\overline{G}(N[v])$;
3. determine a minimum vertex cover of subgraph $\overline{G}(N[v])$;
4. generate the maximum clique by extracting the corresponding maximum independent set;
5. repeat at Step (1) for some fixed number of iterations.

For some instances, the subgraph $G(N[v])$ extracted in Step (2) may be almost as large as the graph G. In such a case, we use modified versions of steps (2), (3), (4), repeated some fixed number of times:

2a. randomly choose a vertex, $z \in G(N(v))$, $\deg(z) > 2$, and extract the subgraph $G(N[v] \cap N[z])$;
2b. generate the complemented subgraph $\overline{G}(N[v] \cap N[z])$;
2c. determine a minimum vertex cover of subgraph $\overline{G}(N[v] \cap N[z])$;
3. generate the maximum clique by extracting the corresponding maximum independent set;
4. repeat at Step (2a) for some fixed number of iterations.

This two-level sampling assumes that if the initial guess of v was correct then we are presented with even better probability of guessing a vertex that is also in the clique, $z \in N(v)$, within the smaller subgraph. The approach accelerates the process of calculating the minimum vertex cover by further restricting the size of the complemented subgraph. In the case of the heuristic algorithm, it helps generate more precise solutions by limiting the subset of vertices that are inspected by the heuristic.

If we iterate through all vertices, solving for the minimum vertex cover in each subgraph, we will find the maximum clique for the graph. On the other hand, one can compromise between accuracy and speed: the more random guesses that are made, the more confident one becomes that the largest clique has been (or will be) found. As we will see, this contrasts with other search algorithms that can fail to find a clique within some allotted time, or can not improve on the largest clique no matter how much time is allowed.

2.2 Heuristic Maximum Clique Algorithm

As with most heuristics, one is willing to trade-off the cost of attempting to find an optimal solution for an algorithm of lower complexity that returns possibly suboptimal solution quickly. Because examining the complemented subgraph still require time exponential in the maximum degree of the graph, it is interesting to consider heuristical solutions to those subproblems.

Algorithm description. In our initial conception, we considered a simple vertex cover heuristic to accelerate the algorithm, in place of the exact k-vertex cover algorithm. The heuristic consisted of removing the vertex with highest degree (and all incident edges), and adding it to the vertex cover. The final implemented version of the algorithm consisted of the same steps as shown in Section 2.1, except that we use the heuristic below to calculate an independent set (cf. ([21, 9])) for subgraph $\overline{G}(N[v] \cap N[z])$:

1. Select vertex, $w \in N[v] \cap N[z]$, where w has minimum degree;
2. add w to the independent set;
3. remove $N[w]$ from $\overline{G}(N[v] \cap N[z])$;
4. repeat at Step (1) until $\overline{G}(N[v] \cap N[z])$ has no edges.

Ideally, any sophisticated heuristic/approximation scheme could be used in this case, producing possibly improved results (within approximation limits, of course).

3   Experimental Results

We use the following methodology in the experimental phase of our research. In the exact parameterized setting, we compare our algorithm that combines k-vertex cover with sampling, to a conventional k-vertex cover-based approach. Both are parallel implementations and so we make a relative comparison between these two FPT algorithms.

We next compare a (sequential) heuristic version of our algorithm to the current best exact implementations for the MAXIMUM CLIQUE problem. We adopt this approach for the following reasons:

1. While there are no current experimental evaluations between heuristic algorithms, Regin [24] recently published an admirable tabulation of the performance of some exact algorithms.
2. We wish to investigate the quality of the solutions obtained by the heuristic approach, while noting the time required to obtain the solutions. Clearly the exact algorithms perform much more work than the heuristic and so run time comparisons, while interesting, are not meant to indicate superiority.

We found that in practice the vertex cover heuristic, described in Section 2.2, generally induced unspectacular clique sizes compared to the known solutions (similar results are observed in [5]). In contrast, we experimented with the simple independent set heuristic and found it produced surprisingly better results. The choice of this heuristic (attributed to Erdös) for our experiments was based on its simplicity and ease of implementation, although ultimately any sophisticated heuristic/approximation would do. As it turns out, this particular heuristic in fact corresponds to an existing approach: select the vertex with highest degree (in the original graph), add it to the clique, delete all its non-neighboring vertices, and repeat until the remaining graph is empty (cf. (Johnson [16])).

In general, algorithms for finding maximum cliques rely on massive experimentation to determine their effectiveness. To expedite our experiments, we make use of two separate platforms. Our first experimental platform consisted of a shared-memory SunFire 6800, configured with 20 900MHz UltraSPARC-III processors and 20GB of memory. Our second experimental platform consisted of a 2.0GHz Intel Xeon processor with 512MB RAM and 60GB of disk storage. Execution time was measured as wall clock time in seconds and includes the time taken to read the input graph from a file and output the solution size.

A *run* consists of a single execution of the code, given some sampling parameters, out of which the size of the largest clique found is output. An *experiment* consists of some number of runs on a given instance and taking the largest clique produced out of all the runs. During a run on instance, $G_i = (V_i, E_i)$, the code generates a random sample set of vertices, $S_1 \subseteq V_i$, $|S_1| = \alpha_1|V_i|$. For each vertex $v \in S_1$, we generate a second random sample set, $S_2 \subseteq N(v)$, $|S_2| = \alpha_2|S_1|$. Finally, for each $z \in S_2$ we extract the subgraph induced by $G(N[v] \cap N[z])$, complement the subgraph, and extract the maximum independent set. This clique is then checked in the original graph to ensure its correctness.

It is difficult to directly compare the algorithm of Section 2.1 with the results in [1, 2, 4], as they have implemented a different kernelization algorithm. Instead we use a readily-available FPT k-vertex cover algorithm described in [7] and compare it against itself, i.e., we measure the performance of our k-vertex cover code running on a complemented graph instance versus our new maximum clique algorithm. This provides a relative performance comparison that carries over to any k-vertex cover algorithm. Because of the time required to calculate an exact minimum vertex cover, we restrict our experiment to consisting of one run for the parallel k-vertex cover algorithm using 9 processors on the first platform. This is compared to one run of the maximum clique algorithm on the same input with sampling parameters $\alpha_1 = 0.10$ and $\alpha_2 = 0.05$. We justify using the parallel

Table 1. Graph instances derived from experimental biological data

Graph	$|K|$	Density	$|V|$	$|E|$
Somatostatin	14	0.22	559	33652
WW	22	0.45	425	40182
Thrombin (10)	24	0.19	646	40717

Table 2. Graph instances derived from experimental biological data

Graph	$|K|$	Density	$|V|$	$|E|$
globin10	12	0.16	972	75,386
globin15	23	0.32	972	149,473
pp_sh2-3	3	0.05	148	494
pp_sh2-10	17	0.27	726	69,982
sh2-5	7	0.08	839	26,612
sh2-10	23	0.37	839	129,697

platform in this series because the minimum vertex cover of a complemented graph can become significantly large and we wish to conduct the experiments in a reasonable time.

Given the shorter execution time of the heuristic algorithm, each experiment is repeated sequentially 50 times on each graph instance using our second platform. We consider an experiment to consist of 10 runs of the heuristic, and the time for an experiment is taken as the average time of the 10 runs. We take as sampling parameters $\alpha_1 = \alpha_2 = 0.10$. We compare the heuristic performance with implementations of the algorithms of Wood [27], Östegard [19], Fahle [14], and Regin [24]. Our platform closely resembles that of Regin[1] so we can draw directly upon his observations to evaluate the accuracy (and to a lesser extent, the run times) of our algorithm against previous work. In Section 3 we present a table adapted from [24] that includes measurements for all four algorithms.

Data sets. We used two different types of graphs for our experiments with the objective of studying the performance of the algorithms using both real-world and public domain benchmark data. The first set of graphs, Table 1, comprise a small subset selected from our experiments in [7, 8]. The second set, Table 2, are those cited in Baldwin, et al. [4] from which the authors derived phylogenetic trees based on proteins domains. The third set comprise the atendentious DIMACS benchmark [10].

FPT k-vertex cover maximum clique algorithm performance. The objective of this experiment is to demonstrate the efficacity of the algorithm from Section 2.1. We selected the instances *WW*, *Somatostatin*, and *Thrombin*, and used the parallel k-vertex cover code described in [7, 8], running on 9 processors.

For each graph instance we measure the total execution time, the time required before a clique matching the optimal size is found, and the number of times

---

[1] His platform was an Intel Pentium IV, 2.0GHz, 512MB RAM.

Table 3. Summary of k-vertex cover-based maximum clique algorithm results using biological datasets

Graph	$|K|$	k-VC: $\overline{G}$ Total time (hr)	k-VCMax. clique: two-level sampling (10%, 5%) Total time (hr)	Largest first found (hr)	No. of largest
Somatostatin	14	02:18:37	00:29:40	00:03:03	7
WW	22	43:34:15	19:33:12	02:45:01	21
Thrombin (10)	24	DNF	198:42:11	116:39:45	10

Table 4. Summary of heuristic algorithm results using biological datasets

Graph	$|K|$	Avg. $|K|$	Avg. run time (s)	Avg. exp. time (s)	Largest found
globin10	12	12.00	4.52	45.22	50
globin15	23	22.57	29.87	298.70	50
pp_sh2-3	3	1.81	0.01	0.10	50
pp_sh2-10	17	16.92	6.81	68.06	50
sh2-5	7	6.26	0.27	2.72	48
sh2-10	23	22.00	28.91	289.10	11

a clique matching the optimal size is found. We emphasize the relative speed with which an optimal clique is found rather than the actual run times. In Table 3 we observe that the sampling algorithm completed in a shorter time than that required to find a minimum vertex cover in a complemented graph. The experiment with the *Thrombin* instance highlights one of the benefits of the new maximum clique algorithm: the new algorithm generated several cliques of size 23 within 48 hours of starting and ultimately found several cliques matching the optimal size before termination. In contrast, the algorithm searching for a minimum vertex cover in the complemented graph had in that same time only processed two cover instances (sizes $k = 623$ and $k = 621$). The execution on the complemented graph was terminated after 2 weeks, never having completed its search.

Maximum clique heuristic performance. The software we developed for these experiments used libraries of graph and set manipulation routines made available by Östegard [20].

We see from Table 4 that the heuristic performed excellently on real-world data, with the exception instance *Sh2-10*. In the latter case the average clique size produced by the heuristic was very close to the optimal.

We follow the methodology of Regin for testing the performance of the heuristic algorithm against the DIMACS clique benchmark set, i.e., code execution on a given instance taking longer than 14,400 seconds is terminated. Table 5 summarizes the performance data for Wood, Östegard, Fahle, and Regin's algorithms for the benchmark (adapted from [24]). The timing measurements have been scaled appropriately to compensate for different processor speeds.

We divide the datasets in two groups, based on the amount of time they require to complete the series of experiments within the 4 hour time limit. For each graph instance we measure the average clique size found during the 50 experiments, the

Table 5. Performance comparison of recent maximum clique algorithms

Graph	DIMACS $\|K\|$	Wood $\|K\|$	Wood Time (s)	Östegard $\|K\|$	Östegard Time (s)	Fahle $\|K\|$	Fahle Time (s)	Regin $\|K\|$	Regin Time (s)
brock400_1	27		fail		fail	≥ 24	fail	27	11,340.80
brock400_2	29		fail		fail	≥ 29	fail	29	7,910.60
brock400_3	31		fail		fail	≥ 24	fail	31	4,477.23
brock400_4	33		fail		fail	≥ 25	fail	33	6,051.77
brock800_1	**23**		**fail**		**fail**	**≥ 21**	**fail**	**≥ 21**	**fail**
brock800_2	**24**		**fail**		**fail**	**≥ 20**	**fail**	**≥ 20**	**fail**
brock800_3	**25**		**fail**		**fail**	**≥ 20**	**fail**	**≥ 20**	**fail**
brock800_4	**26**		**fail**		**fail**	**≥ 20**	**fail**	**≥ 20**	**fail**
hamming10-2	512	512	0	512	0.84	512	5.16	512	1.04
hamming10-4	≥ 40		fail		fail	≥ 32	fail	≥ 40	fail
johnson16-2-4	8	8	13.05	8	0.09	8	7.91	8	3.80
johnson32-2-4	**≥ 16**		**fail**		**fail**	**≥ 16**	**fail**	**≥ 16**	**fail**
johnson8-2-4	4	4	0	4	0	4	0	4	0
johnson8-4-4	14	14	0	14	0	14	0.03	14	0
keller4	11	11	1.23	11	0.17	11	2.53	11	0.50
keller5	**27**		**fail**		**fail**	**≥ 25**	**fail**	**≥ 27**	**fail**
keller6	**≥ 59**		**fail**		**fail**	**≥ 43**	**fail**	**≥ 54**	**fail**
MANN_a9	16	16	0	16	0	16	0	16	0
MANN_a27	126	126	46.95		fail	126	10,348.87	126	18.48
MANN_a81	≥ 1, 100		fail		fail	≥ 996	fail	≥ 1, 100	fail
p_hat500-1	9	9	0.91	9	0.10	9	0.60	9	2.30
p_hat500-2	36	36	17.81	36	142.93	36	203.93	36	32.69
p_hat500-3	50		fail		fail	≥ 48	fail	50	12,744.70
p_hat700-1	11	11	2.69	11	0.22	11	2.67	11	6.01
p_hat700-2	44		fail		fail	44	2,086.63	44	255.79
p_hat700-3	**≥ 62**		**fail**		**fail**	**≥ 54**	**fail**	**≥ 62**	**fail**
p_hat1000-1	10	10	18.88	10	1.95	10	16.43	10	27.80
p_hat1000-2	46		fail		fail	≥ 44	fail	46	16,845.70
p_hat1000-3	≥ 68		fail		fail	≥ 50	fail	≥ 66	fail
p_hat1500-1	12		fail		fail	12	119.77	12	480.84
p_hat1500-2	**≥ 65**		**fail**		**fail**	**≥ 52**	**fail**	**≥ 63**	**fail**
san1000	15	15	43.59	15	0.17	15	3,044.09	15	102.80
san400_0.5_1	13	13	0.75	13	0	13	6.74	13	1.19
san400_0.7_1	40	40	13.25		fail	40	425.99	40	23.28
san400_0.7_2	30	30	415.12	30	168.70	30	159.72	30	67.53
san400_0.7_3	22		fail		fail	22	617.07	22	273.23
san400_0.9_1	**100**		**fail**		**fail**	**100**	**7,219.53**	**100**	**1,700.00**
sanr200_0.7		18	22.50	18	4.70	18	24.99	18	4.30
sanr200_0.9			fail		fail	≥ 41	fail	42	150.08
sanr400_0.5		13	22.55	13	2.21	13	23.09	13	17.12
sanr400_0.7			**fail**		**fail**	**21**	**15,925.00**	**21**	**3,139.11**

average time of all the runs, the average time for the experiments, and a count of the number of times that an experiment found a clique that matched the largest known size. Table 6 consists of instances that were sampled with $\alpha_1 = \alpha_2 = 0.10$. The same methodology is used to test the algorithm on instances in Table 2. As the instances in Table 7 were large, we limited the sampling to $\alpha_1 = 0.01$ and $\alpha_2 = 0.05$. In the tables, we list in boldface entries of particular interest.

In comparing the results in Table 5 and Table 6 we see the heuristic algorithm consistently outperforms the algorithms of Wood, Östegard, and Fahle, and is faster than that of Regin while producing a competitive number of optimal or similar answers. While the shorter run times are not unexpected, the quality of the results are quite surprising. In general, when the heuristic fails it does so for the same instances as does the current best algorithm, while generating comparable suboptimal answers in a fraction of the time. For example, the solution to the *Brock800_1* instance matches the current best suboptimal answer, but required much less time to compute (80 seconds). Another such instance is *P_hat700-3*, where the heuristic found a solution matching the best lower bound in less than a minute. Regin's algorithm exceeded the 4 hour time limit while finding a solution of the same size. There are many cases where the heuristic finds the largest clique extremely quickly (e.g., *Johnson32-2-4*, *Keller5*), *San400_0.9_1*, *Sanr400_0.7*).

Table 6. Summary of heuristic algorithm results using DIMACS benchmark

Graph	$|K|$	Avg. $|K|$	Avg. run time (s)	Avg. exp. time (s)	Largest found
brock400_1	27	24.08	8.17	81.68	0
brock400_2	29	24.35	8.17	81.70	17
brock400_3	31	24.61	8.13	81.32	22
brock400_4	33	26.49	8.19	81.92	47
brock800_1	23	20.81	225.41	2,254.14	0
brock800_2	24	20.78	223.14	2,231.40	0
brock800_3	25	21.13	245.43	2,454.32	0
brock800_4	26	20.55	221.80	2,218.06	0
hamming10-2	512	512.00	1,900.45	19,004.48	50
hamming10-4	≥ 40	36.00	439.04	4,390.44	0
johnson16-2-4	8	8.00	0.09	0.90	50
johnson32-2-4	≥ 16	16.00	25.30	253.04	50
johnson8-2-4	4	4.00	0.01	0.08	50
johnson8-4-4	14	14.00	0.02	0.22	50
keller4	11	11.00	0.20	2.00	50
keller5	27	27.00	106.67	1,066.74	50
MANN_a9	16	16.00	0.01	0.14	50
MANN_a27	126	125.00	24.95	249.48	0
p_hat500-1	9	9.00	0.60	6.00	50
p_hat500-2	36	35.66	6.59	65.92	50
p_hat500-3	50	49.02	21.81	218.10	37
p_hat700-1	11	10.29	1.94	19.38	50
p_hat700-2	44	43.99	24.74	247.44	50
p_hat700-3	≥ 62	60.26	4.50	44.96	25
p_hat1000-1	10	10.00	6.81	68.14	50
p_hat1000-2	46	45.72	91.62	916.20	50
p_hat1000-3	≥ 68	65.08	331.23	3,312.30	0
p_hat1500-1	12	11.22	35.58	355.76	45
p_hat1500-2	65	64.09	514.32	5,143.22	34
san1000	15	10.48	5.09	50.94	29
san400_0.5_1	13	11.60	2.46	24.60	50
san400_0.7_1	40	39.94	6.63	66.26	50
san400_0.7_2	30	29.62	6.46	64.56	50
san400_0.7_3	22	20.31	6.43	64.28	50
san400_0.9_1	100	100.00	16.42	164.18	50
sanr200_0.7	18	17.77	0.45	4.54	50
sanr200_0.9	42	41.37	1.08	10.76	50
sanr400_0.5	13	12.58	1.91	19.12	50
sanr400_0.7	21	20.82	6.46	64.56	50

Table 7. Summary of heuristic algorithm results using large DIMACS instances

Graph	$|K|$	Avg. $|K|$	Avg. run time (s)	Avg. exp. time (s)	Largest found
keller6	≥ 59	55.00	1,007.27	10,072.70	0
MANN_a81	$\geq 1,100$	1,096.00	13,961.40	139,614.00	0

Most surprising, the heuristic improves on the best lower bound for four as yet unsolved instances, and solves a fifth instance optimally. For the instance *Brock800_3*, it determined a clique of size 22. The previous best lower bound for these graphs was 20. For the instances *Brock800_2* and *Brock800_4* the heuristic found cliques of size 21 in each graph. The heuristic found a clique of size 55 in the *Keller6* instance, where previously the best suboptimal clique size was 54. For the instance *P_hat1500-2* the heuristic found a clique matching the optimal size of 65. The best lower bound until now was size 63 as determined by Regin.

We do observe two cases of oversampling. When solving the *Hamming10-2* instance (Table 6), the algorithm found a clique matching the largest known size very quickly, but it continued to search through its rather large sample set, incurring an unnecessarily long run time. In the case of the *MANN_a81* instance (Table 7), the heuristic found a relatively large clique in one run within the time limit, although the experiment exceeded the time limit.

3.1   Discussion of Experimental Results

It is important to note that the effectiveness of the algorithms are a function of the maximum clique size and the sampling rate. Cases where the maximum clique size is small and there are many such cliques, or the maximum clique is large (in relation to the vertex set), are clearly the most favorable circumstances. The contrary situations can be ameliorated with higher sampling at the cost of the longer execution time. Unless we iterate though all the vertices and solve each subgraph exactly, as in the algorithm introduced in Section 2.1, sampling is not guaranteed to find the maximum clique. The motivation for the experimental work is to verify the effectiveness of the approach in practice, particularly using real-world datasets, and to a certain extent the experiments do speak for themselves.

As mentioned in Section 2.2, the vertex cover heuristic generally produced unspectacular clique sizes compared to those generated by the independent set heuristic. The dominant factor determining the quality of a solution is the probability of error at each step, dependent on the number of choices available. The subtle algorithmic difference was significant in that the vertex cover heuristic required many more steps to determine a cover for the subgraph and, at each step in the heuristic, there were many possible choices of vertices of high degree. On the other hand, the independent set heuristic required fewer steps to complete and, at each step, there were generally very few vertices of minimum degree to choose from. (Alternately, in the original subgraph, the vertices of high degree were generally members of the clique.) Again it is important to emphasize that the heuristic, though quicker than the exact algorithms, does not guarantee a solution is optimal. Nonetheless, its degree of accuracy is surprising. Although one may call into question the variety and hardness of the DIMACS instances given the success of the simple heuristic, experiments on the real-world datasets nonetheless demonstrate some practicality of the approach. Finally, given the availability of more sophisticated heuristic/approximation schemes, one can expect higher quality solutions, again at the expense of more time.

4   Conclusion

We describe an improved algorithm for solving the MAXIMUM CLIQUE problem in a graph using a novel sampling technique combined with the FPT k-vertex cover algorithm. Experimental research shows that this approach greatly improves the execution time of the search, and in addition, provides intermediate results during computation. We also experimented with a very effective heuristic for finding a large clique that combines our sampling approach with fast independent set approximation. In experiments using the DIMACS benchmark, the heuristical approach established new lower bounds for four instances and provides the first optimal solution for an instance unsolved until now. The heuristic competitively matched the accuracy of the current best exact algorithm in terms of correct solutions, while requiring a fraction of the run time. Of equal or greater importance, both approaches performed well on real-world datasets.

The research presented here opens several exciting avenues for future exploration, which we are currently pursuing. Because of the speed of the heuristic, it could be used as a preprocessing phase for other clique algorithms, by providing a relatively accurate lower bound for the largest clique in the graph. Indeed, many exact algorithms would benefit from this preprocessing by allowing them to converge more quickly to larger solutions. We are also investigating how our maximum clique algorithms could be augmented with various clique size-bounding heuristics as pruning strategies to improve performance.

References

1. F.N. Abu-Khzam, M.A. Langston, P. Shanbhag. "Scalable parallel algorithms for difficult combinatorial problems: A case study in optimization". In *Proceedings of the International Conference on Parallel and Distributed Computing and Systems*, November, 2003.
2. F.N. Abu-Khzam, R.L. Collins, M.R. Fellows, M.A. Langston, W.H. Suters, C.T. Symons. "Kernelization algorithms for the vertex cover problem: Theory and experiments". In *Proceedings of the ACM-SIAM Workshop on Algorithm Engineering and Experiments*, January, 2004.
3. E. Balas, J. Xue. "Weighted and unweighted maximum clique algorithms with upper bounds from fractional coloring". In *Algorithmica*, Vol.15, pp.397–412, 1996.
4. N.E. Baldwin, R.L. Collins, M.R. Leuze, M.A. Langston, C.T. Symons, B.H. Voy. "High-performance computational tools for motif discovery". In *Proceedings of the IEEE International Workshop on High Performance Computational Biology*, April, 2004.
5. R. Bar-Yehuda, V. Dabholkar, K. Govindarajan, D. Sivakumar. "Randomized local approximations with applications to the MAX-CLIQUE problem". Technical Report 93-30, University at Buffalo, 1993.
6. I.M. Bomze, M. Budinich, P.M. Pardalos, M. Pelillo. "The maximum clique problem". In *Handbook of Combinatorial Optimization*, Du, Pardalos (Eds.), Vol.A, Kluwer, pp.1–74, 1999.
7. J. Cheetham, F. Dehne, A. Rau-Chaplin, U. Stege, P.J. Taillon. "Solving large FPT problems on coarse grained parallel machines". In *Journal of Computer and System Sciences*, Vol.67, No.4, pp.691–706, 2003.
8. J. Cheetham, F. Dehne, A. Rau-Chaplin, U. Stege, P.J. Taillon. "A parallel FPT application for clusters". In *Proceedings of the 3rd IEEE/ACM International Symposium on Cluster Computing and the Grid (CCGrid 2003)*, Tokyo, Japan, 2003, pp.70–77, 2003.
9. J. Chen, I.A. Kanj, W. Jia. "Vertex cover: Further observations and further improvements". In *Proceedings of the 25th International Workshop on Graph-Theoretical Concepts in Computer Science (WG'99)*, LNCS 1665, pp.313–324, 1999.
10. DIMACS clique benchmarks. *ftp://dimacs.rutgers.edu/pub/challenge/graph/*, 1993.
11. R.G. Downey, M.R. Fellows. "Fixed-parameter tractability and completeness." In *Congressus Numerantium*, Vol.87, pp.161–187, 1992.
12. R.G. Downey, M.R. Fellows. *Parameterized Complexity*. Springer-Verlag, 1998.
13. R.G. Downey, M.R. Fellows, K.W. Regan. "Parameterized Circuit Complexity and the *W* Hierarchy". In *Theoretical Computer Science A*, Vol.191, pp.91–115, 1998.

14. T. Fahle. "Simple and fast: Improving a branch-and-bound algorithm for maximum clique". In *10th Annual European Symposium (ESA'02)*, pp.485–498, 2002.
15. F. Focacci, A. Lodi, M. Milano. "Cost-based domain filtering". In *Proceedings of CP'99*, LNCS 1713, pp. 189–203, 1999.
16. D.S. Johnson. "Approximation algorithms for combinatorial problems". In *Journal of Computing and System Science*, Vol.9, pp.256–278, 1974.
17. R.M. Karp. "Reducibility among combinatorial problems". In *Complexity of Computer Computations*, Miller, Thatcher (Eds.), Plenum Press, pp.85–103, 1972.
18. J. Nešetřil, S. Poljak. "On the complexity of the subgraph problem". In *Commentationes Mathematicae Universitatis Carolinae*, Vol.26, pp.415–419, 1985.
19. P.R.J. Östegard. "A new algorithm for the maximum-weight clique problem". In *Nordic Journal of Computing*, Vol.8, No.4, pp.424–436, 2001.
20. P.R.J. Östegard. Private communication, 2004.
21. C.H. Papadimitriou. *Computational Complexity*. Addison-Wesley, 1994.
22. P.M. Pardalos, J. Xue. "The maximum clique problem". In *Journal of Global Optim.*, Vol.4, pp.301–328, 1994.
23. M. Pelillo. "Heuristics for maximum clique and independent set". In *Encyclopedia of Optimization*, Floudas, Pardalos (Eds.), Kluwer Academic, Vol.2, pp.411–423, 2001.
24. J.-C. Regin. "Solving the maximum clique problem with constraint programming". In *Fifth International Workshop on Integration of AI and OR Techniques in Constraint Programming for Combinatorial Optimization Problems*, 2003.
25. J.M. Robson. "Finding a maximum independent set in time $O(2^{n/4})$". Technical Report 1251-01, LaBRI, Université Bordeaux I, 2001.
26. N.Z. Shor. "Dual quadratic estimates in polynomial and Boolean programming". In *Computational Methods in Global Optimization*, Pardalos, Rosen (Eds.), Ann. Oper. Res., Vol.25, pp.163–168, 1990.
27. D.R. Wood. "An algorithm for finding maximum cliques in a graph". In *Operations Research Letters*, Vol.21, pp.211–217, 1997.
28. J. Xue. *Fast Algorithms For Vertex Packing and Related Problems*. Ph.D. Thesis, GSIA, Carnegie Mellon University, 1991.

The Approximability of the Exemplar Breakpoint Distance Problem[*]

Zhixiang Chen[1], Bin Fu[2], and Binhai Zhu[3]

[1] Department of Computer Science, University of Texas-Pan American, Edinburg, TX 78739-2999, USA
chen@cs.panam.edu
[2] Department of Computer Science, University of New Orleans, New Orleans, LA 70148 and Research Institute for Children, 200 Henry Clay Avenue, New Orleans, LA 70118, USA
fu@cs.uno.edu
[3] Department of Computer Science, Montana State University, Bozeman, MT 59717-3880, USA
bhz@cs.montana.edu

Abstract. In this paper we present the first set of approximation and inapproximability results for the Exemplar Breakpoint Distance Problem. Our inapproximability results hold for the simplest case between only two genomes $\mathcal{G}$ and $\mathcal{H}$, each containing only one sequence of genes (possibly with repetitions).

- For the general Exemplar Breakpoint Distance Problem, we prove that the problem does not admit any approximation unless P=NP; in fact, this result holds even when a gene appears in $\mathcal{G}$ ($\mathcal{H}$) at most three times.
- Even on a weaker definition of approximation (which we call weak approximation), we show that the problem does not admit a weak approximation with a factor $m^{1-\epsilon}$, where m is the maximum length of $\mathcal{G}$ and $\mathcal{H}$.
- We present a factor-$2(1 + \log n)$ approximation for an interesting special case, namely, one of the two genomes is a k-span genome (i.e., all genes in the same gene family are within a distance $k = O(\log n)$), where n is the number of gene families in $\mathcal{G}$ and $\mathcal{H}$.

1 Introduction

In the genome comparison and rearrangement area, a standard problem is to compute the number (i.e., genetic distances) and the actual sequence of genetic operations needed to convert a source genome to a target genome. This problem is important in evolutionary molecular biology. Typical genetic distances include edit [15], signed reversal [18, 16, 1] and breakpoint [23], etc. (The idea of signed reversal and, implicitly, breakpoint, was initiated as early as in 1936 by Sturtevant

[*] This research is supported by Louisiana Board of Regents under contract number LEQSF(2004-07)-RD-A-35 and MSU-Bozeman's Short-term Professional Development Leave Program.

S.-W. Cheng and C.K. Poon (Eds.): AAIM 2006, LNCS 4041, pp. 291–302, 2006.
© Springer-Verlag Berlin Heidelberg 2006

and Dobzhansky [21].) Recently, conserved interval distance was also proposed to measure the similarity of multiple sequences of genes [4]. (Interested readers are referred to [11, 12] for a summary of the research performed in this area.)

Until very recently, in genome rearrangement research, it is always assumed that each gene appears in a genome exactly once. Under this assumption, the genome rearrangement problem is in essence the problem of comparing and sorting signed permutations [11, 12]. However, this assumption is very restrictive and is only justified in several small virus genomes. For example, this assumption does not hold on eukaryotic genomes where paralogous genes exist [17, 20]. On the one hand, it is important in practice to compute genomic distances, e.g., Hannenhalli and Pevzner's method [11], when no gene duplications arise; on the other hand, one might have to handle this gene duplication problem as well. In 1999, Sankoff proposed a way to select, from the duplicated copies of genes, the common ancestor gene such that the distance between the reduced genomes (*exemplar genomes*) is minimized [20]. A general branch-and-bound algorithm was also implemented in [20]. Recently, Nguyen, Tay and Zhang proposed to use a divide-and-conquer method to compute the exemplar breakpoint distance empirically [17].

For the theoretical part of research, it was shown that computing the signed reversals and breakpoint distances between exemplar genomes are both NP-complete [2]. Recently, Blin and Rizzi further proved that computing the conserved interval distance between exemplar genomes is NP-complete [3]; moreover, it is NP-complete to compute the minimum conserved interval matching (i.e., without deleting the duplicated copies of genes). Before this work, there has been no formal theoretical results on the approximability of the exemplar genomic distance problems except the NP-completeness proofs [2, 3].

In this paper, we present the first set of inapproximability and approximation results for the Exemplar Breakpoint Distance problem, given two genomes each containing only one sequence of genes drawn from n identical gene families. (Some of the results hold subsequently for the Exemplar Reversal Distance problem.) For the One-sided Exemplar Breakpoint Distance Problem, which is also known to be NP-complete, we obtain a factor-$2(1 + \log n)$, polynomial-time approximation. The approximation algorithm follows the greedy strategy for Set-Cover, but constructing the family of sets is non-trivial and is related to a new problem of *longest constrained common subsequences* which is related to but different from the recently studied *constrained longest common subsequences* [5].

2   Preliminaries

In the genome comparison and rearrangement problem, we are given a set of genomes, each of which is a signed sequence of genes[1]. The order of the genes corresponds to the position of them on the linear chromosome and the signs correspond to which of the two DNA strands the genes are located. While most of the past research are under the assumption that each gene occurs in a genome

---

[1] In general a genome could contain a set of such sequences. The genomes we focus in this paper are typically called *singletons*.

once, this assumption is problematic in reality for eukaryotic genomes or the likes where duplications of genes exist [20]. Sankoff proposed a method to select an *exemplar genome*, by deleting redundant copies of a gene, such that in an exemplar genome any gene appears exactly once; moreover, the resulting exemplar genomes should have a property that certain genetic distance between them is minimized [20].

The following definitions are very much following those in [3]. Given n *gene families* (alphabet) $\mathcal{F}$, a genome $\mathcal{G}$ is a sequence of elements of $\mathcal{F}$ such that each element is with a sign (+ or −). In general, we allow the repetition of a gene family in any genome. Each occurrence of a gene family is called a *gene*, though we will not try to distinguish a gene and a gene family if the context is clear. Given a genome $G = g_1g_2...g_m$ with no repetition of any gene, we say that gene g_i *immediately precedes* g_j if $j = i + 1$. Given genomes G, H, if gene a immediately precedes b in G and neither a immediately precedes b nor $-b$ immediately precedes $-a$ in H, then they constitute a *breakpoint* in G. The *breakpoint distance* is the number of breakpoints in G (symmetrically, it is the number of breakpoints in H).

The number of a gene g appearing in a genome $\mathcal{G}$ is called the cardinality of g in $\mathcal{G}$, written as $card(g, \mathcal{G})$. A gene in $\mathcal{G}$ is called *trivial* if g has cardinality exactly 1; otherwise, it is called *non-trivial*. In this paper, we assume that all the genomes we discuss could contain both trivial and non-trivial genes. A genome $\mathcal{G}$ is called *r-repetitive*, if all the genes from the same gene family appear at most r times in $\mathcal{G}$. A genome $\mathcal{G}$ is called a *k-span* genome, if all the genes from the same gene family are within distance at most k in $\mathcal{G}$. For example, $\mathcal{G} = -adc - bdaeb$ is 2-repetitive and it is a 5-span genome.

Given a genome $\mathcal{G} = g_1g_2 \cdots g_m$, an interval $[g_i, g_j]$ is simply the substring $g_ig_{i+1} \cdots g_j$ (which will also be denoted as $\mathcal{G}[i, j]$). Example: given $\mathcal{G}' = bdc-ag-e-fh, \mathcal{G}'' = bdce-gafh$, between the interval $I_1 = dc-ag-e-f, I_2 = dce-gaf$, there are 2 breakpoints. A *signed reversal* on a genome $\mathcal{G}$ simply reverses the order and signs of all the elements in an interval of $\mathcal{G}$. In the previous example, if a signed reversal operation is conducted on I_1 then we obtain a new genome $\mathcal{G}^* = bfe - ga - c - dh$. (All the reversals concerned in this paper are signed reversals. Henceforth, we simply use *reversal* to make the presentation simpler.) The *reversal distance* between genomes G and H is the minimum number of reversals to transfer G into H.

Given a genome $\mathcal{G}$ over $\mathcal{F}$, an *exemplar genome* of $\mathcal{G}$ is a genome G' obtained from $\mathcal{G}$ by deleting duplicating genes such that each gene family in $\mathcal{G}$ appears exactly once in G'. For example, let $\mathcal{G} = bcaadagef$ there are two exemplar genomes: $bcadgef$ and $bcdagef$.

The Exemplar Breakpoint (Reversal) Distance Problem is defined as follows:

Instance: Genomes $\mathcal{G}$ and $\mathcal{H}$, each is of length $O(m)$ and each covers n identical gene families (i.e., at least one gene from each of the n gene families appears in both $\mathcal{G}$ and $\mathcal{H}$); integer K.

Question: Are there two respective exemplar genomes of $\mathcal{G}$ and $\mathcal{H}$, G and H, such that the breakpoint (reversal) distance between them is at most K?

In the next three sections, we present inapproximability/approximation results for the optimization versions of these problems, namely, to compute or approximate the minimum value K in the above formulation. Given a minimization problem Π, let the optimal solution of Π be OPT. We say that an approximation algorithm $\mathcal{A}$ provides a *performance guarantee* of α for Π if for every instance I of Π, the solution value returned by $\mathcal{A}$ is at most $\alpha \times OPT$. (Usually we say that $\mathcal{A}$ is a factor-α approximation for Π.) Typically we are interested in polynomial time approximation algorithms.

In many biological problems, the optimal solution value OPT could be zero. (For example, in some minimum recombination haplotype reconstruction problems the optimal solution could be zero.) In that case, if computing such a zero optimal solution value is NP-complete then the problem does not admit *any* approximation (unless P=NP). However, in reality one would be happy to obtain a solution with value one or two. Due to this reason, we relax the above (traditional) definition of approximation to a *weak approximation*. Given a minimization problem Π, let the optimal solution of Π be OPT. We say that a weak approximation algorithm $\mathcal{B}$ provides a *performance guarantee* of α for Π if for every instance I of Π, the solution value returned by $\mathcal{B}$ is at most $\alpha \times (OPT + 1)$.

3 Inapproximability Bounds

In this section, we present a series of inapproximability bounds on the Exemplar Breakpoint Distance Problem.

Theorem 1. *If both $\mathcal{G}$ and $\mathcal{H}$ are 2-repetitive genomes, then the Exemplar Breakpoint Distance Problem cannot be approximated within a factor 1.36.*

Proof. We use a reduction from Vertex Cover to the Exemplar Breakpoint Distance Problem in which each gene appears in $\mathcal{G}$ ($\mathcal{H}$) at most twice. Dur and Safra proved that Vertex Cover cannot be approximated within a factor 1.36 [9].

Given a graph $T = (V, E), V = \{v_1, v_2, \cdots, v_n\}, E = \{e_1, e_2, \cdots, e_m\}$, we construct $\mathcal{G}$ and $\mathcal{H}$ as follows. (We assume that the vertices and edges are sorted by their corresponding indices.) Let A_i be the sorted sequence of edges incident to v_i and $-A_i$ be the signed reversal of A_i. (# is not a gene and is used only for the readability purpose.)

$\mathcal{G} : A_1 \# A_2 \# \cdots \# A_{n-1} \# A_n$
$\mathcal{H} : -A_1 \# -A_2 \# \cdots \# -A_{n-1} \# -A_n$

We claim that T has a vertex cover of size K iff the exemplar breakpoint distance between $\mathcal{G}$ and $\mathcal{H}$ is $K - 1$.

If T has a vertex cover of size K, then the claim is trivial. Firstly, construct the exemplar genomes G, H as follows. For all i, if v_i is in the vertex cover, then leave A_i in $\mathcal{G}$ and $-A_i$ in $\mathcal{H}$ and delete all $A_j, -A_j$ in $\mathcal{G}, \mathcal{H}$ for which v_j is not in the vertex cover of T. Finally, if e_i appears twice in the current genomes $\mathcal{G}$ and $\mathcal{H}$, say in A_s, A_t, then delete one copy of e_i in either A_s or A_t arbitrarily (say in A_s), and delete the corresponding copy of $-e_i$ in $-A_s$. The final exemplar genomes obtained, G and H, obviously have a breakpoint distance of $K - 1$. In fact, a breakpoint in G, H can only occur at the # positions—between some A_i and A_j in G ($-A_i$ and $-A_j$ in H).

If the exemplar breakpoint distance between $\mathcal{G}$ and $\mathcal{H}$ is $K - 1$, the first thing to notice is that there is no breakpoint in A_i and $-A_i$; in other words, deleting e_j in A_i inconsistently (say, by deleting e_j in A_i and deleting $-e_j$ in $-A_s$ instead of in $-A_i$) would increase the number of breakpoints in the exemplar genomes G, H. Therefore, we can obtain a pair of exemplar genomes G, H by enforcing the breakpoints to be in between A_i and A_j in G (and symmetrically, $-A_i$ and $-A_j$ in H), with all redundant edges between them deleted. Clearly, the remaining A_i's in G (and $-A_i$'s in H) correspond to a vertex cover of size K in T.

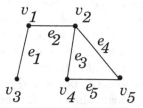

Fig. 1. Illustration of a simple graph for the reduction

In the example shown in Figure 1, we have
$\mathcal{G} : e_1 e_2 \# e_2 e_3 e_4 \# e_1 \# e_3 e_5 \# e_4 e_5$ and
$\mathcal{H} : -e_2 - e_1 \# - e_4 - e_3 - e_2 \# - e_1 \# - e_5 - e_3 \# - e_5 - e_4$.
Corresponding to the optimal vertex cover $\{v_1, v_2, v_5\}$, we have $G : e_1 e_2 \# e_3 e_4 \# e_5$ and $H : -e_2 - e_1 \# - e_4 - v_3 \# - e_5$. □

Corollary 1. *If both $\mathcal{G}$ and $\mathcal{H}$ are 2-repetitive, then the Exemplar Reversal Distance Problem cannot be approximated within a factor 1.36.*

In [17] it was claimed that the Exemplar Breakpoint Distance Problem cannot be approximated within a constant factor. But the proof, which was included in Nguyen's thesis, in fact implies a stronger $c \log n$ inapproximability bound as the reduction was from Set Cover. We extend Theorem 3.1 below to obtain a much simpler and clean proof of the $c \log n$ inapproximability bound, even though this is not the strongest inapproximability bound in this section.

Corollary 2. *The Exemplar Breakpoint Distance Problem cannot be approximated within a factor $c \log n$, for some constant $c > 0$.*

Proof. Similar to the proof of Theorem 3.1, we use a reduction from Dominating Set to the Exemplar Breakpoint Distance Problem in which each gene appears in $\mathcal{G}$ ($\mathcal{H}$) as many as $n - 1$ times. Raz and Safra proved that Dominating Set cannot be approximated within a factor $c \log n$, for some $c > 0$ [19].

Given a graph $T = (V, E), V = \{v_1, v_2, \cdots, v_n\}, E = \{e_1, e_2, \cdots, e_m\}$, we construct $\mathcal{G}$ and $\mathcal{H}$ as follows. (We assume that the vertices and edges are sorted by their corresponding indices.) Let B_i be the sorted sequence of vertices incident to v_i and $-B_i$ be the signed reversal of B_i. (# is not a gene and is again used only for the readability purpose.)

$$\mathcal{G} : v_1 B_1 \# v_2 B_2 \# \cdots \# v_{n-1} B_{n-1} \# v_n B_n$$
$$\mathcal{H} : -B_1 - v_1 \# - B_2 - v_2 \# \cdots \# - B_{n-1} - v_{n-1} \# - B_n - v_n$$

We claim that T has a dominating set of size K iff the exemplar breakpoint distance between $\mathcal{G}$ and $\mathcal{H}$ is $K - 1$.

If T has a dominating set of size K, then the claim is again trivial. Firstly, construct the exemplar genomes G, H as follows. For all i, if v_i is in the dominating set, then leave $v_i B_i$ in $\mathcal{G}$ and $-B_i - v_i$ in $\mathcal{H}$ and delete all other $v_j A_j, -A_j - v_j$ in $\mathcal{G}, \mathcal{H}$ for which v_j is not in the dominating set of T. Finally, if v_i appears x times in the current genomes $\mathcal{G}$ and $\mathcal{H}$, then arbitrarily delete $x - 1$ copies of v_i in all $v_s B_s$ which contains v_i, and delete the corresponding copy of $-v_i$ in $-B_s - v_s$. The final exemplar genomes obtained, G and H, obviously have a breakpoint distance of $K - 1$. In fact, a breakpoint in G, H can only occur at the # positions—between some $v_i B_i$ and $v_j B_j$ in G ($-B_i - v_i$ and $-B_j - v_j$ in H).

If the exemplar breakpoint distance between $\mathcal{G}$ and $\mathcal{H}$ is $K - 1$, the first thing to notice is that there is no breakpoint in $v_i B_i$ and $-B_i - v_i$; in other words, deleting v_j in $v_i B_i$ inconsistently (say, by deleting v_j in $v_i B_i$ and deleting $-v_j$ in $-B_s - v_s$ instead of in $-B_i - v_i$) would increase the number of breakpoints in the exemplar genomes G and H. Therefore, we can obtain a pair of exemplar genomes G and H by enforcing the breakpoints to be in between $v_i B_i, v_j B_j$ in G (and symmetrically, $-B_i - v_i, -B_j - v_j$ in H), with all redundant v_l's deleted. Clearly, the remaining $v_i B_i$'s in G (and $-B_i - v_i$'s in H) correspond to a dominating set of size K in T.

In the example shown in Figure 1, we have
$$\mathcal{G} : v_1 v_2 v_3 \# v_2 v_1 v_4 v_5 \# v_3 v_1 \# v_4 v_2 v_5 \# v_5 v_2 v_4 \text{ and}$$
$$\mathcal{H} : -v_3 - v_2 - v_1 \# - v_5 - v_4 - v_1 - v_2 \# - v_1 - v_3 \# - v_5 - v_2 - v_4 \# - v_4 - v_2 - v_5.$$
Corresponding to the optimal dominating set $\{v_1, v_4\}$, we have $G : v_1 v_2 v_3 \# v_4 v_5$ and $H : -v_3 - v_2 - v_1 \# - v_5 - v_4$. $\qquad\square$

Corollary 3. *The Exemplar Reversal Distance Problem cannot be approximated within a factor $c \log n$, for some constant $c > 0$.*

Proof. Construction is the same as above. The claim that T has a dominating set of size K iff the exemplar reversal distance between $\mathcal{G}$ and $\mathcal{H}$ is K can be proved similarly. $\qquad\square$

Next, we show an even stronger negative result for the Exemplar Breakpoint Distance Problem; namely, deciding whether the exemplar distance between $\mathcal{G}$ and $\mathcal{H}$ is zero is NP-complete. This implies that for the Exemplar Breakpoint Distance Problem there is no approximation unless P=NP. From now on we simply call this problem the *zero breakpoint distance (ZBD)* problem.

Theorem 2. *Deciding if two genomes $\mathcal{G}$ and $\mathcal{H}$ have zero breakpoint distance is NP-complete.*

Proof. We construct a reduction from the SAT problem [10] to the ZBD problem.

Let $F = f_1 \wedge f_2 \wedge \cdots \wedge f_q$ be a conjunctive normal form, where each sub-formula f_i is a disjunctive clause like $(x_2 \vee x_5 \vee \neg x_7)$. We construct a pair of sequences $\mathcal{G}$ and $\mathcal{H}$ such that F is satisfiable iff $\mathcal{G}$ and $\mathcal{H}$ have breakpoint distance zero.

Assume that $x_1, x_2, \cdots, x_n$ are the boolean variables in the formula F. For each variable x_i, we construct two sequences S_i and S_i^*. Let $f_{i_1}, \cdots, f_{i_u}$ be the sub-formulas in F that contains x_i, and let $f_{j_1}, \cdots, f_{j_v}$ be the sub-formulas of F that contains $\neg x_i$. Let $S_i = f_{i_1} \cdots f_{i_u} f_{j_1} \cdots f_{j_v}$ and $S_i^* = f_{j_1} \cdots f_{j_v} f_{i_1} \cdots f_{i_u}$, where $f_1, \cdots, f_q$ are considered as the names of q genes in $\mathcal{G}$ and $\mathcal{H}$.

Let $\mathcal{G} = S_1 g_1 S_2 g_2 \cdots g_{n-1} S_n$ and $\mathcal{H} = S_1^* g_1 S_2^* g_2 \cdots g_{n-1} S_n^*$, where $g_1, \cdots, g_n$ are (peg) genes that occur only once in $\mathcal{G}$ or $\mathcal{H}$.

Assume that $x_1 = b_1, \cdots, x_n = b_n$ are assignments that make F true. If $b_i = 1$, adjust both S_i and S_i^* to $S_i' = f_{i_1} \cdots f_{i_u}$ and $S_i^{*'} = f_{i_1} \cdots f_{i_u}$, respectively. If $b_i = 0$, adjust both S_i and S_i^* to $S_i' = f_{j_1} \cdots f_{j_v}$ and $S_i^{*'} = f_{j_1} \cdots f_{j_v}$, respectively. It is easy to see that $G' = S_1' g_1 S_2' \cdots S_{n-1}' g_{n-1} S_n'$ is the same as $H' = S_1^{*'} g_1 S_2^{*'} \cdots S_{n-1}^{*'} g_{n-1} S_n^{*'}$. Since the assignments make F true, each sub-formula $f_t \in \{f_1, \cdots, f_q\}$ is true due to $x_i = b_i$ for some i. That is, f_t must occur in S_i and S_i^*. If f_t occurs more than once in G' and H' then we can delete their corresponding occurrences in G' and H'. Finally, notice that both G' and H' contain all $q + n - 1$ genes in $\{f_1, \cdots, f_q, g_1, \cdots, g_{n-1}\}$.

Assume that $\mathcal{G}$ is converted into G'' and $\mathcal{H}$ is converted into H'' via removing some genes such that $G'' = H''$ and they contain all genes in the set $\{f_1, \cdots, f_q, g_1, \cdots, g_{n-1}\}$. Let S_i'' and $S_i^{*''}$ be the substrings in G'' and H'' with respect to S_i and S_i^* in $\mathcal{G}$ and $\mathcal{H}$ respectively. This implies that S_i'' and $S_i^{*''}$ are the common subsequence of either $f_{i_1} \cdots f_{i_u}$ or $f_{j_1} \cdots f_{j_v}$, because $S_i = f_{i_1} \cdots f_{i_u} f_{j_1} \cdots f_{j_v}$ and $S_i^* = f_{j_1} \cdots f_{j_v} f_{i_1} \cdots f_{i_u}$. If S_i'' is empty then we can assign a value to x_i arbitrarily. If S_i'' is not empty and it is a subsequence of $f_{i_1} \cdots f_{i_u}$ then we assign $x_i = 1$. If S_i'' is not empty and it is a subsequence of $f_{j_1} \cdots f_{j_v}$ then we assign $x_i = 0$. As each $f_t \in \{f_1, \cdots, f_q\}$ occurs in G'', H'' once, it must occur in a non-empty S_i''. It is easy to see that F is true by the assignments to those variables $x_1, \cdots, x_n$.

The reduction takes linear (in the length of F, $|F|$) time. A sub-formula f_j with y literals appears in $\mathcal{G}$ ($\mathcal{H}$) exactly y times and there are $n - 1$ additional peg genes in $\mathcal{G}$ ($\mathcal{H}$). Therefore, the length of $\mathcal{G}$ and $\mathcal{H}$ are both bounded by $c|F|$ for some constant $c > 1$.   □

The above theorem implies that the Exemplar Breakpoint Distance problem does not admit any approximation unless P=NP—if such a polynomial-time approximation existed then it would be able to decide whether $\mathcal{G}$ and $\mathcal{H}$ have zero breakpoint distance in polynomial time hence contradicting Theorem 3.5. If we parameterize the ZBD problem to kZBD, which is to decide if two k-repetitive sequences have zero break point distance, then the above theorem can be further strengthened as follows.

Theorem 3. *Deciding if two 3-repetitive genomes have zero breakpoint distance is NP-complete.*

Proof. Using the same reduction, a 3SAT sub-formula f_j with three literals appears in $\mathcal{G}$ ($\mathcal{H}$) exactly three times. Therefore, we can reduce 3SAT to 3ZBD in linear time.   □

We now have the following corollary.

Corollary 4. *Unless P=NP, the Exemplar Reversal Distance Problem cannot be approximated even if both $\mathcal{G}$ and $\mathcal{H}$ are 3-repetitive.*

4 Weak Inapproximability Bounds

In this section, we try to generalize Theorem 3.5 to obtain some inapproximability bound under a weak approximation model. Let $opt(\mathcal{G}, \mathcal{H})$ be the optimal exemplar breakpoint distance between $\mathcal{G}$ and $\mathcal{H}$. (We also use $d(X, Y)$ to denote the minimum breakpoint distance between two genomes X and Y, where X and Y do not have to be exemplar.) We obtain the following inapproximability bounds under a much weaker model of approximation.

Theorem 4. *Let $\epsilon > 0$ and $g(x) : N \to N$ be a function computable in polynomial time. If there is a polynomial time algorithm such that given $\mathcal{G}$ and $\mathcal{H}$ of length at most m it can return exemplar genomes G and H satisfying $d(G, H) \le g(m)opt(\mathcal{G}, \mathcal{H}) + m^{1-\epsilon}$, then P=NP.*

Proof. Let f be a SAT formula. Let $G(f), H(f)$ be the sequences as constructed in Theorem 3.5 such that f is satisfiable if and only if $d(G(f), H(f)) = 0$.

Let u be the length of f. Then $|G(f)| = |H(f)| \le cu$ for some positive constant $c > 1$. Let x be a number such that $u^x > u^{(1+x)(1-\frac{\epsilon}{2})}$. Let $M = u^x$.

Let $\Sigma(S)$ be the alphabet of a sequence S. If Σ_i is a different set of letters with $|\Sigma_i| = |\Sigma(S)|$, we define $S(\Sigma_i)$ to be a new sequence obtained by replacing all letters in S, in one to one fashion, by those in Σ_i.

Let $\Sigma_1, \Sigma_2, \cdots, \Sigma_M$ be M disjoint sets of letters of size $|\Sigma(G(f))|$. Let $G_1 = G(f)(\Sigma_1), G_2 = G(f)(\Sigma_2), \cdots, G_M = G(f)(\Sigma_M)$ be the sequences derived from $G(f)$. Let $H_1 = H(f)(\Sigma_1), H_2 = G(f)(\Sigma_2), \cdots, H_M = G(f)(\Sigma_M)$ be the sequences derived from $H(f)$.

Define $\mathcal{G} = G_1 s_1 G_2 s_2 \cdots G_M s_M$ and $\mathcal{H} = H_1 s_1 H_2 s_2 \cdots H_M s_M$, where s_i is a peg gene appearing only once in $\mathcal{G}$ and $\mathcal{H}$. The total length of $\mathcal{G}, \mathcal{H}$ is bounded by $c(u+1)M \le 2cu^{x+1}$. Let m be the maximum length of $\mathcal{G}$ and $\mathcal{H}$, then $m \le c'u^{x+1}$ for some $c' > 2$.

Assume that some polynomial time algorithm $\mathcal{A}$ outputs G, H such that G is an exemplar genome of $\mathcal{G}$ and H is an exemplar genome of $\mathcal{H}$, and $d(G, H) \le g(m)d(\mathcal{G}, \mathcal{H}) + m^{1-\epsilon}$, we can then decide if f is satisfiable by checking whether $d(G, H) \le m^{1-\epsilon}$. If f is satisfiable, it is easy to see that $d(\mathcal{G}, \mathcal{H}) = 0$ then $d(G, H) \le m^{1-\epsilon}$. If f is not satisfiable, then from Theorem 3.5 $d(G_i, H_i) \ge 1$. As no letter is shared by G_i, G_j, we have $d(\mathcal{G}, \mathcal{H}) \ge M = u^x > u^{(1+x)(1-\frac{\epsilon}{2})} \ge (\frac{m}{c'})^{1-\frac{\epsilon}{2}} > m^{1-\frac{3}{4}\epsilon}$ when m is sufficiently large. Since G, H are exemplar genomes of $\mathcal{G}, \mathcal{H}$, $d(G, H) > m^{1-\frac{3}{4}\epsilon}$.   □

Corollary 5. *Let $\epsilon > 0$. If there is a polynomial time algorithm such that given $\mathcal{G}$ and $\mathcal{H}$ of length at most m it can return exemplar genomes G and H satisfying $d(G, H) \leq m^{1-\epsilon}[opt(\mathcal{G}, \mathcal{H}) + 1]$, then P=NP.*

This negative result shows that even under a much weaker model, it is not possible to obtain a good approximation unless P=NP. In next section, we will present a factor-$2(1+\log n)$ approximation for the One-Sided Exemplar Reversal Distance Problem in which one of the two genomes is a k-span genome. It is not surprising that this problem is also known to be NP-complete, in fact, it is NP-complete even when $k = 1$ [2].

5   A $2(1 + \log n)$−Approximation for the One-Sided Case

Given a *k-span* genome $\mathcal{G}_k$ and a general genome $\mathcal{H}$, each is a sequence containing $O(m)$ signed or unsigned genes (drawn from the n gene families and genes from the same family in $\mathcal{G}_k$ are at most k positions away and there are possibly any kind of repetitions in $\mathcal{H}$), the problem is to compute the minimum exemplar breakpoint distance between two exemplar genomes G, H (obtained by deleting redundant genes in $\mathcal{G}_k$ and $\mathcal{H}$). Let $\mathcal{G}_k = a_1 a_2 \cdots a_{n_1}, \mathcal{H} = b_1 b_2 \cdots b_{m_1}$. Throughout this section we assume that $k = O(\log n)$.

Let $opt(\mathcal{G}_k, \mathcal{H})$ be the size of the optimal solution of the above One-sided Exemplar Breakpoint Distance Problem.

Let $A = [a_i, a_{i+s_{p-1}}] \in \mathcal{G}_k$ and $B = [b_j, b_{j+t_{p-1}}] \in \mathcal{H}$. If a gene family, which is a multi-set of genes in $\mathcal{G}_k(\mathcal{H})$, all appear in A (B) then it is called a multi-set of *whole-family* genes in A (B). Example: Let $\mathcal{G}_3 = ga - fgedbedc - e$ and $\mathcal{H} = acefgac - fbebdach - g$. Consider the interval $I_G = a - fgedbed$ in $\mathcal{G}_3$ and the interval $I_H = gac - fbebdc$ in $\mathcal{H}$. The multi-set of whole-family genes in I_H is $\{\{b, b\}, \{d\}\}$.

Given $A = [a_i, a_{i+s_{p-1}}] \in \mathcal{G}_k$ and $B = [b_j, b_{j+t_{p-1}}] \in \mathcal{H}$, an interval $I = c_1 c_2 ... c_p$ or its signed reversal $-I$ is called a *Non-Breaking Interval* (NB-interval for short) if I contains no repetition of any gene, for each multi-set of whole-family genes in A and B one of them must appear in I, and I appears in $\mathcal{G}_k$ with $c_1 = a_i, c_2 = a_{i+s_1}, \cdots, c_p = a_{i+s_{p-1}}$ (or $c_1 = -a_{i+s_{p-1}}, c_2 = -a_{i+s_{p-2}}, \cdots, c_p = -a_i$) and in $\mathcal{H}$ with $c_1 = b_j, c_2 = b_{j+t_1}, \cdots, c_p = b_{j+t_{p-1}}$ (or $c_1 = -b_{j+t_{p-1}}, c_2 = -b_{j+t_{p-2}}, \cdots, c_p = -b_j$) for some $s_{p-1} > s_{p-2} > \cdots > s_1 > 0$ and some $t_{p-1} > t_{p-2} > \cdots > t_1 > 0$. The length p is called the *size* of I. Given $A = [a_i, a_{i+s_{p-1}}] \in \mathcal{G}_k$ and $B = [b_j, b_{j+t_{p-1}}] \in \mathcal{H}$, we are interested in computing a NB-interval of maximum size (length). Notice that a maximum NB-interval is very much a *longest constrained common subsequence* of A and B, it is related to but different from the recently studied *constrained longest common subsequence* [5]. From now on, we will only talk about maximum NB-intervals, which we will simply use NB-intervals if the context is clear.

Now let $A = g_1 g_2 \cdots g_N, B = h_1 h_2 \cdots h_M$ be strings on z identical gene families, and $g_1 = h_1, g_M = h_N$. We assume that both A, B are long enough, say, at least of length $20k$ (otherwise we can simply use a brute-force method). Let $W(A[i, j]), W(B[s, t])$ be the whole-family gene sets in $A[i, j]$ and $B[s, t]$

respectively. We show below a polynomial time dynamic programming algorithm to compute the NB-interval between strings A, B. Let $A[i, j] = P_a H P_b$, where $|P_a| = |P_b| = k$. Since A is a k-span genome, P_a, P_b have no common genes when $|H| \geq k$. Let H_a, H_b be exemplar genomes selected from P_a, P_b respectively. In the dynamic programming table, $table(i, j, H_a, H_b, s, t)$ stores a longest constrained common subsequence $H_a V H_b$ of $A[i, j]$ and $B[s, t]$ such that $W(A[i, j]), W(B[s, t])$ all appear in $H_a V H_b$ and there is no repetition of any gene in $H_a V H_b$.

Let $A[i, j] = P_a H P_b$, with $H = H_1 P_c P_d H_2$ and $|P_c| = |P_d| = k$. Assume that $A[i, j_1] = P_a H_1 P_c$ and $A[j_1, j] = P_d H_2 P_b$, we can merge $table(i, j_1, U_a, U_b, s, t_1)$ and $table(j_1+1, j, T_a, T_b, t_1+1, t)$ into $table(i, j, H_a, H_b, s, t)$—if $U_b T_a$ is exemplar and selected from $P_c P_d$ then all whole family genes in $P_c P_d$ must be in $U_b T_a$ and no gene is repeated in $U_b T_a$; moreover, among all such candidates we select the longest one as $U_b T_a$. So when j_1, t_1 is fixed this merge takes $O(k^2 + n) = O(n)$ time. As we need to try different combinations j_1 and t_1, the final content in $table(i, j, H_a, H_b, s, t)$, which should be the longest, can be computed in $O(n^3)$ time, provided that $table(i, j_1, U_a, U_b, s, t_1)$ and $table(j_1 + 1, j, T_a, T_b, t_1 + 1, t)$ are already available.

There are at most 2^k ways to select H_a from P_a (H_b from P_b). Therefore, this dynamic programming algorithm uses $O(2^{2k} n^5)$ space (there are $O(2^{2k} n^4)$ cells in the table, each could store a sequence of length $O(n)$) and it takes $O(2^{2k} n^7)$ time to compute the (maximum) NB-interval between A and B, which is stored in $table(1, N, -, -, 1, M)$. Finally, notice that each signed/unsigned gene in $\mathcal{G}_k$ or $\mathcal{H}$ is a degenerate (maximum) NB-interval of length one.

This dynamic programming algorithm will be used as a subroutine in our final approximation for the One-sided Exemplar Breakpoint Distance Problem. Now consider the problem of covering all genes in $\mathcal{G}_k$ and $\mathcal{H}$ using the minimum number of (disjoint) NB-intervals. Let $C^*(\mathcal{G}_k, \mathcal{H})$ be the size of the optimal solution for this covering problem.

Lemma 1. $C^*(\mathcal{G}_k, \mathcal{H}) \leq opt(\mathcal{G}_k, \mathcal{H}) + 1$.

Proof. Trivial, as each breakpoint in the exemplar genomes G, H can only occur between two NB-intervals. ☐

We now show how to obtain a factor $2(1 + \log n)$ approximation for $C^*(\mathcal{G}_k, \mathcal{H})$ by converting it to a set-cover problem $(X, \mathcal{F})$. In this case, each (degenerate and non-degenerate) NB-interval is a set $S \in \mathcal{F}$. X contains all of the n genes. The problem is to compute the minimum number of (disjoint) NB-intervals which cover all the genes. The algorithm follows the greedy method [7, 13, 14].

(1) Start with $\mathcal{G}_k, \mathcal{H}$. Enumerate all pairs of intervals $A = [a_i, a_{i+s}]$ and $B = [b_j, b_{j+t}]$ with $a_i = b_j, a_{i+s} = b_{j+t}$. For each such pair (A, B), use the above dynamic programming algorithm to compute a maximum length NB-interval.

(2) Among all the maximum NB-intervals computed at Step (1), select one with the maximum size, I, and put it in the approximation solution.

(3) Delete all the (signed/unsigned) genes in I to have two updated versions of $\mathcal{G}_k, \mathcal{H}$. Repeat Step (1)-(2) until all the genes are covered.

Let $App(\mathcal{G}_k, \mathcal{H})$ be the number of the NB-intervals obtained in the above approximation solution. Following [7, 13, 14], we have the following lemma.

Lemma 2. $App(\mathcal{G}_k, \mathcal{H}) \leq (1 + \log n) \cdot C^*(\mathcal{G}_k, \mathcal{H})$.

We have the following theorem.

Theorem 5. $App(\mathcal{G}_k, \mathcal{H}) \leq 2(1 + \log n) \cdot opt(\mathcal{G}_k, \mathcal{H})$.

Proof. By Lemmas 5.1 and 5.2, $App(\mathcal{G}_k, \mathcal{H}) \leq (1 + \log n) \cdot opt(\mathcal{G}_k, \mathcal{H}) + \log n + 1$. When $opt(\mathcal{G}_k, \mathcal{H}) > 0$, $App(\mathcal{G}_k, \mathcal{H}) \leq (1 + \log n) \cdot opt(\mathcal{G}_k, \mathcal{H}) + \log n + 1 \leq (1 + \log n) \cdot opt(\mathcal{G}_k, \mathcal{H}) + (1 + \log n) \cdot opt(\mathcal{G}_k, \mathcal{H}) = 2(1 + \log n) \cdot opt(\mathcal{G}_k, \mathcal{H})$. When $opt(\mathcal{G}_k, \mathcal{H}) = 0$, which can be identified by the above dynamic programming algorithm, we can ignore using this approximation algorithm.   □

The running time of the above approximation algorithm is as follows: There could be $O(n)$ rounds in the greedy selection process. At each round we could have enumerated $O(n^2)$ intervals and each call to the dynamic programming procedure takes $O(2^{2k}n^7)$ time. Therefore, the overall running time of the approximation algorithm is $O(2^{2k}n^{10})$. The approximation algorithm uses $O(2^{2k}n^5)$ space.

We comment that for this problem, when $k = 1$, the above factor-$2(1 + \log n)$ approximation can be greatly simplified. The complex dynamic programming method can be replaced by a Longest Common Subsequence computation [6] and the algorithm runs in $O(n^5)$ time and $O(n^2)$ space, which is clearly much more efficient.

6   Concluding Remarks

We present the first set of inapproximability/approximation results for the Exemplar Breakpoint Distance Problem. Although it seems that the general problem does not admit any approximation, for a special one-sided case, decent approximation does exist. This also partially conforms with the real-life dataset that repetitions of genes are typically pegged and not very far away [17]. It would be interesting to study some meaningful special cases. For example, can be obtain a good approximation when $\mathcal{G}$ is 2-repetitive and $\mathcal{H}$ is a 3-span genome?

References

1. V. Bafna and P. Pevzner, Sorting by reversals: Genome rearrangements in plant organelles and evolutionary history of X chromosome, *Mol. Bio. Evol.*, 12:239-246, 1995.
2. D. Bryant. The complexity of calculating exemplar distances. *In D. Sankoff and J. Nadeau, editors, Comparative Genomics: Empirical and Analytical Approaches to Gene Order Dynamics, Map Alignment, and the Evolution of Gene Families*, pp. 207-212. Kluwer Acad. Pub., 2000.

3. G. Blin and R. Rizzi. Conserved interval distance computation between non-trivial genomes. *Proc. 11th Intl. Ann. Comput. and Combinatorics (COCOON'05)*, LNCS 3595, pp. 22-31, 2005.
4. A. Bergeron and J. Stoye. On the similarity of sets of permutations and its applications to genome comparison. *Proc. 9th Intl. Ann. Comput. and Combinatorics (COCOON'03)*, LNCS 2697, pp. 68-79, 2003.
5. S. Bereg and B. Zhu. RNA multiple structural alignment with longest common subsequences. *Proc. 11th Intl. Ann. Comput. and Combinatorics (COCOON'05)*, LNCS 3595, pp. 32-41, 2005.
6. T. Cormen, C. Leiserson, R. Rivest, C. Stein. *Introduction to Algorithms*, second edition, MIT Press, 2001.
7. V. Chvátal, "A greedy heuristic for the set-covering problem," *Math. Oper. Res.*, vol. 4, 1979, pp. 233-235.
8. T. Cormen, C. Leiserson and R. Rivest, *Introduction to Algorithms*, The MIT Press, 1990.
9. I. Dur and S. Safra. The importance of being biased. In *Proc. 34th ACM Symp. on Theory Comput. (STOC'02)*, pages 33-42, 2002.
10. M. Garey and D. Johnson. *Computers and Intractability: A Guide to the Theory of NP-completeness*. Freeman, San Francisco, CA, 1979.
11. S. Hannenhalli and P. Pevzner. Transforming cabbage into turnip: polynomial algorithm for sorting signed permutations by reversals. *J. ACM*, 46(1):1-27, 1999.
12. O. Gascuel, editor. *Mathematics of Evolution and Phylogeny.*Oxford University Press, 2004.
13. D. Johnson, "Approximation algorithms for combinatorial problems," *J. Comput. System Sci.*, vol. 9, 1974, pp. 256-278.
14. L. Lovász, "On the ratio of optimal integral and fractional covers," *Discrete Math.*, vol. 13, 1975, pp. 383-390.
15. M. Marron, K. Swenson and B. Moret. Genomic distances under deletions and insertions. *Theoretical Computer Science*, 325(3):347-360, 2004.
16. C. Makaroff and J. Palmer. Mitochondrial DNA rearrangements and transcriptional alternatives in the male sterile cytoplasm of Ogura radish. *Mol. Cell. Biol.*, 8:1474-1480, 1988.
17. C.T. Nguyen, Y.C. Tay and L. Zhang. Divide-and-conquer approach for the exemplar breakpoint distance. *Bioinformatics*, 21(10):2171-2176, 2005.
18. J. Palmer and L. Herbon. Plant mitochondrial DNA evolves rapidly in structure, but slowly in sequence. *J. Mol. Evolut.*, 27:87-97, 1988.
19. R. Raz and S. Safra. A sub-constant error-probability low-degree test, and sub-constant error-probability PCP characterization of NP. In *Proc. 29th ACM Symp. on Theory Comput. (STOC'97)*, pages 475-484, 1997.
20. D. Sankoff. Genome rearrangement with gene families. *Bioinformatics*, 16(11):909-917, 1999.
21. A. Sturtevant and T. Dobzhansky. Inversions in the third chromosome of wild races of *drosophila pseudoobscura*, and their use in the study of the history of the species. *Proc. Nat. Acad. Sci. USA*, 22:448-450, 1936.
22. E. Tannier and M-F. Sagot. Sorting by reversals in subquadratic time. *Proc. 15th Symp. Combinatorial Pattern Matching (CPM'04), Istanbul, Turkey, Pages 1-13, July, 2004 (LNCS series, 3109)*.
23. G. Watterson, W. Ewens, T. Hall and A. Morgan. The chromosome inversion problem. *J. Theoretical Biology*, 99:1-7, 1982.

Computing the λ-Seeds of a String

Qing Guo[1], Hui Zhang[2], and Costas S. Iliopoulos[2]

[1] Department of Computer Science and Engineering, Zhejiang University, Hangzhou, Zhejiang 310027, China
guoqing@tiansign.com
[2] Department of Computer Science, King's College London, Strand, London WC2R 2LS, England
{hui, csi}@dcs.kcl.ac.uk

Abstract. We study the λ-seed problem of a string in this paper. Given a string x of length n and an integer λ, the λ-seed problem is to find all the sets of λ substrings of x that cover a superstring of x, assuming that each element of the set is of equal length. We present an efficient algorithm that can compute all the λ-seeds of x in $O(n^2)$ time.

1 Introduction

The most common regularities in strings, notably periods, covers and seeds, correspond to those repetitive structures of strings. Among them, a substring w is a *period* of a nonempty string x if x is a prefix of a string constructed by concatenations of w, which grasps the typical and classical regularity. The other two related categories are generalized by periods in the way that superpositions as well as concatenations are considered, whereas only concatenations are allowed for periods.

A substring w of x is called a *cover* of x if and only if x can be constructed by concatenations and superpositions of w, so that every position of x lies within some occurrence of w in x. Apostilico, Farach and Iliopoulos first introduced the notion of covers in [1], where a linear-time algorithm for computing the shortest cover of x was presented. A series of linear-time algorithms that improved on this result then followed: Breslauer [3] described a linear-time online algorithm for computing the shortest cover of every prefix of x. Moore and Smyth [13] gave a solution to find all the covers of x, and recently Li and Smyth [12] solved both the all-covers problem and the shortest-cover problem by constructing the cover array for the string. As to parallel computation, Iliopoulos and Park [10] gave a work-time optimal $O(\log \log n)$ algorithm for finding all the covers of x.

Extending the idea of covers in the sense that a set of substrings of x are considered instead of a single string, Iliopoulos and Smyth [11] introduced the idea of k-covers and studied the minimum k-cover problem. This problem is to compute the minimum set of substrings of x each of length k that covers x, which was then proved to be NP-complete in [6]. We [14] proposed an associated problem, that is, the λ-cover problem to find all the sets of λ substrings each of

S.-W. Cheng and C.K. Poon (Eds.): AAIM 2006, LNCS 4041, pp. 303–313, 2006.
© Springer-Verlag Berlin Heidelberg 2006

equal length that cover x, then gave a general algorithm to solve this problem in $O(n^2)$ time.

A *seed* can be thought of as a generalized cover, since it aims at a substring of x that covers a superstring of x. Iliopoulos, Moore and Park [8] introduced this notion for the first time and gave an $O(n \log n)$-time algorithm for computing all the seeds of x. A parallel algorithm that requires $O(\log n)$ time and $O(n \log n)$ work was presented [5] for the same problem.

Inspired by the λ-cover problem, we now intend to study the λ-seed problem. Formally speaking, given a string x of length n and an integer λ, the λ-seed problem is to find all the sets $W = \{w_1, w_2, \ldots, w_\lambda\}$ of substrings of x such that:

(1) $|w_1| = |w_2| = \cdots = |w_\lambda|$;
(2) There exists a superstring $y = uxv$ of x with $|u|, |v| < |w_i|$ such that y can be constructed by concatenating or overlapping copies of the strings w_1, w_2, $\ldots$, w_λ.

If $\lambda = 1$, this problem is actually to compute all the seeds of x, which has been solved in $O(nlogn)$ time [8]. Thus, this paper simply considers the case $\lambda > 1$ and focuses on solving the λ-seed problem of a string. The motivation comes from the efforts to analyze a DNA sequence in the hybridization approach, where we require a set of sample substrings of fixed length to determine the base pairs.

This paper is organized as follows. Section 2 gives preliminaries used throughout the paper. In Section 3, we present an optimal algorithm that finds all the λ-seeds of a string in $O(n^2)$ time. Finally we conclude and discuss our future research interests in Section 4.

2   Preliminaries

Let Σ be a finite alphabet consisting of a set of characters. A *string* over the given alphabet is a sequence of zero or more symbols of the alphabet. A string x of length n is represented by $x[1..n] = x[1]x[2] \cdots x[n]$, where $x[i] \in \Sigma$ for $1 \le i \le n$. The *empty string* is the empty sequence (of zero length) denoted by ε. The set of all nonempty strings over Σ is denoted by Σ^+, then the set of all strings over the alphabet Σ including the empty string is denoted by $\Sigma^* = \Sigma^+ \cup \{\varepsilon\}$.

A string w is a *substring* of x if $x = uwv$ for $u, v \in \Sigma^*$. Equivalently, x is a *superstring* or an *extension* of w; uw is a *left extension* of w; wv is a *right extension* of w. A substring of length p is called a *p-substring* for short. For a nonempty substring $w = x[i..j]$, we say that w occurs at position i and i is an occurrence of w in x. A string w is a *prefix* of x if $x = wu$ for $u \in \Sigma^*$. Similarly, w is a *suffix* of x if $w = uw$ for $u \in \Sigma^*$.

The string xy is a *concatenation* of two strings x and y. The concatenation of k copies of x is denoted by x^k. For two strings $x = x[1..n]$ and $y = y[1..m]$ such that $x[n - i + 1..n] = y[1..i]$ for some $i \ge 1$, the string $x[1..n]y[i + 1..m]$ is a *superposition* of x and y with i overlaps, we say that x and y are *overlapping*.

A *repeat* of x is a substring $r \in \Sigma^+$ such that $x = u_1 r u_2 = v_1 r v_2$ with $u_1, u_2, v_1, v_2 \in \Sigma^*$ and $|u_1| \neq |v_1|$. A substring w is said to be a *cover* of x if x can be constructed from concatenated and/or overlapped copies of w. In other words, w *covers* x. A substring w is called a *seed* of x if there exists an extension of x (possibly x itself) that can be constructed by concatenations and superpositions of w. For example, the string $x = ababab a$ has a cover aba and a seed bab.

Definition 1. *Let* w_1, w_2, ..., w_k *be the substrings of* x, *we say that a set* $W = \{w_1, w_2, \ldots, w_k\}$ *is a combination of* k *substrings drawn from* x, *called for short a* k-*combination of* x. *And,* W *is called a* (k, p)-*combination of* x *if each* w_i *(*$1 \leq i \leq k$*) is of length* p. *We also say that, a* k-*combination* W *occurs at position* j *if* j *is an occurrence of any* w_i *in* x.

Definition 2. *Given a string* x *of length* n *and a certain integer* p, *we say that a* (λ, p)-*combination* $W = \{w_1, w_2, \ldots, w_\lambda\}$ *is a* (λ, p)-*cover of* x *if and only if every position of* x *lies within an occurrence of some* w_i *(*$1 \leq i \leq \lambda$*), and a* (λ, p)-*seed of* x *if and only if* W *is a* (λ, p)-*cover of a superstring* $y = uxv$ *of* x *with* $|u|, |v| < p$.

The λ-seed problem is actually to determine all the (λ, p)-seeds of x. To avoid triviality, we assume that $1 < p < n/\lambda$. The reason for this assumption is because we can trivially find a set of (λ, p)-covers of x if $p \geq n/\lambda$, then easily obtain the (λ, p)-seeds of x. It is evident that any (λ, p)-combination serves as a candidate for both (λ, p)-covers and (λ, p)-seeds. Consequently, we need to record all the occurrences of the (λ, p)-combinations of x, then check which ones are true (λ, p)-seeds.

We use Crochemore's partitioning [7] to find all exact repeats of a string. The main idea of this well-known algorithm is based on the following definition of the equivalence relation over the positions of the string:

Definition 3. *Given a string* $x[1..n] \in \Sigma^*$, *two positions* $i, j \in \{1, \ldots, n-p+1\}$ *of* x, *then* $(i, j) \in E_p$ *iff* $x[i...i + p - 1] = x[j...j + p - 1]$, *noted* $iE_p j$.

That is, two positions i and j of x belong to the same one E_p-class when two p-substrings of x starting at i and j are identical. Clearly each class of E_p of cardinality not less than two records the occurrences of a repeat of length p. Since E_{p+1} is a refinement of E_p, the algorithm first computes E_1, then iteratively builds E_2, E_3, ... until all classes are singleton. Utilizing the "smaller-half trick" [2] [7], the partitioning takes $O(n \log n)$ time to detect the sets of the starting positions of all the repeating substrings of x. More details of this algorithm can be found in the original paper [7].

3   Computing the λ-Seeds of x

Trivially, a (λ, p)-cover is always a (λ, p)-seed. But more commonly, the first (resp. last) appearance of a (λ, p)-seed in x might be incomplete, shown as the structure of a suffix (resp. prefix) of an element in the set.

Definition 4. *Given a* (λ, p)-*combination* $W = \{w_1, w_2, \ldots, w_\lambda\}$ *of* x, *we say that* W *is a candidate* (λ, p)-*seed if there exists a substring* x' *of* $x = ux'v$ *such that* W *covers* x' *and* $|u|, |v| < p$. *For maximal such* x', *we call* u (*resp.* v) *the head* (*resp. tail*) *of* x *with respect to* W.

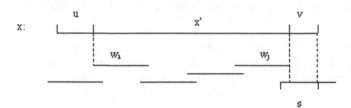

Fig. 1. A candidate (λ, p)-seed of x

Fig. 1 shows a candidate (λ, p)-seed W of the given string x. Assume that the first and the last appearance of W in x' is w_i and w_j respectively($1 \leq i, j \leq \lambda$), in order for a candidate (λ, p)-seed to be a true one, it must suffice to cover a left extension of the string uw_i as well as a right extension of the string w_jv. If it does, a (λ, p)-seed of x can be reported. Relying on this fact, we give the following algorithm for computing the λ-seeds of x.

Step 1. *For a certain* p $(1 < p < n/\lambda)$, *list all the* (λ, p)-*combinations and record their occurrences in* x.

This step is the same with Step 1 of our general algorithm for solving the λ-cover problem [14], since a (λ, p)-combination can serve as a candidate for both (λ, p)-covers and (λ, p)-seeds as we mentioned earlier. We now simply give a brief outline of this algorithm.

The algorithm initially finds valid (λ, p)-combinations for induction in the base case, which is the first p such that the number of distinct substrings of x is greater than λ. Then it works iteratively to deduce all the (λ, p)-combinations and their occurrences in x according to the $(\lambda, p-1)$-combinations.

The iterative deduction rests on the construction of the Equivalence Class Tree (ECT), which expresses the relationship between each E_{p-1}-class and corresponding E_p-classes. Let $\{C_1, \ldots, C_k\}$ be the E_{p-1}-classes, we can create the ECT as follows: The root of the ECT has label 0. There are k nodes of depth $p - 1$, each of which is denoted by the index of $C_i(1 \leq i \leq k)$. The sons of the node corresponding to C_i are the indices of the E_p-classes partitioned by C_i by scanning one more character to the right. Note that, C_i is a union of these E_p-classes. For the convenience of explanation, we label every node by the substring itself instead of the index of its equivalence class.

To achieve an efficient induction, we build a position string to store all the occurrences of a k-combination in x in the following way:

Definition 5. *Let* C_{w_i} *be the equivalence class corresponding to a substring* w_i *of* x, $C_{\{w_1, w_2, \ldots, w_k\}} = \{i_1, i_2, \ldots, i_m\}$ *for a* k-*combination* $W = \{w_1, w_2, \ldots, w_k\}$

be the occurrences of W in x, then a position string $L_{\{w_1,w_2,\ldots,w_k\}}$ for W is a string of length n such that for every position i in L:

$$L_{\{w_1,w_2,\ldots,w_k\}}[i] = \begin{cases} i_1, & i = i_1; \\ i_t - i_{t-1}, & i = i_t \ (2 \leq t \leq m); \\ 0, & \text{otherwise.} \end{cases}$$

The position string L is a string composed of numbers, which identifies each position where a k-combination of x appears by the distance between its current and previous occurrence, and other positions by 0.

We consider the iterative computation of the (λ, p)-combinations and their occurrences in x from the $(\lambda, p - 1)$-combinations. Given a certain $(\lambda, p - 1)$-combination $S = \{s_1, s_2, \ldots, s_\lambda\}$, it might produce a series of (λ, p)-combinations as a result of the partitioning of each E_{p-1}-class C_{s_i} ($1 \leq i \leq \lambda$) according to the ECT. Let the number of sons of s_i in the ECT be r_i for $1 \leq i \leq \lambda$, then the relevant p-substrings can be denoted by $w_i^1, w_i^2, \ldots, w_i^{r_i}$, clearly $r_i \leq |\Sigma|$. From $i = 1$ to $i = \lambda$, we successively replace s_i by l_i among these r_i p-substrings respectively. Every current combination obtained after s_i being replaced is saved to be further updated, denoted by W_i. In other words, we first compute $\{W_1\}$, then update each W_1 to obtain $\{W_2\}$, etc., until $\{W_\lambda\}$ is iteratively created. Obviously, $\{W_\lambda\}$ consists of all the (λ, p)-combinations associated with the given $(\lambda, p - 1)$-combination S.

The induction of $\{W_j\}$ ($2 \leq j \leq \lambda$) from $\{W_{j-1}\}$ can be stated as follows: Consider any W_{j-1},

1. if W_{j-1} contains s_j: Keep all those members of W_{j-1} located before s_j, namely, $l_1 + l_2 + \ldots + l_{j-1}$ p-substrings produced respectively by $s_1, s_2, \ldots, s_{j-1}$ unchanged. The following cases need to be discussed by checking s_j.
   (a) s_j is not the last element of W_{j-1}: replace s_j by l_j among r_j sons of s_j according to ECT, then remove $l_j - 1$ from the remaining s_i's ($i > j$) in the case of $l_j \neq 0$; or delete s_j and reserve all the remaining s_i's ($i > j$) in the case of $l_j = 0$, which leads to one W_j. Let $\lambda' = \lambda - (l_1 + l_2 + \ldots + l_{j-1})$.
      i. $r_j < \lambda'$: As $0 \leq l_j \leq r_j$ and $r_j \leq |\Sigma|$, in this case there are at most $\sum_{l_j=1}^{\min(|\Sigma|,\lambda'-1)} C_{\min(|\Sigma|,\lambda'-1)}^{l_j} C_{\lambda'-1}^{\lambda'-l_j} + 1$ W_j's, which is a constant dependent only on λ' and $|\Sigma|$, or on λ and $|\Sigma|$ since each l_i ($1 \leq i \leq j - 1$) reckons on them.
      ii. $r_j \geq \lambda'$: In this case, $0 \leq l_j \leq \lambda'$, $r_j \leq |\Sigma|$. The cardinality of $\{W_j\}$ is at most $\sum_{l_j=1}^{\lambda'} C_{|\Sigma|}^{l_j} C_{\lambda'-1}^{\lambda'-l_j} + 1$, a constant independent of p.
   (b) s_j is the last element of W_{j-1}: Replace s_j with exactly λ' among r_j p-substrings partitioned by s_j in the case of $r_j \geq \lambda'$; or delete this W_{j-1} otherwise.
2. if W_{j-1} does not contain s_j: $\{W_j\} \leftarrow W_{j-1}$.

The position string for any W_λ can also be computed by iteratively updating the position string for S as shown in the procedure UPDATE. We create a doubly linked list to store all the nonzero-value positions i of a position string, where

$i.left$ and $i.right$ points to the previous and the next one respectively. Initially, for all $2 \leq i \leq n - 1$, $i.left = i - 1$, $i.right = i + 1$; for $i = 1$, $i.left =$NULL, $i.right = i + 1$; for $i = n$, $i.left = i - 1$, $i.right =$NULL. The list for a position string L_{W_λ} can be iteratively updated along with the induction of W_λ from S (see while loop). Observe that the occurrences of W_λ in x can be viewed as removing all the occurrences of w_i^j's $(1 \leq i \leq \lambda, 1 \leq j \leq r_i)$ that are not included in W_λ from those of S in x. When an occurrence i' needs to be removed, the distance between adjacent nonzero-value positions, etc. $i'.left$ and $i'.right$, should be correspondingly updated (line 2-6 of while loop). The list is as well updated by directly linking $i'.left$ with $i'.right$ (line 10-11 of while loop). After all the "removed" positions are examined, we get the position string L_{cur} for W_λ.

Step 2. *Filter all the (λ, p)-combinations to determine candidate (λ, p)-seeds.*

We define three variables to represent the maximal differences between adjacent occurrences, the first and the last occurrence of a λ-combination in x, denoted by MAX-GAP, pos_0, pos_t respectively.

Initially, MAX-GAP=1, $pos_0 = 1$, $pos_t = n$. The values of these three variables of W_λ can also be iteratively calculated along with the induction of W_λ from S, as shown in the procedure UPDATE. Suppose that MAX-GAP, pos_0, pos_t for the $(\lambda, p - 1)$-combination $S = \{s_1, s_2, \ldots, s_\lambda\}$ are g_0, e_0 and e_t respectively. As we mentioned earlier, when a position i' is removed, the distance between adjacent nonzero-value positions are correspondingly updated (line 2-6 of while loop), then MAX-GAP is maintained as the larger one between this updated distance and the current MAX-GAP (line 7 of while loop). Similarly, if the current pos_0 (resp. pos_t) is removed, then the next (resp. previous) nonzero-value position $pos_0.right$ (resp. $pos_t.left$) is updated as the new one (line 8-9 of while loop).

procedure UPDATE
  REMOVE $\leftarrow \emptyset$;
for $i \leftarrow 1$ **to** λ **do**
  for $j \leftarrow 1$ **to** r_i **do**
    if $w_i^j \notin W_\lambda$ **then** REMOVE $\leftarrow$ every element of $C_{w_i^j}$;
  $L_{cur} \leftarrow L_S$;
  $L_{cur}[n - p + 2] \leftarrow 0$;
  MAX-GAP $\leftarrow g_0$;
  $pos_0 \leftarrow e_0$;
  $pos_t \leftarrow e_t$;
while REMOVE not empty **do**
  $pos\_remove \leftarrow$ extract an element from REMOVE;
  $temp \leftarrow L_{cur}[pos\_remove]$;
  $pl \leftarrow pos\_remove.left$;
  $pr \leftarrow pos\_remove.right$;
  $L_{cur}[pos\_remove] \leftarrow 0$;

$L_{cur}[pr] \leftarrow L_{cur}[pr] + temp;$

MAX-GAP $\leftarrow \max(\text{MAX-GAP}, L_{cur}[pr]);$

if $pos\_remove = pos_0$ **then** $pos_0 \leftarrow pr;$

if $pos\_remove = pos_t$ **then** $pos_t \leftarrow pl;$

$pr.left \leftarrow pl;$

$pl.right \leftarrow pr;$

end do

end

Clearly, when the position string for a certain (λ, p)-combination W_λ is obtained, the corresponding values of MAX-GAP, pos_0, pos_t are also acquired. By Definition 4, any W_λ with MAX-GAP$> p$ or the length of the head or tail with respect to W_λ is larger than or equal to p should be eliminated. As $|u| = pos_0 - 1$, $|v| = n - pos_t + 1 - p$, we follow that any (λ, p)-combination with $p \geq \max(pos_0, \lceil (n - pos_t + 2)/2 \rceil, \text{MAX-GAP})$ is qualified to become a candidate (λ, p)-seed.

For example, suppose that $x = abbabbaababbab$ and $\lambda = 3$, we can obtain the position string for a $(3, 3)$-combination $L_{\{bba,baa,aba\}} = 02003102003000$ following Step 1. Furthermore, we get: MAX-GAP$=3$, $pos_0 = 2$ and $pos_t = 11$. Therefore, We determine $\{bba, baa, aba\}$ to be a candidate $(3, 3)$-seed.

Step 3. *Test each candidate (λ, p)-seed if it is a true one.*

Consider a candidate (λ, p)-seed $W = \{w_1, w_2, \ldots, w_\lambda\}$, suppose that the first and the last appearance of W in x is the element w_i and w_j separately ($1 \leq i, j \leq \lambda$), then $uw_i = x[1..pos_0 + p - 1]$, $w_j v = x[pos_t..n]$. We use the ECT to test if a candidate (λ, p)-seed can cover a right extension of $w_j v$, while we construct the Reversed Equivalence Class Tree (RECT) to help checking whether a candidate can cover a left extension of uw_i.

The RECT is built similar to the ECT, with the major distinction that the refinement of E_p from E_{p-1} scans characters in the contrary direction, that is, from left to right in the ECT, backward in the RECT. Next we illustrate the construction of the RECT and the ECT.

Consider the example $x = abbabbaababbab$, both trees have label 0 as root. Let C_w be the equivalence class associated with a substring w of x. Since there are two E_1-classes: $C_a = \{1, 4, 7, 8, 10, 13\}$, $C_b = \{2, 3, 5, 6, 9, 11, 12, 14\}$, we can add two nodes a and b of depth 1 into both trees. At stage 2, scanning one more character to the right, class C_a is partitioned into two subclasses: $C_{ab} = \{1, 4, 8, 10, 13\}$, $C_{aa} = \{7\}$, both bearing node a as their parent in the ECT. If C_a reads one character to the left, two E_2-classes C_{ba} and C_{aa} are produced. We hence put nodes ba and aa into the RECT as a's sons. Note that, it does not imply that C_a itself is partitioned into C_{ba} and C_{aa}, but that we partition with respect to C_a the class C_b obtaining C_{ba} and partition with respect to C_a the class C_a obtaining C_{aa}. The same work can be done on C_b, getting two nodes bb and ba of depth 2 in the ECT and two sons bb and ab in the RECT. Sequentially, as soon as the length p increases and the new E_p-classes are acquired, the nodes at this level are created and added into these two trees. The construction of both

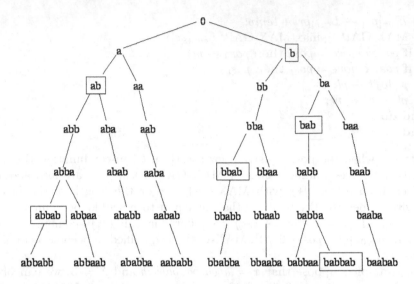

Fig. 2. The ECT of $x = abbabbaababbab$

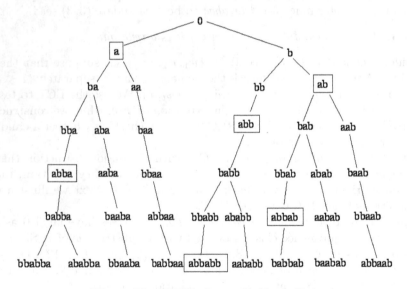

Fig. 3. The RECT of $x = abbabbaababbab$

the ECT and the RECT is finished once the computation of equivalence classes stops, that is, all classes are composed of only one element or $p = n/2$.

Following this method we build the ECT and the RECT of the given example, respectively shown in Fig. 2 and Fig. 3. We label those special substring in the ECT (resp. RECT) such that each of them aligns with the end (resp. start) of x, called *end-aligned* substrings (resp. *start-aligned* substrings) in the following context. In other words, the ending position of the last appearance of an end-aligned substring in x is position n, while the first occurrence of a start-aligned substring in x is position 1. For example, b, ab, bab, $bbab$, $abbab$, $babbab$ are all end-aligned, while the substrings a, ab, abb, $abba$, $abbab$, $abbabb$ are start-aligned, thus boxes are drawn to mark them respectively in the ECT and the RECT. Furthermore, it is easily noticed that:

Fact 1. *In any path from the leaf to the root of the ECT (resp. RECT), an internal node represents a prefix (resp. suffix) of the leaf, in order of decreasing length.*

This fact allows us to test a candidate (λ, p)-seed using the following approach, where two parts are involved.

Case 1. Check the tail v with respect to a candidate (λ, p)-seed W.

Recall that if the last appearance of $W = \{w_1, w_2, \ldots, w_\lambda\}$ in x is w_j ($1 \leq i, j \leq \lambda$), W should be able to cover the right extension of the string $w_j v$.

Consider any element w_k in W ($1 \leq k \leq \lambda$, $|w_k| = p$), we can observe from the ECT that if w_k is a marked node, then the tail v with respect to W is the empty string, which allows us to be relieved of the further testing. If w_k is unmarked, we can trace up in the branch where w_k is located to the nearest marked node s. By Fact 1, s identifies the longest prefix of w_k that is end-aligned.

(i) if $|s| < |v|$, $w_j v$ cannot be constituted by concatenations or superpositions of w_j and any prefix of w_k.
(ii) if $|s| \geq |v|$, $w_j v$ is concatenated or overlapped by w_j and s, $s = x[n-|s|+1..n]$ is the last incomplete appearance of w_k in x. We call such s a *valid prefix* of w_k, as shown in Fig. 1. Note that w_j and w_k might equal or differ.

Obviously, our task is to examine every element of W if it has a valid prefix. If no such substring exists, we can immediately claim that W is not a (λ, p)-seed. Once a valid prefix is found, we realize that W succeeds in covering the right extension of $w_j v$. Consider the candidate $(3, 3)$-seed $W = \{bba, baa, aba\}$ of the example string acquired in Step 2. Observe from Fig. 2 that bba has a valid prefix b such that $|b| = |v| = 1$. Actually, baa and aba both have valid prefixes as well. Therefore, this candidate wins its way to the further testing for the head covering. To efficiently implement the above tail testing, our algorithm redefines each node in the ECT as below:

Definition 6. *A node in the ECT is denoted by a triple: Node(w)=(w, P-Align, E-Align), where w stands for the equivalence class for a string w or simply w itself. P-Align points to the nearest ancestor of w that is marked in the ECT. E-Align is a boolean value for w, where E-Align=TRUE if w itself is end-aligned, and E-Align=FALSE otherwise.*

For instance, a substring $bbabb$ of $x = abbabbaababbab$ can be denoted in the ECT as: Node($bbabb$)=($bbabb$, $bbab$, FALSE). When a new E_p-class is computed, the corresponding node triple of depth p is updated and inserted into the ECT. For any node w of depth p, there are two possibilities regarding updating its values:

(i) if w is a marked node, that is, $pos_t + p - 1 = n$ where pos_t is the last occurrence of w in x, then Node(w).E-Align= TRUE, Node(w).P-Align= w.
(ii) if w is an unmarked node, then Node(w).E-Align=FALSE, Node(w).P-Align= Node(pw).P-Align where pw is the parent of w in the ECT.

Given this node triple, checking for a node w if it has a valid prefix is a direct one-step operation.

Case 2. Check the head u with respect to a candidate (λ, p)-seed W.

With a slight modification, checking the head is symmetric to the case of checking the tail. As we discussed before, W should be able to cover the left extension of the string uw_i.

We utilize the RECT to help checking the head. Recall that in the RECT, the start-aligned substrings are marked. We then examine every element w_k in W $(1 \leq k \leq \lambda)$ if it has a *valid suffix* s such that s is the closest marked ancestor of w_k in the RECT, and $|s| \geq |u|$. Once a valid suffix is found, we declare that W succeeds in checking the head.

Every node in the the RECT is also denoted by a triple: Node(w)=(w, P-Align, E-Align), with the distinction that E-Align stands for a boolean symbol identifying if w itself is start-aligned or not, and P-Align points to the nearest ancestor of w in the RECT that is marked. Above computation for the values of the nodes holds in this case.

Theorem 1. *The above algorithm can solve the λ-seed problem in $O(n^2)$ time.*

Proof. As we discussed earlier, the ECT and the RECT are both constructed along with the partitioning of equivalence classes, when the values of every node triple are updated simultaneously. The construction for each tree requires $O(n \log n)$ time.

At a certain stage p, recall that the cardinality of $\{W_j\}$ is dependent only on $|\Sigma|$ and λ. That is, the number of (λ, p)-combinations associated with a given $(\lambda, p-1)$-combination is a constant independent of p. The position string for any W_λ can be computed using the procedure UPDATE, which takes $O(1)$ time for each "removed" position. Taking all the (λ, p)-combinations into account, every position is examined for constant times in the procedure UPDATE. Therefore, Step 1 costs $O(n)$ time for all the n positions.

The values of MAX-GAP, pos_0, pos_t are updated along with updating position strings, which does not take more time than computing the position strings. Thus Step 2 requires $O(1)$ time to determine whether a (λ, p)-combination is a candidate (λ, p)-seed or not, then $O(n)$ time for all the $O(n)$ (λ, p)-combinations.

Step 3 simply executes some arithmetic operations, which runs in constant time for any element in the candidate $\{\lambda, p\}$-seed, and constant time for all the λ elements.

To sum up, our algorithm takes $O(n)$ time for computing the (λ, p)-seeds for a certain p. Since there are at most n/λ stages, the overall complexity of this algorithm is $O(n^2)$.

4   Conclusions

In this paper we introduce the λ-seed problem of a string and present an $O(n^2)$-time algorithm for solving this problem.

Our future direction is focused on a valid definition for the approximate λ-cover and the approximate λ-seed problem of a string and seeking solutions to these problems. We are also interested in the general λ-seed problem that does not limit the size of each element in the set to be equal.

References

1. Apostolico, A., Farach, M., Iliopoulos, C.S.: Optimal superprimitivity testing for strings. Information Processing Letters, Vol. 39. (1991) 17–20.
2. Apostolico, A., Preparata, F.P.: Optimal off-line detection of repetitions in a string. Theoretical Computer Science, Vol. 22, (1983) 297–315.
3. Breslauer, D.: An on-line string superprimitivity test. Information Processing Letters, Vol. 44, (1992) 345–347.
4. Breslauer, D.: Testing string superprimitivity in parallel. Information Processing Letters, Vol. 49, (1994) 235–241.
5. Ben-Amram, A.M., Berkman, O. Iliopoulos, C.S., Park, K.: The subtree max gap problem with application to parallel string covering. In: Proc. of 5th ACM-SIAM Symp. on Discrete Algorithmsocessing Letters, Arlington, VA, (1994) 501–510.
6. Cole, R., Iliopoulos, C.S., Mohamed, M., Smith, W.F., Yang, L.: Computing the minimum k-cover of a string. In: Proc. of the 2003 Prague Stringology Conference (PSC'03), (2003) 51–64.
7. Crochemore, M.: An Optimal Algorithm for Computing the Repetitions in a Word. Information Processing Letters, Vol. 12 (5), (1981) 244–250.
8. Iliopoulos, C.S., Moore, D.W.G., Park, K.: Covering a string. Algorithmica, Vol. 16, (1996) 288–297.
9. Iliopoulos, C.S., Park, K.: An optimal $O(\log \log n)$ time algorithm for parellel superprimitivity testing. J. of the Korean Information Science Society, Vol. 21(8), (1994) 1400–1404.
10. Iliopoulos, C.S., Park, K.: A work-time optimal algorithm for computing all string covers. Theoretical Computer Science, Vol. 2(164), (1996) 299–310.
11. Iliopoulos, C.S., Smith, W.F.: An on-line algorithm of computing a minimum set of k-covers of a string. Proc. of the Ninth Australian Workshop on Combinatorial Algorithms (AWOCA), (1998) 97–106.
12. Yin Li, Smyth, W.F.: Computing the cover array in linear time. Algorithmica, Vol. 32(1), (2002) 95–106.
13. Moore, D.W.G., Smyth, W.F.: A correction to "Computing the covers of a string in linear time". Information Processing Letters, Vol. 54, (1995) 101–103.
14. Zhang, H., Guo, Q., Iliopoulos, C.S.: The λ-cover problem of a string. Submitted to CIAC'06.

Subsequence Packing: Complexity, Approximation, and Application

Minghui Jiang

Department of Computer Science, Utah State University,
Logan, Utah 84322-4205, USA
mjiang@cc.usu.edu

Abstract. We study the subsequence packing problem: given a string T and a collection of strings $\{S_i\}$, find disjoint subsequences $\{T_i\}$ of T with maximum total length such that each T_i is a subsequence of S_i. We prove the NP-completeness of the decision problem, present the first non-trivial deterministic approximation, and show its applications to DNA sequencing verification and preemptive job shop scheduling with two machines.

1 Introduction

Given a string T and a collection of strings $C = \{S_1, S_2, \ldots, S_n\}$ over the same alphabet Σ, a *subsequence packing* (or *packing* for short), $P = \{T_1, T_2, \ldots, T_n\}$, is a collection of n disjoint subsequences of T such that, for $1 \leq i \leq n$, each T_i is also a subsequence of S_i (that is, T_i is a common subsequence of T and S_i). Given an instance (Σ, T, C), the *subsequence packing optimization problem* is to find a packing of maximum size, where the *size* of a packing is the total length of its disjoint subsequences: $\|P\| = \sum_{i=1}^{n} |T_i|$. Given an instance (Σ, T, C, x), where x is a positive integer, the *subsequence packing decision problem* is to decide whether there exists a packing of size x.

We can visualize the subsequence packing problem as a game in which each of the $n + 1$ strings is represented by a stack of characters with the first character at the top. In each step of the game, a player can either pop one of the $n + 1$ stacks or, if the top of the stack T is identical to the top of a stack S_i, pop the two stacks and declare a *match*. The game continues until all stacks are empty; the goal is to maximize the number of matches.

The subsequence packing problem is a direct generalization of the longest common subsequence problem: when the collection C contains only a single string S_1, the maximum subsequence packing is simply the longest common subsequence of two strings S_1 and T. The longest common subsequence problem is a classic string matching problem with numerous applications to bioinformatics and computational biology [6]. The subsequence packing problem itself has applications to both DNA sequencing verification and preemptive job shop scheduling.

1.1 Application to DNA Sequencing Verification

The most common strategy for DNA sequencing [1] today is the shotgun method. In this method, copies of a genome are first broken apart randomly into many

S.-W. Cheng and C.K. Poon (Eds.): AAIM 2006, LNCS 4041, pp. 314–323, 2006.
© Springer-Verlag Berlin Heidelberg 2006

short fragments. These fragments together cover the whole genome, that is, for each base pair location in the genome, there is at least one fragment that includes it. If, on average, each base pair location is included in k fragments, we say these fragments are a k-fold coverage of the genome. Each fragment of the genome is short enough that its sequence can be determined by experimental methods. Given the short sequences corresponding to the fragments, a computer then tries to arrange them into the whole genome, using the overlap between the sequences to deduce their order and to guide the assembly.

The shotgun method has been successfully applied in sequencing the genomes of Haemophilus influenzae (a bacterium, the first organism to have its complete genome sequence determined by the shotgun method), mouse, and even human [9]; it is the technique of choice for sequencing small genomes. However, the sequencing of highly repetitive genomes still poses an intriguing problem for scientists. If a genome is highly repetitive, fragments with the same sequence may come from different locations of the genome—to resolve these ambiguities is challenging and often requires a post-sequencing verification process to guarantee the accuracy of the deduced sequence. We next sketch a verification method based on subsequence packing.

We first introduce some notations. $S[i]$ denotes the i'th character of the string S; $S[i..j] = S[i]S[i+1]\ldots S[j]$ denotes the substring of S between indices i and j. $S_1 S_2$ denotes the concatenation of the two strings S_1 and S_2; S^k denotes the concatenation of k copies of the string S. The empty string is denoted by ϵ.

Given a collection of strings $C = \{S_1, \ldots, S_n\}$ that is a k-fold coverage of a longer string $S = S[1]\ldots S[\mu]$ of length μ, we compose a string $T = (S[1])^k \ldots (S[\mu])^k$ of length $k\mu$, in which each character of S is repeated k times. If C is a *perfect* k-fold coverage of S, that is, if each character of S is covered by exactly k strings in C, then the subsequence packing instance (Σ, T, C) has optimal packing size exactly $k\mu$. To illustrate our idea, we give the following example where $C = \{S_1, S_2, S_3, S_4\}$ is a perfect two-fold coverage of the string S:

```
S:   CGTACGTACGTA     T:   CCGGTTAACCGGTTAACCGGTTAA
S1:  CGTA                  C G T A
S2:          CGTA                             C G T A
S3:  CGTACGTA              C G T A C G T A
S4:      CGTACGTA                C G T A C G T A
```

The observation above implies that, if the sequence S deduced from C by the shotgun method is close to the real sequence S^*, then the optimal packing size for the constructed instance (Σ, T, C) must be close to $k\mu$. Therefore, an algorithm for the subsequence packing problem can be used to indirectly verify the accuracy of a DNA sequencing.

1.2 Application to Preemptive Job Shop Scheduling

We have shown a biological application of the subsequence packing problem. The earliest motivation of subsequence packing, however, is not biology but scheduling. In fact, subsequence packing over the binary alphabet is closely related to preemptive job shop scheduling with two machines. This relation was first discovered by Bansal et al. [3]. They designed a randomized approximation algorithm for subsequence packing to indirectly approximate the notoriously difficult job shop scheduling problem.

The problem of preemptive job shop scheduling with two machines, as noted by Anderson et al. [2], arises naturally in the scheduling of a computer system in which the CPU and the IO processing powers are modeled by two preemptible machines. In this scheduling problem, there is a set $J = \{J_i\}_{i=1..n}$ of n jobs that must be processed on two machines M_0 and M_1. Each job J_i consists of a sequence of operations of unit processing time that must be processed in order (the unit processing time implicitly models the preemption). Each operation of a job must be scheduled on a particular machine, either M_0 or M_1. At any time, a machine can process at most one operation, and a job may be processed by at most one machine. For a given schedule, let C_i be the completion time of the job J_i; the optimization goal is to find a schedule that minimizes the maximum completion time $C_{\max} = \max_i C_i$. The standard notation for this scheduling problem is $J2|pmtn|C_{\max}$, where $J2$ means job shop with two machines, $pmtn$ means preemption, and $C_{\max}$ refers to the optimization goal.

Let ℓ be the maximum length, that is, the maximum number of operations, of a job. Let L_i be the load of the machine M_i, and define $L = \max\{L_0, L_1\}$. The makespan $C_{\max}$ of any schedule is at least L. If $L_i < L$, then M_i has $L - L_i$ idle time units in which a dummy job of $L - L_i$ operations can be inserted without changing the optimal makespan. Without loss of generality, we assume that $L_0 = L_1 = L$. We have $C_{\max} \geq \max\{L, \ell\}$.

$J2|pmtn|C_{\max}$ was shown to be NP-hard [5] in 1978. The first non-trivial 1.5 approximation was presented by Sevastianov and Woeginger [8] in 1998. Sevastianov and Woeginger used $\max\{L, \ell\}$ as the lower bound for $C_{\max}$ in their approximation, and noted that any approximation ratio better than 1.5 would require a new lower bound. In 2000, an expected 1.5 approximation for the online version was achieved by Kimbrel and Saia [7]. In 2001, Anderson et al. [2] presented a linear-time (linear in the total number of operations) algorithm that finds a schedule with makespan $C_{\max} \leq L + \ell/2$. This algorithm has the same worst-case approximation ratio 1.5, but for certain cases it is superior to the previous 1.5 approximation; for example, if all n jobs are of equal length, then this algorithm achieves a $1 + \frac{1}{2n}$ approximation. Moreover, Anderson et al.'s result revealed the most difficult case of the problem: one long job is amidst many short jobs. An effective algorithm must schedule the short jobs in parallel to the long job in as many time units as possible.

This difficult case was recently modeled by Bansal et al. [3] as a subsequence packing problem over the binary alphabet: each short job J_i is represented by a binary string S_i in which a character 0 corresponds to an operation on M_0 and a character 1 to an operation on M_1; the long job is represented by a binary string T with the inverted 0/1 correspondence. The number of matches in a packing for the subsequence packing instance $(\{0, 1\}, T, \{S_i\})$ is equal to the number of time units in which an operation of a short job is scheduled in parallel to an operation of the long job. For this problem, Bansal et al. presented a randomized algorithm based on linear programming relaxation that achieves an expected $\frac{1}{1-1/e} \approx 1.58$ approximation. This randomized algorithm was then combined with Anderson et al.'s algorithm into an expected $1 + \frac{e}{3e-2} \approx 1.44$ approximation for $J2|pmtn|C_{\max}$. It is not obvious how their algorithm can be derandomized; the current best deterministic approximation ratio is still 1.5 to our knowledge. We found that, using Bansal et al.'s technique, we can derive the first deterministic bound for $J2|pmtn|C_{\max}$ that is less than 1.5.

In this paper, we further the study of the subsequence packing problem. First, we prove the NP-completeness of the decision problem. Next, we present the first non-trivial deterministic approximation for the optimization problem. Finally, we use Bansal et al.'s technique to derive an improved deterministic approximation for the problem of preemptive job shop scheduling with two machines.

2   The Hardness Result

Theorem 1. *The subsequence packing decision problem is NP-complete.*

Proof. The decision problem is clearly in NP: in polynomial time, we can assign each character of the string T non-deterministically to one of the $n + 1$ subsequences $\{T_i\}_{i=1..n+1}$, use the first n subsequences $\{T_i\}_{i=1..n}$ as the packing (T_{n+1} is the subsequence of unmatched characters in T), check whether each T_i is also a subsequence of S_i, for $1 \leq i \leq n$, and whether $\sum_{i=1}^{n} |T_i| \geq x$. We next prove that the decision problem is NP-hard.

Our proof is based on a reduction from the NP-hard bin packing problem [4]. Given a set of m positive integers $I = \{t_i\}_{i=1..m}$ (the items), a positive integer a (the capacity), and a positive integer b, the bin packing decision problem is to decide whether I can be partitioned into b disjoint subsets $\{I_1, \ldots, I_b\}$ (the bins) such that, for each I_j, $1 \leq j \leq b$, the sum of the integers in I_j does not exceed a. Bin packing is strongly NP-hard, that is, it remains NP-hard even if the integers in the input instance are encoded in unary.

Given a bin packing instance (I, a, b), we construct a subsequence packing instance (Σ, T, C, x) as follows: Σ is the binary alphabet $\{0, 1\}$; T is the string $(0^a 1^a)^b$; C contains a string $S_i = 0^{t_i} 1^{t_i}$ for each integer t_i in I, $1 \leq i \leq m$, and, in addition, $ab - \sum_{i=1}^{m} t_i$ copies of the string 01 for padding (we assume that $a \geq t_i$, for $1 \leq i \leq m$, and that $ab \geq \sum_{i=1}^{m} t_i$ to exclude the trivial negative cases); and $x = 2ab$. The parameter x is both the length of the string T and the total length of the $n = m + ab - \sum_{i=1}^{m} t_i$ strings in C; therefore, in any solution for the constructed subsequence packing instance (Σ, T, C, x), each character of the strings in C is matched to a distinct character in the string T, that is, the strings in C are interwoven into the string T. Assuming that the input bin packing instance is encoded in unary, the size of the constructed subsequence packing instance is clearly polynomial in the input size; our reduction is therefore polynomial. To complete the proof, we need to show that (I, a, b) is a positive instance of bin packing if and only if (Σ, T, C, x) is a positive instance of subsequence packing.

We first prove the "only if" direction. Let $\{I_j\}_{j=1..b}$ be a solution for the bin packing instance (I, a, b). We show how to pack all the strings in C into the string T. Intuitively, each of the b substrings of T in the form of $0^a 1^a$ simulates a bin to be filled. Let $C_j \subseteq C$ be the collection of strings that correspond to the items in I_j, and let $C'_j \subseteq C$ be $a - \sum_{t_i \in I_j} t_i$ copies of the padding string 01. The strings in $C_j + C'_j$ have exactly a 0s and a 1s. To pack these strings, all of the pattern $0^z 1^z$, into the j'th substring $0^a 1^a$, first match all the 0s, then match the 1s.

We next prove the "if" direction. We show that, in any solution for the subsequence packing instance (Σ, T, C, x), each substring $0^a 1^a$ of T must be filled by complete strings in C, that is, no string in C is packed into more than one "bin." Let $C_j \subseteq C$ be the collection of strings that are used to fill the j'th substring $0^a 1^a$. We examine how the first substring $0^a 1^a$ is filled by the strings in C_1. Before the 1s in a string $0^z 1^z$ in C_1 can be matched to the 1s in the substring $0^a 1^a$, the preceding z 0s in $0^z 1^z$ must all be matched. If some strings in C_1 are not completely packed into the substring $0^a 1^a$, then the number of matched 0s in C_1 must exceed the number of matched 1s in C_1, which is impossible because the substring $0^a 1^a$ contains equal numbers of 0s and 1s. Therefore, the strings in C_1 are completely packed into the first substring $0^a 1^a$ and their total length is exactly $2a$. It follows by induction that $\{C_j\}_{j=1..b}$ is a partition of C. For each non-padding string S_i in C_j, we put the corresponding item t_i into I_j. The resulting partition $\{I_j\}_{j=1..b}$ is a solution for the bin packing instance (I, a, b). This completes the proof. $\square$

3   The Algorithms

3.1   Dynamic Programming Algorithm

Given a subsequence packing instance (Σ, T, C), where $C = \{S_1, \ldots, S_n\}$, we denote by $P(t; s_1, \ldots, s_n)$ an optimal packing for the sub-instance

$$(\Sigma, T[1..t], \{S_1[1..s_1], \ldots, S_n[1..s_n]\}).$$

An optimal packing for the complete instance

$$(\Sigma, T, \{S_1, \ldots, S_n\})$$

is therefore $P(|T|; |S_1|, \ldots, |S_n|)$ and can be computed by a dynamic programming algorithm with the base conditions

$$P(t; 0, \ldots, 0) = \{\epsilon, \ldots, \epsilon\}$$
$$P(0; s_1, \ldots, s_n) = \{\epsilon, \ldots, \epsilon\}$$

and the recurrence

$$P(t; s_1, \ldots, s_n) = \max \begin{cases} P(t - 1; s_1, \ldots, s_n) \\ \max_i P(t; s_1, \ldots, s_i - 1, \ldots, s_n) \\ \max_i P(t - 1; s_1, \ldots, s_i - 1, \ldots, s_n) \oplus^i_{t, s_i}, \end{cases}$$

where the operation

$$P(t-1; s_1, \ldots, s_i - 1, \ldots, s_n) \oplus^i_{t, s_i}$$

extends the subsequence T_i of the partial packing $P(t-1; s_1, \ldots, s_i - 1, \ldots, s_n)$ with the character $S_i[s_i]$ if it matches the character $T[t]$. Intuitively, the recurrence means that either the character $T[t]$ is unmatched, or a character $S_i[s_i]$ is unmatched, or the character $T[t]$ is matched to a character $S_i[s_i]$.

Let $\ell_T = |T|$ and $L_S = \sum_i |S_i|$. The dynamic programming table has

$$(|T| + 1) \cdot (|S_1| + 1) \cdots (|S_n| + 1) = O\left(\ell_T \left(\frac{L_S}{n} + 1\right)^n\right)$$

cells; each cell takes $O(n)$ time to fill. The total running time of the algorithm is $O(n\ell_T(\frac{L_S}{n} + 1)^n)$. We have the following theorem.

Theorem 2. *The dynamic programming algorithm solves the subsequence packing optimization problem and runs in $O(n\ell_T(\frac{L_S}{n} + 1)^n)$ time.*

3.2   Greedy Algorithm

For a sequence S and a character c, we define $|S|_c$ to be the number of cs in S. Given a subsequence packing instance $(\Sigma, T, \{S_i\})$, $M_c = \min\{|T|_c, \sum_i |S_i|_c\}$ is the maximum number of c-c matches that can appear in any packing for the instance. It follows that the optimal packing size is at most $\sum_{c \in \Sigma} M_c$. On the other hand, the Pigeonhole Principle implies that

$$\max_{c \in \Sigma} M_c \geq \sum_{c \in \Sigma} M_c / |\Sigma|. \tag{1}$$

Using this bound, we can design a greedy algorithm: first find the character $c \in \Sigma$ that maximizes M_c, then compute a packing of size M_c containing only c-c matches. This algorithm has a simple implementation that runs in time linear in the total length of the strings and achieves a $|\Sigma|$ approximation for the subsequence packing problem. We have the following theorem.

Theorem 3. *The greedy algorithm approximates the subsequence packing optimization problem with ratio $|\Sigma|$ and runs in $O(\ell_T + L_S)$ time.*

In some cases, the greedy algorithm can indeed find the optimal packing; for example, when $T = 1^t 0^t$ and $S_i = 0^{s_i} 1^{s_i}$, a packing may contain either 0-0 matches or 1-1 matches, but not both. However, using only (1) as the bound, the approximation ratio $|\Sigma|$ of the greedy algorithm is the best possible, as we can see from the example $T = 0^t 1^t$ and $S_i = 0^{s_i} 1^{s_i}$.

3.3   Hybrid Algorithm

Combining the dynamic programing algorithm and the greedy algorithm, we design a hybrid algorithm that trades better approximation ratio with increased time complexity:

1. Find the character $c \in \Sigma$ that maximizes M_c.
2. Partition the collection $C = \{S_i\}_{i=1..n}$ into $\lceil \frac{n}{k} \rceil$ sub-collections $\{C_j\}$ each of size at most k.
3. For each C_j, $1 \leq j \leq \lceil \frac{n}{k} \rceil$,
   (a) Gather all the cs in the strings in $C - C_j$ into a single string S'_j. If $|S'_j| > |T|$, truncate S'_j to length $|T|$. Extend C_j to C'_j: $C'_j \leftarrow C_j + \{S'_j\}$.
   (b) Use the dynamic programming algorithm to find an optimal packing O'_j for the instance (Σ, T, C'_j).
   (c) Break the subsequence in O'_j that corresponds to the string S'_j into $|C - C_j|$ smaller subsequences, one for each string in $C - C_j$. The result is a packing O_j for the instance (Σ, T, C).
4. Find the index j that maximizes $\|O_j\|$. Output O_j.

Theorem 4. *The hybrid algorithm approximates the subsequence packing optimization problem with ratio $|\Sigma|/(1 + \frac{|\Sigma|-1}{\lceil \frac{n}{k} \rceil})$ and runs in $O(n\ell_T^2(\frac{L_S}{k} + 1)^k)$ time.*

Proof. Given a packing P for the instance (Σ, T, C) and a sub-collection $C_j \subseteq C$, we denote by $P_j \subseteq P$ the induced packing for the sub-instance (Σ, T, C_j).

Let P^* be an optimal packing for the instance (Σ, T, C). Let P be a greedy packing for (Σ, T, C) that maximizes the number of c-c matches and, at the same time, satisfies the additional constraint that, for each string $S_i \in C$, the subsequence $T_i \in P$ contains at least as many cs as the subsequence $T_i^* \in P^*$, that is $|T_i|_c \geq |T_i^*|_c$. Such a packing P always exists because we can start from P^*, discard all the non-c-c matches, "shift" the c-c matches in the packing to the front of each string S_i and the string T as far as possible, then extend the packing by greedily matching the remaining cs.

The optimal packing P^* contains $\|P^*\| - \|P\|$ more matches than the greedy packing P. According to the Pigeonhole Principle, there is at least one sub-collection C_j among the $\lceil \frac{n}{k} \rceil$ sub-collections of C such that, for the sub-instance (Σ, T, C_j), the packing P_j^* contains $(\|P^*\| - \|P\|)/\lceil \frac{n}{k} \rceil$ more matches than the packing P_j, that is, $\|P_j^*\| - \|P_j\| \geq (\|P^*\| - \|P\|)/\lceil \frac{n}{k} \rceil$.

For each string $S_i \in C_j \subseteq C$, the subsequence $T_i^* \in P_j^* \subseteq P^*$ contains as most as many cs as the subsequence $T_i \in P_j \subseteq P$. The total number of cs in the subsequences in P_j^* is at most the total number of cs in the subsequences in P_j, which is $\|P_j\|$. This implies that, even after we extract all the subsequences in P_j^* from the string T, the remaining part of T still contains enough cs to form $\|P\| - \|P_j\|$ additional c-c matches with the strings in $C - C_j$. That is, we can extend the packing P_j^* for the sub-instance (Σ, T, C_j) to a packing for the instance (Σ, T, C) by greedily adding at least $\|P\| - \|P_j\|$ c-c matches. Based on the observation that the non-c characters in the strings in $C - C_j$ are essentially ignored by this extension, our hybrid algorithm simply gathers all the cs in these strings into a temporary string S'_j, then computes an optimal packing for the extended sub-instance (Σ, T, C'_j) where $C'_j = C_j + \{S'_j\}$. The size of the packing output by the hybrid algorithm is at least

$$\|P_j^*\| + \|P\| - \|P_j\|$$

$$\geq \|P_j\| + \frac{\|P^*\| - \|P\|}{\lceil \frac{n}{k} \rceil} + \|P\| - \|P_j\|$$

$$= \|P\| \left(1 - \frac{1}{\lceil \frac{n}{k} \rceil}\right) + \frac{\|P^*\|}{\lceil \frac{n}{k} \rceil}$$

$$\geq \frac{\|P^*\|}{|\Sigma|} \left(1 - \frac{1}{\lceil \frac{n}{k} \rceil}\right) + \frac{\|P^*\|}{\lceil \frac{n}{k} \rceil}$$

$$= \|P^*\| \frac{1 + \frac{|\Sigma| - 1}{\lceil \frac{n}{k} \rceil}}{|\Sigma|}.$$

We now analyze the running time of the hybrid algorithm, which is clearly dominated by the dynamic programming algorithm for the $\lceil \frac{n}{k} \rceil$ sub-problems. The sub-problem for C_j takes $O(k\ell_T^2(\frac{L_j}{k} + 1)^k)$ time, where L_j is the total length of the strings in C_j and is at most L_S. The total running time for the $\lceil \frac{n}{k} \rceil$ sub-problems is $O(n\ell_T^2(\frac{L_S}{k} + 1)^k)$. □

When the alphabet is binary, and when the parameter k is chosen to be 2, we have the following corollary.

Corollary 1. *The subsequence packing optimization problem over the binary alphabet has a $\frac{2}{1 + 1/\lceil \frac{n}{2} \rceil}$ approximation that runs in $O(n\ell_T^2 L_S^2)$ time.*

4   Preemptive Job Shop Scheduling with Two Machines

We show that, using Bansal et al.'s technique, our approximation for the subsequence packing problem over the binary alphabet can used to derive the first deterministic bound for $J2|pmtn|C_{\max}$ that is less than 1.5. We first prove the following lemma.

Lemma 1. *For $0 < \delta \leq \frac{5 - \sqrt{13}}{3} \approx 0.465$, a $2 - \delta$ approximation for the subsequence packing problem over the binary alphabet implies a $1.5 - \delta'$ approximation for $J2|pmtn|C_{\max}$, where $\delta' = \frac{\delta}{8 - 2\delta}$.*

Proof. From $0 < \delta < \frac{5 - \sqrt{13}}{3}$ and $\delta' = \frac{\delta}{8 - 2\delta}$, a calculation shows that $1 - 2\delta' < 2 - 2\delta$. We generalize Bansal et al.'s proof and consider the following three cases:

1. $\ell \leq (1 - 2\delta')L$: Anderson et al.'s algorithm finds a schedule of makespan at most $L + \ell/2$. A lower bound L gives the approximation ratio at most

$$1 + \frac{\ell}{2L} \leq 1 + \frac{1 - 2\delta'}{2} = 1.5 - \delta'.$$

2. $\ell \geq (2 - 2\delta)L$: Again, Anderson et al.'s algorithm finds a schedule of makespan at most $L + \ell/2$. A lower bound ℓ gives the approximation ratio at most

$$\frac{L}{\ell} + 0.5 \leq \frac{1}{2 - 2\delta} + 0.5 = 1.5 - \frac{1 - 2\delta}{2 - 2\delta}.$$

A calculation shows that, when $\delta \leq \frac{5 - \sqrt{13}}{3}$, we have $\frac{1 - 2\delta}{2 - 2\delta} \geq \frac{\delta}{8 - 2\delta} = \delta'$. The approximation ratio is at most $1.5 - \delta'$ in this case.

3. $(1 - 2\delta')L \leq \ell < (2 - 2\delta)L$: Let V denote the optimal packing size for the corresponding subsequence packing problem. The number of operations in the long job is ℓ; let ℓ' be the total number of operations in the short jobs. At least $\ell' - V$ operations of the short jobs cannot be scheduled in parallel to the long job; they take at least $\frac{\ell'-V}{2}$ time units. The optimal makespan is at least

$$\ell + \frac{\ell' - V}{2} = \frac{\ell + \ell'}{2} + \frac{\ell - V}{2} = L + \frac{\ell - V}{2}.$$

On the other hand, the makespan of our approximation is at most $2L - \frac{V}{2-\delta}$. Therefore, the approximation ratio is at most

$$f(V) = \frac{2L - \frac{V}{2-\delta}}{L + \frac{\ell-V}{2}} = \frac{2L(2 - \delta) - V}{2L + \ell - V} \cdot \frac{2}{2 - \delta}.$$

Since $\ell < (2-2\delta)L$, we have $2L(2-\delta) > 2L+\ell$. Therefore, $f(V)$ is monotonically increasing in V. Since $V \leq \ell$, $f(V)$ reaches the maximum

$$g(\ell) = \frac{2L(2 - \delta) - \ell}{2L} \cdot \frac{2}{2 - \delta} = 2 - \frac{\ell}{L(2 - \delta)}$$

at $V = \ell$, and $g(\ell)$ reaches the maximum

$$2 - \frac{1 - 2\delta'}{2 - \delta}$$

at $\ell = (1 - 2\delta')L$. A calculation shows that, when $\delta' = \frac{\delta}{8-2\delta}$, we have $2 - \frac{1-2\delta'}{2-\delta} = 1.5 - \delta'$. Again, the approximation ratio is at most $1.5 - \delta'$. $\square$

Anderson et al.'s algorithms runs in $O(L)$ time; our hybrid approximation algorithm runs in $O(n\ell_T^2 L_S^2)$ time. Note that $\ell_T = \ell$ and $L_S = O(L)$. Lemma 1 and Corollary 1 together imply the following theorem.

Theorem 5. $J2|pmtn|C_{\max}$ has a $1.5 - \Theta(\frac{1}{n})$ approximation that runs in $O(n\ell^2 L^2)$ time.

It is interesting to note that, for the online version of $J2|pmtn|C_{\max}$, Kimbrel and Saia [7] proved that no algorithm can achieve a competitive ratio less than $1.5 - \frac{1}{2n}$ against an oblivious adversary.

5 Future Work

We believe the subsequence packing problem is a fundamental combinatorial problem with important real-world applications. We intend to continue our work in the following directions:

1. $J2|pmtn|C_{max}$ is a very difficult problem whose upper bound 1.5 had remained intact for nearly two decades. Both the recent breakthrough by Bansal et al. [3] and our modest improvement showed that the subsequence packing problem is at the core of its complexity. Can we design a deterministic algorithm for subsequence packing over the binary alphabet with an approximation ratio that is a constant strictly less than 2? Lemma 1 shows that this will immediately imply a deterministic algorithm for $J2|pmtn|C_{max}$ with a constant approximation ratio strictly less than 1.5.

2. For the application of subsequence packing to DNA sequencing verification, we can assume that the *span* of each subsequence T_i, that is, the difference of its starting and ending indices in T, is a constant times its length. Does this assumption change the complexity of the subsequence packing problem? Can we design a better approximation algorithm with this assumption?

3. Bansal et al. presented a randomized approximation algorithm [3] for subsequence packing over the binary alphabet. Can we generalize this algorithm from binary alphabet to any constant-size alphabet (in particular, alphabet of size four for DNA sequences)?

Acknowledgment

The author thanks Dr. Zhixiang Chen, Dr. Guohui Lin, Dr. Bin Ma, and anonymous referees for helpful comments.

References

1. Bruce Alberts, Alexander Johnson, Julian Lewis, Martin Raff, Keith Roberts, and Peter Walter. *Molecular Biology of the Cell, Fourth Edition*. Garland Science, 2002.
2. Eric J. Anderson, T. S. Jayram, and Tracy Kimbrel. Tighter bounds on preemptive job shop scheduling with two machines. *Computing*, 67(1):83–90, 2001.
3. Nikhil Bansal, Tracy Kimbrel, and Maxim Sviridenko. Job shop scheduling with unit processing times. In *Proc. 16th Annual ACM-SIAM Symposium on Discrete Algorithms (SODA'05)*, pages 207–214, 2005.
4. Michael R. Garey and David S. Johnson. *Computers and Intractability: A Guide to the Theory of NP-Completeness*. W. H. Freeman and Company, 1979.
5. Teofilo Gonzalez and Sartaj Sahni. Flowshop and jobshop schedules: complexity and approximation. *Operations Research*, 26:36–52, 1978.
6. Dan Gusfield. *Algorithms on Strings, Trees, and Sequences: Computer Science and Computational Biology*. Cambridge University Press, 1997.
7. Tracy Kimbrel and Jared Saia. Online and offline preemptive two-machine job shop scheduling. *Journal of Scheduling*, 3:355–364, 2000.
8. Sergey V. Sevastianov and Gerhard J. Woeginger. Makespan minimization in preemptive two machine job shops. *Computing*, 60(1):73–80, 1998.
9. J. Craig Venter, Mark D. Adams, Eugene W. Myers, et al. The sequence of the human genome. *Science*, 291(5507):1304–1351, 2001.

Decomposition Based Heuristic Approach to Frequency Reassignment Problem

Junghee Han

College of Business Administration, Kangwon National University
Hyoja-2Dong, Chunchon-Shi, Kangwon-Do, Korea
jhhan@kangwon.ac.kr

Abstract. In this paper, I present a frequency reassignment problem (FRP) arising from the installation of new base stations for capacity expansion in mobile telecommunication systems, and develop an integer programming (IP) formulation along with some valid inequalities. Also, I develop a novel decomposition based heuristic algorithm. Computational results show that the developed valid inequalities are quite strong, and that the developed heuristic algorithm finds a feasible solution of good quality within reasonable time bound.

1 Introduction

This paper deals with a frequency reassignment problem (FRP) arising from the installation of new base stations (BSs) in a mobile telecommunication network. When we add new BSs to the current mobile telecommunication network in order to expand the capacity or service area, we need to assign frequencies to new BSs. However, if we cannot find frequencies to assign to new BSs that do not incur interference with the frequencies used by the existing BSs, we need to change the frequencies that are already assigned to the existing BSs in order to avoid (or minimize) interference between frequencies. Interference may occur between frequencies assigned to adjacent BSs if the frequencies are not separated enough. Interference between frequencies may degrade the service quality, and may reduce the capacity of a code division multiple access (CDMA) network (see [9]. As addressed in [6], pseudo noise (PN) code (corresponding to frequency in general term) reassignment process in a CDMA network requires quite long time, during which service quality may degrade significantly. Also, if two or more adjacent BSs change their frequencies in parallel, large geographical area may become out-of-service. This may cause heavy traffic load to the adjacent BSs. Consequently, this may make the CDMA network in this local area unstable [6]. Thus, we need to determine a subset of existing BSs to perform frequency reassignment, the sequence of frequency reassignments and the new frequencies to reassign to these existing BSs along with the frequencies to assign to new BSs carefully. Also, during the frequency reassignment process, we may observe interference between new frequencies assigned to the existing BSs and the current frequencies assigned to the existing BSs to perform frequency reassignment. Thus, the interference that may occur during the frequency reassignment process should be

S.-W. Cheng and C.K. Poon (Eds.): AAIM 2006, LNCS 4041, pp. 324–333, 2006.

taken into account when we design frequency reassignment process even if we can transform the existing frequency assignment matrix to an optimized frequency assignment matrix incurring minimum interference.

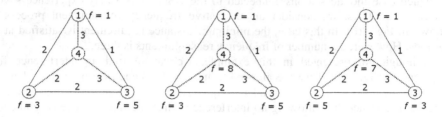

Fig. 1. Frequency reassignment example: (a) initial assignment, (b) reassignment example 1 and (c) reassignment example 2

BS	Frequency		
	$t = 0$	$t = 1$	$t = 2$
1	1	1	1
2	3	5^*	5
3	5	5	3^*
4	–	–	7

BS	Frequency			
	$t = 0$	$t = 1$	$t = 2$	$t = 3$
1	1	1	1	1
2	3	3	5^*	5
3	5	7^*	7	3^*
4	–	–	–	7

Fig. 2. Frequency reassignment process of Fig. 1(c): (a) reassignment process example 1 and (b) reassignment process example 2

For example, suppose that we have a network with three BSs, 1, 2 and 3, having initial frequency 1, 3 and 5, respectively, and we add a new BS, indexed by 4. This is illustrated in Fig. 1(a), where the minimum distance between frequencies assigned to adjacent BSs to avoid interference is denoted by the number on links. Here, note that the current frequency assignment incurs no interference. As shown in Fig. 1(b), we can assign frequency 8 to the new BS 4 without changing the initial frequencies assigned to the existing BSs. However, if we are allowed to use frequencies from 1 to 7, we have to reassign frequencies to the existing BSs in order to obtain an interference free frequency assignment. Fig. 1(c) illustrates an example of frequency reassignment using the frequencies from 1 to 7. Now, suppose that simultaneous frequency reassignments at the BSs 2 and 3 are not allowed. Then, we have to change the frequencies assigned to the BSs 2 and 3 in order or in reverse order. Based on the frequency assignment of Fig. 1(c), two alternative frequency reassignment processes are illustrated in Fig. 2. As shown in Fig. 2(a), we can change the frequency of BS 2 from 3 to 5 first. Then, we change the frequency of BS 3 from 5 to 3, and assign frequency 7 to the BS 4. Here, note that the number of frequency reassignments is two. Also, note that at period $t = 1$ both the BSs 2 and 3 use the same frequency of 5, in which case mobile stations connected to the BSs 2 and 3 may experience severe interference. Thus, we consider an alternative frequency reassignment process as shown in Fig. 2(b). In this case, the minimum distance requirement is satisfied at all periods. However, the number of frequency reassignments is three.

Although not presented in this example, we cannot find an interference free frequency assignment if we are allowed to use the frequencies from 1 to 6. Thus, it may be more practical to consider frequency reassignment allowing minimum interference rather than allowing no interference when the frequency reassignment is completed unlike the work by Han [6] that seeks to find an interference free frequency assignment at the end of frequency reassignment. In this context, this paper considers the problem FRP that minimizes the total cost consisting of frequency reassignment cost and interference cost that may occur when the frequency reassignment is completed as well as the interference cost that may occur during the frequency reassignment process.

On the frequency reassignment problem, Han [6] showed that this problem belongs to NP-hard class, and developed two integer programming (IP) formulations for the problem FRP that does not allow interference when the frequency reassignment is completed. Although there are numerous studies on frequency assignment problem (FAP) such as [2], [3], [5], [7], [8], [10]-[12], to the best of my knowledge, frequency reassignment is considered only by Han [3]. Aardal et al. [1] summarized the results of quite may research papers on FAP.

This paper is organized as follows. In Section 2, a mathematical formulation and some valid inequalities are developed. Also, a compact formulation is derived by applying some preprocessing rules. In Section 3, an effective heuristic procedure based on a decomposition principle is proposed. Computational results are provided in Section 4, and Section 5 concludes this paper.

2 Formulation

Let us define some notations in order to formulate the problem FRP. Let N be a set of existing BSs, and let V be a set of new BSs. Also, let K be a set of frequencies, and let T be a set of time periods for frequency reassignments. Let $E = \{(i, j): r(i, j) > 0$ for i, $j(> i) \in N \cup V\}$, where $r(i, j)$ is the minimum distance to avoid interference between the frequencies assigned to the BSs i and $j(> i) \in N \cup V$. In particular, we define $E(N) = \{(i, j): r(i, j) > 0$ for $i, j(> i) \in N\}$. And, let A be a set of pairs of adjacent BSs $(i, j) \in E(N)$ such that simultaneous frequency reassignments at the BSs i and $j(> i) \in N$ are not allowed. Now, let us define decision variables and input parameters. Let $x_{tik} = 1$ if frequency $k \in K$ is assigned to the BS $i \in N \cup V$ at $t \in T$, and 0 otherwise. Also, let $y_{ti} = 1$ if a new frequency is assigned to the BS $i \in N$ at $t \in T$, and 0 otherwise. Let $v_t = 1$ if frequency reassignment is completed at $t \in T$, and 0 otherwise. And, let us denote the frequency reassignment cost by c_i for BS $i \in N$. Let $u_{tij} = 1$ if the distance between the frequencies assigned to the BSs i and $j(> i) \in N \cup V$, respectively, is less than

$r(i, j)$, and 0 otherwise. When $u_{tij} = 1$ for $t = 1,\ldots, |T| - 1$ and $(i, j) \in E(N)$, interference cost p_{ij} occurs. Also, if $u_{tij} = 1$ for $t = |T|$ and $(i, j) \in E$, interference cost q_{ij} occurs. Here, note that (q) needs to be set quite large compared to (p) since (q) indicates the interference cost that may occur after the frequency reassignment is completed. Let (h_{ik}) be the input vector indicating the current frequency assignment to the BS $i \in N$ at $t = 0$. That is, if frequency $k \in K$ is assigned to the BS $i \in N$ at $t = 0$, we set $h_{ik} = 1$, and 0 otherwise.

Using the above notations, we can formulate the problem FRP as follows, denoted by P1.

P1: Minimize $\sum_{t \in T} \sum_{i \in N} c_i y_{ti} + \sum_{t = 1,\ldots, |T| - 1} \sum_{(i, j) \in E} p_{ij} u_{tij} + \sum_{(i, j) \in E} q_{ij} u_{|T|ij}$

Subject to

$\sum_{k \in K} x_{tik} = 1$	$t \in T, i \in N,$	(1)
$\sum_{k \in K} x_{tik} = v_t$	$t \in T, i \in V,$	(2)
$\sum_{t \in T} v_t \geq 1$		(3)
$x_{tik} + x_{tjl} \leq 1 + u_{tij}$	$t \in T, (i, j) \in E, k, l \in K: |k - l| < r(i, j),$	(4)
$y_{ti} \geq x_{tik} - x_{(t-1)ik}$	$t \in T, i \in N, k \in K,$	(5)
$y_{ti} \geq x_{(t-1)ik} - x_{tik}$	$t \in T, i \in N, k \in K,$	(6)
$y_{ti} + y_{tj} \leq 1$	$t \in T, (i, j) \in A,$	(7)
$x_{tik} \in \{0,1\}$	$t \in T, i \in N \cup V, k \in K,$	
$y_{ti} \in \{0,1\}$	$t \in T, i \in N,$	
$u_{tij} \in \{0,1\}$	$t \in T, (i, j) \in E,$	
$v_t \in \{0,1\}$	$t \in T,$	

where $x_{0ik} = h_{ik}$ for $i \in N$ and $k \in K$.

Constraint (1) forces that a frequency should be assigned to each existing BS at all time periods in T. Constraint (2) forces that a frequency should be assigned to each new BS when the frequency reassignment is completed. Constraint (3) forces that frequency reassignment should be completed within $|T|$ time periods. Constraint (4) expresses the interference between frequencies assigned to a pair of adjacent BSs. Constraints (5) and (6) express the frequency reassignment at existing BSs. Constraint (7) prohibits simultaneous frequency reassignments for a pair of adjacent BSs in A.

Remark 1. Note that if the initial frequency assignment (h) is interference free, there exists an optimal solution such that $v_{|T|} = 1$ and $v_t = 0$ for all $t = 1,\ldots, |T| - 1$. Thus, assuming that the (h) is interference free, we can delete constraint (3), which in turn enables us to set $x_{tik} = 0$ for all $t = 1,\ldots, |T| - 1$, $i \in V$ and $k \in K$ and to set $u_{tij} = 0$ for all $t = 1,\ldots, |T| - 1$, $(i, j) \in E - E(N)$. Then, we can rewrite the P1 as follows, denoted by P2.

P2: Minimize $\sum_{t \in T} \sum_{i \in N} c_i y_{ti} + \sum_{t = 1,\ldots, |T| - 1} \sum_{(i, j) \in E(N)} p_{ij} u_{tij} + \sum_{(i, j) \in E} q_{ij} u_{|T|ij}$

Subject to

$\sum_{k \in K} x_{tik} = 1$	$t \in T, i \in N,$	(8)
$\sum_{k \in K} x_{|T|ik} = 1$	$i \in V,$	(9)
$x_{tik} + x_{tjl} \leq 1 + u_{tij}$	$t = 1,\ldots, |T| - 1, (i, j) \in E(N), k, l \in K: |k - l| < r(i, j),$	(10)
$x_{|T|ik} + x_{|T|jl} \leq 1 + u_{|T|ij}$	$(i, j) \in E, k, l \in K: |k - l| < r(i, j),$	(11)

$$y_{ti} \geq x_{tik} - x_{(t-1)ik} \qquad\qquad t \in T, i \in N, k \in K, \tag{12}$$
$$y_{ti} \geq x_{(t-1)ik} - x_{tik} \qquad\qquad t \in T, i \in N, k \in K, \tag{13}$$
$$y_{ti} + y_{tj} \leq 1 \qquad\qquad t \in T, (i,j) \in A, \tag{14}$$
$$x_{tik} \in \{0,1\} \qquad\qquad t \in T, i \in N, k \in K,$$
$$x_{|T|ik} \in \{0,1\} \qquad\qquad i \in V, k \in K,$$
$$y_{ti} \in \{0,1\} \qquad\qquad t \in T, i \in N,$$
$$u_{tij} \in \{0,1\} \qquad\qquad t = 1,\ldots, |T| - 1, (i,j) \in E(N),$$
$$u_{|T|ij} \in \{0,1\} \qquad\qquad (i,j) \in E.$$

Next, we develop some valid inequalities based on the constraints (10), (11) and (14) in order to enhance the lower bound of the P2.

Remark 2. For $t = 1,\ldots, |T| - 1$, $(i,j) \in E(N)$ and $k \in K$, we obtain the following valid inequalities by lifting the constraint (10).

$$x_{tik} + \sum_{f = \max\{1, k - (r(i,j)-1)\},\ldots,\, \min\{|K|, k + (r(i,j)-1)\}} x_{tjf} \leq 1 + u_{tij}, \tag{15a}$$

and

$$x_{tjk} + \sum_{f = \max\{1, k - (r(i,j)-1)\},\ldots,\, \min\{|K|, k + (r(i,j)-1)\}} x_{tif} \leq 1 + u_{tij}. \tag{15b}$$

Here, note that the valid inequalities (15a) and (15b) dominate the constraints (10). Similarly, we can derive valid inequalities based on the constraint (11). For $(i,j) \in E$ and $k \in K$ at $t = |T|$, we can derive the following valid inequalities.

$$x_{|T|ik} + \sum_{f = \max\{1, k - (r(i,j)-1)\},\ldots,\, \min\{|K|, k + (r(i,j)-1)\}} x_{|T|jf} \leq 1 + u_{|T|ij}, \tag{16a}$$

and

$$x_{|T|jk} + \sum_{f = \max\{1, k - (r(i,j)-1)\},\ldots,\, \min\{|K|, k + (r(i,j)-1)\}} x_{|T|if} \leq 1 + u_{|T|ij}. \tag{16b}$$

Also, note that the valid inequalities (16a) and (16b) dominate the constraints (11). Now, let us consider constraint (14), from which we can derive a clique inequality as follows.

$$\sum_{i \in C} y_{ti} \leq 1 \qquad\qquad t \in T, \tag{17}$$

where C is a clique of the graph G(A) induced by the edge set A. Separating the inequality (17) amounts to finding a maximal clique in a graph G(A). This problem is known to be NP-hard. Thus, we implement heuristic procedure for separating the inequality (17) from G(A). Our approach is as follows. First, we detect a triangle subgraph G(S), where S (A. Then, we expand G(S) by simple greedy procedure that adds a new BS i (N not in G(S) until we cannot find any new BS to add. Then, we let C = {i (G(S)}.

3 Heuristic Algorithm

In this section, we develop an efficient heuristic algorithm based on a decomposition principle. Let $\tilde{x}$ be a frequency assignment at $t = |T|$. If we let $|T| = 1$, we can delete index $t \in T$ from the P2. Then, the P2 reduces to as follows, denoted by P3. Thus, by solving the P3, we can obtain a frequency assignment at $t = |T|$, denoted by $\tilde{x}$.

P3: Minimize $\sum_{i \in N} \sum_{k \in K} c_i (1 - h_{ik}) x_{ik} + \sum_{(i,j) \in E} q_{ij} u_{ij}$
Subject to

$$
\begin{array}{ll}
\sum_{k \in K} x_{ik} = 1 & i \in N \cup V, \\
x_{ik} + x_{jl} \leq 1 + u_{ij} & (i, j) \in E, k, l \in K, |k - l| < r(i, j), \\
x_{ik} \in \{0,1\} & i \in N \cup V, k \in K, \\
u_{ij} \in \{0,1\} & (i, j) \in E.
\end{array}
$$

If $\tilde{x}$ is given, we can define a set $\Omega = \{i \in N: \tilde{x}_{ik} \neq h_{ik} \text{ for some } k \in K\}$. Thus, to find a feasible solution to the P2 it is sufficient to determine at what time period $t \in T$ to perform frequency reassignment for the BSs in Ω considering the constraint (14). Letting $\Omega(t) \subseteq \Omega$ be a set of BSs to perform frequency reassignment at $t \in T$, any partition of Ω over T, $\{\Omega(t): t \in T\}$, defines a feasible solution to the P2 if $\bigcup_{t \in T} \Omega(t)$ $= \Omega$ and $(i, j) \notin A$ for $i, j(> i) \in \Omega(t)$ and $t \in T$. Then, the outline of the heuristic algorithm can be described as follows. First, we find a frequency assignment at $t = |T|$, $\tilde{x}$, which automatically defines Ω. Then, we find a feasible solution $\{\Omega(t): t \in T\}$ such that $\Omega(s) \cap \Omega(t) = \varnothing$ for all s and $t(> s) \in T$. That is, we change the initial frequency assigned to the BS in Ω only once. Then, we resume this process by finding an alternative feasible solution to the P3, $\tilde{x}$. The above process is repeated for a given time bound.

3.1 Algorithm for Finding an Initial Solution

Below, we describe an heuristic algorithm to find an initial feasible solution $\{\Omega(t): t \in T\}$ such that $\Omega(s) \cap \Omega(t) = \varnothing$ for all s and $t(> s)$ in T.

Initialize. Let $\Omega(t) = \varnothing$ for $t \in T$, and let $t = 1$.

Step 1. Find a frequency assignment $\tilde{x}$ at $t = |T|$ and get Ω. If $\Omega = \varnothing$, stop. Otherwise, go to Step 2.
Step 2. If $\{(i, j) \in A: i, j \in \Omega\} = \varnothing$, stop. Otherwise, go to Step 3.
Step 3. Pick an arbitrary BS $i \in \Omega$. Then, let $\Omega(t) = \{i\}$ and $\Omega = \Omega - \{i\}$. If $\Omega = \varnothing$, stop. Otherwise, go to Step 4.
Step 4. Pick an arbitrary BS $j \in \Omega$ such that $(i, j) \notin A$ for all $i \in \Omega(t)$. Then, let $\Omega(t) = \Omega(t) + \{j\}$ and $\Omega = \Omega - \{j\}$. If $\Omega = \varnothing$, stop. Otherwise, resume Step 4. If such BS is not found, let $t = t + 1$ and go to Step 3.

Remark 3. For finding a frequency assignment at $t = |T|$, $\tilde{x}$, at Step 1, we solved the P3 using a commercial optimization software, CPLEX Version 9.0 [4]. If the above heuristic algorithm terminates at Step 1 or at Step 2, $\tilde{x}$ defines an optimal solution to the P2 provided that we solved the P3 optimally. However, if the above heuristic procedure terminates at Step 3 or at Step 4, we need to calculate the interference cost incurred by the current partition $\{\Omega(t): t \in T\}$. If no interference is observed, we see that the current partition $\{\Omega(t): t \in T\}$ defines an optimal solution to the P2 provided that we solved the P3 optimally. Otherwise, we improve the initial solution, which is described in the following.

3.2 Improving Algorithm

First, we seek to minimize the interference cost (p) by finding an optimal partition of Ω over T, $\{\Omega(t): t \in T\}$. If we cannot reduce the interference cost (p) for given Ω, we generate an alternative $\tilde{x}$, which defines a new set Ω. Then, we try to minimize the interference cost (p) again based on a new Ω. This process is repeated for a given time limit. Another termination criterion is that we fail to find a new $\tilde{x}$. In Section 3.2.1, we describe a procedure that minimizes the interference cost (p) based on an incumbent Ω. And, in Section 3.2.2, we describe a procedure that finds an alternative frequency assignment at $t = |T|$ in order to derive a new Ω.

3.2.1 Minimizing the Interference Cost (p)
For given Ω, minimizing the interference cost (p) is equivalent to optimally solving the following problem, denoted by P4.

P4: Minimize $\sum_{t=1,\ldots,|T|-1} \sum_{i,j(\neq i) \in \Omega} p_{ij} u_{tij}$
   Subject to

$x_{tif(i)} + x_{tig(i)} = 1$	$t = 1,\ldots, |T|-1, i \in \Omega,$
$x_{tig(i)} + x_{tif(j)} \leq u_{tij} + 1$	$t = 1,\ldots, |T|-1, i, j(\neq i) \in \Omega, |f(i) - g(j)| < r(i,j),$
$y_{ti} = x_{(t-1)if(i)} + x_{tig(i)} - 1$	$t \in T, i \in \Omega,$
$y_{ti} + y_{tj} \leq 1$	$t \in T, i, j(> i) \in \Omega$ if $(i,j) \in A,$
$\sum_{t \in T} y_{ti} = 1$	$i \in \Omega,$
$x_{tif(i)} \in \{0,1\}$	$t = 1,\ldots, |T|-1, i \in \Omega,$
$x_{tig(i)} \in \{0,1\}$	$t \in T, i \in \Omega,$
$y_{ti} \in \{0,1\}$	$t \in T, i \in \Omega,$
$u_{tij} \in \{0,1\}$	$t = 1,\ldots, |T|-1, i, j(\neq i) \in \Omega, |f(i) - g(j)| < r(i,j),$

where $f(i) = arg_{k \in K}\{h_{ik} = 1\}$ and $g(i) = arg_{k \in K}\{\tilde{x}_{ik} = 1\}$ for $i \in \Omega$.

Note that the P4 determines the optimal time period $t \in T$ at which we change the initial frequency $f(i)$ to a new frequency $g(i)$ for each BS $i \in \Omega$ in order minimize the interference cost (p). Here, note that the P4 can be obtained from the P2 by

- setting $x_{|T|ik} = \tilde{x}_{ik}$ for $i \in N \cup V, k \in K,$
- setting $x_{tik} = \tilde{x}_{ik}$ for $t = 1,\ldots, |T|-1, i \in N - \Omega, k \in K,$ and
- adding $\sum_{t \in T} y_{ti} = 1$ for $i \in \Omega$ to the P2.

3.2.2 Finding an Alternative Frequency Assignment
We already have a frequency assignment at $t = |T|$. Thus, we try to find an alternative frequency assignment at $t = |T|$, which can be obtained by solving the modified P3. For the sake of notational convenience, let us denote the incumbent $\tilde{x}$ by w. Then, we can find a new $\tilde{x}$ by solving the P3 after adding the following constraint to the P3:

$$\sum_{i \in N \cup V} \sum_{k \in K} w_{ik} \, x_{ik} \leq |N \cup V| - 1.$$

If the updated P3 returns no feasible solution, or if the optimal objective value of the updated P3 is greater than the total cost of the best solution to the P2, we terminate the heuristic procedure. Otherwise, we minimize the interference cost (p) based on a new $\tilde{x}$ by performing the procedure described in Section 3.2.1.

4 Computational Results

In order to evaluate the performance of the proposed heuristic procedure and the effectiveness of the developed valid inequalities, we randomly generated test problems as follows.

Step 1. Generate $|N| + |V|$ BSs at random on a rectangle with scale 1,000 by 1,000, and calculate the distance for all pairs of BSs.

Step 2. If the distance d_{ij} between a pair of BSs i and $j(> i) \in N \cup V$ is greater than a threshold R, we set $r(i, j) = 0$, otherwise, we set $r(i, j) = \lceil R/d_{ij} \rceil$. In our test, we set $R = 300$ or 400 depending on the number of BSs.

Step 3. Find an interference free initial frequency assignment (h_{ik}) for $i \in N$ and $k \in K$, using the following procedure.

    Step 3.1 Pick an arbitrary BS and assign frequency of 1.

    Step 3.2 Find a combination of BS and frequency $\{i, f\}$, where i denotes the BS that is not assigned with any frequency and is adjacent to at least one of the BSs that are already assigned with frequencies, and f denotes the lowest index frequency that incurs no interference with adjacent BSs. Then, assign frequency f to the BS i. If there exists any BS that is not assigned with any frequency, resume Step 3.2. Otherwise, go to Step 4.

Step 4. Let $|K|$ be the integer value of the highest frequency index used in Step 3 multiplied by 0.9. For each BS, if the assigned frequency is less than or equal to $|K|$, let this BS belong to the set N. Otherwise, let this BS belong to the set V, and delete the initial frequency assigned to it.

Step 5. Let $A = \{(i, j) \in E(N): d_{ij} < 0.7 \times R\}$. And, let c_i be a random integer value in the rage [10, 20] for $i \in N$. Also, let $p_{ij} = (c_i + c_j) \times r(i, j)$ for $(i, j) \in E(N)$ and let $q_{ij} = 10 \times (c_i + c_j) \times r(i, j)$ for $(i, j) \in E$.

The developed heuristic algorithm was coded in Visual Basic 6.0 coupled with CPLEX 9.0 (see [4]), and was tested on a Pentium IV PC (CPU: 2.8GHz, RAM: 512Mbytes). Computational results are presented in Table 1, where P2V denotes the P2 enhanced by the valid inequalities (15), (16) and (17) at the root node of branch-and-bound tree. We terminated the CPLEX optimization procedure and the heuristic algorithm after 10,000 seconds and after 1,000 seconds, respectively. Also, we limited computation time for solving the P3 to 100 seconds. Computation time is presented in parenthesis. Also, the mark "NA" represents that CPLEX optimization procedure failed to find an initial feasible solution in 10,000 seconds. From Table 1, we see that the valid inequalities (15), (16) and (17) significantly improves the lower bound of the P2. Also, we see that the proposed heuristic algorithm finds a feasible solution of good quality to the most test problems within 1,000 seconds. In particular, the heuristic algorithm finds an optimal solution to the 7 test problems out of 30 test problems, which is indicated by asterisk mark (*) at the fifth column "Opt." of Table 1. Also, the marks "b", "w" and "e" indicate that the heuristic algorithm found a "better", "worse" and "equally good" feasible solution, respectively, compared with that obtained by the P2V. Another observation is that P3 provides tight lower bound.

Table 1. Computational results of test problems

No	Size $|N|, |V|, |E|, |K|, |T|$	Lower bound $P2_{LP}, P2V_{LP}, P3$	Upper bound P2V, Heuristic	Opt.
1	19, 1, 76, 17, 19	48, 115, 578(2)	598(10K), 598(21)	e
2	19, 1, 75, 13, 19	85, 217, 770(1)	770(10K), 770(1)	e
3	18, 2, 51, 8, 18	86, 314, 637(1)	637(106), 637(1)	*
4	19, 1, 62, 15, 19	50, 106, 219(1)	219(746), 219(1)	*
5	18, 2, 58, 11, 18	107, 362, 965(2)	1026(6.7K), 1072(1K)	w
6	18, 2, 50, 10, 18	107, 256, 1003(1)	1100(7.3K), 1121(1K)	w
7	19, 1, 85, 16, 19	55, 113, 339(1)	750(10K), 387(1K)	b
8	19, 1, 96, 16, 19	62, 152, 402(1)	402(10K), 402(1)	*
9	16, 4, 74, 12, 16	116, 279, 1246(11)	1350(10K), 1442(1K)	w
10	19, 1, 69, 14, 19	63, 130, 530(2)	530(6.6K), 530(2)	*
11	29, 1, 158, 15, 29	109, 251, 387(1)	1450(10K), 410(1K)	b
12	27, 3, 164, 17, 27	159, NA, 993(28)	NA(10K), 1002(1K)	b
13	28, 2, 136, 19, 28	68, 177, 290(1)	1020(10K), 311(1K)	b
14	28, 2, 168, 17, 28	72, 237, 627(25)	3181(10K), 632(1K)	b
15	29, 1, 181, 17, 29	69, 154, 492(6)	1150(10K), 580(133)	b
16	28, 2, 143, 14, 28	137, 313, 826(6)	2982(10K), 826(6)	b
17	27, 3, 154, 17, 27	97, 252, 695(19)	3894(10K), 773(1K)	b
18	28, 2, 141, 20, 28	103, NA, 783(19)	NA(10K), 863(1K)	b
19	29, 1, 131, 17, 29	47, 121, 395(1)	1335(10K), 395(1)	*
20	29, 1, 157, 18, 29	41, 132, 264(1)	810(10K), 264(1)	*
21	37, 3, 276, 19, 18	135, 289, 676(90)	NA(10K), 831(1K)	b
22	38, 2, 308, 24, 19	79, 246, 373(48)	901(10K), 513(1K)	b
23	39, 1, 266, 23, 19	89, 156, 172(1)	172(81), 172(1)	*
24	37, 3, 288, 20, 18	151, NA, NA(100)	NA(10K), 920(1K)	b
25	38, 2, 302, 22, 19	76, NA, 576(23)	NA(10K), 590(1K)	b
26	38, 2, 267, 20, 19	50, 152, NA(100)	1740(10K), 1021(1K)	b
27	37, 3, 271, 20, 18	175, 374, 768(15)	2700(10K), 921(1K)	b
28	38, 2, 308, 26, 19	87, 140, 376(9)	1346(10K), 399(1K)	b
29	38, 2, 287, 23, 19	82, 215, 576(20)	2089(10K), 829(1K)	b
30	38, 2, 233, 20, 19	94, 255, 395(6)	1205(10K), 443(1K)	b

5 Conclusion

In this paper, we addressed a new frequency reassignment problem arising from the installation of new BSs in a CDMA based mobile radio network. And, we developed mathematical formulations for this problem along with some valid inequalities. For solving large size problems, we developed heuristic algorithm that decomposes the original problem into two sub-problems and solves the two sub-problems in turn repeatedly. Computational results show that the developed valid inequalities are quite strong and that the proposed heuristic algorithm finds a feasible solution of good quality within reasonable time limit.

Further research task is to develop a meta-heuristic procedure in order to handle larger problem instances. Also, as an extension of this study, frequency reassignment for the global system for mobile communications (GSM) based telecommunication system, allocating multiple frequencies to each BS, should be followed.

References

1. Aardal, K., Hoesel, S., Koster, A., Mannino, C. and Sassano, A.: Models and Solution Techniques: Frequency Assignment Problems, ZIB Report 01-40 (2001).
2. Allen, S. M., Smith, D. H. and Hurley, S.: Lower Bounding Techniques for Frequency Assignment, Discrete Mathematics, Vol. 197, 41-52 (1999).
3. Borndörfer, R., Eisenblätter, A. Grötschel, M and Martin, A.: The Orientation Model for Frequency Assignment Problem, Tech. Report TR 98-01, Konrad-Zuse-Zentrum für Informationstechnik Berlin (1998).
4. CPLEX Division: CPLEX 9.0 Users' Manual, ILOG Inc. (2004).
5. Fischetti, M., Lepschy, C., Minerva, G., Romanin-Jacur, G. and Toto, E.: Frequency Assignment in Mobile Radio Systems Using Branch-and-cut Techniques, European Journal of Operational Research, Vol. 123, 241-255 (2000).
6. Han, J. H.: Frequency Reassignment Problem in Mobile Communication Networks, Computers and Operations Research, accepted.
7. Hao, J., Dorne, R. and Galinier, P.: Tabu Search for Frequency Assignment in Mobile Radio Networks, Journal of Heuristics, Vol. 4, 47-62 (1990).
8. Koster, A. M. C. A.: Frequency Assignment: Models and Algorithms, Ph.D. thesis, Maastricht University (1999).
9. Lee, S. and Bang, H.: IMT-2000 CDMA Technology, Sehwa Publishing (2001).
10. Montemanni, R., Smith, D. H. and Allen, S. M.: Lower Bounds for Fixed Spectrum Frequency Assignment, Annals of Operations Research, Vol. 107, 237-250 (2001).
11. Sung, W. and Wong, W.: Sequential Packing Algorithm for Channel Assignment Under Cochannel and Adjacent Channel Interference Constraint, IEEE Trans. on Veh. Tech., Vol. 46, 676-686 (1997).
12. Tiourine, R., Hurkens, C. and Lenstra, J.: Local Search Algorithm for the Radio Link Frequency Assignment Problem, Telecommunication Systems, Vol. 13, 293-314 (2000).

Approximation Algorithms for Minimum Span Channel Assignment Problems

Yuichiro Miyamoto[1] and Tomomi Matsui[2]

[1] Sophia University, Kioicho 7-1, Chiyoda city, Tokyo 102-8554, Tokyo, Japan
y-miyamo@sophia.ac.jp
[2] The University of Tokyo, Hongo 7-3-1, Bunkyo city, Tokyo 113-8656, Japan
tomomi@mist.i.u-tokyo.ac.jp

Abstract. We propose polynomial time approximation algorithms for minimum span channel (frequency) assignment problems, which is known to be NP-hard. Let α be the approximation ratio of our algorithm and $W \geq 2$ be the maximum of numbers of channels required in vertices. If an instance is defined on a perfect graph G, then $\alpha \leq 1 + (1 + \frac{1}{W-1})H_{\omega(G)}$, where H_i denotes the i-th harmonic number. For any instance defined on a unit disk graph G, α is less than or equal to $(1 + \frac{1}{W-1})(3H_{\omega(G)} - 1)$. If a given graph is 4 or 3 colorable, α is bounded by $(2.5 + \frac{1.5}{W-1})$ and $(2 + \frac{1}{W-1})$, respectively. We also discuss well-known practical instances called Philadelphia instances and propose an algorithm with $\alpha \leq 12/5$.

1 Introduction

In this paper, we denote the set of non-negative (positive) integers by $\mathbb{Z}_+$ ($\mathbb{N}$), respectively. Let $G = (V, E)$ be an undirected simple graph. We introduce a non-negative integer vertex weight function (vector) $w : V \to \mathbb{Z}_+$, a non-negative integer edge length function (vector) $l : E \to \mathbb{Z}_+$ and a non-negative integer $k \in \mathbb{Z}_+$. A *channel assignment* of (G, w, l, k) is an assignment $\phi : V \to 2^{\mathbb{N}}$ such that

$$|\phi(v)| \geq w(v) \qquad (\forall v \in V),$$
$$|c_1 - c_2| \geq k \qquad (\forall c_1, \forall c_2 \in \phi(v), \ \forall v \in V),$$
$$|c_1 - c_2| \geq l(\{u,v\}) \quad (\forall c_1 \in \phi(u), \ \forall c_2 \in \phi(v), \ \forall\{u,v\} \in E).$$

Given a channel assignment ϕ, a *span* of the channel assignment ϕ, denoted by $\mathrm{span}(\phi)$, is defined by $\mathrm{span}(\phi) \overset{\mathrm{def}}{=} \max\{c_1 - c_2 + 1 \mid c_1, c_2 \in \cup_{v \in V}\phi(v)\}$. A minimum span channel assignment problem (G, w, l, k) finds a channel assignment ϕ of (G, w, l, k) which minimizes $\mathrm{span}(\phi)$.

The minimum span channel assignment problem is a kind of discrete versions of channel assignment problems which originated from wireless communication networks [1, 2]. Channel assignment problems are also called frequency assignment problems and/or radio channel assignment problem. For problems related to radio channel assignment problems, see McDiarmid's survey papers [3, 4]. The minimum span channel assignment problem includes the ordinary vertex coloring problem as a special case. It is known that if there exists a polynomial time

S.-W. Cheng and C.K. Poon (Eds.): AAIM 2006, LNCS 4041, pp. 334–342, 2006.

algorithm for the vertex coloring problem which provides a performance guarantee of $O(n^\epsilon)$ for a constant $\epsilon > 0$, then P=NP [5]. When a given graph is perfect, we can solve the vertex coloring problem in polynomial time [6].

In this paper, we propose approximation algorithms for minimum span channel assignment problems. Let α be the approximation ratio of our algorithm and $W \geq 2$ be the maximum of numbers of vertex weights. If an instance is defined on a perfect graph G, then $\alpha \leq 1 + (1 + \frac{1}{W-1})H_{\omega(G)}$, where H_i denotes the i-th harmonic number. For any instance defined on a unit disk graph G, α is less than or equal to $(1 + \frac{1}{W-1})(3H_{\omega(G)} - 1)$. If a given graph is 4 or 3 colorable, α is bounded by $(2.5 + \frac{1.5}{W-1})$ and $(2 + \frac{1}{W-1})$, respectively. We also discuss well-known practical instances called Philadelphia instances and propose an algorithm with $\alpha \leq 12/5$.

2   Algorithms

In this section, we describe our algorithm for the case that $0 < \exists l \leq k, \forall e \in E, l(e) = l$. In this case, the minimum span channel assignment problem remains NP-hard, since the problem still includes the vertex coloring problem. First, we introduce a naive algorithm. Given a graph G and a vertex weight w, $(G, w)_+$ denotes a subgraph of G induced by a set of vertices with positive weights. Throughout this paper, W denotes the maximum of vertex weights.

Algorithm 1
Step 1: Find a coloring of graph $(G, w)_+$. Let c be a number of used colors.
Step 2: For each vertex v whose color is $i \in \{1, \ldots, c\}$, we assign channels

$$\{1 + (i - 1)l, 1 + (i - 1)l + \max\{cl, k\},$$
$$\ldots, 1 + (i - 1)l + (w(v) - 1)\max\{cl, k\}\}.$$

It is clear that Algorithm 1 gives a channel assignment to the problem (G, w, l, k) whose span is less than or equal to

$$1 + (c - 1)l + (W - 1)\max\{cl, k\} \leq cl + (W - 1)\max\{cl, k\} \leq W\max\{cl, k\}.$$

Since $l \leq k$, the span is bounded by $1 + (W - 1)k$, if $c = 1$. The sequence

$$(1 + (i - 1)l, 1 + (i - 1)l + \max\{cl, k\}, \ldots, 1 + (i - 1)l + (w(v) - 1)\max\{cl, k\})$$

is an arithmetic sequence with the first term $1 + (i - 1)l$, common difference of $\max\{cl, k\}$ and length of $w(v)$. Thus we can output the set of channels by a triplet $(1 + (i - 1)l, \max\{cl, k\}, w(v))$. If we use a polynomial time coloring algorithm in Step 1 and output a channel assignment by a set of triplets representing arithmetic sequences, then the time complexity of Algorithm 1 is bounded by a polynomial of the input size of the problem (G, w, l, k). Here we note that the input size of the problem (G, w, l, k) is $O(|V|\lceil \log W + 1\rceil + |E|\lceil \log L + 1\rceil + \lceil \log k \rceil)$, where L denotes the maximum of edge lengths.

Next, we propose our algorithm. Let $(G, w)_{\geq d}$ be a subgraph of G induced by $\{v \in V \mid w(v) \geq d\}$. For any graph G', $\omega(G')$ and $\chi(G')$ denote the clique number and the chromatic number of G', respectively. Clearly, $d \leq d'$ implies that $\omega((G, w)_{\geq d}) \geq \omega((G, w)_{\geq d'})$. Throughout this paper, n denotes the number of vertices in G. We define a sequence $(W(1), W(2), \ldots, W(n))$ by

$$W(q) \stackrel{\text{def.}}{=} \begin{cases} 0 & (\text{ if } q > \omega(G)), \\ \max\{d \mid \omega((G, w)_{\geq d}) \geq q\} & (\text{ if } q \leq \omega(G)). \end{cases}$$

The definition above implies that $W(1) = W$. Let $(w^1, w^2, \ldots, w^n)$ be a sequence of vertex weight vectors defined by

$$w^n(v) + w^{n-1}(v) + \cdots + w^q(v) = \min\{W(q), w(v)\} \quad (\forall v \in V, \ 1 \leq \forall q \leq n).$$

Clearly, the equality $w^1 + w^2 + \cdots + w^n = w$ holds. Our algorithm finds a channel assignment ϕ^q for each problem (G, w^q, l, k) by applying Algorithm 1, independently. Lastly, we output a channel assignment ϕ defined by

$$\phi(v) = \cup_{q=1}^{n}\{c + p^0 + p^1 + \cdots + p^{q-1} \mid c \in \phi^q(v)\} \ (\forall v \in V)$$

where
$$p^q = \begin{cases} 0 & (\text{ if } q = 0 \text{ or span}(\phi^q) = 0), \\ \text{span}(\phi^q) + k - 1 & (\text{ if } q \geq 1 \text{ and span}(\phi^q) > 0). \end{cases}$$

The span of an assignment ϕ is bounded by $\sum_{q=1}^{n} \text{span}(\phi^q) + (k-1)(W-1)$, since the inequality $|\{q \in \{1, 2, \ldots, n\} \mid \text{span}(\phi^q) > 0\}| \leq W$ holds. We briefly describe our algorithm below.

Algorithm 2
Step 1: When $W = 1$, solve the problem by Algorithm 1 and stop.
Step 2: Obtain the sequence $(W(1), \ldots, W(n))$.
Step 3: Construct the sequence of vertex weights $(w^1, w^2, \ldots, w^n)$ defined by
$w^q(v) := \min\{W(q), w(v)\} - \min\{W(q+1), w(v)\} \ (\forall v \in V, \ 1 \leq \forall q < n)$
and $w^n(v) := \min\{W(n), w(v)\} \ (\forall v \in V)$.
Step 3: For each $q \in \{1, 2, \ldots, n\}$, we solve the problem (G, w^q, l, k) by Algorithm 1 and output an assignment obtained by merging n assignments.

3   Approximation Ratios

In this section, we discuss the approximation ratio of our algorithm. Throughout this section, we assume that $0 < \exists l \leq k, \forall e \in E, l(e) = l$.

3.1   Lower Bounds

First, we give some lower bounds which plays an important role to analyze the approximation ratio.

Lemma 1. *The optimal span Z^* of the problem (G, w, l, k) satisfies that*

$$k(W - 1) \leq Z^* - 1, \text{ and } qlW(q) \leq Z^* + l - 1 \ (\forall q \in \{2, 3, \ldots, n\}).$$

Proof. The existence of a vertex v satisfying $w(v) = W$ implies that $1 + k(W - 1) \leq Z^*$. From the definition of $W(q)$ ($2 \leq q \leq n$), the graph $(G, w)_{\geq W(q)}$ contains a clique $Q \subseteq V$ satisfying $|Q| = q$ and $w(v) \geq W(q)$ ($\forall v \in Q$). Thus $1 + l(qW(q) - 1) \leq 1 + l(\sum_{v \in Q} w(v) - 1) \leq Z^*$ holds.   □

When $\omega(G) = 1$, we can use the trivial 1-coloring at Step 1 of Algorithm 1, and Algorithm 2 finds an optimal solution. Now we discuss the approximation ratio of Algorithm 2 in case $\omega(G) \geq 2$ and $W = 1$.

Lemma 2. *When $W = 1$ and $\omega(G) \geq 2$, the approximation ratio of Algorithm 2 is less than or equal to $(c - 1)/(\omega(G) - 1)$.*

Proof. In case $W = 1$, Algorithm 2 is essentially equivalent to Algorithm 1. When $W = 1$, Algorithm 1 finds a channel assignment whose span is bounded by $1 + (c - 1)l$. Lemma 1 shows that $Z^* \geq 1 - l + lqW(q)$ ($2 \leq \forall q \leq n$). By setting $q = \omega(G)$, the inequality $c \geq \omega(G) \geq 2$ implies that the approximation ratio is bounded by

$$\frac{1 + (c - 1)l}{Z^*} \leq \frac{1 + (c - 1)l}{1 - l + l\omega(G)W(\omega(G))} = \frac{1 + (c - 1)l}{1 + (\omega(G) - 1)l} \leq \frac{c - 1}{\omega(G) - 1}.$$ □

Let $w^1, w^2, \ldots, w^n$ be the decomposition of vertex weight vector w defined in the previous section. The following lemma gives a lower bound of the number of colors required for coloring $(G, w^q)_+$.

Lemma 3. *The clique number of the graph $(G, w^q)_+$ is less than or equal to q.*

Proof. Assume that $(G, w^q)_+$ has a clique Q whose size is $q + 1$. For any vertex $v \in Q$, $0 < w^q(v) = \min\{W(q), w(v)\} - \min\{W(q + 1), w(v)\}$ and thus $w(v) > W(q + 1)$. It implies that clique Q is contained in the graph $(G, w)_{\geq W(q+1)+1}$. Contradiction.   □

3.2   Perfect Graphs

A graph G is perfect if and only if every induced subgraph of G satisfies $\chi(G) = \omega(G)$. In Step 1 of Algorithm 1, we can find a coloring of $(G, w^q)_+$ with $\chi((G, w^q)_+) = \omega((G, w^q)_+)$ colors in polynomial time [6]. We can obtain the sequence $(W(1), W(2), \ldots, W(n))$ by using ordinary binary search technique in polynomial time. Thus, the total time complexity of Algorithm 2 is bounded by a polynomial of the input size of the problem, where we output a channel assignment by a set of triplets for representing arithmetic sequences.

Theorem 1. *Let G be a perfect graph and $\mathcal{I}_G$ a class of instances of minimum span channel assignment problems defined on G.*

For any instance $I \in \mathcal{I}_G$ with $W = 1$, Algorithm 2 finds an optimal solution of I. For any instance $I \in \mathcal{I}_G$ with $W \geq 2$, the approximation ratio of Algorithm 2 is bounded by $1 + (1 + \frac{1}{W-1})\mathrm{H}_{\omega(G)}$, where H_i denotes the i-th harmonic number.

Proof. We omit the trivial case that $\omega(G) = 1$. When $W = 1$ and $\omega(G) \geq 2$, Lemma 2 implies that the approximation ratio is bounded by $(\chi(G)-1)/(\omega(G)-1) = 1$ and thus Algorithm 2 finds an optimal solution.

Next, we consider the case that $W \geq 2$. Lemma 3 and perfectness of G implies that $\chi((G, w^q)_+) = \omega((G, w^q)_+) \leq q$. For each $q \in \{1, 2, \ldots, n\}$, Algorithm 1 finds an optimal coloring of the graph $(G, w^q)_+$ and outputs a channel assignment whose span is bounded by $W^q \max\{ql, k\}$, where W^q denotes the maximum of vertex weights $w^q(v)$ $(v \in V)$. In the following, we define $W(n+1) = 0$. Let A be the span of a channel assignment obtained by Algorithm 2 and Z^* the optimal span. Then the following inequalities hold;

$$
\begin{aligned}
A &\leq (k-1)(W-1) + \sum_{q=1}^{n} W^q \max\{ql, k\} \\
&\leq k(W-1) + \sum_{q=1}^{n}(W(q) - W(q+1)) \max\{ql, k\} \\
&= k(W-1) + \sum_{q=1}^{\omega(G)}(W(q) - W(q+1)) \max\{ql, k\} \\
&\leq Z^* + W(1)\max\{l, k\} + \sum_{q=2}^{\omega(G)} W(q)(\max\{ql, k\} - \max\{(q-1)l, k\}) \\
&\leq Z^* + Wk + \sum_{q=2}^{\omega(G)} W(q)l \leq Z^* + (Z^* + k - 1) + (Z^* + l - 1)\sum_{q=2}^{\omega(G)}(1/q) \\
&\leq Z^* + (Z^* + k) + (Z^* + k)\sum_{q=2}^{\omega(G)}(1/q) \\
&\leq Z^* + (Z^* + \tfrac{Z^*-1}{W-1})\sum_{q=1}^{\omega(G)}(1/q) \leq (1 + (1 + \tfrac{1}{W-1})H_{\omega(G)})Z^* \qquad \square
\end{aligned}
$$

3.3   Unit Disk Graphs

A graph $G = (V, E)$ is called a *unit disk graph* if the vertex set V is a point-set on 2-dimensional space and a pair of vertices is adjacent if and only if the Euclidean distance between them is less than or equal to 1. There exists a polynomial time algorithm for finding a vertex coloring of a given unit disk graph G satisfying that the required number of colors is bounded by $3\omega(G) - 2$ (see [7, 8]). We can employ the algorithm in Step 1 of Algorithm 1. Since the maximum clique problem defined on a unit disk graph G is solvable in polynomial time (see [9]), we can find the sequence $(W(1), \ldots, W(n))$ in polynomial time by employing ordinary binary search technique. If we adopt the procedures above, Algorithm 2 becomes a polynomial time algorithm.

Theorem 2. *Let G be a unit disk graph and $\mathcal{I}_G$ a class of instances of minimum span channel assignment problems defined on G.*

For any $I \in \mathcal{I}_G$ with $W = 1$, the approximation ratio of Algorithm 2 is bounded by 3. For any instance $I \in \mathcal{I}_G$ with $W \geq 2$, the approximation ratio of Algorithm 2 is bounded by $(1 + \tfrac{1}{W-1})(3H_{\omega(G)} - 1)$.

Proof. We omit the trivial case that $\omega(G) = 1$. In case $W = 1$ and $\omega(G) \geq 2$, Lemma 2 implies that the approximation ratio is bounded by $(c-1)/(\omega(G)-1) \leq (3\omega(G) - 2 - 1)/(\omega(G) - 1) = 3$.

Next, we consider the case that $W \geq 2$. For each $q \in \{1, 2, \ldots, n\}$, Lemma 3 implies that Algorithm 1 finds a coloring of graph $(G, w^q)_+$ with at most $3q - 2$

colors and outputs an assignment whose span is bounded by $W^q \max\{(3q - 2)l, k\}$, where W^q denotes the maximum of vertex weights $w^q(v)$ $(v \in V)$. In the following, we define $W(n+1) = 0$. The span A of a channel assignment obtained by Algorithm 2 and the optimal span Z^* satisfy the following,

$$
\begin{aligned}
A &\leq (k-1)(W-1) + \textstyle\sum_{q=1}^{n} W^q \max\{(3q-2)l, k\} \\
&\leq k(W-1) + \textstyle\sum_{q=1}^{n}(W(q) - W(q+1)) \max\{(3q-2)l, k\} \\
&= k(W-1) + \textstyle\sum_{q=1}^{\omega(G)}(W(q) - W(q+1)) \max\{(3q-2)l, k\} \\
&= Z^* + W(1)\max\{l, k\} + \textstyle\sum_{q=2}^{\omega(G)} W(q)(\max\{(3q-2)l, k\} - \max\{(3q-5)l, k\}) \\
&\leq Z^* + Wk + \textstyle\sum_{q=2}^{\omega(G)} W(q)3l \leq Z^* + (Z^* + k - 1) + (Z^* + l - 1)\textstyle\sum_{q=2}^{\omega(G)}(3/q) \\
&\leq 2(Z^* + k) + (Z^* + k)\textstyle\sum_{q=2}^{\omega(G)}(3/q) \leq (Z^* + k)3\textstyle\sum_{q=1}^{\omega(G)}(1/q) - (Z^* + k) \\
&\leq (Z^* + \tfrac{Z^*-1}{W-1})(3\mathrm{H}_{\omega(G)} - 1) \leq (1 + \tfrac{1}{W-1})(3\mathrm{H}_{\omega(G)} - 1)Z^* \qquad \square
\end{aligned}
$$

3.4   General Cases

In the previous subsections, we dealt with perfect graphs and unit disk graphs. Here, we extend our results to general cases.

First, we consider a graph class $\mathcal{G}$ satisfying following three properties:

1. for any $G \in \mathcal{G}$, every induced subgraph H of G is in $\mathcal{G}$,
2. there exists a polynomial time α-approximation algorithm for the coloring problem instances defined on graphs in $\mathcal{G}$,
3. there exists a polynomial time exact algorithm for the maximum clique problem instances defined on graphs in $\mathcal{G}$.

Then Algorithm 2 is a polynomial time $\left(1 + \left(1 + \frac{1}{W-1}\right)\alpha\mathrm{H}_{\omega(G)}\right)$-approximation algorithm for a class of instances of minimum span channel assignment problem defined on graphs in $\mathcal{G}$. If $\mathcal{G}$ is the class of perfect graphs and unit disk graphs, then $\alpha = 1$ and $\alpha = 3$ hold, respectively.

Next, we consider a case that we have a c-coloring of a given (general) graph G. In the following, we describe a modified version of Algorithm 2. The definition of $W(2)$ implies that $W(2)$ attains the minimum value of W' subject to the condition that the graph $G_{\geq W'+1}$ is a set of isolated vertices. Thus, we can calculate the value $W(2)$ in polynomial time. We introduce vertex weight vectors $\widetilde{w}^1, \widetilde{w}^2$ defined by $\widetilde{w}^2(v) = \min\{W(2), w(v)\}$ and $\widetilde{w}^1 = w - \widetilde{w}^2$. Our algorithm applies Algorithm 1 to problems $(G, \widetilde{w}^1, l, k)$, $(G, \widetilde{w}^2, l, k)$, and obtain two assignments ϕ^1 and ϕ^2, independently. When we solve $(G, \widetilde{w}^2, l, k)$, we use a given c-coloring of G in Step 1 of Algorithm 1. When we solve $(G, \widetilde{w}^1, l, k)$, we use the trivial 1-coloring of $(G, \widetilde{w}^1)_+$. Then we output a channel assignment defined by $\phi(v) = \phi^1(v) \cup \{c + (k-1) + \mathrm{span}(\phi^1) \mid c \in \phi^2(v)\}$ $(\forall v \in V)$.

Theorem 3. *Let G be a c-colorable graph with $c \geq 3$ and $\mathcal{I}_G$ a class of instances of minimum span channel assignment problems defined on G.*

For any $I \in \mathcal{I}_G$ with $W = 1$, the approximation ratio of Algorithm 2 is bounded by $c - 1$. For any $I \in \mathcal{I}_G$ with $W \geq 2$, the approximation ratio of Algorithm 2 is bounded by $(1/2)(1 + c + \frac{c-1}{W-1})$.

Proof. We omit the trivial case that $W(2) = 0$. If $W(2) > 0$, then $\omega(G) \geq 2$. In case that $W = 1$ and $W(2) > 0$, Lemma 2 implies that the approximation ratio is bounded by $(c - 1)/(\omega(G) - 1) \leq c - 1$.

Next, we consider the case that $W \geq 2$ and $W(2) > 0$. Algorithm 1 finds a channel assignment of $(G, \widetilde{w}^1, l, k)$ whose span is bounded by $1 + (\widetilde{W}^1 - 1)k$ where $\widetilde{W}^1$ is the maximum of vertex weights $\widetilde{w}^1(v)$ $(v \in V)$. For the problem $(G, \widetilde{w}^2, l, k)$, Algorithm 1 outputs a channel assignment whose span is bounded by $1 + (c - 1)l + (\widetilde{W}^2 - 1)\max\{cl, k\}$ where $\widetilde{W}^2$ denotes the maximum of vertex weights $\widetilde{w}^2(v)$ $(v \in V)$. The span A of a channel assignment obtained by Algorithm 2 and the optimal span Z^* satisfy that

$$
\begin{aligned}
A &\leq (1 + (\widetilde{W}^1 - 1)k) + (k - 1) + (1 + (c - 1)l + (\widetilde{W}^2 - 1)\max\{cl, k\}) \\
&\leq (1 + (W - W(2) - 1)k) + (k - 1) + (1 + (c - 1)l + (W(2) - 1)\max\{cl, k\}) \\
&= (1 + Wk - k) + (c - 1)l + (W(2) - 1)(\max\{cl, k\} - k) \\
&= (1 + Wk - k) + (c - 1)l + (W(2) - 1)(\max\{cl, k\} - \max\{l, k\}) \\
&\leq Z^* + (c - 1)l + (W(2) - 1)(cl - l) = Z^* + W(2)l(c - 1) \\
&\leq Z^* + \tfrac{Z^* + l - 1}{2}(c - 1) = Z^* + (1/2)(c - 1)Z^* + (1/2)(c - 1)(l - 1) \\
&\leq (1/2)(1 + c)Z^* + (1/2)(c - 1)k \leq (1/2)\left(1 + c + \tfrac{c-1}{W-1}\right)Z^*. \qquad \square
\end{aligned}
$$

When a given graph G is planar, G is 4-colorable. Some instances of the minimum span channel assignment problem are defined on a triangular lattice, which is 3-colorable. The above theorem gives bounds of the approximation ratio of our algorithm for these cases.

Corollary 1. *Assume that $W \geq 2$. When a given graph G is 4-colorable, the approximation ratio of Algorithm 2 is bounded by $2.5 + 1.5/(W - 1)$. When we have a 3-coloring of a given graph, the approximation ration of Algorithm 2 is bounded by $2 + 1/(W - 1)$.*

4   Philadelphia Instances

In this section, we discuss a class of practical instances in *Philadelphia Instances* [10]. A triangular lattice graph $T_{m,n}$ has a vertex set $\{(x e_1 + y e_2) \mid x \in \{0, 1, 2, \ldots, m-1\}, y \in \{0, 1, 2, \ldots, n-1\}\}$ where $e_1 \overset{\text{def.}}{=} (1, 0)$, $e_2 \overset{\text{def.}}{=} (1/2, \sqrt{3}/2)$, and an edge set consists of pairs of vertices with unit distance. In this section, we denote $T_{m,n}$ by T for simplicity. We denote the 3rd power of the graph T by T^3. Here, we consider a class of Philadelphia instances, which is defined on T^3. Let $l^*(e)$ be a weight of an edge e in T^3 defined by

$$
l^*(e) \overset{\text{def.}}{=} \begin{cases} 2 & (e \text{ is an edge in } T), \\ 1 & (\text{otherwise}). \end{cases}
$$

The minimum span channel assignment problem $(T^3, w, l^*, 5)$ includes a class of the Philadelphia instances [10]. The minimum span channel assignment problem $(T^3, w, l^*, 5)$ is NP-hard, since the problem includes the multicoloring problem defined on T^3 which is known to be NP-hard.

Figure 1 shows a coloring c^* of T^3 defined by

$$c^*(v) \stackrel{\text{def.}}{=} \begin{cases} 1, & v \in \{(2x + 4y)e_1 + (2x - 2y)e_2 + (0,0) \mid x, y \in \mathbb{Z}\}, \\ 2, & v \in \{(2x + 4y)e_1 + (2x - 2y)e_2 + (0,2) \mid x, y \in \mathbb{Z}\}, \\ 3, & v \in \{(2x + 4y)e_1 + (2x - 2y)e_2 + (2,0) \mid x, y \in \mathbb{Z}\}, \\ 4, & v \in \{(2x + 4y)e_1 + (2x - 2y)e_2 + (-1,0) \mid x, y \in \mathbb{Z}\}, \\ 5, & v \in \{(2x + 4y)e_1 + (2x - 2y)e_2 + (-1,2) \mid x, y \in \mathbb{Z}\}, \\ 6, & v \in \{(2x + 4y)e_1 + (2x - 2y)e_2 + (1,0) \mid x, y \in \mathbb{Z}\}, \\ 7, & v \in \{(2x + 4y)e_1 + (2x - 2y)e_2 + (0,-1) \mid x, y \in \mathbb{Z}\}, \\ 8, & v \in \{(2x + 4y)e_1 + (2x - 2y)e_2 + (2,-1) \mid x, y \in \mathbb{Z}\}, \\ 9, & v \in \{(2x + 4y)e_1 + (2x - 2y)e_2 + (0,1) \mid x, y \in \mathbb{Z}\}, \\ 10, & v \in \{(2x + 4y)e_1 + (2x - 2y)e_2 + (1,-1) \mid x, y \in \mathbb{Z}\}, \\ 11, & v \in \{(2x + 4y)e_1 + (2x - 2y)e_2 + (-1,1) \mid x, y \in \mathbb{Z}\}, \\ 12, & v \in \{(2x + 4y)e_1 + (2x - 2y)e_2 + (-1,-1) \mid x, y \in \mathbb{Z}\}. \end{cases}$$

Since T^3 has a clique of size 12, the coloring c^* is an optimal coloring of T^3.

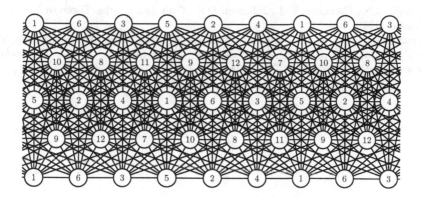

Fig. 1. The coloring c^* of a graph $T^3_{m,5}$

It is clear that Algorithm 1 gives a channel assignment to the problem $(T^3, w, l^*, 5)$ whose span is at most $12 + 12(W - 1) = 12W$, when the coloring c^* is used in Step 1 of Algorithm 1. Since Lemma 1 holds for the problem $(T^3, w, l^*, 5)$, the optimal span of the problem is at least $1 + 5(W - 1) = 5W - 4$. Thus we have the following.

Lemma 4. *Algorithm 1 (with c^*) finds a channel assignment of $(T^3, w, l^*, 5)$ whose span A and the optimal span Z^* satisfies*

$$A \leq 12W = \frac{12}{5}(5W - 4) + \frac{48}{5} \leq \frac{12}{5}Z^* + \frac{48}{5}.$$

In a similar way, we can estimate approximation ratio of Algorithm 1 for each 'Philadelphia instance'.

References

1. Aardal, K.I., van Hoesel, S.P.M., Koster, A.M.C.A., Mannino, C., Sassano, A.: Models and solution techniques for frequency assignment problems. Technical Report 01-40, ZIB (2001)
2. Koster, A.M.C.A.: Frequency Assignment—Models and Algorithms. PhD thesis, Maastricht University (1999)
3. McDiarmid, C.: Graph imperfection and channel assignment. In Alfonsín, J.L.R., Reed, B.A., eds.: Perfect Graphs. Wiley (2001)
4. McDiarmid, C.: Discrete mathematics and radio channel assignment. In: Recent Advances in Algorithms and Combinatorics. Springer (2003)
5. Arora, S., Lund, C., Motowani, R., Sudan, M., Szegedy, M.: Proof verification and intractability of approximation problems. In: Proceedings of the 33rd IEEE Symposium on Foundation of Computer Science. (1992) 13–22
6. Grötschel, M., Lovász, L., Schrijver, A.: Geometric Algorithms and Combinatorial Optimization. Springer-Verlag (1993)
7. Marathe, M.V., Breu, H., Hunt, H.B., Ravi, S.S., Rosenkrantz, D.J.: Simple heuristics for unit disk graph. Networks **25** (1995) 59–68
8. Hochbaum, D.S.: Efficient bounds for the stable set, vertex cover and set packing problems. Discrete Applied Mathematics **6** (1983) 243–254
9. Clark, B.N., Colbourn, C.J., Johnson, D.S.: Unit disk graphs. Discrete Mathematics **86** (1990) 165–177
10. Anderson, L.G.: A simulation study of some dynamic channel assignment algorithms in a high capacity mobile telecommunications system. IEEE Transactions on Communications **21** (1973) 1294–1301

Weighted Broadcast in Linear Radio Networks

Gautam K. Das and Subhas C. Nandy

Indian Statistical Institute, Kolkata - 700 108, India

Abstract. The non-homogeneous version of the range assignment problem in Ad-Hoc wireless network is studied in the context of information *broadcast* and *accumulation*. Efficient algorithms are presented for the unbounded and bounded-hop *broadcast* problems for a set of radio-stations when they are placed on a straight line. This improves time complexity of the existing results for the same two problems by a factor of n, where n is the number of radio-stations [2]. An easy to implement algorithm for the unbounded version of *accumulation* problem is also presented for the non-homogeneous version of the range assignment problem. Its worst case running time complexity is $O(n^2)$. The same algorithm works when the radio-stations are placed in R^d.

1   Introduction

In Ad-Hoc wireless network, the *range assignment* problem is studied extensively in the context of information broadcast, accumulation and all-to-all communication [15]. Here, a set of radio-stations $S = \{s_1, s_2, \ldots, s_n\}$ is assumed to be placed in $I\!\!R^d$, $d \geq 1$. If a radio-station is assigned a range ρ, it can communicate with any other radio-station(s) located in the hyper-sphere of radius ρ. The cost of assigning a range ρ_i to a radio-station s_i is $w_i \times \rho_i^\mu$, where μ is a fixed constant, which is assumed to be 2 for all practical applications, and w_i is a constant specified apriori for each radio-station s_i. In the *broadcast* problem, a source node $s^* \in S$ is specified, and we need to assign ranges to the radio-stations in S such that s^* can broadcast message to all other nodes in S using at most h hops $(1 \leq h \leq n-1)$. In *accumulation* problem, all the members of S need to send the message to the source station s^*. In both cases, the objective is to minimize the total power required for all the members in S. In each of these problems, if the restriction on the number of hops h is not mentioned, (i.e., $h = n-1$ in the worst case), then it is referred to as the *unbounded* version of that problem. There is another classification of the broadcast/accumulation range assignment problems, namely homogeneous and non-homogeneous cases. In the homogeneous case, w_i values are assumed to be same for all i. But, the practical environment insists studying the *non-homogeneous* version, where w_i values differ due to several locality parameters of the radio-stations.

The homogeneous h-hop broadcast range assignment problem $(h > 2)$ in $I\!\!R^d$ is known to be NP-complete even for $d = 2$ [4]. A dynamic programming based algorithm is proposed in [1] for the 2-hop broadcast range assignment in $I\!\!R^2$ which needs $O(n^7)$ time in the worst case. Approximation algorithms are

S.-W. Cheng and C.K. Poon (Eds.): AAIM 2006, LNCS 4041, pp. 343–353, 2006.

available for the unbounded broadcast problem (i.e., $h = n - 1$) in $I\!\!R^2$, with approximation factor equal to 12 [16]. In [7], the homogeneous h-hop broadcast range assignment problem is considered for $d = 1$. Here the radio-stations are placed on a straight line, and the proposed algorithm runs in $O(hn^2)$ time. The time complexity of this problem is improved to $O(n^2)$ in [9]. For a detailed survey in the broadcast range assignment problem, see [5, 14].

Though there is a long history of the homogeneous broadcast range assignment problem, the non-homogeneous version is studied very little. The first work on this problem appeared in [2]. A number of variations of the problem studied and algorithms are proposed using dynamic programming, as stated below.

For the unbounded case (i.e., $h = n - 1$), the time and space complexities of the proposed algorithm are $O(n^3)$ and $O(n^2)$ respectively.

For the h-hop broadcast, the time and space complexities of the proposed algorithms are $O(hn^4)$ and $O(hn^2)$ respectively.

For the unbounded multisource broadcast, the time and space complexities are $O(n^6)$ and $O(n^2)$ respectively.

In higher dimension (i.e., $d > 2$) and $\mu = 1$, the problem is formulated as a shortest path problem in a graph, and the proposed algorithm produces a 3-approximation result in $O(n^3)$ time.

For $\mu > 1$, the proposed algorithm works for some special (mentioned as q-spread) instances and produces q-approximation algorithm for all-to-all communication.

For the linear radio-network, the homogeneous h-hop *accumulation* range assignment problem is studied in [3], and an $O(hn^3)$ time algorithm is proposed using dynamic programming. Next, the proposed algorithm is used to design a 2-approximation algorithm for h-hop all-to-all communication. Heuristic algorithms are proposed in [6] for the homogeneous h-hop accumulation problem in $I\!\!R^2$. The performance analysis of the algorithms are done assuming the distribution of points (radio-stations) from different bivariate statistical distributions.

We will consider the non-homogeneous version of the broadcast /accumulation problem in linear radio network. Using graph-theoretic formulation, we propose algorithms for unbounded and bounded hop broadcast range assignment problem with time complexities $O(n^2)$ and $O(hn^3)$ respectively. This is an improvement on the existing results (in [2]) by a factor of n in both the cases. Next, we consider the unbounded version of the accumulation problem in the non-homogeneous case. Our proposed algorithm runs in $O(n^2)$ time and it works for any arbitrary dimension (i.e., $d \geq 1$). This result is important due to the fact that, the range assignment problem for the unbounded version of both broadcast and all-to-all communication are NP-complete for $d \geq 2$, but for the accumulation problem, it is solvable in polynomial time.

In spite of the fact that the model considered in this paper is simple, it is very much useful in studying road traffic information system where the vehicles follow roads and messages are to be broadcasted along lanes. Typically, the curvature of the road is small in comparison to the transmission range so that we can consider that the vehicles are moving on a line [3]. Linear radio networks have been observed to be important in several recent studies [3, 7, 11, 12, 13].

2   Preliminaries

Let $S = \{s_1, s_2, \ldots, s_n\}$ be the set of n radio-stations on a straight line from left to right. The radio-station s_j is assigned with an weight $w_j \geq 0$, for $j = 1, 2, \ldots, n$. Let $s_\alpha \in S$ be the source radio-station where from the message needs to be broadcast. Let $\mathcal{R} = \{\rho(s_1), \rho(s_2), \ldots, \rho(s_n)\}$ be a range assignment, where $\rho(s_i)$ be the range assigned to s_i. The graph $G = (V, E)$ with $V = S$ and $E = \{(s_i, s_j), \rho(s_i) \geq \delta(s_i, s_j)\}$ is referred to as the broadcast communication graph for the range assignment $\mathcal{R}$, where $\delta(s_i, s_j)$ denotes the Euclidean distance between s_i and s_j.

Definition 1. *An edge $e = (s_i, s_j)$ is said to be* functional *in the h-hop broadcast communication graph G if the removal of this edge indicates that there exists a radio-station $s_k \in S$ which is not reachable from s_α using a h-hop path.*

Consider a path $\Pi = \{s_\alpha = s_{i_1}, s_{i_2}, \ldots, s_{i_k} = s_j\}$ from source station s_α to a station s_j $(j > \alpha)$ in graph G corresponding to a range assignment $\mathcal{R}$. An edge $(s_{i_\alpha}, s_{i_{\alpha+1}})$ is said to be a *right back edge* if $i_{\alpha-1} < i_\alpha$ and $i_{\alpha+1} < i_\alpha$. Similarly, on a path from s_α to a node s_j $(j < \alpha)$ a *left back edge* can be defined.

Lemma 1. *If $\alpha = 1$ (resp. $\alpha = n$) (i.e. the source station is at one end of the linearly ordered radio-stations) then in the optimum (minimum cost) h-hop non-homogeneous broadcast range assignment, there is no functional right (resp. left) back edge.*

Proof. Let $\alpha = 1$. Suppose there exists a functional right back edge $e = (s_i, s_j)$ on the path Π in the communication graph corresponding to the optimum broadcast range assignment (see the dashed edge in Fig. 1(a)). Note that, $j < i$, and there are paths from s_α to both s_i and s_j without using that back edge. In order to communicate with a radio-station s_k, $k > i$, there is an path from s_j to s_k. Note

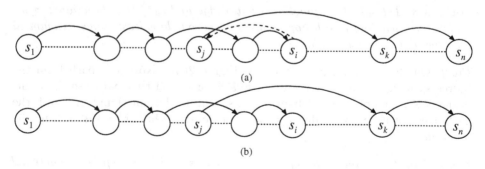

(a)

(b)

Fig. 1. Proof of Lemma 1

that the broadcast is still possible from s_α to all the members in S if we remove the edge e from graph G by setting $\rho(s_i) = 0$ (see Fig. 1(b)). Thus, we have a contradiction as the total power consumption gets reduced.

Definition 2. *In a h-hops broadcast range assignment $\mathcal{R}$ a left-bridge $\overrightarrow{s_a s_b}$ corresponds to a pair of radio-stations (s_a, s_b) such that $a < \alpha$, $b > \alpha$ and $\delta(s_a, s_b) \leq \rho(s_a) < \delta(s_a, s_{b+1})$.*

The bridge $\overrightarrow{s_a s_b}$ is said to be functional *if there exists a radio-station $s_i \in S$ such that no h-hop path from s_α to s_i exists which can avoid the direct (1-hop) communication $\overrightarrow{s_a s_b}$.*

Similarly, we can define a *right-bridge* ($\overleftarrow{s_d s_c}$ as in Fig. 2(b)) and a *functional right-bridge* in a broadcast range assignment $\mathcal{R}$.

Lemma 2. *If $\overrightarrow{s_a s_b}$ and $\overrightarrow{s_{a'} s_{b'}}$ are two functional left-bridges corresponding to a h-hops weighted broadcast range assignment $\mathcal{R}$ with $a < a'$, then $b' < b$.*

Proof. On the contrary, let $b' \geq b$ (see Figure 2(a)). Now any path from source s_α to s_a implies that there is also a path from s_α to $s_{a'}$. Since $b' \geq b$, all the radio-stations s_k ($\alpha < k \leq b$) is reachable using the left-bridge $\overrightarrow{s_{a'} s_{b'}}$. Thus, the left bridge $\overrightarrow{s_a s_b}$ no longer remains functional.

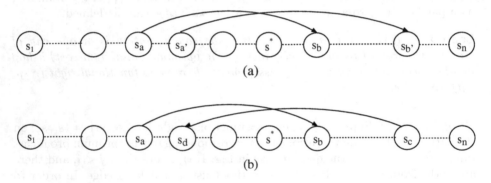

Fig. 2. (a) Contrary of Lemma 2 and (b) Contrary of Lemma 3

Lemma 3. *Let $\overrightarrow{s_a s_b}$ be a functional left-bridge and $\overleftarrow{s_d s_c}$ be a functional right-bridge corresponding to a h-hops non-homogeneous broadcast range assignment $\mathcal{R}$. Now, if $a < d$ then $c < b$.*

Proof. On the contrary, let $c \geq b$ (see Figure 2(b)). Now, the path from the source s_α to s_a can use the right-bridge $\overleftarrow{s_d s_c}$ or not. If that path use the right-bridge $\overleftarrow{s_d s_c}$, then obviously left-bridge $\overrightarrow{s_a s_b}$ will not be functional. Again, if the path does not use right-bridge $\overleftarrow{s_d s_c}$, then obviously right-bridge $\overleftarrow{s_d s_c}$ will not be functional.

Definition 3. *A functional left-bridge $\overrightarrow{s_{a^*} s_{b^*}}$ is said to be leftmost functional left-bridge in a h-hops non-homogeneous broadcast range assignment $\mathcal{R}$, if the* *following conditions hold: (i) there is no other functional left-bridge $\overrightarrow{s_a s_b}$ with $a < a^*$ and $b > b^*$ (see Fig. 3(a)), (ii) there is no functional right-bridge $\overrightarrow{s_d s_c}$ with $c > b^*$ and $d \leq a^*$ (see Fig. 3(b)), and (iii) there is no functional right-bridge $\overrightarrow{s_d s_c}$ with $c = b^*$ and $d < a^*$ (see Fig. 3(c)). Similarly, one can define a rightmost functional right-bridge $\overrightarrow{s_{d^*} s_{c^*}}$.*

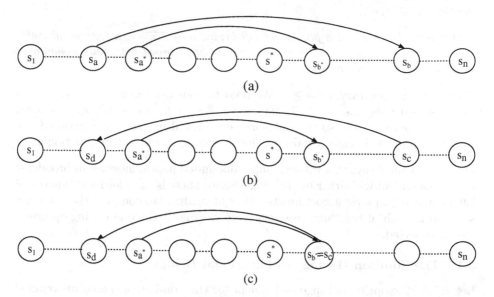

Fig. 3. Impossible configurations of leftmost functional left bridge

3   Unbounded Non-homogeneous Broadcast Problem

In this section, we consider unbounded version ($h = n - 1$) of the *non-homogeneous broadcast problem*. First we solve this problem when the source radio-station is at one end (i.e., $\alpha = 1$ or n).

Let M be an array of size n, where $M[i]$ is the cost of broadcasting from s_i to all the radio-stations in the set $S_i^+ = \{s_i, s_{i+1}, \ldots, s_n\}$. The following lemma helps in easy computation of the optimum range assignment.

Lemma 4. *If $i < n$ then $M[i] = \min_{k=i}^{n}(w_i \times (\delta(s_i, s_k))^2 + M[k])$, and if $i = n$ then $M[i] = 0$.*

Proof. The case where $i = n$ is trivial. So, we consider the case where $i < n$. It is clear that, if there is a path from s_i to s_n in the communication graph corresponding to a range assignment, then that range assignment is a feasible solution for the broadcast from s_i to all the members in S_i^+. By Lemma 1, in the optimum range assignment there is no functional back edge. Thus, there exists some $s_k \in S_i^+$ such that s_i first reaches s_k in 1-hop, and then reaches s_n in a minimum cost path. This proves the lemma.

Similarly, we can compute another array N of size n where $N[i]$ contains the cost of minimum broadcast range assignment from s_i to all the nodes in $S_i^- = \{s_1, s_2, \ldots, s_i\}$. By Lemma 3, we have the following result.

Lemma 5. *The computation of the arrays M and N needs $O(n^2)$ time.*

Next, we will consider the case where the source radio-station is in an arbitrary position α ($\in [1, n]$).

Lemma 6. *Let $\overrightarrow{s_a s_b}$ be a functional left-bridge and $\overrightarrow{s_d s_c}$ be a functional right-bridge corresponding to a unbounded $((n-1)$-hop) broadcast range assignment $\mathcal{R}$. Now, if $a > d$ then $c > b$.*

Proof. On the contrary, let $b \geq c$. We need to consider the following two cases : (i) $\delta(s_a, s_d) \leq \delta(s_b, s_c)$ and (ii) $\delta(s_a, s_d) > \delta(s_b, s_c)$. In case (i), s_d is covered by the range of s_a, and so the right-bridge $\overrightarrow{s_d s_c}$ will not remain functional. By similar argument, in case (ii) also the left-bridge $\overrightarrow{s_a s_b}$ will not be functional.

Lemmata 3 and 6 says that the optimum unbounded non-homogeneous broadcast range assignment will either be (i) bridge-free or there is (ii) a leftmost functional left-bridge or (iii) a rightmost functional right-bridge. We compute the optimum solution in each of the three cases separately. Finally, the one having optimum cost is reported.

3.1 The Solution Having No Functional Bridge

Let R_i^+ be the optimum range assignment for the non-homogeneous unbounded broadcast to the radio-stations $S_i^+ = \{s_i, s_{i+1}, \ldots, s_n\}$ from source station s_i, and R_i^- be the optimum range assignment for the non-homogeneous unbounded broadcast to the radio-stations $S_i^- = \{s_1, s_2, \ldots, s_i\}$ from source s_i. We can adopt the technique mentioned earlier using the array M to compute R_i^+ and R_i^-. Now, our algorithm executes as follows:

Step 1: Compute two arrays M and N.
Step 2: Assign $\rho(s_\alpha) = \max(\delta(s_\alpha, s_{\alpha-1}), \delta(s_\alpha, s_{\alpha+1}))$. Let $\rho(s_\alpha)$ corresponds to $s_{\alpha-1}$, and it covers $s_{\alpha+1}, s_{\alpha+2}, \ldots, s_j$ to the right.
Step 3: The cost of the optimum solution for assigning $\rho(s_\alpha) = \delta(s_\alpha, s_{\alpha-1})$ is obtained as

$$C = N[\alpha - 1] + w_\alpha \times (\rho(s_\alpha))^2 + \min_{i=\alpha+1}^{j} M[i]$$

Initialize *opt_cost* with C. In the last part of the expression for C, we have chosen the minimum among $M[i]$ values for $i = \alpha + 1, i = \alpha + 2 \ldots j$ because the weights assigned to the radio-stations may be different.
Step 4: Consider each element $s \in S_{\alpha-2}^- \cup S_{j+1}^+$ in order of their distances from s_α. Assign $\rho(s_\alpha) = \delta(s_\alpha, s)$, and use Step 3 for computing the cost of the optimum range assignment with the current range of s_α. If the present value of C is less than *opt_cost*, then update *opt_cost*.

3.2   The Solution Having Left-Most Functional Left-Bridge

We first compute a complete digraph G with the vertices corresponding to radio-stations S. The weight of a directed edge (s_a, s_b) is $w_a \times (\delta(s_a, s_b))^2$. Let P be an array with $P[a]$ containing the cost of the shortest path from s_α to s_a in the graph G. The array P can be computed using Dijkstra's single source shortest path algorithm in $O(n^2)$ time [8].

We consider the optimum broadcast range assignment with $\overline{s_d s_c}$ as the right-most functional right-bridge, for each pair (s_c, s_d), $s_c \in S^+_{\alpha+1}$ and $s_d \in S^-_{\alpha-1}$. Let this covers $s_{c+1}, s_{c+2}, \ldots, s_j$. The cost of this range assignment is computed as follows:

$$C = P[c] + w_c \times (\delta(s_c, s_d))^2 + N[d] + \min_{i=\alpha+1}^{j} M[i]$$

In order to expedite this computation, we may apply a linear pass prior to execution of this step. This computes an array M', where $M'[i] = \min_{i=\alpha+1}^{j} M[i]$.

The process is repeated to compute the cost of optimum broadcast range assignment with $\overline{s_a s_b}$ as the left-most functional left-bridge, for each pair (s_a, s_b), $s_a \in S^-_{\alpha-1}$ and $s_b \in S^+_{\alpha+1}$. The total time complexity of this pass is $O(n^2)$. Thus we have the following theorem:

Theorem 1. *The worst case time complexity of computing the non-homogeneous unbounded broadcast range assignment is $O(n^2)$.*

4   Non-homogeneous h-Hop Broadcast Problem

If the number of hops is restricted to a specified integer h ($\in [1, n-1]$), the graph-theoretic approach, described above, does not works. We apply the dynamic programming approach for solving this problem. We first compute three $n \times h$ matrices, namely A, B and C, whose each entry is a tuple (χ, γ), as mentioned below. These are used for computing the optimum cost range assignment for the non-homogeneous h-hop broadcast range assignment problems.

(a) $A[i, j].\chi =$ the minimum cost for communicating message from s_i to s_n using at most j hops, and $A[i, j].\gamma =$ index k of the radio-station in S where the first hop takes place for the minimum cost j-hops path from s_i to s_n, i.e., $\rho(s_i) = \delta(s_i, s_k)$). Note that, this broadcasts the message from s_i to all the radio-stations $S^+_{i+1} = \{s_{i+1}, s_{i+2}, \ldots, s_n\}$.

(b) $B[i, j].\chi =$ the minimum cost for communicating message from s_i to s_1 using at most j hops, and $B[i, j].\gamma =$ index k of the radio-station in S where the first hop takes place for the minimum cost j-hops path from s_i to s_1 as in (a) part. Note that, this broadcasts the message from s_i to all the radio-stations $S^-_{i-1} = \{s_{i-1}, s_{i-2}, \ldots, s_1\}$.

(c) $C[i, j].\chi =$ the minimum cost of communicating message from s_α to s_i using at most j hops, and $C[i, j].\gamma =$ index k of the radio-station in S where the last hop (to s_i) takes place in the minimum cost j-hops path from s_α to s_i.

We explain the computation of matrices A and C. The computation of the matrix B is similar to that of A.

The elements of the first row of matrix A are $A[i, 1] = (w_i \times (\delta(s_i, s_n))^2, n)$ for $i = 1, 2, \ldots, n$. After computing the $(j - 1)$-th row, the computation of the j-th row is as follows:

$A[i, j].\chi = \min_{k=i}^{n}(w_i \times (\delta(s_i, s_k))^2 + A[k, j-1].\chi)$. If the minimum is achieved for $k = \theta$, then we set $A[i, j].\gamma = \theta$.

The elements of the first row of matrix C are $C[i, 1] = (w_\alpha \times (\delta(s_\alpha, s_i))^2, \alpha)$, for $i = 1, 2, \ldots, n$. After computing the $(j - 1)$-th row, the computation of the j-th row is as follows:

$C[i, j].\chi = \min_{k=1}^{n}(C[k, j - 1].\chi + w_k \times (\delta(s_k, s_i))^2)$. If the minimum is achieved for $k = \theta$, then we set $C[i, j].\gamma = \theta$.

It is clear from the above discussion that the time required for computing the matrices A, B and C is $O(hn^2)$.

Lemma 7. *Let $\overrightarrow{s_a s_b}$, $\overrightarrow{s_{a'} s_{b'}}$ are two left-bridges and $\overleftarrow{s_d s_c}$, $\overleftarrow{s_{d'} s_{c'}}$ are two right-bridges such that (i) $a' < a$, (ii) $c' > c$, (iii) $d < a$ and $b > c$, and (iv) $d' < a'$ and $b' > c'$. Now, if $\overrightarrow{s_a s_b}$ and $\overleftarrow{s_d s_c}$ are functional then both of the $\overrightarrow{s_{a'} s_{b'}}$ and $\overleftarrow{s_{d'} s_{c'}}$ are not functional in the optimum non-homogeneous h-hop broadcast range assignment.*

Proof. Since $\overrightarrow{s_a s_b}$ and $\overleftarrow{s_d s_c}$ are functional, therefore, in the optimum solution the path from s_α to $s_{a'}$ and $s_{c'}$ do not contain the bridge $\overleftarrow{s_{d'} s_{c'}}$ and $\overrightarrow{s_{a'} s_{b'}}$ respectively. Let in the optimum solution $s_{a'}$ and $s_{c'}$ are reached from s_α using ℓ and k hop respectively. Now, if $\ell > k$, then reaching $s_{a'}$ using $\overleftarrow{s_{d'} s_{c'}}$ is sufficient, in this case $\overleftarrow{s_d s_c}$ is no more functional. By similar argument, if $\ell < k$, then $\overrightarrow{s_a s_b}$ is no more functional. Now, let $\ell = k$. In this case, either $s_{d'}$ is covered by the range of $s_{a'}$ or $s_{b'}$ is covered by the range of $s_{c'}$. Therefore, both of the $\overrightarrow{s_{a'} s_{b'}}$ and $\overleftarrow{s_{d'} s_{c'}}$ are not functional.

In the following two subsections we describe the method of computing the optimum range assignment for the h-hop broadcast with (i) no functional bridge and (ii) with the left-most functional left-bridge. The optimum solution with rightmost functional right-bridge is similarly computed.

4.1   The Solution Having No Functional Bridge

The algorithm for the non-homogeneous h-hop broadcast range assignment problem having no functional bridge is more or less similar to the algorithm with unbounded version of the same problem described in section 3. The only change is in the arrays M and N. The elements of the arrays M and N Changed is as follows such that $M[i] = A[i, h - 1]$ and $N[i] = B[i, h - 1]$.

4.2   The Solution Having Functional Bridge

In this subsection, we solve the problem with functional bridge. To solve this problem, we consider each of the left-bridge $\overrightarrow{s_a s_b}(a < \alpha < b)$ as the leftmost and right bridge $\overleftrightarrow{s_d s_c}(d < \alpha < c)$ as the rightmost functional bridge and find feasible range assignment. Here, we solve the problem having leftmost functional left-bridge. By similar argument, we can compute the solution having rightmost functional right-bridge.

The Solution with Leftmost Functional Left-Bridge

Let $\overrightarrow{s_i s_j}$ be the last functional left-bridge. We consider each such left-bridge as a last functional left-bridge. Now, we want to compute the optimum weighted broadcast range assignment with last functional left-bridge $\overrightarrow{s_i s_j}$. Let it reaches s_ℓ to the left of s_i. The algorithm is as follows :

For each $j(\ell \leq j \leq h)$, we compute the cost of following range assignment, and choose one having minimum cost.

Take the range assignment from s_α to s_i in k-hops with minimum cost and let for this range assignment $s_\ell, s_{\ell+1}, \ldots, s_{j-1}$ is reachable using $h_\ell (= k + 1), h_{\ell+1}, \ldots, h_{j-1}$ hops. It needs to mention that, $h_\ell = h_{\ell+1} = \ldots = h_{i-1} = k+1$. Now, consider the minimum cost range assignment among s_p to s_ℓ using at most $(h - h_p)$-hops, where $\ell \leq p \leq j - 1$. Then take the range assignment from s_j to s_n in $(h - k - 1)$-hops with minimum cost.

5   Weighted Accumulation Problem

In this section, we solve the accumulation problem for $h = n - 1$ in $O(n^2)$ time for the radio-stations located on d-dimensional $(d > 0)$ Euclidean space. And if the radio-stations are located on a line, then we can solve h-hops accumulation problem in $O(hn^3)$ time using the method in [3].

Let $S = \{s_1, s_2, \ldots, s_n\}$ be a set of n radio-stations located on a d-dimensional Euclidean space and $W = \{w_1, w_2, \ldots, w_n\}$ be the weight set of the radio-station in S, where weight $w_i(1 \leq i \leq n)$ correspond to radio-stations s_i. Without loss of generality, assume s_n be the sink radio-station. To solve the accumulation problem, we construct a weighted complete directed graph $G = (V, E)$, where vertex set V of G corresponds to set of radio-stations S and weight of edge $e = (s_i, s_j)$ is equal to $w_i \times (\delta(s_i, s_j))^2$. Here, we solve the accumulation problem by solving the minimum-cost *arborescence* problem in the augmented graph $G' = (V, E')$ (say) of the graph G.

Arborescence Problem. Given a weighted directed graph G^* and a root node r, the minimum cost arborescence problem is to find the minimum-cost spanning tree in G^* directed out of r. Here, tree cost is the sum of the tree arc cost.

Computation of G' from G. The only change in G' from G is edge weights and the change of edge weights is as follows : interchange the edge weights between edges $(u, v), (v, u) \in G$, for each $u, v \in V$. For each $u, v \in V$, there are arcs $(u, v), (v, u) \in G$, as G is a complete directed graph.

From the definition of arborescence problem, it follows that the solution of minimum cost accumulation problem in G is equivalent to the solution of minimum cost arborescence problem in G'. Gabow et. al. in [10], propose an algorithm for solving arborescence problem optimally in $O(n \log n + m)$ time in a weighted directed graph with n vertices and m edges. Therefore, the weighted accumulation problem can be solved in $O(n^2)$ time for the unbounded version of the problem.

References

1. C. Ambhl, A. E. F. Clementi, M. Di Ianni, N. Lev-Tov, A. Monti, D. Peleg, G. Rossi, R. Silvestri, *Efficient Algorithms for Low-Energy Bounded-Hop Broadcast in Ad-Hoc Wireless Networks*, Proc. 21st Annual Symposium on Theoretical Aspects of Computer Science, LNCS 2996, pp. 418-427, 2004.
2. C. Ambuhl, A.E.F. Clementi, M. Di Ianni, A. Monti, G. Rossi, and R. Silvestri, *The range assignment problem in non-homogeneous static ad-hoc networks*, In Proc. of 18th International Parallel and Distributed Precessing Symposium (IPDPS'04), 2004.
3. A. E. F. Clementi, A. Ferreira, P. Penna, S. Penennes and R. Silvestri, *The minimum range assignment problem on linear radio networks*, Algorithmica, vol. 35, pp. 95 - 110, 2003.
4. M. Cagalj, J. -P. Hubaux and C. Enz, *Minimum energy broadcast in all wireless networks: NP-completeness and distribution issues*, Proc. of the 8th. Annual Int. Conf. on Mobile Computing and Networking, pp. 172-182, 2002.
5. A. E. F. Clementi, G. Huiban, P. Penna, G. Rossi and Y. C. Verhoeven, *Some recent theoretical advances and open questions on energy consumption in Ad Hoc wireless networks*, Proc. 3rd. Workshop on Approximation and Randomization Algorithms in Communication Networks, pp. 23 - 38, 2002.
6. A. E. F. Clementi, M. Di Ianni, A. Monti, G. Rossi and R. Silvestri, *Experimental analysis of practically efficient algorithms for bounded-hop accumulation in Ad-Hoc wireless networks*, Proc. 19th International Parallel and Distributed Precessing Symposium (IPDPS '05), 2005.
7. A.E.F. Clementi, M. Di Ianni and R. Silvestri, *The minimum broadcast range assignment problem on linear multihop wireless networks*, Theoretical Computer Science, vol. 299, pp. 751-761, 2003.
8. T. H. Cormen, C. E. Leiserson, and R. L. Rivest, *Introduction to Algorithms*, Prentice Hall (India) Pvt. Ltd., New Delhi, 1998.
9. G. K. Das, S. Das and S. C. Nandy, *Efficient Algorithm for Energy Efficient Broadcasting in Linear Radio Networks*, Proc. 11th. Int. Conf. on High Performance Computing, LNCS 3296, pp. 420-429, 2004.
10. H. N. Gabow, Z. Galil, T. Spencer and R. E. Tarjan, *Efficient algorithm for finding minimum spanning trees in undirected and directed graphs*, Combinatorica, vol. 6, no. 2, pp. 109 - 122, 1986.
11. C. Gaibisso, G. Proietti, and R. Tan, *Efficient management of transient station failures in linear radio communication networks with bases*, 2nd Internatonal Workshop on Approximation and Randomized Algorithms in Communication Networks (ARACNE'01), vol. 12 of Proceedings in Informatics, Carleton Scientific, pp. 37-54, 2001.
12. L. Kirousis, E. Kranakis, D. Krizanc and A. Pelc, *Power consumption in packet radio networks*, Theoretical Computer Science, vol. 243, pp. 289-305, 2000.

13. R. Mathar and J. Mattfeldt, *Optimal transmission ranges for mobile communication in linear multihop packet radio networks*, Wireless Networks, vol. 2, pp. 329-342, 1996.
14. J. Park and S. Sahni, *Maximum lifetime broadcasting in wireless networks*, IEEE Trans. on Computers, vol. 54, pp. 1081-1090, 2005.
15. P. Santi, Topology Control in Wireless Ad Hoc and Sensor Networks, John Wiley & Sons, 2005.
16. P. -J. Wan, G. Calinescu, X. -Y. Li and O. Frieder, *Minimum energy broadcast in static ad hoc wireless networks*, Wireless Networks, vol. 8, pp. 607-617, 2002.

Secure Overlay Network Design

Li (Erran) Li[1], Mohammad Mahdian[2], and Vahab S. Mirrokni[3]

[1] Bell Laboratories
erranlli@dnrc.bell-labs.com
[2] Microsoft Research
mahdian@microsoft.com
[3] MIT Computer Science and Artificial Intelligence Lab
mirrokni@theory.csail.mit.edu

Abstract. Due to the increasing security threats in the Internet, new overlay network architectures have been proposed to secure privileged services. In these architectures, the application servers are protected by a defense perimeter where only traffic from entities called servelets are allowed to pass. End users must be authorized and can only communicate with entities called access points (APs). APs relay authorized users' requests to servelets, which in turn pass them to the servers. The identity of APs are publicly known while the servelets are typically secret. All communications are done through the public Internet. Thus all the entities involved forms an overlay network. The main component of this distributed system consists of n APs. and m servelets. A design for a network is a bipartite graph with APs on one side, and the servelets on the other side. If an AP is compromised by an attacker, all the servelets that are connected to it are subject to attack. An AP is *blocked*, if all servelets connected to it are subject to attack. We consider two models for the failures: In the *average case model*, we assume that each AP i fails with a given probability p_i. In the *worst case model*, we assume that there is an adversary that knowing the topology of the network, chooses at most k APs to compromise. In both models, our objective is to design the connections between APs and servelets to minimize the (expected/worst-case) number of blocked APs. In this paper, we give a polynomial-time algorithm for this problem in the average-case model when the number of servelets is a constant. We also show that if the probability of failure of each AP is at least $1/2$, then in the optimal design each AP is connected to only one servelet (we call such designs *star-shaped*), and give a polynomial-time algorithm to find the best star-shaped design. We observe that this statement is not true if the failure probabilities are small. In the worst-case model, we show that the problem is related to a problem in combinatorial set theory, and use this connection to give bounds on the maximum number of APs that a perfectly failure-resistant design with a given number of servelets can support. Our results provide the *first* rigorous theoretical foundation for practical secure overlay network design.

Keywords: network design, network security, optimization, combinatorics.

S.-W. Cheng and C.K. Poon (Eds.): AAIM 2006, LNCS 4041, pp. 354–366, 2006.
© Springer-Verlag Berlin Heidelberg 2006

1   Introduction

Providing secure and highly available services using the shared Internet infrastructure is very challenging due to security threats in the Internet. Distributed Denial of Service (DDoS) attacks are a major threat to Internet security. Attacks against high-profile web sites such as Yahoo, CNN, Amazon and E*Trade in early 2000 [7] rendered the services of these web sites unavailable for hours or even days. During the hour long attack against root Domain Name Servers (DNS) in Oct, 2002, only four or five of the 13 servers were able to withstand the attack and remain available to legitimate Internet traffic throughout the strike [13]. Internet service would have started degrading if the attack had been sustained long enough for the information contained in the secondary DNS caches to start expiring—a process that usually takes from a few hours to about two days. A recent attack on June 15, 2004 against Akamai's DNS servers caused several major customers of Akamai's DNS hosting services, including Microsoft Corp., Yahoo Inc., and Google Inc. to suffer brief but severe slowdown [22] in their web performance. The event was marked by being a step beyond "simple bandwidth attacks" on individual web sites to more sophisticated targeting of core upstream Internet routers, DNS servers and bandwidth bottlenecks.

To defend against DDoS attacks, one can trace the attack sources and punish the perpetrators [3, 5, 19, 21, 4, 20, 8, 1, 11]. Due to the large number of compromised hosts (known as Zombies) used in the attack, finding the attack origin can be very difficult. Techniques to prevent DDoS attacks and/or to mitigate the effect of such attacks while they are raging on have been proposed [12, 6, 17, 9, 14, 16, 15]. These mechanisms alone do not prevent DDoS attacks from disrupting Internet services as they are reactive in nature. Recent research efforts [9, 2] have focused on designing overlay network architectures where certain critical elements are hidden from the attackers. The key entities in these architectures are access points (APs), servelets and end application servers. The end application servers are protected by a defense perimeter. Routers at the boundary are installed with filters which only allow traffic from the servelets in. The servelets are hidden from the attackers. Only a subset of access points are allowed to access each servelet. User requests must be authorized by access points and the requests are tunneled to their corresponding servelets via access points. The servelets then communicate with the end application servers. The access points can be geographically well placed to service the end users. The number of access points is assumed to be much larger than the number of secret servelets. All communications go through the public Internet. Thus all the entities involved form an overlay network.

The ability of such distributed systems to service their users is characterized by how many access points can still communicate to the end application servers, should an attack happens. This depends on how the access points are connected to the servelets. Intuitively, if a vulnerable access point connects to all the servelets, once it is compromised, all the servelets will be subject to DDoS attacks. In the worse case, this in turn denies all other access points from accessing the servelets. The network must be designed to resist such attacks. However, how the network should be designed has not been rigorously analyzed. In this paper, we formalize the problem as a combinatorial optimization problem with the objective to maximize the number of surviving access points. We first define our problem settings.

Definition 1. *A design for a network with n APs and m servelets is a bipartite graph with APs on one side, and the servelets on the other side. If an AP fails (or is compromised), it attacks all the servelets that are connected to it and we say that these servelets are* attacked. *If all servelets connected to an AP are attacked, we say that the AP is* blocked. *By definition, we say any compromised AP is blocked.*

We are interested in designing secure networks in which the number of blocked APs is minimized. We consider two models of failures:

- In the *average case model*, we assume that each AP i fails with a given probability p_i. Our objective is to design the connections between APs and servelets to minimize the expected number of blocked APs[1].
- In the *worst case model*, we assume that there is an adversary that knowing the topology of the network, chooses at most a given number k of APs to compromise. Our objective is to design the connections between APs and servelets to minimize the worse-case number of blocked APs.

This paper presents the *first* theoretical study of secure overlay network design. Our results provide guidelines for practical design of such networks.

The rest of this paper is organized as follows. In Section 2, we study the problem in the average case model. We first prove a lemma on the structure of the optimal design. This lemma restricts the number of possible solutions and gives a polynomial-time algorithm for the problem where the number of servelets is constant. It also implies a polynomial-time algorithm for the case that each AP can be connected to at most one servelet. We prove that if all failure probabilities are large enough (namely, greater than $\frac{1}{2}$), then the optimal design is of this form, and therefore can be found in polynomial time. At the end of Section 2, we give an example that if failure probabilities are not small, then the optimal design is not necessarily star shaped, and in fact, the best star-shaped design can be worse than the optimal design by an arbitrary factor. Finally, in Appendix A, we show hardness results for computing the expected number of blocked APs for a given network. In Section 3, we study the worst case model. We establish a connection between the secure network design problem and a problem in combinatorial set theory, and use this to give the optimal design for one failed AP. For constant number of failed APs, we use the probabilistic method to bound the maximum number of APs that we can support using a fixed number of servelets without blocking any other APs. We conclude in Section 4 with several open questions.

2   The Average-Case Model

In this section, we study the average case model. We give polynomial-time algorithms for this problem in two cases: when the number of servelets is a constant,

---

[1] For a detailed justification of this model, please see [2].

and when the probability of failure of each AP is at least $1/2$. We also demonstrate the difficulty of the problem in Appendix A by showing that even when a design is given, computing the probability that a given AP will be blocked or the expected number of APs that will be blocked is $\#P$-complete.

Our algorithms are based on the following lemma about the structure of the optimal design.

Lemma 1. *Assume that APs are ordered in decreasing order of their failure probabilities, i.e., $p_1 \geq p_2 \geq \ldots \geq p_n$. For an AP i, let S_i be the set of servelets connected to i. There exists an optimal design in which for all $i < j < k$, if $S_i = S_k$, then $S_j = S_i$.*

Proof. Assume that there is no optimal design with the desired property. Let $S_1, \ldots, S_n$ be an optimal solution in which for some $i < j < k$, $S_i = S_k$ but $S_j \neq S_i$. Note that $S_i = S_k$ implies that if either i or k fails, then both i and k are blocked. In particular, the expected number of blocked APs given that i fails is equal to the expected number of blocked APs given that k fails and is equal to the expected number of blocked APs given that i and k fail. Let B_{11} be the expected number of blocked APs given that j fails and at least one of i and k fail. Let B_{10} be the expected number of blocked APs given that at least one of i and k fail and j does not fail. Similarly, let B_{01} be the expected number of blocked APs given that j fails but neither i nor k fails and B_{00} be the expected number of blocked APs given that none of i and k and j fails. From this definitions, it is straightforward to see that $B_{11} \geq B_{01}$. The expected number $\mathcal{P}^*$ of blocked APs in an optimal design can be expressed as follows.

$$
\begin{aligned}
\mathcal{P}^* &= \mathbf{E}[\#\text{blocked APs}] \\
&= p_j(p_i + p_k - p_i p_k)B_{11} + (1 - p_j)(p_i + p_k - p_i p_k)B_{10} \\
&\quad + (1 - p_i)p_j(1 - p_k)B_{01} + (1 - p_i)(1 - p_j)(1 - p_k)B_{00}
\end{aligned}
$$

Now we prove that the set of servelets of j can be exchanged with the set of servelets of either i or k without increasing the expected number of blocked APs. For contradiction, assume that both these exchanges increase the expected number of blocked APs. The expected number of blocked APs after exchanging i and j can be written as

$$
\begin{aligned}
\mathcal{P}_1 &= \mathbf{E}[\#\text{blocked APs}] \\
&= p_i(p_j + p_k - p_j p_k)B_{11} + (1 - p_i)(p_j + p_k - p_j p_k)B_{10} \\
&\quad + (1 - p_j)p_i(1 - p_k)B_{01} + (1 - p_j)(1 - p_i)(1 - p_k)B_{00}
\end{aligned}
$$

Similarly, the expected number of blocked APs after exchanging j and k is

$$
\begin{aligned}
\mathcal{P}_2 &= \mathbf{E}[\#\text{blocked APs}] \\
&= p_k(p_i + p_j - p_i p_j)B_{11} + (1 - p_k)(p_i + p_j - p_i p_j)B_{10} \\
&\quad + (1 - p_i)p_k(1 - p_j)B_{01} + (1 - p_i)(1 - p_k)(1 - p_j)B_{00}
\end{aligned}
$$

By our assumption, we have $\mathcal{P}^* < \mathcal{P}_1$ and $\mathcal{P}^* < \mathcal{P}_2$. Therefore,

$$p_j p_k B_{11} + p_i B_{10} + p_j (1 - p_k) B_{01} < p_i p_k B_{11} + p_j B_{10} + p_i (1 - p_k) B_{01}$$

Thus,

$$(p_i - p_j)(p_k B_{11} - B_{10} - (1 - p_k) B_{01}) > 0$$

Since $p_i \geq p_j$, this implies

$$p_k B_{11} - B_{10} - (1 - p_k) B_{01} > 0 \tag{1}$$

Similarly, $\mathcal{P}^* < \mathcal{P}_2$ implies

$$p_i B_{11} - B_{10} - (1 - p_i) B_{01} < 0 \tag{2}$$

By subtracting (1) from (2), we get $(p_i - p_k) B_{11} < (p_i - p_k) B_{01}$, and hence $B_{11} < B_{01}$. However, this is impossible by the definition of B_{11} and B_{01}. $\square$

Using Lemma 1, we can prove the following result.

Theorem 1. *There is a polynomial-time algorithm that constructs the optimal design in the average case model when the number of servelets is at most a constant.*

Proof Sketch. Assume that APs are ordered in the decreasing order of their failure probabilities, i.e., $p_1 \geq p_2 \geq \ldots \geq p_n$. Let S_i denote the set of servelets connected to the AP i. From Lemma 1, we know that there are indices $1 = \alpha_0 < \alpha_1 < \alpha_2 < \cdots < \alpha_s = n + 1$ such that for each $j \in [\alpha_i, \alpha_{i+1})$, $S_j = S_{\alpha_i}$, and the sets $S_{\alpha_0}, S_{\alpha_1}, \ldots, S_{\alpha_{s-1}}$ are pairwise distinct. Since the total number of distinct sets of servelets is 2^m, there are at most $\binom{n+2^m}{2^m}(2^m)!$ ways to pick the indices $\alpha_0, \ldots, \alpha_s$ and the corresponding S_i's. This number is bounded by a polynomial in n if m is a constant. Therefore, the algorithm can check all such configurations. Computing the expected number of blocked APs for each configuration can also be done in polynomial time when m is a constant. $\square$

If we can connect each AP to at most one servelet, the resulting graph is a union of stars. We say that the design is *star-shaped* in this case. The following theorem proves that the optimal star-shaped design can be found in polynomial time.

Theorem 2. *The optimal star-shaped design can be computed in polynomial time.*

Proof. Let the failure probabilities of the APs be $p_1 \leq p_2 \leq \ldots \leq p_n$. It is easy to see that the proof of Lemma 1 holds even if the design is restricted to a star-shaped design. This shows that in the optimal star-shaped design we should partition the APs $1, \ldots, n$ into at most $m + 1$ consecutive parts each of which is connected to no servelet or to one of the servelets. This can be done by dynamic programming in polynomial time. We observe that the subset of APs that are connected to none of the servelets should be among the APs with larger failure probability. Let $A[k, t]$ be the minimum (over the choice of the star-shaped design) of the expected number of blocked APs when the set of APs consists of $1, 2, \ldots, k$ and there exists t servelets. Let $B(a, b)$ be the expected number of blocked APs among the APs $a, a + 1 \ldots, b$, if they are all connected to the same servelet (and no other AP is connected to this servelet). Note that $B(a, b)$ can be easily computed in polynomial time for each a and b. It is not hard to see that $A[k, t] = \min\{\min_{1 \leq l \leq k}\{A[l, t - 1] + B(l + 1, k)\}, \min_{1 \leq l \leq k}\{A[l, t] + k - l\}\}$ and $A[k, 0] = k$. Using this recurrence, the values of $A[k, t]$ can be computed in polynomial time. The value of the best star-shaped design is given by $A[n, m]$. $\square$

It might appear that star-shaped designs are weaker than general designs. The following theorem shows that if all failure probabilities are at least $\frac{1}{2}$, there is an optimal design that is star-shaped.

Theorem 3. *If all failure probabilities are at least $\frac{1}{2}$ then there is a star-shaped optimal design and therefore an optimal design can be found in polynomial time.*

Proof. We start from an optimal design, $\mathcal{D}$, and prove that we can change this design to a star-shaped design without increasing the expected number of blocked APs.

First we prove that we can get rid of all the cycles in the optimal design $\mathcal{D}$. If there is a cycle in $\mathcal{D}$, then there is a chordless cycle C in $\mathcal{D}$ as well. The length of cycle C is even and is at least 4. We consider two cases:

Case 1: $|C| \geq 6$.   In this case, let cycle C be $s_1 c_1 s_2 c_2 \ldots s_k c_k s_1$, where c_i's are APs and s_i's are servelets. We claim that removing one of the matchings $c_1 s_1, c_2 s_2, \ldots, c_k s_k$ or $c_1 s_2, c_2 s_3, \ldots, c_{k-1} s_k, c_k s_1$ will not increase the expected number of blocked APs. Let $\mathcal{D}_1$ be the design $\mathcal{D}$ after removing the matching $c_1 s_1, c_2 s_2, \ldots, c_k s_k$ and $\mathcal{D}_2$ be the design after removing the matching $c_1 s_2, c_2 s_3, \ldots, c_{k-1} s_k, c_k s_1$. Removing a matching from C will not increase the blocking probability of any AP other than $c_1, c_2, \ldots, c_k$. So it is enough to argue that the expected number of blocked APs in $c_1, c_2, \ldots, c_k$ decreases as we remove one of these two matchings. Let E_{c_i} for all $1 \leq i \leq k$ be the event that all of servelets that are connected to c_i and are not in the set $\{s_1, s_2, \ldots, s_k\}$ are attacked. Let E_{s_i} be the event that at least one of the APs that are connected to servelet s_i fails. The probability of E_{c_i} is denoted by P_{c_i}, and the probability of E_{c_i} and not E_{s_j} is denoted by $P_{c_i \bar{s}_j}$. Similarly, the probability of E_{c_i} and E_{s_j} and not E_{s_l} is denoted by $P_{c_i s_j \bar{s}_l}$, etc. Let $\mathcal{P}_{\mathcal{T}}(c_i)$ be the blocking probability of c_i in design $\mathcal{T}$. Then,

$$\mathcal{P}_{\mathcal{D}}(c_i) = p_i + (1 - p_i)\big(P_{c_i} - P_{c_i \bar{s}_i}(1 - p_{i-1})$$
$$- P_{c_i \bar{s}_{i+1}}(1 - p_{i+1}) + P_{c_i \bar{s}_i \bar{s}_{i+1}}(1 - p_{i-1})(1 - p_{i+1})\big).$$

Furthermore $\mathcal{P}_{\mathcal{D}_1}(c_i) = p_i + (1 - p_i)P_{c_i s_i}$ and $\mathcal{P}_{\mathcal{D}_2}(c_i) = p_i + (1 - p_i)P_{c_i s_{i+1}}$.

In order to prove that the expected number of blocked APs is not more in one of the designs $\mathcal{D}_1$ and $\mathcal{D}_2$, it is enough to prove that $\mathcal{P}_{\mathcal{D}}(c_i) \geq \frac{1}{2}(\mathcal{P}_{\mathcal{D}_1}(c_i) + \mathcal{P}_{\mathcal{D}_2}(c_i))$. In order to prove this, it is enough to show the following:

$$\mathcal{P} := P_{c_i} - P_{c_i \bar{s}_i}(1 - p_{i-1}) - P_{c_i \bar{s}_{i+1}}(1 - p_{i+1}) + P_{c_i \bar{s}_i \bar{s}_{i+1}}(1 - p_{i-1})(1 - p_{i+1})$$

$$\geq \frac{1}{2}(P_{c_i s_i} + P_{c_i s_{i+1}})$$

Using $P_{c_i} = P_{c_i s_i} + P_{c_i \bar{s}_i} = P_{c_i s_{i+1}} + P_{c_i \bar{s}_{i+1}}$, we have:

$$P \geq \frac{1}{2}(P_{c_i s_i} + P_{c_i \bar{s}_i} + P_{c_i s_{i+1}} + P_{c_i \bar{s}_{i+1}}) - P_{c_i \bar{s}_i}(1 - p_{i-1}) - P_{c_i \bar{s}_{i+1}}(1 - p_{i+1})$$

$$\geq \frac{1}{2}(P_{c_i s_i} + P_{c_i s_{i+1}})$$

where we use the fact that $p_{i-1} \geq \frac{1}{2}$ and $p_{i+1} \geq \frac{1}{2}$.

Case 2: $|C| = 4$.   Let cycle C be $c_1 s_1 c_2 s_2 c_1$. The analysis of this case is very similar to the that of $|C| > 4$. We use the same notation as in the previous case. Again we prove that removing one the matchings $c_1 s_1, c_1 s_2$ or $c_1 s_2, c_2 s_1$ will not increase the expected number of blocked APs. Let $\mathcal{D}$, $\mathcal{D}_1$ and $\mathcal{D}_2$ be an optimal design, and this design after removing matchings $c_1 s_1, c_1 s_2$ and $c_1 s_2, c_2 s_1$, respectively.

$$\mathcal{P}_{\mathcal{D}}(c_i) = p_i + (1 - p_i)(P_{c_i} - (1 - p_{i+1})(P_{c_i \bar{s}_1} + P_{c_i \bar{s}_2} - P_{c_i \bar{s}_1 \bar{s}_2}))$$

$$\geq \frac{1}{2}(p_i + (1 - p_i)P_{c_i s_1} + p_i + (1 - p_i)P_{c_i s_2})$$

$$+ (1 - p_i)(p_{i+1} - \frac{1}{2})(P_{c_i \bar{s}_1} + P_{c_i \bar{s}_2})$$

$$\geq \frac{1}{2}(\mathcal{P}_{\mathcal{D}_1}(c_i) + \mathcal{P}_{\mathcal{D}_2}(c_i)) + (1 - p_i)(p_{i+1} - \frac{1}{2})(P_{c_i \bar{s}_1} + P_{c_i \bar{s}_2})$$

$$\geq \frac{1}{2}(\mathcal{P}_{\mathcal{D}_1}(c_i) + \mathcal{P}_{\mathcal{D}_2}(c_i))$$

Thus, in at least one of the designs $\mathcal{D}_1$ and $\mathcal{D}_2$, the expected number of blocked APs is less than or equal to the expected number of blocked APs in $\mathcal{D}$.

After getting rid of all cycles, $\mathcal{D}$ is a tree. Next, we show that it is possible to change this tree to a star-shaped design without increasing the expected number of blocked APs. Again, we consider two cases:

Case 1: There is a leaf s in tree $\mathcal{D}$ that is a servelet.
In this case, let c be the AP connected to servelet s. Removing all edges of c to servelets other than s will decrease the expected number of blocked APs among APs other than c. Furthermore, the blocking probability of c will not increase, since c has a private servelet s.

Case 2: All leaves of $\mathcal{D}$ are APs.
Consider a connected component of $\mathcal{D}$ which is not a star. Now consider a leaf AP c in this component. AP c is connected to servelet s. Servelet s must have a neighboring AP c' which is connected to at least one other servelet s', for otherwise the component would be a star. We claim that removing the edge $c's'$ decreases the expected number of blocked APs. Let $\mathcal{D}'$ be the tree after removing $c's'$.

The blocking probability of all APs except c' decrease in $\mathcal{D}'$. In the following, we prove that removing $c's'$ also decreases the sum of blocking probabilities of the APs c and c'. Let $P_{c'}$ be the probability that all servelets connected to c', except possibly s, are attacked. Let P_s be the probability that one AP other than c' and c in the neighborhood of s fails. As before, let $\mathcal{P}_\mathcal{D}(c)$ be the blocking probability of AP c in the design $\mathcal{D}$. Using the fact that $\mathcal{D}$ is a tree, we have

$$\mathcal{P}_\mathcal{D}(c) = p_c + (1 - p_c)(p_{c'} + P_s - p_{c'}P_s)$$
$$\mathcal{P}_\mathcal{D}(c') = p_{c'} + (1 - p_{c'})P_{c'}(p_c + P_s - p_cP_s)$$
$$\mathcal{P}_{\mathcal{D}'}(c) = p_c + (1 - p_c)P_s$$
$$\mathcal{P}_{\mathcal{D}'}(c') = p_{c'} + (1 - p_{c'})P_{c'}.$$

Therefore,

$$\begin{aligned}
\mathcal{P}_\mathcal{D}(c) + \mathcal{P}_\mathcal{D}(c') &= p_c + (1 - p_c)(p_{c'} + P_s - p_{c'}P_s) + p_{c'} \\
&\quad + (1 - p_{c'})P_{c'}(p_c + P_s - p_cP_s) \\
&= \mathcal{P}_{\mathcal{D}'}(c) + \mathcal{P}_{\mathcal{D}'}(c') + (p_{c'} - (1 - p_{c'})P_{c'})(1 - p_c)(1 - P_s) \\
&\geq \mathcal{P}_{\mathcal{D}'}(c) + \mathcal{P}_{\mathcal{D}'}(c'),
\end{aligned}$$

where in the last inequality we use the fact that $p_{c'} \geq \frac{1}{2}$ and $P_{c'} \leq 1$, and hence $p_{c'} - (1 - p_{c'})P_{c'} \geq 0$. This completes the proof of this case.

Using the above operations, we can change the tree-shaped design $\mathcal{D}$ to a star-shaped design without increasing the expected number of blocked APs. Hence, we can change any optimal design to an optimal tree-shaped design and then to an optimal star-shaped design. $\square$

Another case for which we can show that there is an optimal star-shaped design is when the number of servelets is two.

Theorem 4. *If the number of servelets is two, then there is an optimal design that is star-shaped.*

Proof. For simplicity, we prove the theorem assuming all APs have the same failure probability p. The proof in the general case is similar. Let $q = 1 - p$. Let A_{00}, A_{10}, A_{01}, and A_{11} be the set of APs connected to none of the servelets, to servelet 1, to servelet 2, and to both servelets in an optimal solution. Let $n_{uv} = A_{uv}$ for $0 \leq u, v \leq 1$ and $n = n_{01} + n_{10} + n_{11}$. Let P_1 be the probability that servelet 1 is not attacked. For $i \in A_{10}$, $P_1 = \mathbf{Pr}[i \text{ is blocked}] = 1 - q^{n_{10}+n_{11}}$. For $i \in A_{01}$, $P_2 = \mathbf{Pr}[i \text{ is blocked}] = 1 - q^{n_{01}+n_{11}}$. For $i \in A_{11}$, $P_3 = \mathbf{Pr}[i \text{ is blocked}] = 1 - q^{n_{01}+n_{11}} - q^{n_{10}+n_{11}} + q^{n_{01}+n_{10}+n_{11}}$. Thus, the expected number of blocked APs is equal to $\mathcal{P}^* = n_{10}P_1 + n_{01}P_2 + n_{11}P_3$. WLOG, assume that $n_{01} \geq n_{10}$. We prove that moving one of the APs from A_{11} to A_{10} decreases the expected number of blocked APs. Before moving this AP from A_{11} to A_{10},

$$\mathcal{P}^* = n - (n_{10} + n_{11})q^{n_{10}+n_{11}} - (n_{01} + n_{11})q^{n_{01}+n_{11}} + n_{11}q^{n_{01}+n_{10}+n_{11}}$$

After this movement, the expected number of blocked APs is

$$\mathcal{P} = n - (n_{10} + n_{11})q^{n_{10}+n_{11}} - (n_{01} + n_{11} - 1)q^{n_{01}+n_{11}-1} + (n_{11} - 1)q^{n_{01}+n_{10}+n_{11}}$$

Now we have,

$$
\begin{aligned}
\mathcal{P}^* - \mathcal{P} &= q^{n_{01}+n_{11}-1}[(n_{01}+n_{11})(1-q)-1+q^{n_{10}+1}] \\
&\geq q^{n_{01}+n_{11}-1}[(n_{01}+n_{11})p-1+(1-p)^{n_{10}+1}] \\
&\geq q^{n_{01}+n_{11}-1}(n_{01}+n_{11}-n_{10}-1)p \\
&\geq 0
\end{aligned}
$$

where the last two inequalities are from $(1-p)^{n_{10}+1}-1 > -p(n_{10}+1)$ and $n_{01}+n_{11} \geq n_{10}+1$. Therefore, we can move all APs from A_{11} to either A_{10} or A_{01} without increasing the expected number of blocked APs. Thus, there is a star-shaped optimal solution. □

The above proof was based on a local operation that removes one of the edges attached to an AP of degree more than one. However, this local operation can increase the expected number of blocked APs when the number of servelets is more than two. For example, consider a cycle of size six with three APs and three servelets. It is not hard to show that removing any of the edges of this design will increase the expected number of blocked APs. In the following theorem, we show that without an assumption on the failure probabilities or the number of servelets, the optimal design need not be star shaped.

Theorem 5. *There is an instance of the secure network design problem in which the expected number of blocked APs in the optimal design is larger than that of the optimal star-shaped design by an arbitrary factor.*

Proof. Choose a sufficiently large number m, and let $n = \binom{m}{m/2}$ and $p = 1/n^2$. We first analyze the expected number of blocked APs in the best star-shaped design with these parameters. Let n_i denote the number of APs connected to the ith servelet in such a design, and n_0 denote the number of APs not connected at all. The expected number of blocked APs can be expressed as

$$
n_0 + \sum_{i=1}^{m} n_i \left(1-(1-p)^{n_i}\right) \geq \sum_{i=0}^{m} n_i \left(1-(1-p)^{n_i}\right).
$$

There is at least one i, $0 \leq i \leq m$, with $n_i \geq n/(m+1)$. Thus, the above expression is at least

$$
\frac{n}{m+1}\left(1-(1-p)^{n/(m+1)}\right) \geq \frac{n}{m+1}\left(\frac{pn}{m+1}-\frac{p^2n^2}{(m+1)^2}\right) \geq \frac{pn^2}{2(m+1)^2},
$$

where the first inequality follows from $(1-p)^s \leq 1-ps+p^2s^2$.

Now, we propose a different design and analyze the expected number of blocked APs in such a design. For each of the $n = \binom{m}{m/2}$ APs, we pick a distinct subset of $m/2$ servelets, and connect the AP to the servelets in this set. This design guarantees that if only one AP is attacked, then no other AP will be blocked. We use this to bound the expected number of blocked APs. By the union bound, the probability that more than one AP is attacked can be bounded by n^2p^2. In this case, we bound the number of blocked APs by n. Similarly, with probability at most np, exactly one AP is attacked, and in this case only one AP (the one that is attacked) is blocked. Thus, the expected number of blocked APs is at most $n^2p^2 \times n + np \times 1 = 2/n$.

Therefore, the ratio of the expected number of blocked APs in the latter design to the one in the best star-shaped design is at most $4(m+1)^2/n$, which tends to zero as m tends to infinity.   □

3 The Worst-Case Model

In this section, we study a model where an adversary selects at most a given number k of APs to compromise, and the objective is to minimize the number of blocked APs in the worst case. We observe that the worst-case model is closely related to the following problem in extremal combinatorics.

Definition 2. Let $\mathcal{A} = (A_1, A_2, \ldots, A_n)$ be a family of subsets of the universe $U = \{1, 2, \ldots, m\}$. We call the family $\mathcal{A}$ k-union free if for any $A_{i_0}, \ldots, A_{i_k} \in \mathcal{A}$ such that $i_j \neq i_t$ for $j \neq t$, we have $A_{i_0} \not\subseteq \cup_{1 \leq j \leq k} A_{i_j}$. In particular, a family $\mathcal{A}$ is 1-union free if none of the elements of $\mathcal{A}$ is a subset of another. Let $\mathcal{L}_k(m)$ be the maximum number of subsets in a k-union free family of subsets of the universe $\{1, 2, \ldots, m\}$.

We call a design *perfect* for k failures, if no matter which k APs fail, no other AP is blocked. It is not difficult to see that there exists a perfect design for k failures with m servelets and n APs if and only if $n \leq \mathcal{L}_k(m)$. The following theorem gives lower and upper bounds on the value of $\mathcal{L}_k(m)$. The lower bound in this theorem is proved by Kleitman and Spencer [10] for a more general problem. We include the proof here for the sake of completeness. We also give an upper bound based on Sperner's theorem. Sperner's theorem gives a tight bound on the maximum number of subsets in a 1-union free family of subsets. See also Ruszinkó [18] for an upper bound for a related problem.

Theorem 6. *For every k and m,*

$$(1 - \frac{k^k}{(k+1)^{k+1}})^{-m/(k+1)} \leq \mathcal{L}_k(m) \leq k\left(1 + \left(\frac{m}{\frac{m}{2}}\right)^{\frac{1}{k}}\right) = O(k2^{m/k}m^{-1/(2k)}).$$

Proof. We start by proving the upper bound. Let $\mathcal{A} = (A_1, A_2, \ldots, A_n)$ be a k-union-free family of subsets of $\{1, 2, \ldots, m\}$. Consider unions of k distinct sets from $\mathcal{A}$. We claim that no two such unions, say $A_{i_1} \cup \cdots \cup A_{i_k}$ and $A_{j_1} \cup \cdots \cup A_{j_k}$, are equal unless $\{i_1, \ldots, i_k\} = \{j_1, \ldots, j_k\}$. The reason for this is that if two such unions are equal and there is an index i_l not contained in $\{j_1, \ldots, j_k\}$, then we have $A_{i_l} \subseteq A_{j_1} \cup \cdots \cup A_{j_k}$, contradicting the assumption that $\mathcal{A}$ is k-union-free. Therefore, the collection of sets that are obtained by taking the union of k distinct sets in $\mathcal{A}$ contains exactly $\binom{n}{k}$ distinct sets. Furthermore, similar reasoning shows that no set in this collection is contained in another. Therefore, by Sperner's theorem, this collection can contain at most $\binom{m}{m/2}$ sets. Thus,

$$\binom{n}{k} \leq \binom{m}{m/2} \Rightarrow n \leq k\left(1 + \binom{m}{\frac{m}{2}}^{\frac{1}{k}}\right) = O(k2^{m/k}m^{-1/(2k)}),$$

completing the proof of the upper bound.

To prove the lower bound, we use the probabilistic method to construct a k-union-free collection of sets of the required size. Fix $p = \frac{1}{k+1}$, and pick each of the n sets in the collection by picking each element in $\{1, \ldots, m\}$ independently with probability p. Therefore, for a given set of indices $i_0, i_1, \ldots, i_k$, the probability that $A_{i_0} \subseteq A_{i_1} \cup \cdots \cup A_{i_k}$ is precisely $(1-p(1-p)^k)^m = (1 - \frac{k^k}{(k+1)^{k+1}})^m$. Therefore, by the union bound, the probability that the collection is not k-union-free is less than $n^{k+1}(1 - \frac{k^k}{(k+1)^{k+1}})^m$. Hence, if we pick $n \leq (1 - \frac{k^k}{(k+1)^{k+1}})^{-m/(k+1)}$, there is a nonzero probability that the resulting collection is k-union-free. This completes the proof of the lower bound.   □

Note that the above theorem suggests a randomized algorithm for our network design problem: put each edge in the graph with probability $\frac{1}{k+1}$. We can bound the expected number of blocked APs resulting from this randomized algorithm using similar ideas of the proof of the above theorem. For small values of k, this algorithm works exponentially better than the optimal star-shaped design.

The only case where we know the exact value of $\mathcal{L}_k(m)$ is when $k = 1$. In this case, we can prove the following stronger theorem.

Theorem 7. *If $k = 1$, then there is a design in which the maximum number of APs an adversary can block is at most $\lceil n/\binom{m}{\lfloor m/2 \rfloor} \rceil$. Conversely, for every design for such a network, there is a strategy for the adversary to block at least $\lceil n/\binom{m}{\lfloor m/2 \rfloor} \rceil$ APs.*

Proof Sketch. We can obtain a design for $k = 1$ by duplicating each of the $\binom{m}{\lfloor m/2 \rfloor}$ subsets of size $\lfloor m/2 \rfloor$ of the set of servelets $\lceil n/\binom{m}{\lfloor m/2 \rfloor} \rceil$ times, and associate an AP to each subset. To prove the other direction, we use the fact that the collection of all subsets of a set of size m can be partitioned into $\binom{m}{\lfloor m/2 \rfloor}$ chains. Therefore, in every design there are at least $\lceil n/\binom{m}{\lfloor m/2 \rfloor} \rceil$ APs that are connected to sets of servelets belonging to the same chain. Hence, if the adversary compromises the AP connected to the subset at the top of this chain, all other APs connected to the subsets in this chain will fail.   □

4   Conclusion

In this paper, we presented the first theoretical study of the secure network design problem. We showed that in the average case model, when failure probabilities are large (greater than $\frac{1}{2}$), there is an optimal star-shaped design, and such a design can be computed in polynomial time. On the other hand, there are instances with small failure probabilities where the optimal star-shaped design is arbitrarily worse than the optimal design. The case of small failure probabilities seems to be related to the stronger model where an adversary is allowed to select at most k APs to compromise. We observed that in this model, a random design performs considerably better than the optimal star-shaped design.

We still do not know of any hardness result or a polynomial-time algorithm for the general case of the secure network design problem, although the connection between this problem and the problem of finding a tight bound on the size of the largest k-union-free family of sets (which is a long-standing open problem) suggests that computing the exact optimum is difficult. Even an approximation algorithm for this problem, or tighter bounds for the k-union-free problem, would be interesting. Lovasz Local Lemma gives us a small improvement in the lower bound, but more significant improvement seem to require new techniques. Finally, it would be interesting to prove Theorem 3 with a weaker assumption (e.g., that probabilities are greater than a small constant), or show that such a generalization is not true.

References

1. M. Adler. Tradeoffs in probabilistic packet marking for ip traceback. In *Proc. ACM Symposium on Theory of Computing (STOC)*, May 2002.
2. T. Bu, S. Norden, and T. Woo. Trading resiliency for security: Model and algorithms. In *Proc. IEEE International Conference on Network Protocols(ICNP)*, 2004.
3. H. Burch and B. Cheswick. Tracing anonymous packets to their approximate source. In *Proc. USENIX LISA*, pages 319–327, December 2000.
4. D. Dean, M. Franklin, and A. Stubblefield. An algebraic approach to IP traceback. In *Proc. NDSS*, pages 3–12, February 2001.
5. T. Doeppner, P. Klein, and A. Koyfman. Using router stamping to identify the source of IP packets. In *Proc. ACM CCS*, pages 184–189, November 2000.
6. P. Ferguson. *Network Ingress Filtering: Defeating Denial of Service Attacks Which Employ IP Source Address Spoofing.* RFC 2267, January 1998.
7. L. Garber. Denial-of-service attacks rip the Internet. *IEEE Computer*, 33(4):12–17, April 2000.
8. M. T. Goodrich. Efficient packet marking for large-scale IP traceback. In *Proc. ACM CCS*, pages 117–126, November 2002.
9. A. D. Keromytis, V. Misra, and D. Rubenstein. SOS: Secure overlay services. In *Proc. ACM SIGCOMM*, pages 61–72, August 2002.
10. D. Kleitman and J. Spencer. Families of k-independent sets. *Discrete Mathematics*, 6:255–262, 1973.
11. J. Li, M. Sung, J. Xu, and L.E. Li. Large-scale ip traceback in high-speed internet: Practical techniques and theoretical foundation. In *Proc. IEEE Symposium on Security and Privacy*, pages 115–129, 2004.
12. R. Mahajan, S. Bellovin, S. Floyd, J. Ioannidis, V. Paxson, and S. Shenker. Controlling high bandwidth aggregates in the network. *ACM Computer Communication Review*, 32(3):62–73, July 2002.

13. D. McGuire and B. Krebs.   Attack on internet called largest ever.
    http://www.washingtonpost.com/wp-dyn/articles/A828-2002Oct22.html,   Oc-
    tober 2002.
14. Jelena Mirkovic, Gregory Prier, and Peter Reiher. Attacking DDoS at the source.
    In *Proc. IEEE ICNP*, pages 312–321, November 2002.
15. Jelena Mirkovic, Max Robinson, Peter Reiher, and Geoff Kuenning. Alliance forma-
    tion for ddos defense. In *Proc. New Security Paradigms Workshop, ACM SIGSAC*,
    August 2003.
16. Christos Papadopoulos, Robert Lindell, John Mehringer, Alefiya Hussain, and
    Ramesh Govidan. COSSACK: coordinated suppression of simultaneous attacks.
    In *DISCEX III*, pages 22–24, April 2003.
17. K. Park and H. Lee. On the effectiveness of route-based packet filtering for dis-
    tributed DoS attack prevention in power-law Internets. In *Proc. ACM SIGCOMM*,
    pages 15–26, August 2001.
18. M. Ruszinkó. On the upper bound of the size of the *r*-cover-free families. *Journal
    of Combinatorial Theory, Series A*, 66:302–310, 1994.
19. S. Savage, D. Wetherall, A. Karlin, and T. Anderson. Practical network support
    for IP traceback. In *Proc. ACM SIGCOMM*, pages 295–306, August 2000.
20. A. Snoeren, C. Partridge, et al. Hash-based IP traceback. In *Proc. ACM SIG-
    COMM*, pages 3–14, August 2001.
21. D. Song and A. Perrig.   Advanced and authenticated marking schemes for IP
    traceback. In *Proc. IEEE INFOCOM*, pages 878–886, April 2001.
22. J. Vijayan. Akamai attack reveals increased sophistication: Host's DNS servers were
    DDoS targets, slowing large sites. http://www.computerworld.com/securitytopics/
    security/story/0,10801,93977p2,00.html, June 2004.

A   Expected Number of Blocked APs: A Hardness Result

In terms of hardness, we can show that given a particular design, it is hard to
compute the probability that a given AP is blocked, and the expected number
of APs that will be blocked.

Theorem 8. *The following two problems are #P-complete:*

- *Given a design and assuming uniform failure probabilities of $p = 1/2$, compute the probability that a given AP i will be blocked.*
- *Given a design and assuming uniform failure probabilities of $p = 1/2$, compute the expected number of APs that will be blocked.*

Proof Sketch. For the first problem, we can give a reduction from the problem of
computing the number of solutions of a set-cover instance. The second problem
can be reduced to the first by adding a "private servelet" for each AP except
one.   □

Even though finding the exact expected number of blocked APs is hard, it is
not hard to approximate within a factor of $1 + \epsilon$ for any positive constant ϵ by
sampling polynomially many times and taking the average. Note that the above
theorem does not show any hardness result for finding the optimal network. The
complexity of this problem for general failure probabilities is still open.

A Portfolio Selection Method Based on Possibility Theory

Wei-Guo Zhang[1,2,*], Qianqin Chen[1], and Hai-Lin Lan[1]

[1] School of Business Administration, South China University of Technology,
Guangzhou, 510641, P.R. China
wgzhang@scut.edu.cn, zhwg61@263.net
[2] School of Management, Xi'an Jiaotong University, Xi'an,710049, P.R. China

Abstract. This paper discusses the portfolio selection problem based on the possibilistic theory. The possibilistic portfolio model with general constraints to investment is proposed by means of possibilistic mean value and possibilistic variance. The conventional probabilistic mean-variance model can be simplified under the assumption that the returns of assets are triangular fuzzy numbers. Finally, a numerical example of the portfolio selection problem is given to illustrate our proposed effective means and approaches.

1 Introduction

The mean-variance methodology for the portfolio selection problem, posed originally by Markowitz [1], has played an important role in the development of modern portfolio selection theory. It combines probability and optimization techniques to model the behavior investment under uncertainty. In the mean-variance portfolio selection problem, previous research includes Perold [2], Pang [3], VÖRÖS [4] and Best [5], etc.. The key principle of the mean-variance model is to use the expected return of a portfolio as the investment return and to use the variance of the expected returns of the portfolio as the investment risk. The basic assumption for using Markowitz's mean-variance model is that the situation of asset markets in future can be correctly reflected by asset data in the past, that is, the mean and covariance of assets in future is similar to the past one. It is hard to ensure this kind of assumption for real ever-changing asset markets.

Recently, a few of authors such as Watada [6], Tanino and Guo [7], Inuiguchi and Tanino [8], Zhang and Nie [9] etc., studied the fuzzy portfolio selection problem. Watada [6] presented portfolio selection models where he used fuzzy numbers to represent the decision maker's aspiration levels for the expected rate of return and a certain degree of risk. Inuiguchi and Tanino [8] introduced a novel possibilistic programming approach to the portfolio selection problem: their approach, which prefers a distributive investment solution, is based on the minimax regret criterion (the regret which the decision maker is ready to undertake). Tanaka and Guo [7] proposed the portfolio selection models based on

---

* Corresponding author.

S.-W. Cheng and C.K. Poon (Eds.): AAIM 2006, LNCS 4041, pp. 367–374, 2006.

fuzzy probabilities and possibilistic distributions. Zhang and Nie [9] introduced the admissible efficient portfolio model under the assumption that the expected returns and risks of assets have admissible errors.

Zadeh [10] proposed possibility theory based on possibilistic distributions. Dubois and Prade [11-12] developed it further. Carlsson and Fullér [13] defined the notions of possibilistic mean value and variance of fuzzy numbers. Carlsson [14] introduced a possibilistic approach to selecting portfolios with highest utility score. Zhang et.al [15] considered the portfolio selection problem based on a new crisp possibilistic variance and a new crisp possibilistic covariance of fuzzy numbers. Zhang and Wang [16] discussed the general weighted possibilistic portfolio selection problem. In this paper, we consider the portfolio selection problem under general constraints to investment. The possibilistic mean value corresponds to the return, while the possibilistic variance corresponds to the risk. We obtain a simple programming model to replace Markowitz's mean-variance model under general linear constraints to investment.

2   Possibilistic Mean and Variance

Let us introduce some definitions, which we shall need in the following section. A fuzzy number A is a fuzzy set of the real line $\mathcal{R}$ with a normal, fuzzy convex and continuous membership function of bounded support. The family of fuzzy numbers will be denoted by $\mathcal{F}$.

Let A be a fuzzy number with $\gamma-$ level set $[A]^\gamma = [a_1(\gamma), a_2(\gamma)](\gamma > 0)$. Carlsson and Fullér [13] introduced the possibilistic mean value of A as

$$M(A) = \frac{1}{2}\left[\frac{\int_0^1 a(\gamma)Pos[A \leq a(\gamma)]d\gamma}{\int_0^1 Pos[A \leq a(\gamma)]d\gamma} + \frac{\int_0^1 b(\gamma)Pos[A \geq b(\gamma)]d\gamma}{\int_0^1 Pos[A \geq b(\gamma)]d\gamma}\right]$$
$$= \int_0^1 \gamma[a(\gamma) + b(\gamma)]d\gamma, \tag{1}$$

where Pos denotes possibility, i.e.,

$$Pos[A \leq a_1(\gamma)] = \Pi((-\infty, a_1(\gamma)]) = \gamma,$$

$$Pos[A \geq a_2(\gamma)] = \Pi([a_2(\gamma), \infty)) = \gamma.$$

Let A with $[A]^\gamma = [a_1(\gamma), a_2(\gamma)]$ and B with $[B]^\gamma = [b_1(\gamma), b_2(\gamma)](\gamma \in [0,1])$ be two fuzzy numbers. Carlsson and Fullér [13] also introduced the possibilistic variance and covariance of fuzzy numbers as

$$Var(A) = \frac{1}{2}\int_0^1 \gamma[a_2(\gamma) - a_1(\gamma)]^2 d\gamma \tag{2}$$

and

$$Cov(A, B) = \frac{1}{2}\int_0^1 \gamma[(a_2(\gamma) - a_1(\gamma))(b_2(\gamma) - b_1(\gamma))]d\gamma, \tag{3}$$

respectively.

The following theorem was showed by Carlsson and Fullér [13].

Theorem 1. *Let A and B be two fuzzy numbers and let $\lambda, \mu \in \mathcal{R}$. Then*

$$Var(\lambda A + \mu B) = \lambda^2 Var(A) + \mu^2 Var(B) + 2|\lambda\mu|Cov(A, B),$$

where the addition of fuzzy numbers and the multiplication by a scalar of fuzzy number are defined by the sup-min extension principle (Zadeh [10]).

Theorem 1 means that the possibilistic variance of linear combinations of fuzzy numbers can easily be computed in a similar manner as in probability theory.
   The following theorem is easily obtained.

Theorem 2. *Let $A_1, A_2, \ldots, A_n$ be n fuzzy numbers and let $\lambda_0, \lambda_1, \ldots, \lambda_n$ be $n + 1$ real numbers. Then*

$$M(\lambda_0 + \sum_{i=1}^{n} \lambda_i A_i) = \lambda_0 + \sum_{i=1}^{n} \lambda_i M(A_i),$$

$$Var(\lambda_0 + \sum_{i=1}^{n} \lambda_i A_i) = \sum_{i=1}^{n} \lambda_i^2 Var(A_i) + 2 \sum_{i<j=1}^{n} |\lambda_i \lambda_j| Cov(A_i, A_j),$$

where the addition of fuzzy numbers and the multiplication by a scalar of fuzzy number are defined by the sup-min extension principle (Zadeh [10]).

3   Possibilistic Efficient Portfolio Model

Let us give a brief description of Markowitz's mean-variance model. Assume that there are n risky assets, the return rate of asset j is denoted as a random variable r_j with expected return $\bar{r}_j = E(r_j)$ and the proportion of total investment funds devoted to this asset is denoted as $x_j, j = 1, \ldots, n$. In order to describe conveniently, we set $\mathbf{x} = (x_1, x_2, \ldots, x_n)'$, $\mathbf{r} = (r_1, r_2, \ldots, r_n)'$ and $\mathbf{F} = (1, 1, \ldots, 1)'$. We use prime $(\prime)$ to denote matrix transposition and adopt the convention that all non-primed vectors are column vectors.
Then the return associated with the portfolio $\mathbf{x} = (x_1, x_2, \ldots, x_n)'$ is $r = \mathbf{r}'\mathbf{x}$.
The expected return and variance of r are, respectively, given by

$$E(r) = \bar{\mathbf{r}}'\mathbf{x}, D(r) = \mathbf{x}'\mathbf{V}\mathbf{x},$$

where $\bar{\mathbf{r}} = (\bar{r}_1, \bar{r}_2, \ldots, \bar{r}_n)'$ and $\mathbf{V} = (\sigma_{ij})_{n \times n}$ are the expected return vector and covariance matrix of returns, respectively.
Markowitz's mean-variance model for portfolio selection can be formulated as

$$
\begin{aligned}
\min \quad & \mathbf{x}'\mathbf{V}\mathbf{x} \\
s.t. \quad & \bar{\mathbf{r}}'\mathbf{x} = \mu, \\
& \mathbf{F}'\mathbf{x} = 1, \\
& \mathbf{x} \geq \mathbf{0}.
\end{aligned}
\tag{4}
$$

In order to use the mean-variance model (4), it is necessary to estimate the probability distribution, strictly speaking, a expected return vector and a covariance matrix. It means that all the expected returns, variances, covariances of risky assets can be accurately estimated by an investor. In the mean-variance model (4) uncertainty is equated with randomness, which actually combines both objectively observable and testable random events with subjective judgments of the decision maker into probability assessments. A purist on theory would accept the use of probability theory to deal with observable random events, but would frown upon the transformation of subjective judgments to probabilities. It is well-known that the returns of risky assets are in a fuzzy uncertain economic environment and vary from time to time, the future states of returns and risks of risky assets cannot be predicted accurately. Fuzzy number is a powerful tool used to describe an uncertain environment with vagueness and ambiguity. In many important cases, it might be easier to estimate the possibility distributions of rates of return on risky assets, rather than the corresponding probability distributions. Based on these facts, we discuss the portfolio selection problem under the assumption that the returns of assets are fuzzy numbers.

In next section, we consider a financial market with n risky assets and a risk-less asset. Let r_0 be the interest rate of the risk-less asset. Analogous to Markowitz's mean-variance methodology for the portfolio selection problem, the possibilistic mean value is termed measure of investment return and the possibilistic variance is termed measure of investment risk. Probabilic mean and variance in the model (4) may be replaced by possibilistic mean and possibilistic variance. The general possibilistic mean-variance model for the portfolio selection problem may be described by

$$
\begin{aligned}
\min \quad & Var[\mathbf{r}'\mathbf{x} + r_0(1 - \mathbf{F}'\mathbf{x})] \\
s.t. \quad & M[\mathbf{r}'\mathbf{x} + r_0(1 - \mathbf{F}'\mathbf{x})] \geq \mu, \\
& \mathbf{x} \in \mathbf{H},
\end{aligned} \tag{5}
$$

where $\mathbf{H}$ is a convex set that represents additional constraints on the choice of $\mathbf{x}$.

Let $r_i = (a_i, \alpha_i, \beta_i), i = 1, \ldots, n$ be triangular fuzzy numbers with center a_i, left-width $\alpha_i > 0$ and right-width $\beta_i > 0$.
Then a $\gamma-$level of r_i is computed by

$$
[r_i]^\gamma = [a_i - (1 - \gamma)\alpha_i, a_i + (1 - \gamma)\beta_i],
$$

for all $\gamma \in [0.1], i = 1, \ldots, n$.
According to (1), (2) and (3), we easily obtain

$$
M(r_i) = a_i - \frac{\alpha_i}{6} + \frac{\beta_i}{6},
$$

$$
Var(r_i) = \frac{(\alpha_i + \beta_i)^2}{24},
$$

$$
Cov(r_i, r_j) = \frac{(\alpha_i + \beta_i)(\alpha_j + \beta_j)}{24}.
$$

Then the possibilistic mean value of the return associated with the portfolio $\mathbf{x}$ is given by

$$M[\mathbf{r}'\mathbf{x} + r_0(1 - \mathbf{F}'\mathbf{x})] = M(\sum_{i=1}^{n} r_i x_i) + r_0(1 - \sum_{i=1}^{n} x_i)$$

$$= \sum_{i=1}^{n} M(r_i)x_i + r_0(1 - \sum_{i=1}^{n} x_i)$$

$$= \sum_{i=1}^{n} (a_i + \frac{\beta_i - \alpha_i}{6})x_i + r_0(1 - \sum_{i=1}^{n} x_i).$$

The possibilistic variance of the return associated with the portfolio $\mathbf{x}$ is given by

$$Var[\mathbf{r}'\mathbf{x} + r_0(1 - \mathbf{F}'\mathbf{x})] = \frac{1}{24}[\sum_{i=1}^{n}(\alpha_i + \beta_i)^2 x_i^2 + \sum_{i \neq j=1}^{n}(\alpha_i + \beta_i)(\alpha_j + \beta_j)|x_i||x_j|]$$

$$= \frac{1}{24}[\sum_{i=1}^{n}(\alpha_i + \beta_i)|x_i|]^2.$$

Thus, the possibilistic mean-variance model of portfolio selection problem may be described by

$$\min Var(r) = \frac{1}{24}[\sum_{i=1}^{n}(\alpha_i + \beta_i)|x_i|]^2$$

$$s.t. \quad \sum_{i=1}^{n}(a_i + \frac{\beta_i - \alpha_i}{6})x_i + r_0(1 - \sum_{i=1}^{n} x_i) \geq \mu, \tag{6}$$

$$\mathbf{x} \in \mathbf{H},$$

where $\mathbf{H}$ is a convex set that represents additional constraints on the choice of $\mathbf{x}$.

Furthermore, the possibilistic mean-variance model (6) is equal to the following programming:

$$\min \sum_{i=1}^{n}(\alpha_i + \beta_i)|x_i|$$

$$s.t. \quad \sum_{i=1}^{n}(a_i - r_0 + \frac{\beta_i - \alpha_i}{6})x_i \geq \mu - r_0, \tag{7}$$

$$\mathbf{x} \in \mathbf{H},$$

where $\mathbf{H}$ is a convex set that represents additional constraints on the choice of $\mathbf{x}$.

It should be noted that the model (7) contains $3n$ unknown parameters, but the conventional probabilistic mean-variance model (5) contains $(n^2 + 3n + 2)/2$ unknown parameters. Clearly, the unknown parameters are greatly decreased to compare the model (7) with conventional probabilistic mean-variance methodology.

Especially, if $r_i = (a_i, \alpha_i), i = 1, \ldots, n$ are symmetric triangular fuzzy numbers, that is $\alpha_i = \beta_i$, then the possibilistic mean-variance model (7) may be described by

$$\min \sum_{i=1}^{n} \alpha_i |x_i|$$

$$s.t. \quad \sum_{i=1}^{n} (a_i - r_0)x_i \geq \mu - r_0, \tag{8}$$

$$\mathbf{x} \in \mathbf{H},$$

where $\mathbf{H}$ is a convex set that represents additional constraints on the choice of $\mathbf{x}$.

If $\mathbf{H}$ only contains linear constraints and $x_i \geq 0$ for all $i = 1, \ldots, n$, the possibilistic mean-variance model (7) is a simple linear programming, using some related algorithms for solving linear programming problem easily obtain the possibilistic efficient portfolios.

4   Numerical Example

In order to illustrate our proposed effective means and approaches of the efficient portfolios in this paper, we considered a real portfolio selection example. In this example, we selected five stocks from Shanghai Stock Exchange, their returns $r_i, i = 1, \ldots, 5$ were regarded as symmetric triangular fuzzy numbers. Based on both the historical data and the future information, A $\gamma-$ level set of $r_i, i = 1, \ldots, 5$ are given by

$$[r_1]^\gamma = [0.069 + 0.021\gamma, 0.111 - 0.021\gamma],$$

$$[r_2]^\gamma = [0.105 + 0.03\gamma, 0.165 - 0.03\gamma],$$

$$[r_3]^\gamma = [0.123 + 0.042\gamma, 0.207 - 0.042\gamma],$$

$$[r_4]^\gamma = [0.15 + 0.06\gamma, 0.27 - 0.06\gamma],$$

$$[r_5]^\gamma = [0.174 + 0.081\gamma, 0.336 - 0.081\gamma].$$

We selected a risk-less asset which is current deposit of bank, the lending interest rate r_0 is 2%. The lower bound vector and upper bound vector of $\mathbf{x}$ are given by

$$\mathbf{L} = (0.1, 0.1, 0.1, 0.1, 0.1)'$$

and

$$\mathbf{U} = (0.4, 0.4, 0.5, 0.6, 0.7)',$$

respectively.

Thus, we obtain the possibilistic mean-variance model:

$$\min 0.021x_1 + 0.03x_2 + 0.042x_3 + 0.06x_4 + 0.081x_5$$
$$s.t. \quad 0.07x_1 + 0.115x_2 + 0.145x_3 + 0.19x_4 + 0.235x_5 + 0.02 \geq \mu,$$
$$0.1 \leq x_1 \leq 0.4,$$
$$0.1 \leq x_2 \leq 0.4,$$
$$0.1 \leq x_3 \leq 0.5,$$
$$0.1 \leq x_4 \leq 0.6,$$
$$0.1 \leq x_5 \leq 0.7,$$
$$x_1 + x_2 + x_3 + x_4 + x_5 \leq 1.$$

Table 1. The possibilistic efficient portfolios

μ	0.12	0.17	0.19	0.21
x_1	0.1	0.1	0.1	0.1
x_2	0.3130	0.3733	0.1067	0.1
x_3	0.1	0.1	0.1	0.1
x_4	0.1	0.3267	0.5933	0.1667
x_5	0.1	0.1	0.1	0.5333

We give the possibilistic efficient portfolios as $\mu = 0.12, 0.17, 0.19, 0.21$ in Table 1.

Table 1, representing the possibilistic efficient portfolio, shows that the investor does need to invest total capital to risky assets except $\mu = 0.12$. It is 71.30% that the proportion of total investment funds devoted to five risky assets associated with the possibilistic efficient portfolio $\mathbf{x} = (0.1, 0.3130, 0.1, 0.1, 0.1)'$ at $\mu = 0.12$.

5   Conclusions

In this paper, we have considered the portfolio selection problem based on possibilistic mean and variance under assumption that the returns of assets are fuzzy numbers. We have used the possibilistic means, variances and covariances to replace the probabilistic means, variances and covariances in Markowitz's mean-variance model, respectively. We have obtained the possibilistic mean-variance model for portfolio selection under general linear constraints to investment, which can better integrate the experts' knowledge and the managers' subjective opinions to compare with conventional probabilistic mean-variance methodology. Our proposed possibilistic approaches to selecting portfolios can better describe an uncertain decision problem with vagueness and ambiguity.

Acknowledgements

This research is supported by the National Natural Science Foundation of China (No.70571024) and China Postdoctoral Science Foundation (No.2005037241).

References

1. Markowitz, H.: Portfolio selection: efficient diversification of Investments. Wiley, New York.1959
2. Perold, A.F.: Large-scale portfolio optimization. Management Science, 30 (1984) 1143-1160
3. Pang, J.S.: A new efficient algorithm for a class of portfolio selection problems. Operational Research, 28(1980) 754-767
4. Vörös, J.: Portfolio analysis-An analytic derivation of the efficient portfolio frontier. European journal of operational research, 203(1986) 294-300

5. Best, M.J.,and Hlouskova,J.: The efficient frontier for bounded assets. Math.Meth.Oper.Res. 52(2000) 195-212
6. Watada, J.: Fuzzy portfolio selection and its applications to decision making. Tatra Mountains Mathematical Publication, 13(1997) 219-248
7. Tanaka, H.,and Guo, P.:. Portfolio selection based on upper and lower exponential possibility distributions. European Journal of Operational Research, 114(1999) 115-126
8. Inuiguchi M., and Tanino T.: Portfolio selection under independent possibilistic information, Fuzzy Sets and Systems 115(2000) 83-92
9. Zhang W.G., and Z.K Nie.: On admissible efficient portfolio selection problem, Applied mathematics and computation 159(2004) 357-371
10. Zadeh, L. A.: Fuzzy Sets. Inform. and Control 8(1965) 338-353
11. Dubois, D., and Prade, H.: The mean value of a fuzzy number. Fuzzy Sets and Systems 24(1987) 279-300
12. Dubois, D., and Prade, H.: Fuzzy sets and systems: Theory and applications. New York: Academic Press,1980
13. Carlsson, C., and Fullér, R.: On possibilistic mean value and variance of fuzzy numbers. Fuzzy Sets and Systems 122(2001) 315-326
14. Carlsson, C.,and Fullér, R. and Majlender. P.: A possibilistic approach to selecting portfolios with highest utility score. Fuzzy Sets and Systems, 131(2002) 13-21
15. Zhang, W.G., Liu, W.A , and Wang, Y.L.: A Class of Possibilistic Portfolio Selection Models and Algorithms. Lecture Notes in Computer Science, 3828(2005) 464-472
16. Zhang, W.G.,and Wang, Y.L.: Portfolio selection: Possibilistic mean-variance model and possibilistic efficient frontier. Lecture Notes in Computer Science, 3521 (2005) 203-213

Branch on Price: A Fast Winner Determination Algorithm for Discount Auctions

S. Kameshwaran and Lyés Benyoucef

INRIA Lorraine
ISGMP Bat: A, Ile du Saulcy
Metz 57000 France
{kamesh, lyes.benyoucef}@loria.fr

Abstract. Discount auction is a market mechanism for buying hetero-geneous items in a single auction. The suppliers submit discount bids, which consist of two parts: the individual cost for each of the items and discounts based on the number of items procured. The winner determi-nation problem faced by the buyer is to determine the winning suppliers and their corresponding winning items, such that the total cost of pro-curement is minimized. This problem is $\mathcal{NP}$-hard and in this paper we propose a novel branch and bound algorithm called as *branch on price*, which uses a tight integer programming formulation with valid inequal-ities. Computational experiments show that the proposed algorithm is many folds faster than the existing algorithm.

1 Introduction

Online auction mechanisms over the Internet pose several algorithmic challenges. The design of Internet auctions [1] is inherently a multidisciplinary activity re-quiring expertise from disciplines like economics, game theory, operations re-search, and computer science. One of the algorithmic aspects in auction design is the design of algorithms for the *winner determination problem* (WDP). The WDP is an optimization problem faced by the auctioneer to select a set of win-ning bidders and their corresponding winning quantity of items, such that the total profit (cost) is maximized (minimized). The WDP need to be solved once or several times in an auction depending on the auction dynamics. The solution time of the WDP plays a significant role in the auction implementation.

Discount auction (DA) is an auction mechanism for procuring heterogeneous items in a single auction [2]. In procurement auctions, buyer is the auctioneer and the suppliers (sellers) are the bidders. In DA, each supplier submits a dis-count bid, which consists of two parts: individual costs for each of the items and a discount function defined over the number of items. Through the discount function, the supplier conveys to the buyer the discount that he can avail on the original cost, depending on the number of items procured. The WDP to be solved by the buyer is to determine the winning suppliers and their respective winning items such that the total cost of procurement is minimized. It is easy to see that without the discount function the WDP is polynomially solvable

S.-W. Cheng and C.K. Poon (Eds.): AAIM 2006, LNCS 4041, pp. 375–386, 2006.

(for each item, choose the minimum cost supplier), but with the discount function, the WDP is $\mathcal{NP}$-hard [3]. A branch and bound algorithm was proposed in [3] which used the embedded network structure in the WDP to generate the bounds. These bounds were tighter than that of the linear programming (LP) relaxation of an integer programming (IP) formulation for the WDP. In this paper, we propose a tighter formulation by adding valid inequalities to the original IP formulation. Using this formulation we develop a novel branch and bound algorithm, which branches on the price of the items rather than on the fractional decision variables. The improved formulation, together with the novel branching scheme, results in a superior algorithm that is many fold faster than the existing algorithm.

The rest of the paper is organized as follows. In Sect. 2, we introduce the DA and define the WDP. The existing IP formulation and the proposed tighter formulation are presented in Sect. 3. The *branch on price* algorithm is systematically developed in Sect. 4. Section 5 presents the results of computational experiments comparing the proposed algorithm with the existing algorithm. We conclude the paper in Sect. 6.

2   Discount Auctions

The buyer is interested in procuring M different items and there are N suppliers. Each of the item is indivisible, *i.e.* it can be supplied by only one supplier. An *item* need not refer to a single unit. It can be *a computer* or *a computer and a printer* or *hundred computers*, but it cannot be supplied by multiple suppliers. An item is denoted by index m and a supplier by index j. Each supplier can submit only one discount bid and hence the index j denotes both the supplier and his bid. The discount bid j consists of two parts: (1) cost Q_j^m for each item m and (2) discount θ_j^i for i ($= 1, \ldots, M$) items. The bid can be compactly expressed as an ordered pair of M-tuples: $((Q_j^1, \ldots, Q_j^m, \ldots, Q_j^M), (\theta_j^1, \ldots, \theta_j^i, \ldots, \theta_j^M))$. Note that m denotes a particular item and i denotes any i number of items. If the buyer procures items 2, 4, and 7 from bid j, then the cost of procurement is $(1-\theta_j^3)(Q_j^2+Q_j^4+Q_j^7)$. All the Q_j^m are positive (possibly infinite for an unavailable item) and the θ_j^i are non-decreasing over i (the discount cannot decrease with the number of items bought).

The DA is related to two well known auctions: *volume discount auctions* and *combinatorial auctions*. In volume discount auctions [4], the suppliers provide discount based on the quantity procured. However, the auction is for procuring *multiple* units of *an* item, in contrast to the DA which is for procuring *single* unit of *different* items. The discount in DA is for the *number* of different items procured. Combinatorial auctions [5] is for trading different items in a single auction like DA. The bidder can submit several combinatorial bids, one for each of the possible subsets of the items. This is applicable in scenarios where the cost of a combination of items can be more or less than the sum of individual costs of items. By quoting a single price for a combination of items, a combinatorial bid can express complementarity or substitutability among the items in the combination. The DA can be considered as a special instance of the combinatorial auctions where the discount function enables to determine the bid price of a combination of items [3]. The winner determination algorithms for combinatorial auctions are well studied [6] and hence one can use them for DA. However, the DA is only a special instance of combinatorial auctions, and hence algorithms that exploit the structure of DA will be more effective. A branch and bound algorithm that exploits the network structure of the WDP was proposed in [3]. In the following, we first propose a tight IP formulation with valid inequalities, followed by a novel branch and bound algorithm called as *branch on price*.

3   IP Formulations and LP Relaxations

3.1   An IP Formulation

An IP formulation for the WDP was proposed in [2]. The formulation uses *effective cost* p_j^{im}, which is the cost of item m from bid j, when i items are procured from j.

$$p_j^{im} = (1 - \theta_j^i) Q_j^m \tag{1}$$

The WDP can be restated as choosing the items with minimal sum of effective costs subject to the demand and supply constraints. The IP formulation is:

$$\min \sum_j \sum_i \sum_m p_j^{im} w_j^{im} \tag{2}$$

subject to

$$\sum_i v_j^i \leq 1 \quad \forall j \tag{3}$$

$$\sum_m w_j^{im} = i v_j^i \ \forall i, j \tag{4}$$

$$\sum_j \sum_i w_j^{im} = 1 \ \forall m \tag{5}$$

$$w_j^{im}, v_j^i \in \{0, 1\} \ \forall j, i, m \tag{6}$$

The binary decision variable w_j^{im} is to choose an item m from bid j with effective cost p_j^{im} and v_j^i is to choose i items from j. If $\sum_i v_j^i = 0$, then no items are chosen from j. Constraints (4) ensures that the items chosen from j are consistent with their effective cost: if i items are chosen, then they have the effective cost with respect to i. Constraints (5) ensures that every item is procured from only one supplier. Thus the constraint sets (3) and (4) are for supply constraints and constraint set (5) is for the demand constraints. The binary decision variables $\{w_j^{im}\}$ can be relaxed to be continuous, as for any feasible allocation of $\{v_j^i\}$ variables, there exists an allocation for $\{w_j^{im}\}$ with binary values.

3.2   LP Relaxation

The LP relaxation of the above IP formulation has interesting parallels with network optimization problems, in particular the transportation problem. Consider the following transportation problem with N supply nodes and M demand nodes. Each supplier corresponds to a supply node who can supply M items and each demand node corresponds to an item with a demand of one unit. The cost of transporting a unit from source j to sink m is p_j^{Mm}. Let $\{x_j^m\}$ denote the optimal flow in the above transportation network. The following solution is optimal to the LP relaxation of the WDP:

- $w_j^{im} = 0$, $\forall j, m$, and $i < M$
- $v_j^i = 0$, $\forall j$ and $i < M$
- $w_j^{Mm} = x_j^m$ $\forall j, m$
- $v_j^i = \frac{\sum_m w_j^{Mm}}{M}$ $\forall j, i$

The optimality of the above solution was proved in [2] using duality theory. The LP relaxation, in essence, violates the discount constraint by procuring less number of items from a supplier but with a maximum discount for M items. It is worth noting that only variables $\{v_j^i\}$ take continuous values.

3.3   A Tight Formulation with Valid Inequalities

To improve the lower bound generated by the LP relaxation, we add some *valid inequalities* to the IP formulation. Valid inequalities [7] are essentially redundant constraints to the original IP formulation. However, they may serve as *cuts* for the LP relaxation by reducing the continuous solution space thereby resulting in tighter bounds. The following valid inequalities serve as cuts in the LP relaxation for the WDP:

$$w_j^{im} \leq v_j^i \ \forall j, i, m \tag{7}$$

They are obviously valid for the IP formulation. However, they exclude the optimal solution of the LP relaxation of the original IP. The LP solution had binary values for $\{w_j^{im}\}$ and fractional values for $\{v\}$. With the valid inequalities added as cuts, this optimal solution is no longer feasible and hence one can expect a tighter lower bound.

The optimal solution to the new LP relaxation with cuts satisfies the following properties:

1. If $v_j^i > 0$, then number of non-zero w_j^{im}s are greater than or equal to i.
2. The $\{w_j^{im}\}$ take binary values only if $\{v_j^i\}$ are binary.

The first property is a direct consequence of the valid inequalities and the constraint set (4). The second property follows from the first. In the following, we develop a branch and bound algorithm using the improved IP formulation with cuts.

4   Branch on Price

In this section we propose *branch on price* (BoP), an optimal branch and bound (B&B) algorithm for solving the WDP. B&B is an exact intelligent enumerative technique that attempts to avoid enumerating a large portion of the feasible integer solutions [8, 9]. It is a widely used approach for solving discrete optimization, combinatorial optimization, and integer programming problems in general. The B&B approach first partitions the overall set of feasible solutions into two or more sets and as the algorithm proceeds the set is partitioned into many simpler and smaller sets, which are explored for the optimal solution. Each such set is represented by a *candidate problem* (CP). A typical iteration of B&B consists of:

- **Selection/Removal** of a CP from the list of CPs
- Determining the **lower bound** of the selected CP
- **Fathoming** or **pruning**, if possible, the selected CP
- Determining and updating the **incumbent** solution, if possible
- **Branching strategy**: If the CP is not fathomed, branching creates sub-problems which are added to the list of CPs

The algorithm first starts with the original IP as the only CP in the list, that is, the entire feasible set of solutions is considered at this point. The above steps are repeated until the list of CPs is empty. We use the LP relaxation of the tight IP formulation to generate lower bounds for BoP. The details of other steps of the BoP are explained below.

4.1   Branching Criteria and the Candidate Problem

The common branching technique in B&B algorithms using LP relaxation as the bounding technique is *variable dichotomy*: to branch on a fractional integer variable by imposing bounds on the variables. According to the properties of the optimal LP solution mentioned in Sect. 3.3, either all variables are binary (in which case optimal to the IP) or many variables are fractional. The variable dichotomy branching is to chose a particular v_j^i or a particular w_j^{im} that is fractional, to create two CPs by imposing the variable to equal to 0 and 1, respectively. The branching on a v_j^i is more generic than that on a w_j^{im}. The former splits the solution space based on the number of items supplied by a bid, whereas the latter is more specific about an item, supplied by a particular bid that supplies a certain number of other items. In essence, the variable dichotomy branching is about branching on the fractional decision variables, which may be more than one for the LP relaxation of the tight formulation. We propose here a novel technique to create candidate problems by branching on the price of an item.

Most of the integer variables in an IP formulation of combinatorial or discrete optimization problems are auxiliary variables created to impose logical constraints or linear constraints. Hence branching on such variables directly may not have logical or natural implication on the way the subsequent candidate problems are created. Further, the size of the solution set of these candidate problems may not be balanced, thus resulting in a unbalanced search tree. For example, a candidate problem created by fixing $w_j^{im} = 1$ may have very less number of solutions compared with the CP created by fixing $w_j^{im} = 0$. Instead

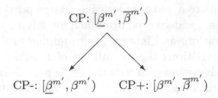

Fig. 1. Branching Strategy of BoP

of creating CPs by branching on fractional variables, we create by branching on the *price* of an item, which is fractionally supplied by more than one supplier.

Let w_j^{im} be fractional. Due to the constraint (5), there exists at least one another $w_{j'}^{i'm}$ with both $i \neq i'$ and $j \neq j'$, which is fractional. Let β^m be the price of the item m as dictated by the LP solution:

$$\beta^m = \sum_j \sum_i p_j^{im} w_j^{im} \tag{8}$$

We create two CPs, CP- and CP+, by branching on the above price. The CP- is created by adding the following constraints:

$$w_j^{im} = 0 \text{ if } p_j^{im} \geq \beta^m, \ \forall j, i, m \tag{9}$$

The CP+ is created by adding constraints

$$w_j^{im} = 0 \text{ if } p_j^{im} < \beta^m, \ \forall j, i, m \tag{10}$$

The two CPs partition the IP feasible solution space. The optimal LP solution (which is infeasible) does not belong to the solution space of the relaxations of the either of the CPs. To facilitate this branching, we represent a CP by using bounds on the prices of each of the items.

The CP essentially represents the feasible solution space it contains. We present a compact representation using an allowable price range $[\underline{\beta}^m, \overline{\beta}^m)$ for each item m. Algebraically, this is achieved by imposing the following bounds in the IP formulation: $w_j^{im} = 0$ if p_j^{im} is outside the above range. The initial CP contains all the solutions and hence $\underline{\beta}^m = \min_j\{p_j^{Mm}\}$ and $\overline{\beta}^m = \max_j\{p_j^{1m}\} + \epsilon$, for some $\epsilon > 0$. If an item m' is chosen for branching with price $\beta^{m'}$, then CP- has $[\underline{\beta}^{m'}, \beta^{m'})$ and CP+ has $[\beta^{m'}, \overline{\beta}^{m'}]$, as the respective price range for m'. This is illustrated in Fig. 1. Note that the price range of other items will remain the same as that of the parent CP.

The above branching scheme imposes many variables to be zero, across several bids, rather than just fixing one variable to 0 or 1, as in the variable dichotomy branching. Further, such a branching is more meaningful in terms of the WDP. The β_m can be considered as the price of the item m and the branch and bound algorithm is searching for the optimal price from the set of $\{p_j^{im}\}$, subject to the discount and demand constraints. Violation of the discount and the demand constraints leads β^m to be a convex combination of some of the prices from the set $\{p_j^{im}\}$. The proposed branching scheme partitions the set such that the same convex combination cannot be encountered again, thereby removing the violation in the constraints. In this way, one can expect that the algorithm will converge fast towards the optimal prices. It is worth noting that even though the β^m is a real number and thus the branching could be infinitely divisible, the possible optimal values it can take is NM and hence the number of branches is finite.

If there are more than one item which has fractional allocations, the algorithm has to choose one item to branch upon. Let β^m be the price of item m defined by the convex combination (8). The item m' to branch on is chosen according the following rule:

$$m' = \arg\min_m\{\beta^m - \underline{\beta}^m : \beta^m > \underline{\beta}^m\} \tag{11}$$

The above rule chooses the item whose price is closest to its lower bound, but not equal to it. According to the rules of the creation of CPs, branching on an item with $\beta^m = \underline{\beta}^m$ will create an infeasible CP- with range $[\underline{\beta}^m, \beta^m)$ and CP+ with range $[\beta^m, \overline{\beta}^m)$, which is same as its parent CP. To avoid an infinite loop, we consider only items whose price is strictly greater than its lower bound. However, a pathological case might arise with all items having their prices equal to their respective lower bounds. In this situation, we chose an item m randomly and create only CP+ with range $[\beta^m + \epsilon, \overline{\beta}^m)$. The $\epsilon > 0$ is chosen such that it is small enough to exclude just the price β^m. Note that there is no CP- created and with this new CP+ creation, it is possible that a feasible IP solution with prices $\{\beta^m\}$ might be excluded in the search. However, this does not happen as the BoP will find such an IP solution, if it exists, using the heuristic algorithm to be explained next.

4.2   Heuristic Algorithm to Determine Incumbent Solutions

Incumbent solutions are feasible IP solutions to the WDP that are obtained in the course of the algorithm. Usually in B&B, an incumbent solution is obtained if the LP relaxation provides optimal IP solutions. For BoP, we propose a rounding heuristic to obtain a feasible IP solution from the optimal LP solution for each CP. Let $\{w^{im}\}$ be the optimal LP solution. If all are binary, then it is a feasible solution to the IP. The following heuristic constructs a feasible solution from the fractional LP solution.

1. (Initialize) $S_j = 0$, $\forall j$
2. **do** $\forall m$: $k = \arg_j \max_{j,i}\{w_j^{im}\}$; $S_k \leftarrow S_k + 1$;
3. Construct a transportation network with winning bids as sources and items as sinks. The source corresponding to winning bid j has a supply of $S_j > 0$ and each sink has a unit demand. The cost of flow from j to m is $p_j^{S_j,m}$. Let $\{x_j^m\}$ be the optimal flow. Assign $v_j^{S_j} = 1$ and $w_j^{S_j,m} = x_j^m$.

The winning bid for an item m is chosen as the one with the largest w_j^{im} value. This is used to determine the number of winning items S_j for each winning bid. This is in turn used to determine the winning items with the consistent discount prices $p_j^{S_j,m}$. This will provide a better IP solution than directly rounding the largest w_j^{im} to 1. Using this heuristic, an incumbent solution is obtained whenever a new CP is created and the best known solution is updated and stored. Thus BoP is an *anytime* algorithm that can be terminated anytime with a feasible solution available.

4.3   Fathoming and Pruning

If the lower bounding technique provides an optimal solution to the CP, then the CP is said to be fathomed, that is, no further probe into the CP is required. A CP can also be fathomed if the problem is infeasible with the given price ranges. Pruning is another technique by which a CP can removed from further analysis. If the lower bound of a CP is greater than the objective value of the best incumbent solution, then the CP can be removed from further analysis as it cannot guarantee a better solution than what is already obtained. BoP has an incumbent solution right from the first node in the search and hence prunes lot of futile nodes, reducing the solution time.

4.4   Search Strategy

In each iterative step of B&B, a CP is selected and removed from the list of CPs for further analysis. We used *best first search* as the search strategy, which explores the best CP from the current list of CPs. This reduces the search space, but it has to store all the unexplored CPs in memory. One need not store the entire LP formulation for the CP, but just the information about the price range. Based on this price range, variables bounds are imposed on the LP formulation, to determine the lower bound. Once the lower bound is determined, the bounds are removed and the same LP model could be used by the other CPs. BFS is implemented by creating a priority queue that holds the list of CPs. At every iteration, the root of the queue, which is the CP with the least lower bound, is deleted from the queue and explored. If it is not fathomed or pruned, then two new CPs are generated and added to the queue.

4.5   The Algorithm

We present here the high level pseudocode of BoP. The list of CPs is stored in a priority queue pq. The deletemin operation of the priority queue returns the CP with the least lower bound. If the selected CP is not pruned, two child CPs are generated and added to the priority queue if they are not infeasible/fathomed/pruned. The priority queue is implemented using the binary heap data structure. Every insertion and deletion is of complexity $O(\log P)$, where P is the current size of the queue [10]. The is and bs store the incumbent and the best incumbent solution, and ls stores the LP solution. The $Z(\cdot)$ denotes the optimal objective value of the solution $(\cdot)$. At the end of the algorithm, bs contains the optimal solution to the WDP.

1. (Initialize)
   $cp \leftarrow \{[\underline{\beta}^m, \overline{\beta}^m)\}$; $pq = \emptyset$; $ls = \emptyset$; $is = \emptyset$; $bs = \emptyset$;
2. $ls \leftarrow$ LowerBound(cp); $is \leftarrow$ Heuristic(ls); $bs \leftarrow is$;
3. if $Z(ls) = Z(is)$ end;
4. pq.insert(cp);
5. while $pq \neq \emptyset$ do:
   (a) $cp \leftarrow pq$.DeleteMin();
   (b) if $Z($LowerBound$(cp)) \geq Z(bs)$ end;
   (c) Create$(cp-)$;
   (d) $ls \leftarrow$ LowerBound$(cp-)$; $is \leftarrow$ Heuristic(ls);
   (e) if $Z(is) < Z(bs)$ $bs \leftarrow is$;
   (f) if $Z(ls) < Z(bs)$ pq.insert$(cp-)$;
   (g) Create$(cp+)$;
   (h) $ls \leftarrow$ LowerBound$(cp+)$; $is \leftarrow$ Heuristic(ls);
   (i) if $Z(is) < Z(bs)$ $bs \leftarrow is$;
   (j) if $Z(ls) < Z(bs)$ pq.insert$(cp+)$;

5   Computational Experiments

In this section we present the computational experiments comparing the proposed BoP with the existing winner determination algorithm [3], which we henceforth refer as *branch on supply* (BoS). First, we present a brief review of BoS, to point out the differences in the design philosophy of BoS and BoP.

5.1   Branch on Supply

BoS is also a B&B algorithm, which used the embedded network structure in DA. The WDP in DA can be considered as a transportation network with N supply nodes (bids) and M demand nodes (one for each item), as shown in Fig. 2. Each supply node has a supply of M units and each demand node has a unit demand. A flow in the network connecting node j to node m indicate that bid j is supplying item m. The complicating feature of the network is the cost $c(j, m)$ of the flow in the arc (j, m), which is the function of number of units supplied from node j. This is different from the conventional nonlinear cost network models, where the cost will vary based on the flow through the arc, whereas in this case the cost varies on the total flow from the supply node. The LP relaxation of the original IP formulation, discussed in Sect. 3.2 is indeed the above network with the cost on arc $c(j, m)$ fixed as p_j^{Mm}, which is the least among p_j^{im}, having the highest discount.

The BoS does not use the IP formulation or its LP relaxation directly, but rather uses the above network structure. The CP is represented by a supply range $[\underline{S}_j, \overline{S}_j]$ for each bid j. This means that the CP contains all solutions in which, bid j can supply items in range $[\underline{S}_j, \overline{S}_j]$. The lower bound is determined by solving an *interval transportation problem* with supply for each node in the above interval and the cost on link (j, m) as $p_j^{\overline{S}_j, m}$. The search nodes are created by branching on the number of items supplied by a supplier. This is in contrast with the BoP, which branches on the price of an item rather than the supply of a bid.

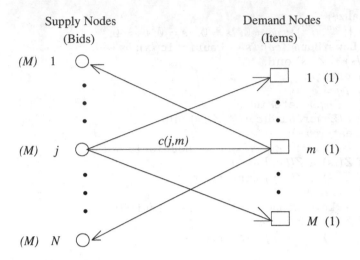

Fig. 2. The transportation network structure of DA

5.2   Experimental Setup

The two algorithms BoS and BoP were compared on the basis of the CPU time to solve the problem to optimality. The sample problems were randomly generated for $M = 5$ and $M = 10$, with N varying from 10 to 50, in intervals of 10. The other parameters of the problem were generated as follows. A relative cost rc^m for each of the item m was chosen randomly in range $[0.2, 1]$, with at least two items each taking one of the boundary values. The individual cost of m for each of the bids are chosen in the following way:

$$Q_j^m = \text{Random}[0.5, 1] \times rc^m \tag{12}$$

The individual costs are chosen in such a way that at least one bid has the maximum relative cost rc^m and one has the minimum cost $0.5 \times rc^m$. For the discount functions, $\theta_j^1 = 0$ and θ_j^M was chosen randomly in range $[0.15, 0.2]$. With the above values chosen, the discount type for each of the bids were randomly chosen from the following types: *linear, marginally decreasing, marginally increasing, step,* and *arbitrary.* All of them are increasing functions and based on the type, the intermediate value for θ_j^m were determined. The experiments were carried out on a Windows XP based PC equipped with a 2.8GHz Intel P4 processor with 1GB RAM. The algorithms were coded in Java, and for the model building and solving of LP relaxations and transportation problems, ILOG Concert Technology of CPLEX 9.0 [11] was used.

5.3   Experimental Results

For $M = 5$ and $M = 10$, with N varying from 10 to 50, 50 DA problem instances were randomly generated for each of the M and N combinations. The BoS and BoP were applied to each of the problem instances. The average computational

Table 1. Average Computational Time in milliseconds

N	M = 5		M = 10	
	BoS	BoP	BoS	BoP
10	73.12	11.76	2478.20	31.76
20	223.76	24.44	20846.44	115.04
30	545.00	71.80	76083.36	337.48
40	961.44	66.12	285967.96	944.48
50	1773.64	129.56	638697.20	1456.24

time (over the 50 problem instances) for each of the algorithms are shown in
Table 1. The BoP is many folds faster than BoS on all problem sizes.

6   Conclusions and Future Research

In this paper we presented a branch and bound algorithm BoP to solve the WDP
in discount auctions to optimality. It uses a tight IP formulation with valid inequalities, which provides better lower bounds. The BoP uses a novel branching
strategy instead of the traditional variable dichotomy branching. The price of
each of the items as dictated by the LP solution are determined and are used
to decide the branching. The computational experiments show that the BoP is
many folds faster than the existing algorithm BoS. Though the BoP is clearly
better than BoS, its performance could be improved further in many ways. The
strategy used to choose an item for branching was the item with price close to
its lower bound. There are other possible strategies: choose an item from a bid
that has highest number of fractional solutions, choose an item with minimum
weighted distance between its price and the effective costs of the fractional solutions, etc. The best first search was used as the search strategy. Other strategies
like depth first search and breadth first search, along with the branching strategies needs to be computationally tested to find the best combination of strategies.
The computational experiments should take into account different kinds of data
instances to study the effect of the algorithm on the instance type.

References

1. Rothkopf, M.H., Park, S.: An elementary introduction to auctions. Interfaces **31**
   (2001) 83–97
2. Kameshwaran, S., Benyoucef, L., Xie, X.: Discount auctions for procuring heterogeneous items. In: Proceedings of Seventh International Conference on Electronic
   Commerce (ICEC 2005). (2005) 244 – 249
3. Kameshwaran, S., Benyoucef, L., Xie, X.: Winner determination in discount auctions. In: Proceedings of Workshop on Internet and Network Economics (WINE
   2005). Lecture Notes in Computer Science 3828, Springer (2005) 868 – 877
4. Eso, M., Ghosh, S., Kalagnanam, J., Ladanyi, L.: Bid evaluation in procurement
   auctions with piece-wise linear supply curves. Research Report RC 22219, IBM
   Research, Yorktown Heights, NJ, USA (2001)

5. Cramton, P., Shoham, Y., Steinberg, R.: Combinatorial Auctions. MIT Press, Cambridge (2006)
6. Sandholm, T., Suri, S., Gilpin, A., Levine, D.: CABOB: A fast optimal algorithm for winner determination in combinatorial auctions. Management Science **51** (2005) 374–390
7. Wolsey, L.A.: Integer Programming. John Wiley and Sons, New York (1998)
8. Chandru, V., Rao, M.R.: Integer programming. In Atallah, M.J., ed.: Handbook of Algorithms and Theory of Computing. CRC Press (1999) 32.1–32.45
9. Murty, K.G.: Linear and Combinatorial Programming. R. E. Krieger (1985)
10. Cormen, T., Leiserson, C., Rivest, R.: Introduction to Algorithms. MIT Press, Cambridge (1990)
11. ILOG CPLEX: ILOG Concert Technology 1.1 User's Manual. (2001)

Note on an Auction Procedure for a Matching Game in Polynomial Time

Winfried Hochstättler[1], Hui Jin[2], and Robert Nickel[1]

[1] Department of Mathematics FernUniversität in Hagen, D-58084 Hagen
[2] Department of Mathematics Brandenburg Technical University Cottbus,
D-03044 Cottbus

Abstract. We derive a polynomial time algorithm to compute a stable solution in a mixed matching market from an auction procedure as presented by Eriksson and Karlander [5]. As a special case we derive an $\mathcal{O}(nm)$ algorithm for bipartite matching that does not seem to have appeared in the literature yet.

1 Introduction

In past years scientists from different fields such as game theory, economics, computer science, and combinatorial optimization have focused on the problem of two-sided matching markets where there are two finite and disjoint sets of agents, P and Q, that are to be matched in pairs consisting of one P-agent and one Q-agent. Two famous models of two-sided matching markets are the marriage model of Gale and Shapley [9] and the assignment game of Shapley and Shubik [17].

In the marriage model (see e.g. [13, 11]) the two sets of agents are usually referred to as the eligible marriage candidates in some small village. Each agent has preferences over the agents of the opposite set. A marriage is called *stable* when there is no pair which is not matched but prefers each other over their partners. Using the algorithm named "men propose – women dispose" Gale and Shapley [9] proved the existence of such a stable marriage when the preference lists are strict.

In the assignment game money plays a prominent role. It is modeled as a continuous variable. A matching and an allocation of its weight to the players compose a solution of the assignment game which is called *outcome*. By a stable outcome we mean a solution where no pair gets allocated less than the weight of its connecting edge. Shapley and Shubik [17] observed that stable outcomes coincide with the primal-dual pairs of solutions of the maximum weighted matching linear program, thereby showing the existence of a stable outcome. However, algorithms and complexity issues of game theoretical solution concepts have raised attention only recently (see e.g. Deng and Papadimitriou [3], Faigle et al. [6], Deng et al. [4]) and the classical algorithm for weighted bipartite matching, namely the Hungarian Method of Kuhn [14], is not as prominent in game theory as it is in combinatorial optimization.

S.-W. Cheng and C.K. Poon (Eds.): AAIM 2006, LNCS 4041, pp. 387–394, 2006.

Quite similar results such as the non-emptiness of the set of stable matchings and the lattice structure of the core have been established for these two models. To find a satisfactory explanation for the similarities in behavior between the two models Roth and Sotomayor [16] themselves offered a first model containing both the two old models as special cases and showed that its set of stable solutions, if it is non-empty, also has the lattice property under certain conditions. Eriksson and Karlander [5] modified this model to the *RiFle assignment game*, another common generalization of the two old models, and gave an algorithmic proof of the non-emptiness of its set of stable solutions. This algorithm computes a stable solution not in polynomial time but in pseudopolynomial time. For the classical special cases, it coincides with "men propose – women dispose", respectively with the "exact" auction procedure of [2]. The existence of stable solutions in presence of irrational data is proved by Eriksson and Karlander only via arguments from non-standard analysis.

We consider the model of Eriksson and Karlander [5]. A careful analysis of their algorithm reveals that a proper implementation solves the problem in $\mathcal{O}(n^4)$. This implementation was developed in parallel with [12] where we derive another polynomial time algorithm, to compute a stable solution for the same model, from the key lemma in Sotomayor [18]. Both algorithms run in $\mathcal{O}(n^4)$, where $2n$ is the number of players and n^2 is the size of a problem instance.

In the next section we briefly introduce the model and its notion of stability. Then we design a polynomial time algorithm to compute a stable solution in Section 3. Finally, we discuss the behavior of the algorithm in the special cases of Stable Matching, Assignment Game and cardinality matching and summarize differences from and similarities to the algorithm from [12].

2   Notation

We have two sets of players P (firms indexed by i) and Q (workers indexed by j) w.l.o.g. satisfying $|P| = |Q| =: n$. Let furthermore $P \cup Q$ be partitioned into *flexible* players (F) and *rigid* players (R). Consider the complete bipartite graph on $P \dot\cup Q$. An edge (i, j) is called *rigid* if one of i or j is in R and *flexible*, otherwise. For each edge (i, j) there are nonnegative numbers a_{ij} and b_{ij}. The sum $a_{ij} + b_{ij}$ is the *productivity* of a cooperation between i and j. A pair of functions $u : P \to \mathbb{R}$ and $v : Q \to \mathbb{R}$ is called a *payoff*. If i cooperates with j and (i, j) is a free edge the productivity can be freely divided into payoffs u_i and v_j while $u_i = a_{ij}$ and $v_j = b_{ij}$ must hold if (i, j) is a rigid edge.

Definition 1. *A payoff (u, v) is called* stable *if for any edge $(i, j) \in P \times Q$ we have*

(i) $u_i + v_j \geq a_{ij} + b_{ij}$ *if (i, j) is a free edge and*
(ii) $u_i \geq a_{ij}$ *or $v_j \geq b_{ij}$ if (i, j) is a rigid edge.*

A stable outcome is a stable payoff (u, v) together with a bijective map $\mu : P \to Q$ (denoted by $(u, v; \mu)$) so that

(iii) $u_i \geq 0$ *and* $v_j \geq 0$ *for all* $(i,j) \in P \times Q$.
(iv) $u_i + v_j = a_{ij} + b_{ij}$ *for* $\mu(i) = j$ *and* $\{i,j\} \subseteq F$.
(v) $u_i = a_{ij}$ *and* $v_j = b_{ij}$ *for* $\mu(i) = j$ *and* $\{i,j\} \cap R \neq \emptyset$.

Let $\mu : P \to Q$ be a map. If $\mu(i) = j$ then we say i *proposes to* j. A proposal is called *free* or *rigid* if the corresponding edge is free resp. rigid. A firm i (a worker j) is called *mapped* if $i \in \mu^{-1}(Q)$ (resp. $j \in \mu(P)$) and *unmapped*, otherwise. If there are firms i_1, i_2 so that $\mu(i_1) = \mu(i_2) = j$ then j is called *doubly mapped*. We denote by

$\quad\quad Q_U$ the set of unmapped workers,

$\quad\quad Q_{2\mu}$ the set of doubly mapped workers,

$\quad\quad Q_R$ the set of workers that have a rigid proposal, and by

$\quad\quad Q_{2R}$ the set of workers with at least 2 rigid proposals.

Let furthermore

$$f_{ij}^{(v,\mu)} := \begin{cases} a_{ij} + b_{ij} - v_j & \text{if } (i,j) \text{ is a free edge} \\ a_{ij} & \text{if } (i,j) \text{ is rigid and } v_j < b_{ij} \\ a_{ij} & \text{if } (i,j) \text{ is rigid and } v_j = b_{ij} \text{ and } \mu(i) = j \\ 0 & \text{otherwise} \end{cases}$$

define the possible profit of i from j if j receives v_j.

The strategy of the algorithm is the following: The map μ always defines stable relations but is not necessarily injective. In the course of the algorithm we will try and increase $|\mu(P)|$, keeping stability of the relations, until the map is injective. The procedure to increase $|\mu(P)|$ acts on the *augmentation digraph* $G^{(v,\mu)} = (P \cup Q, A)$ with backward arcs (j,i) for $\mu(i) = j$ and forward arcs (i,j) for $j \in D_i^{(v,\mu)}$ where

$$D_i^{(v,\mu)} = \{j \in Q \mid f_{ij}^{(v,\mu)} = \max_k f_{ik}^{(v,\mu)}\}$$

is the set of workers that maximize the potential benefit of firm i. A directed path $\mathcal{P}$ in $G^{(v,\mu)}$ that connects a doubly mapped worker $j_1 \in Q_{2\mu}$ with another worker j_s is called *(μ-)alternating* resp. *(μ-)augmenting* if j_s is not mapped.

3   An Algorithm to Find a Stable Outcome

Eriksson and Karlander [5] assume integer data and in one step increase a free payoff by at most one. We modify this approach in such a way that we increase the payoff by the smallest possible amount that changes the augmentation digraph. Our strategy to make the map $\mu : P \to Q$ bijective is as follows: As in the classical "men propose – women dispose" algorithm from Gale and Shapley [9] workers with more than one rigid proposal choose the best one and dispose the rest. This way some firms become temporarily unmapped. Each of these unmapped firms has to place another proposal until every worker has at most one rigid proposal. Next, we search the graph $G^{(v,\mu)}$ for alternating paths that reach a worker in $Q_U \cup Q_R$ and alternate the map μ along the path. If none of the above is possible, we increase the payoffs v of workers which are reachable by an alternating path until $G^{(v,\mu)}$ receives a new edge and the process is repeated until the map becomes injective.

The algorithm uses several sub-procedures:

PROPOSE(i): Places a proposal from i to a worker in $D_i^{(v,\mu)}$, i.e. chooses $\mu(i) \in D_i^{(v,\mu)}$.

DISPOSE(j, i^*): Disposes all firms $i \neq i^*$ that made a rigid proposal to j, i.e. sets $\mu(i)$ to be undefined for all $i \in \mu^{-1}(j) \setminus \{i^*\}$.

ALTERNATE($\mathcal{P}$): μ is alternated along the alternating path $\mathcal{P}$, i.e. all arcs are reoriented and μ is modified such that it uses the new backward arcs. If $\mathcal{P}$ is augmenting then the size of the image of μ increases by 1.

BFS($G, Q_{2\mu}$): Returns all vertices reachable from $Q_{2\mu}$ in G.

PLACERIGIDPROPOSALS: This procedure is the "men propose – women dispose" algorithm of Gale and Shapley [9]. Here, we denote by P_U the set of temporarily unmapped firms. See Algorithm 2.

HUNGARIANUPDATE: Increases the payoffs of all workers reachable from a doubly mapped worker. See Algorithm 3 for details.

Algorithm 1. An Algorithm to Find a Stable Outcome

$v \leftarrow 0$
PLACERIGIDPROPOSALS
while $Q_{2\mu} \neq \emptyset$
    while $\exists$ μ-alternating path to $j \in (Q \setminus \mu(P)) \cup Q_R$ **do**
        ALTERNATE($\mathcal{P}$)
        PLACERIGIDPROPOSALS
    end while
    HUNGARIANUPDATE
end while

Algorithm 2. PLACERIGIDPROPOSALS

while $P_U \neq \emptyset$ **do**
    for all $i \in P_U$ **do**
        PROPOSE(i)
    end for
    for all $j \in Q_{2R}$ **do**
        Let i^* be the favorite proposal in $\mu^{-1}(j)$
        DISPOSE(j, i^*)
        $v_j \leftarrow b_{i^* j}$
    end for
end while

Theorem 1. *Algorithm 1 eventually finishes with a stable outcome and can be implemented to run in $\mathcal{O}(n^4)$ time.*

Algorithm 3. HUNGARIANUPDATE

$\bar{P} \dot{\cup} \bar{Q} \leftarrow \text{BFS}(G^{(v,\mu)}, Q_{2\mu})$

$u_i \leftarrow \max_j f_{ij}^{(v,\mu)}$

$\Delta \leftarrow \min\{u_i - f_{ik}^{(v,\mu)} \mid i \in \bar{P}, k \in Q \setminus \bar{Q}\}$

for all $j \in \bar{Q}$ **do**

   $v_j \leftarrow v_j + \Delta$

end for

Proof. In any iteration of the inner loop of line 4 in Algorithm 1 $|\mu(P)|$ is increased or a rigid proposal is disposed. If there is a path to $Q \setminus \mu(P)$ then $|\mu(P)|$ increases. If the path ends in $j \in Q_R$ then PLACERIGIDPROPOSALS is called and disposes at least one rigid edge. Note, that a rigid edge once disposed will never be proposed again. If no path exists at all then v is increased by HUNGARIANUPDATE until this is the case and in each call of HUNGARIANUPDATE at least one new arc shows up in $G^{(v,\mu)}$. Thus, the procedure is finite.

The while-loop in line 4 of Algorithm 1 might be iterated more than once without finding a path as desired. Anyway, HUNGARIANUPDATE can be implemented so that its consecutive calls until a path is found need $\mathcal{O}(n^2)$ time in sum including an update of the augmentation graph by reusing the BFS-structure from the previous call and storing a minimum distance Δ_j from unmapped vertices and vertices in Q_R to the current BFS forest (see e.g. Galil [10] or Hochstättler et al. [12] for details). Hence, after $\mathcal{O}(n^2)$ time steps we can augment μ or dispose a rigid edge which can happen at most $\mathcal{O}(n^2)$ times. Hence, without considering the complexity of PLACERIGIDPROPOSALS the algorithm runs in $\mathcal{O}(n^4)$.

We also can implement PLACERIGIDPROPOSALS at a total cost of $\mathcal{O}(n^4)$ without any effort. For the first call we have to place n proposals taking $\mathcal{O}(n^2)$ time including the time to find a favorite partner for any $i \in P$. Note that the preference lists may change during the course of the algorithm thus, sorting the lists in a preprocessing does not suffice to speed up the procedure. For each discarded rigid edge we have to find a new favorite partner and after each call of PLACERIGIDPROPOSALS in the inner while-loop we may freely use $\mathcal{O}(n^2)$ to update the augmentation graph which happens at most $\mathcal{O}(n^2)$ times taking the number of rigid edges into account. Thus, the overall complexity of PLACERIGID-PROPOSALS is $\mathcal{O}(n^4)$. As the total cost of ALTERNATE is bounded by $\mathcal{O}(n^3)$ we get a total complexity of $\mathcal{O}(n^4)$.

Next we will show that the algorithm produces a stable outcome. In any stage of the algorithm let $\bar{u}_i := \max_j f_{ij}^{(v,\mu)}$. Then $(\bar{u}, v)$ is stable and $(\bar{u}, v; \mu)$ satisfies (iv) and (v) of Definition 1 since $\mu(i) = j$ implies $j \in D_i^{(v,\mu)}$. As v monotonically increases we also have $v \geq 0$. A worker with no proposer always has payoff zero and is therefore of non-negative value to all firms. Hence together with (iv) and (v) this implies $u \geq 0$. When the algorithm terminates μ is bijective and thus, $(\bar{u}, v; \mu)$ is a stable outcome. $\square$

4 Special Cases and Remarks

Cardinality Matching. If $R = \emptyset$ and $a_{ij} + b_{ij} \in \{0, 1\}$ for any edge (i, j) the problem reduces to finding a matching of maximum cardinality among edges with productivity 1 (referred to as *1-edges*). The presented algorithm (see Algorithm 4 for the reduced version) does not seem to have appeared in the literature yet and differs from the standard approach which starts with an empty matching M and searches the graph of 1-edges G_M^1 for an M-augmenting path. The algorithm presented here starts with a total but not surjective (and therefore not injective) map μ on the set of nodes in P with at least one 1-edge. A μ-alternating path in the graph of 1-edges G_μ^1 is a path from a doubly mapped worker to an unmapped worker using only 1-edges (forward) and μ-edges (backward) and is used to modify the map in a similar fashion as the augmentation of matchings is done in more classical algorithms. Here, the size of the image of μ increases. If no such μ-augmenting path exists, then the set of doubly mapped workers $Q_{2\mu}$ together with the set of firms which are mapped to a worker not in $Q_{2\mu}$ form a vertex cover of G_μ^1 with the same cardinality as the image of μ resulting in a maximum matching constructed from μ as in Algorithm 4 (e.g. [8]). If a perfect matching exists, we turn a total (not necessarily injective) map into an injective map instead of making a partial injective map (i.e. a matching) total.

Algorithm 4. Cardinality Matching by Increasing the Image of a Map

for all $i \in P$ **do**
   $\mu(i) \leftarrow j$ $((i, j)$ is a 1-edge)
end for
while $\exists \; \mu$-augmenting path in G_μ^1 to $j \in (Q \setminus \mu(P))$ **do**
   ALTERNATE$(\mathcal{P})$
end while
for all $j \in P, \; \mu^{-1}(j) \neq \emptyset$ **do**
   $M \leftarrow M \cup \{(i, j)\}$ $(i \in \mu^{-1}(j))$
end for

While the standard approach is essentially due to Ford and Fulkerson [7] the approach presented here reminds of the preflow-push algorithm (see e.g. [1]), as in the first step we send as much flow as possible from nodes in P to nodes in Q. Then, nodes in $Q_{2\mu}$ correspond to *excess nodes*, i.e. nodes that violate Kirchhoff's law. However, the strategy of lifting node potentials in preflow-push in successive steps does not seem to have anything in common with the augmenting path procedure used here.

A naive implementation of Algorithm 4 leads to an $\mathcal{O}(nm)$ algorithm. Note, that the main difference to the classical approach is in the orientation of the arcs in the search graph. While in the standard approach backward arcs are matchings, here we have exactly one backward arc ending in each non-isolated vertex of P. Thus, the ratio of forward to backward arcs decreases and the search tree in average should be shorter. We wonder if this approach might lead to more efficient implementations for cardinality matching.

Weighted Bipartite Matching. If $R = \emptyset$ the algorithm reminds of the Hungarian Method. Like the latter our method is a primal-dual algorithm and can be viewed to start with a weighted vertex cover (u, v) if we set $u_i \leftarrow \max_j f_{ij}^{(v,\mu)}$. We then search for alternating paths or update the payoffs if no such path can be found. Up to a different notion of an augmenting path (i. e. a different algorithm for cardinality matching) and a different orientation of the search graph this strategy is identical with that of the Hungarian Method (see e. g. Frank [8] for a transparent presentation).

Stable Marriage. When $F = \emptyset$ our model coincides with the Stable Marriage Model, since the a_{ij} at firm i resp. b_{ij} at worker j may as well be replaced by preference lists. The algorithm is identical to the classical "men propose – women dispose" algorithm of Gale and Shapley [9], that proceeds in rounds.

Comparison with the Algorithm in [12]. The algorithm in [12] to find a stable outcome differs from the algorithm presented here in various ways. In [12] (especially rigid) proposals are made asynchronously and not in rounds as in the present implementation. Furthermore, this algorithm is a direct extension of the Hungarian Method as introduced in Kuhn [15, Variant 2], while the algorithm presented here is a direct extension of the original "men propose – women dispose" algorithm of Gale and Shapley [9]. Also the concepts of augmenting paths differ as described above.

Bibliography

[1] Ravindra K. Ahuja, Thomas L. Magnanti, and James B. Orlin. *Network Flows.* Prentice Hall, 1993.

[2] Gabrielle Demange, David Gale, and Marilda Sotomayor. Multi-item auctions. *Journal of Political Economy*, 94(4):863–872, 1986.

[3] Xiaotie Deng and Christos H. Papadimitriou. On the complexity of cooparative game solution concepts. *Mathematics of Operations Research*, 19:257–266, 1994.

[4] Xiaotie Deng, Toshihide Ibaraki, and Hiroshi Nagamochi. Combinatorial optimization games. In *Proceedings of the 8th Annual ACM-SIAM Symposium on Discrete Algorithms*, pages 720–729, New Orleans, LA, 1997.

[5] Kimmo Eriksson and Johan Karlander. Stable matching in a common generalization of the marriage and assignment models. *Discrete Mathematics*, 217(1-3): 135–156, 2000.

[6] Ulrich Faigle, Sandor P. Fekete, Winfried Hochstättler, and Walter Kern. On the complexity of testing membership in the core of min cost spanning tree games. *International Journal of Game Theory*, 26:361–366, 1997.

[7] Lester R. Ford and Delbert R. Fulkerson. A simple algorithm for finding maximal network flows and an application to the hitchcock problem. *Canadian Journal of Mathematics*, 9:210–218, 1957.

[8] András Frank. On Kuhn's Hungarian method – A tribute from Hungary. Technical report, Egerváry Research Group on Combinatorial Optimization, October 2004.

[9] David Gale and Lloyd S. Shapley. College admissions and the stability of marriage. *American Mathematical Monthly*, 69:9–15, 1962.

[10] Zvi Galil. Efficient algorithms for finding maximum matchings in graphs. *ACM Computing Surveys*, 18(1):23–38, 1986.

[11] Dan Gusfield and Robert W. Irving. *The stable marriage problem: Structure and algorithms*. MIT Press, Cambridge, MA, USA, 1989.

[12] Winfried Hochstättler, Hui Jin, and Robert Nickel. The hungarian method in a mixed matching market. Technical report, FernUniversität in Hagen, Germany, October 2005.

[13] Donald E. Knuth. Stable marriage and its relation to other combinatorial problems. In *CRM Proceedings and Lecture Notes*, volume 10. American Mathematical Society, 1997.

[14] Harold W. Kuhn. The Hungarian method for the assignment problem. *Naval Research Logistics Quaterly*, 2:83–97, 1955.

[15] Harold W. Kuhn. Variants of the Hungarian method for the assignment problem. *Naval Research Logistics Quaterly*, 3:253–258, 1956.

[16] Alvin E. Roth and Marilda Sotomayor. Stable outcomes in discrete and continuous models of two-sided matching: A unified treatment. *Revista de Econometria, The Brazilian Review of Econometrics*, 16(2), November 1996.

[17] Lloyd S. Shapley and Martin Shubik. The assignment game I: The core. *International Journal of Game Theory*, 1:111–130, 1972.

[18] Marilda Sotomayor. Existence of stable outcomes and the lattice property for a unified matching market. *Mathematical Social Sciences*, 39:119–132, 2000.

Author Index

Vol. 4044: P. Abrahamsson, M. Marchesi, G. Succi (Eds.), Extreme Programming and Agile Processes in Software Engineering. XII, 230 pages. 2006.

Vol. 4041: S.-W. Cheng, C.K. Poon (Eds.), Algorithmic Aspects in Information and Management. XI, 395 pages. 2006.

Vol. 4039: M. Morisio (Ed.), Reuse of Off-the-Shelf Components. XIII, 444 pages. 2006.

Vol. 4038: P. Ciancarini, H. Wiklicky (Eds.), Coordination Models and Languages. VIII, 299 pages. 2006.

Vol. 4037: R. Gorrieri, H. Wehrheim (Eds.), Formal Methods for Open Object-Based Distributed Systems. XVII, 474 pages. 2006.

Vol. 4034: J. Münch, M. Vierimaa (Eds.), Product-Focused Software Process Improvement. XVII, 474 pages. 2006.

Vol. 4027: H.L. Larsen, G. Pasi, D. Ortiz-Arroyo, T. Andreasen, H. Christiansen (Eds.), Flexible Query Answering Systems. XVIII, 714 pages. 2006. (Sublibrary LNAI).

Vol. 4025: F. Eliassen, A. Montresor (Eds.), Distributed Applications and Interoperable Systems. XI, 355 pages. 2006.

Vol. 4024: S. Donatelli, P. S. Thiagarajan (Eds.), Petri Nets and Other Models of Concurrency - ICATPN 2006. XI, 441 pages. 2006.

Vol. 4021: E. André, L. Dybkjær, W. Minker, H. Neumann, M. Weber (Eds.), Perception and Interactive Technologies. XI, 217 pages. 2006. (Sublibrary LNAI).

Vol. 4011: Y. Sure, J. Domingue (Eds.), The Semantic Web: Research and Applications. XIX, 726 pages. 2006.

Vol. 4010: S. Dunne, B. Stoddart (Eds.), Unifying Theories of Programming. VIII, 257 pages. 2006.

Vol. 4007: C. Àlvarez, M. Serna (Eds.), Experimental Algorithms. XI, 329 pages. 2006.

Vol. 4006: L.M. Pinho, M. González Harbour (Eds.), Reliable Software Technologies – Ada-Europe 2006. XII, 241 pages. 2006.

Vol. 4004: S. Vaudenay (Ed.), Advances in Cryptology - EUROCRYPT 2006. XIV, 613 pages. 2006.

Vol. 4003: Y. Koucheryavy, J. Harju, V.B. Iversen (Eds.), Next Generation Teletraffic and Wired/Wireless Advanced Networking. XVI, 582 pages. 2006.

Vol. 4001: E. Dubois, K. Pohl (Eds.), Advanced Information Systems Engineering. XVI, 560 pages. 2006.

Vol. 3999: C. Kop, G. Fliedl, H.C. Mayr, E. Métais (Eds.), Natural Language Processing and Information Systems. XIII, 227 pages. 2006.

Vol. 3998: T. Calamoneri, I. Finocchi, G.F. Italiano (Eds.), Algorithms and Complexity. XII, 394 pages. 2006.

Vol. 3997: W. Grieskamp, C. Weise (Eds.), Formal Approaches to Software Testing. XII, 219 pages. 2006.

Vol. 3996: A. Keller, J.-P. Martin-Flatin (Eds.), Self-Managed Networks, Systems, and Services. X, 185 pages. 2006.

Vol. 3995: G. Müller (Ed.), Emerging Trends in Information and Communication Security. XX, 524 pages. 2006.

Vol. 3994: V.N. Alexandrov, G.D. van Albada, P.M.A. Sloot, J. Dongarra (Eds.), Computational Science – ICCS 2006, Part IV. XXXV, 1096 pages. 2006.

Vol. 3993: V.N. Alexandrov, G.D. van Albada, P.M.A. Sloot, J. Dongarra (Eds.), Computational Science – ICCS 2006, Part III. XXXVI, 1136 pages. 2006.

Vol. 3992: V.N. Alexandrov, G.D. van Albada, P.M.A. Sloot, J. Dongarra (Eds.), Computational Science – ICCS 2006, Part II. XXXV, 1122 pages. 2006.

Vol. 3991: V.N. Alexandrov, G.D. van Albada, P.M.A. Sloot, J. Dongarra (Eds.), Computational Science – ICCS 2006, Part I. LXXXI, 1096 pages. 2006.

Vol. 3990: J. C. Beck, B.M. Smith (Eds.), Integration of AI and OR Techniques in Constraint Programming for Combinatorial Optimization Problems. X, 301 pages. 2006.

Vol. 3989: J. Zhou, M. Yung, F. Bao, Applied Cryptography and Network Security. XIV, 488 pages. 2006.

Vol. 3987: M. Hazas, J. Krumm, T. Strang (Eds.), Location- and Context-Awareness. X, 289 pages. 2006.

Vol. 3986: K. Stølen, W.H. Winsborough, F. Martinelli, F. Massacci (Eds.), Trust Management. XIV, 474 pages. 2006.

Vol. 3984: M. Gavrilova, O. Gervasi, V. Kumar, C.J. K. Tan, D. Taniar, A. Laganà, Y. Mun, H. Choo (Eds.), Computational Science and Its Applications - ICCSA 2006, Part V. XXV, 1045 pages. 2006.

Vol. 3983: M. Gavrilova, O. Gervasi, V. Kumar, C.J. K. Tan, D. Taniar, A. Laganà, Y. Mun, H. Choo (Eds.), Computational Science and Its Applications - ICCSA 2006, Part IV. XXVI, 1191 pages. 2006.

Vol. 3982: M. Gavrilova, O. Gervasi, V. Kumar, C.J. K. Tan, D. Taniar, A. Laganà, Y. Mun, H. Choo (Eds.), Computational Science and Its Applications - ICCSA 2006, Part III. XXV, 1243 pages. 2006.

Vol. 3981: M. Gavrilova, O. Gervasi, V. Kumar, C.J. K. Tan, D. Taniar, A. Laganà, Y. Mun, H. Choo (Eds.), Computational Science and Its Applications - ICCSA 2006, Part II. XXVI, 1255 pages. 2006.

Vol. 3980: M. Gavrilova, O. Gervasi, V. Kumar, C.J. K. Tan, D. Taniar, A. Laganà, Y. Mun, H. Choo (Eds.), Computational Science and Its Applications - ICCSA 2006, Part I. LXXV, 1199 pages. 2006.

Vol. 3979: T.S. Huang, N. Sebe, M.S. Lew, V. Pavlović, M. Kölsch, A. Galata, B. Kisačanin (Eds.), Computer Vision in Human-Computer Interaction. XII, 121 pages. 2006.

Vol. 3978: B. Hnich, M. Carlsson, F. Fages, F. Rossi (Eds.), Recent Advances in Constraints. VIII, 179 pages. 2006. (Sublibrary LNAI).

Vol. 3976: F. Boavida, T. Plagemann, B. Stiller, C. Westphal, E. Monteiro (Eds.), Networking 2006. Networking Technologies, Services, and Protocols; Performance of Computer and Communication Networks; Mobile and Wireless Communications Systems. XXVI, 1276 pages. 2006.

Vol. 3975: S. Mehrotra, D.D. Zeng, H. Chen, B.M. Thuraisingham, F.-Y. Wang (Eds.), Intelligence and Security Informatics. XXII, 772 pages. 2006.

Vol. 3973: J. Wang, Z. Yi, J.M. Zurada, B.-L. Lu, H. Yin (Eds.), Advances in Neural Networks - ISNN 2006, Part III. XXIX, 1402 pages. 2006.

Vol. 3972: J. Wang, Z. Yi, J.M. Zurada, B.-L. Lu, H. Yin (Eds.), Advances in Neural Networks - ISNN 2006, Part II. XXVII, 1444 pages. 2006.

Vol. 3971: J. Wang, Z. Yi, J.M. Zurada, B.-L. Lu, H. Yin (Eds.), Advances in Neural Networks - ISNN 2006, Part I. LXVII, 1442 pages. 2006.

Vol. 3970: T. Braun, G. Carle, S. Fahmy, Y. Koucheryavy (Eds.), Wired/Wireless Internet Communications. XIV, 350 pages. 2006.

Vol. 3968: K.P. Fishkin, B. Schiele, P. Nixon, A. Quigley (Eds.), Pervasive Computing. XV, 402 pages. 2006.

Vol. 3967: D. Grigoriev, J. Harrison, E.A. Hirsch (Eds.), Computer Science – Theory and Applications. XVI, 684 pages. 2006.

Vol. 3966: Q. Wang, D. Pfahl, D.M. Raffo, P. Wernick (Eds.), Software Process Change. XIV, 356 pages. 2006.

Vol. 3965: M. Bernardo, A. Cimatti (Eds.), Formal Methods for Hardware Verification. VII, 243 pages. 2006.

Vol. 3964: M. Ü. Uyar, A.Y. Duale, M.A. Fecko (Eds.), Testing of Communicating Systems. XI, 373 pages. 2006.

Vol. 3963: O. Dikenelli, M.-P. Gleizes, A. Ricci (Eds.), Engineering Societies in the Agents World VI. X, 303 pages. 2006. (Sublibrary LNAI).

Vol. 3962: W. IJsselsteijn, Y. de Kort, C. Midden, B. Eggen, E. van den Hoven (Eds.), Persuasive Technology. XII, 216 pages. 2006.

Vol. 3960: R. Vieira, P. Quaresma, M.d.G.V. Nunes, N.J. Mamede, C. Oliveira, M.C. Dias (Eds.), Computational Processing of the Portuguese Language. XII, 274 pages. 2006. (Sublibrary LNAI).

Vol. 3959: J.-Y. Cai, S. B. Cooper, A. Li (Eds.), Theory and Applications of Models of Computation. XV, 794 pages. 2006.

Vol. 3958: M. Yung, Y. Dodis, A. Kiayias, T. Malkin (Eds.), Public Key Cryptography - PKC 2006. XIV, 543 pages. 2006.

Vol. 3956: G. Barthe, B. Grégoire, M. Huisman, J.-L. Lanet (Eds.), Construction and Analysis of Safe, Secure, and Interoperable Smart Devices. IX, 175 pages. 2006.

Vol. 3955: G. Antoniou, G. Potamias, C. Spyropoulos, D. Plexousakis (Eds.), Advances in Artificial Intelligence. XVII, 611 pages. 2006. (Sublibrary LNAI).

Vol. 3954: A. Leonardis, H. Bischof, A. Pinz (Eds.), Computer Vision – ECCV 2006, Part IV. XVII, 613 pages. 2006.

Vol. 3953: A. Leonardis, H. Bischof, A. Pinz (Eds.), Computer Vision – ECCV 2006, Part III. XVII, 649 pages. 2006.

Vol. 3952: A. Leonardis, H. Bischof, A. Pinz (Eds.), Computer Vision – ECCV 2006, Part II. XVII, 661 pages. 2006.

Vol. 3951: A. Leonardis, H. Bischof, A. Pinz (Eds.), Computer Vision – ECCV 2006, Part I. XXXV, 639 pages. 2006.

Vol. 3950: J.P. Müller, F. Zambonelli (Eds.), Agent-Oriented Software Engineering VI. XVI, 249 pages. 2006.

Vol. 3948: H.I Christensen, H.-H. Nagel (Eds.), Cognitive Vision Systems. VIII, 367 pages. 2006.

Vol. 3947: Y.-C. Chung, J.E. Moreira (Eds.), Advances in Grid and Pervasive Computing. XXI, 667 pages. 2006.

Vol. 3946: T.R. Roth-Berghofer, S. Schulz, D.B. Leake (Eds.), Modeling and Retrieval of Context. XI, 149 pages. 2006. (Sublibrary LNAI).

Vol. 3945: M. Hagiya, P. Wadler (Eds.), Functional and Logic Programming. X, 295 pages. 2006.

Vol. 3944: J. Quiñonero-Candela, I. Dagan, B. Magnini, F. d'Alché-Buc (Eds.), Machine Learning Challenges. XIII, 462 pages. 2006. (Sublibrary LNAI).

Vol. 3943: N. Guelfi, A. Savidis (Eds.), Rapid Integration of Software Engineering Techniques. X, 289 pages. 2006.

Vol. 3942: Z. Pan, R. Aylett, H. Diener, X. Jin, S. Göbel, L. Li (Eds.), Technologies for E-Learning and Digital Entertainment. XXV, 1396 pages. 2006.

Vol. 3941: S.W. Gilroy, M.D. Harrison (Eds.), Interactive Systems. XI, 267 pages. 2006.

Vol. 3940: C. Saunders, M. Grobelnik, S. Gunn, J. Shawe-Taylor (Eds.), Subspace, Latent Structure and Feature Selection. X, 209 pages. 2006.

Vol. 3939: C. Priami, L. Cardelli, S. Emmott (Eds.), Transactions on Computational Systems Biology IV. VII, 141 pages. 2006. (Sublibrary LNBI).

Vol. 3936: M. Lalmas, A. MacFarlane, S. Rüger, A. Tombros, T. Tsikrika, A. Yavlinsky (Eds.), Advances in Information Retrieval. XIX, 584 pages. 2006.

Vol. 3935: D. Won, S. Kim (Eds.), Information Security and Cryptology - ICISC 2005. XIV, 458 pages. 2006.

Vol. 3934: J.A. Clark, R.F. Paige, F.A. C. Polack, P.J. Brooke (Eds.), Security in Pervasive Computing. X, 243 pages. 2006.

Vol. 3933: F. Bonchi, J.-F. Boulicaut (Eds.), Knowledge Discovery in Inductive Databases. VIII, 251 pages. 2006.

Vol. 3931: B. Apolloni, M. Marinaro, G. Nicosia, R. Tagliaferri (Eds.), Neural Nets. XIII, 370 pages. 2006.

Vol. 3930: D.S. Yeung, Z.-Q. Liu, X.-Z. Wang, H. Yan (Eds.), Advances in Machine Learning and Cybernetics. XXI, 1110 pages. 2006. (Sublibrary LNAI).

Vol. 3929: W. MacCaull, M. Winter, I. Düntsch (Eds.), Relational Methods in Computer Science. VIII, 263 pages. 2006.

Vol. 3928: J. Domingo-Ferrer, J. Posegga, D. Schreckling (Eds.), Smart Card Research and Advanced Applications. XI, 359 pages. 2006.

Vol. 3927: J. Hespanha, A. Tiwari (Eds.), Hybrid Systems: Computation and Control. XII, 584 pages. 2006.